THE PHYSICS OF MEDICAL IMAGING

P

Medical Science Series

THE PHYSICS OF MEDICAL IMAGING

Edited by

Steve Webb

Joint Department of Physics,
Institute of Cancer Research and
Royal Marsden Hospital, Sutton, Surrey

Institute of Physics Publishing, Bristol and Philadelphia

© IOP Publishing Ltd 1988

Translated into Russian by MIR Publishers 1991
ISBN 5-03-001925-5 (vol 1)
ISBN 5-03-001924-3 (vol 2)

British Library Cataloguing-in-Publication Data

The physics of medical imaging.
  1. Physics – For radiology
  I. Webb. Steve II. Series
  530′024616

  ISBN 0-85274-361-0
  ISBN 0-85274-349-1 (pbk)

Library of Congress Cataloging-in-Publication Data

The physics of medical imaging.
  (Medical science series)
  Includes bibliographies and index.
  1. Diagnostic imaging – Technique. 2. Radiology.
I. Webb. Steve II. Series. [DNLM: 1. Diagnostic
Imaging. 2. Physics. WN 200 P578]
RC78. 7. D53P48 1988 616.07′57 88-4487

  ISBN 0-85274-361-0
  ISBN 0-85274-349-1 (pbk)

First published in 1988
Reprinted with corrections 1990 (pbk)
Reprinted with corrections 1992, 1993 (pbk)
Reprinted 1995, 1996, 1998

Series Editor: **Dr R F Mould**
            Westminster Hospital, London

Published by Institute of Physics Publishing, wholly owned by The Institute of Physics, London
Institute of Physics Publishing, Dirac House, Temple Back, Bristol BS1 6BE
US Office: Institute of Physics Publishing, The Public Ledger Building, Suite 1035, 150 South Independence Mall West, Philadelphia, PA 19106, USA

First typeset by KEYTEC, Bridport, Dorset
Printed in Great Britain at The Bath Press, Avon
Reprinted in Great Britain by J W Arrowsmith Ltd, Bristol

'What is the use of a book' thought Alice 'without pictures'
Alice's Adventures in Wonderland

The Medical Science Series is the official book series of the International Federation for Medical and Biological Engineering (IFMBE) and the International Organization for Medical Physics (IOMP).

IFMBE

The IFMBE was established in 1959 to provide medical and biological engineering with an international presence. The Federation has a long history of encouraging and promoting international cooperation and collaboration in the use of technology for improving the health and life quality of man.

The IFMBE is an organization that is mostly an affiliation of national societies. Transnational organizations can also obtain membership. At present there are 42 national members, and one transnational member with a total membership in excess of 15 000. An observer category is provided to give personal status to groups or organizations considering formal affiliation.

Objectives

- To reflect the interests and initiatives of the affiliated organizations.
- To generate and disseminate information of interest to the medical and biological engineering community and international organizations.
- To provide an international forum for the exchange of ideas and concepts.
- To encourage and foster research and application of medical and biological engineering knowledge and techniques in support of life quality and cost-effective health care.
- To stimulate international cooperation and collaboration on medical and biological engineering matters.
- To encourage educational programmes which develop scientific and technical expertise in medical and biological engineering.

Activities

The IFMBE has published the journal *Medical and Biological Engineering and Computing* for over 34 years. A new journal, *Cellular Engineering*, was established in 1996 in order to stimulate this emerging field in biomedical engineering. In *IFMBE News* members are kept informed of the developments in the Federation. *Clinical Engineering Update* is a publication of our division of Clinical Engineering. The Federation also has a division for Technology Assessment in Health Care.

Every three years, the IFMBE hold a World Congress on Medical Physics and Biomedical Engineering, organized in cooperation with the IOMP, and the IUPESM. In addition, annual, milestone regional conferences are organized in different regions of the world, such as the Asia Pacific, Baltic, Mediterranean, African and South American regions.

The administrative council of the IFMBE meets once or twice a year and is the steering body for the IFMBE. The council is subject to the rulings of the General Assembly which meets every three years.

For further information on the activities of the IFMBE, please contact Jos A E Spaan, Professor of Medical Physics, Academic Medical Center, University of Amsterdam, PO Box 22660, Meibergdreef 9, 1105 AZ, Amsterdam, The Netherlands. Tel: 31 (0) 20 566 5200. Fax: 31 (0) 20 6917233. E-mail: IFMBE@amc.uva.nl. WWW: http://www.iupesm.org/ifmbe.html.

IOMP

The IOMP was founded in 1963. The membership includes 64 national societies, two international organizations and 12 000 individuals. Membership of IOMP consists of individual members of the Adhering National Organizations. Two other forms of membership are available, namely Affiliated Regional Organization and Corporate Member. The IOMP is administered by a Council, which consists of delegates from each of the Adhering National Organizations; regular meetings of Council are held every three years at the International Conference on Medical Physics (ICMP). The Officers of the Council are the President, the Vice-President and the Secretary-General. IOMP committees include: developing countries, education and training; nominating; and publications.

Objectives

- To organize international cooperation in medical physics in all its aspects, especially in developing countries.
- To encourage and advise on the formation of national organizations of medical physics in those countries which lack such organizations.

Activities

Official publications of the IOMP are *Physiological Measurement*, *Physics in Medicine and Biology* and the *Medical Science Series*, all published by Institute of Physics Publishing. The IOMP publishes a bulletin, *Medical Physics World*, twice a year.

Two Council meetings and one General Assembly are held every three years at the ICMP. The most recent ICMPs were held in Kyoto, Japan (1991), Rio de Janeiro, Brazil (1994) and Nice, France (1997). The next conference is scheduled for Chicago, USA (2000). These conferences are normally held in collaboration with the IFMBE to form the World Congress on Medical Physics and Biomedical Engineering. The IOMP also sponsors occasional international conferences, workshops and courses.

For further information contact Gary D Fullerton, Professor, University of Texas HSC – San Antonio, Department of Radiology, 7703 Floyd Curl Drive, San Antonio, TX 78284-7800, USA. Tel: (210) 567-5550. Fax: (210) 567-5549. E-mail: fullerton@uthscsa.edu.

CONTENTS

PROFILES OF CONTRIBUTORS   xi

ACKNOWLEDGMENTS   xv

INTRODUCTION — AND SOME CHALLENGING QUESTIONS   1
    S Webb

1   IN THE BEGINNING   7
    S Webb
    References   16

2   DIAGNOSTIC RADIOLOGY WITH X-RAYS   20
    D R Dance
    2.1   Introduction   20
    2.2   The imaging system and image formation   21
    2.3   Photon interactions   23
    2.4   Important physical parameters   26
    2.5   X-ray tubes   32
    2.6   Image receptors   40
    2.7   Digital radiology   66
    References   71

3   QUALITY ASSURANCE AND IMAGE IMPROVEMENT IN
    DIAGNOSTIC RADIOLOGY WITH X-RAYS   74
    S H Evans
    3.1   Introduction to quality assurance   74
    3.2   Basic quality-assurance tests for x-ray sets   74
    3.3   Specific quality-assurance tests   85
    3.4   Data collection and presentation of the results   88
    3.5   Summary of quality assurance   88
    3.6   Improvement in radiographic quality   90
    3.7   Scatter removal   92
    3.8   Contrast enhancement   94
    3.9   Summary of methods of image enhancement   96
    References   97

4   X-RAY TRANSMISSION COMPUTED TOMOGRAPHY   98
    W Swindell and S Webb
    4.1   The need for sectional images   98

4.2   The principles of sectional imaging   100
4.3   Fourier-based solutions: the method of convolution and
      backprojection   111
4.4   Iterative methods of reconstruction   119
4.5   Other considerations   121
4.6   Appendix   125
References   125

5   CLINICAL APPLICATIONS OF X-RAY COMPUTED
    TOMOGRAPHY IN RADIOTHERAPY PLANNING   128
    H J Dobbs and S Webb
    5.1   X-ray computed tomography scanners and their role in
          planning   128
    5.2   Non-standard computed tomography scanners   135
    References   140

6   THE PHYSICS OF RADIOISOTOPE IMAGING   142
    R J Ott, M A Flower, J W Babich and P K Marsden
    6.1   Introduction   142
    6.2   Radiation detectors   143
    6.3   Radioisotope imaging equipment   147
    6.4   Radionuclides for imaging   181
    6.5   The role of computers in radioisotope imaging   193
    6.6   Static and dynamic planar scintigraphy   204
    6.7   Emission computed tomography   221
    6.8   Quality control and performance assessment of radioisotope
          imaging equipment   245
    6.9   Clinical applications of radioisotope imaging   256
    References   309

7   DIAGNOSTIC ULTRASOUND   319
    J C Bamber and M Tristam
    7.1   Introduction   319
    7.2   Basic physics   320
    7.3   Engineering principles of ultrasonic imaging   337
    7.4   Clinical applications and biological aspects of diagnostic
          ultrasound   365
    7.5   Research topics   377
    References   386

8   SPATIALLY LOCALISED NUCLEAR MAGNETIC
    RESONANCE   389
    M O Leach
    8.1   Introduction   389
    8.2   The development of nuclear magnetic resonance   390
    8.3   Principles of nuclear magnetic resonance   391
    8.4   Nuclear magnetic resonance pulse sequences   402
    8.5   Relaxation processes and their measurement   406

Contents   ix

8.6	Nuclear magnetic resonance image acquisition and reconstruction	412
8.7	Spatially localised spectroscopy	445
8.8	Instrumentation	465
8.9	Nuclear magnetic resonance safety	474
	References	479

9 PHYSICAL ASPECTS OF INFRARED IMAGING   488
C H Jones

9.1	Introduction	488
9.2	Infrared photography	488
9.3	Transillumination	489
9.4	Infrared imaging	489
9.5	Liquid-crystal thermography	505
9.6	Microwave thermography	506
	References	506

10 IMAGING OF TISSUE ELECTRICAL IMPEDANCE 509
S Webb

10.1	The electrical behaviour of tissue	509
10.2	Tissue impedance imaging	511
10.3	Suggested clinical applications of applied potential tomography	517
	References	522

11 IMAGING BY DIAPHANOGRAPHY 524
S Webb

11.1	Clinical applications	524
11.2	Physical basis of transillumination	525
11.3	Experimental arrangements	528
	References	532

12 THE MATHEMATICS OF IMAGE FORMATION AND IMAGE PROCESSING 534
S Webb

12.1	The concept of object and image	534
12.2	The relationship between object and image	536
12.3	The general image processing problem	539
12.4	Discrete Fourier representation and the models for imaging systems	542
12.5	The general theory of image restoration	545
12.6	Image sampling	549
12.7	Two examples of image processing from modern clinical practice	553
12.8	Iterative image processing	559
12.9	Appendix	562
	References	565

13   PERCEPTION AND INTERPRETATION OF IMAGES   567
     C R Hill
     13.1   Introduction   567
     13.2   The eye and brain as a stage in an imaging system   569
     13.3   Spatial and contrast resolution   571
     13.4   Perception of moving images   576
     13.5   Quantitative measures of investigative performance   578
     References   582

14   COMPUTER REQUIREMENTS OF IMAGING SYSTEMS   584
     R E Bentley and S Webb
     14.1   Single- versus multi-user systems   584
     14.2   Generation and transfer of images   585
     14.3   Processing speed   588
     14.4   Display of medical images   590
     14.5   Three-dimensional image display: methodology   593
     14.6   Three-dimensional image display: clinical applications 597
     References   602

15   EPILOGUE   604
     S Webb and C R Hill
     15.1   Introduction   604
     15.2   The impact of radiation hazard on medical imaging
            practice   606
     15.3   Attributes and relative roles of imaging modalities   610
     References   614

INDEX   617

PROFILES OF THE CONTRIBUTORS

All the contributors are on the staff of the Joint Department of Physics of the Institute of Cancer Research (ICR) and Royal Marsden Hospital (RMH) at either Fulham Road or Sutton, London, UK.

JOHN BABICH is Senior Radiopharmaceutical Scientist. His current research interests include the development and evaluation of new radiopharmaceuticals for detection and physiological assessment of tumours. His early research work included radiolabelling of blood cells and the development of an ultra-short-lived radionuclide generator for cardiovascular imaging.

JEFF BAMBER is Lecturer in Physics as Applied to Medicine. His research interests are in ultrasound imaging, where he has pioneered new techniques in speckle reduction. His work has included tissue characterisation, ultrasonic investigation of the breast and mathematical modelling of the interaction of ultrasound waves with tissue.

ROY BENTLEY is Senior Lecturer in Physics as Applied to Medicine and Head of the Digital Computer Service. In addition to applications of minicomputers in medical imaging, his current interests include computer networking, computer systems management and parallel architectures. He has pioneered the use of minicomputers in radiation treatment planning and in digital scintigraphy. His early research experience was in environmental radiation.

DAVID DANCE is a Principal NHS Physicist. His extensive research experience is in the physics of x-radiology and, in particular, mammographic imaging. He has published papers on radiotherapy physics, dosimetry calculation by numerical techniques and imaging by coded aperture techniques. He began research as a high-energy nuclear physicist.

JANE DOBBS is Consultant in Radiotherapy and Oncology at King's College Hospital and formerly Lecturer in Radiotherapy at ICR/RMH. Her research experience has included assessing the role of x-ray computed tomography in radiotherapy treatment planning. She has published a textbook of radiotherapy external beam techniques used at the Royal Marsden Hospital entitled *Practical Radiotherapy Planning*.

STEPHEN EVANS is a Senior NHS Physicist working in diagnostic radiology and radiation protection. His main research interests are in computer applications in diagnostic imaging, especially mammography. He is currently investigating patterns formed by scattered radiation.

MAGGIE FLOWER is Lecturer in Physics as Applied to Medicine. Her research interests have included studies of equipment for emission tomography and quantitative imaging using single-photon emission computed tomography, positron emission tomography and planar imaging. She is currently engaged in evaluating the clinical role of new radiopharmaceuticals with tomographic imaging. Her early experience was in radiotherapy physics.

KIT HILL is Professor of Physics as Applied to Medicine and Head of the Joint Physics Department. He is best known for his research in imaging with ultrasound following research in the physics of environmental radiation. He began as an engineer in industry and is currently Vice-President of the British Institute of Radiology.

COLIN JONES is Senior Lecturer in Physics as Applied to Medicine and Head of Department at Fulham Road. He has a long history of interest in thermographic imaging and has published widely. Among other interests, he has published papers on the physics of radiotherapy and dosimetry, radiation protection and breast imaging.

MARTIN LEACH is Lecturer in Physics as Applied to Medicine. His current research includes nuclear magnetic resonance imaging and spectroscopy and imaging in radiotherapy. He has also published papers on imaging using single-photon emission computed tomography, positron emission tomography and computed tomography. He has previously worked on *in vivo* activation analysis, inert-gas metabolism and cyclotron isotope production.

PAUL MARSDEN is a Research Fellow. He has recently completed a PhD thesis on positron emission tomography and is currently working on hardware and software developments for the prototype positron emission tomography system.

BOB OTT is Reader in Physics as Applied to Medicine, Chairman of the Sutton Physics Department and Head of the Radionuclide Imaging Group. He is currently establishing clinical positron emission tomography as a major research interest. He has published papers on single-photon emission computed tomography, imaging of monoclonal antibodies and dynamic scintigraphy. He converted to medical physics after 15 years of research in high-energy nuclear physics.

BILL SWINDELL is Reader in Physics as Applied to Medicine and Head of the Radiodiagnostic and Radiotherapy Physics Sections. His currently active research includes imaging in radiotherapy and ultrasound hyperthermia. He was Professor of Optical Sciences and Professor of Radiology at the University of Arizona and has published extensively in the physics of optical imaging and image processing. He was in charge of the NASA Pioneer 10 imaging project.

MARIA TRISTAM is a Research Fellow. Her current research interests include the development of new imaging techniques to assess tissue movement using ultrasound. Her early research experience was in cosmic-ray physics detecting high-energy muons deep underwater and underground.

STEVE WEBB is Lecturer in Physics as Applied to Medicine. His current research interests include the technology of single-photon emission computed tomography and positron emission tomography, reconstruction theory, image processing and imaging for radiotherapy planning. He has also published papers on coded aperture imaging, radiation dosimetry and Monte-Carlo modelling, longitudinal tomography and x-ray computed tomography. His early research experience was in cosmic-ray physics.

The contributors together, photographed at the Royal Marsden Hospital in October 1987 when the book was completed. Left to right: Roy Bentley, Colin Jones, Jeff Bamber, Paul Marsden, Jane Dobbs, John Babich, Steve Webb, Maggie Flower, Kit Hill, Maria Tristam, David Dance, Martin Leach, Bill Swindell, Bob Ott and Stephen Evans.

ACKNOWLEDGMENTS

This book has been compiled by members of the Physics Department of the Institute of Cancer Research and Royal Marsden Hospital, and our principal debt of gratitude is to those, too numerous to mention by name, whose wisdom foresaw the importance of medical imaging for cancer and other diseases and whose perseverance established imaging as a central component of our work. We particularly thank three past professors, Val Mayneord, Jack Boag and Ged Adams, who brought most of us here and whose influence still governs what we do and how we do it. We should also like to acknowledge the late Professor Roy Parker, whose own serious illnesses did not weaken his research initiative in imaging. The direction of the Department owes much to the wisdom of Dr Nigel Trott, who has also provided help and support for this book.

The book arose from a course of lectures given initially at Chelsea College, University of London, as part of the MSc in Biophysics and Bioengineering organised by Dennis Rosen. We should like to acknowledge our long association with Dr Rosen and his past and present colleagues.

Joint-author papers in the reference lists give some indication of specific credit due to our own past and present colleagues for some of the material included in this book. We are very grateful to the large number of people who have lent us figures and images and who are mentioned by name at the appropriate places. Our thanks also go to the publishers who have allowed this material to be freely used.

Several illustrations have been drawn from work at the Royal Marsden Hospital and we are grateful to clinical colleagues who have allowed us to use this material. We should also like to acknowledge our principal funding agencies, including the Medical Research Council and the Cancer Research Campaign.

To say that Sheila Dunstan, Rosemary Atkins and Marion Barrell typed the manuscript gives no indication of the organisational effort they have put into this project. They have been typing at times of the day and night when ordinary mortals have been enjoying themselves or asleep. We cannot thank them enough. We should also like to thank Sue Sugden for help with references and Ray Stuckey for help with the production of figures.

Our thanks go to Neville Hankins (commissioning editor) and to Sarah James (desk editor) at Adam Hilger publishers for their most professional work.

INTRODUCTION

AND SOME CHALLENGING QUESTIONS

S WEBB

The role of accurate investigation and diagnosis in the management of all disease is unquestionable. This is especially true for cancer medicine, and central to the diagnostic process are physical imaging techniques. Medical imaging not only provides for diagnosis but also serves to assist with planning and monitoring the treatment of malignant disease. Additionally, the role of imaging in cancer screening is widening to complement imaging symptomatic disease, as the appreciation of the importance of early diagnosis has become established.

When in its infancy, imaging with ionising x-radiation could be and was carried out by general practitioners and other medical specialists, but it has since become the *raison d'être* of diagnostic radiographers and radiologists and is rightly regarded as a speciality in its own right. In the mid-1940s the only satisfactory medical images available were radiographs of several kinds. Within the last 40 years or so, the situation has changed so dramatically that diagnostic imaging is rarely contained within a single hospital department, there being many imaging modalities, some important enough to be regarded as medical specialities in their own right. The creation of these imaging techniques called on the special skills of physicists, engineers and chemists, which collectively complement medical expertise. Diagnostic imaging has become a team activity: it is to the former group of paramedical scientists that this book is mainly directed. The intention has been to explain the physical principles governing medical imaging in a manner that should be accessible to the postgraduate or final-year undergraduate physical scientist. We have attempted to set different techniques in context with each other, to provide some historical perspective and to speculate on future developments. As clinical imaging is the end-point for such physical developments, clinical applications are discussed but only to the level of clarifying intentions and providing examples. The medical specialist should therefore find this a useful adjunct to the several

1

comprehensive volumes that have recently appeared describing state-of-the-art clinical imaging. The practising or training radiographer may also find something of interest.

Fifteen members from one department, the Joint Department of Physics of the Institute of Cancer Research and Royal Marsden Hospital, have contributed to this book. This team has conducted research together in medical imaging for a number of years and the diversity of medical imaging today should serve to explain why so much individual expertise is required even within one department. It is hoped that the advantage of choosing authors in this way has produced a readable text by contributors who were aware of what their colleagues were providing. An attempt has been made substantially to reframe contributions editorially to simulate a single-author text for convenient reading while retaining the separate expertise, rarely available from one author. Mistakes introduced by this process are entirely the responsibility of the editor. The framework of the text arose from a course of lectures on medical imaging given as part of the MSc degree in Biophysics and Bioengineering at Chelsea College, London (now the Intercollegiate London MSc). This course (and hence this book) was designed to introduce the physics of medical imaging to students from widely differing undergraduate science courses.

It is today difficult to imagine how diagnosis was carried out without images. There are, of course, very few people alive whose experience pre-dates the use of x-rays to form images. There are, however, plenty of people who can remember how things were done before the use of ultrasound, radionuclide and nuclear magnetic resonance imaging and other techniques became available, and they have witnessed how these imaging methods were conceived, developed and put into practice. They have observed the inevitable questioning of roles and the way the methods compare with each other in clinical utility. They have lived in a unique and exciting period in which the science of diagnostic medicine has matured and diversified.

Against this background, it is reasonable to raise a number of challenging questions, some of which this book hopes to answer. Why are there so many imaging modalities and does each centre need all of them? (This is a favourite question of hostile critics and financial providers!) Almost without exception, new techniques come to be regarded as complementary rather than replacing existing ones, and resources of finance, staff and space are correspondingly stretched. The essence of an answer to this question is not difficult to find. We shall see how different imaging methods are based on separate physical interactions of energy with biological tissue and thus provide measurements of different physical properties of biological structures. By some fortunate provision of nature, two (or more) tissues that are similar in one physical property may well differ widely in another. It does, of course, become necessary to interpret what these physical properties may reflect in terms of normal tissue function or abnormal pathology, and for some of the new imaging modalities these matters are by no means completely resolved.

The reverse of the question concerning the multiplicity of imaging methods is to ask whether one should expect the number of (classes of) imaging techniques to remain finite and in this sense small. Stated differently, will the writer of a text on medical imaging in 10 years' time or 100 years' time be expected to be able to include grossly different material from that of today? Certainly one may expect the hardware of existing classes of imaging to become less expensive, more compact, computationally faster and more widely available. With these changes, procedures that can be perfectly well specified today but cannot yet be reasonably done routinely will be achieved and included in the armament of practising radiologists (for example, image reconstruction in three dimensions by maximum entropy or simulated annealing mathematics). None of this, however, really amounts to crystal-ball gazing for quite new methodology. It would be a foolish person who predicts that no major developments will arise, but one must, of course, observe the baseline principle that governs the answer to this question. All imaging rests on the physics of the interaction of energy and matter. It is necessary for the energy to penetrate the body and be *partially* absorbed or scattered. The body must be semitransparent and there is a limited (and dare one say finite?) number of interactions for which such a specification exists. This requirement for *semi*-opacity becomes obvious from considering two extremes. The body is completely opaque to long-wavelength optical electromagnetic radiation, which therefore cannot be used as an internal probe. Equally, neutrinos, to which the body is totally transparent, are hardly likely to be the useful basis for imaging! The windows currently available for *in vivo* probing of biological tissue are all described in this volume. External probes lead to either resonant or non-resonant interaction between matter and energy. When the wavelength of the energy of the probe matches a scale size of the tissue, a resonant interaction leads to inelastic scattering, energy absorption and re-emission; attenuation of the probe is the key to forming images from the transmitted intensity. Such is the case, for example, for x-rays interacting with inner and outer electron shells or γ-rays interacting with atomic nuclei. Infrared and optical radiation similarly interacts with outer electron shells. When the frequency of matter and energy differ widely, elastic scattering, which is isotropic in a homogeneous material, leads to classical Huygens-optics behaviour. The scattering becomes anisotropic at tissue boundaries and in inhomogeneous matter, and the reflections and refractions generated form the basis of imaging. This is, for example, the essence of how ultrasound waves may be used to form images. With this in mind, the reader should consider whether the interactions described in this volume comprise a complete set of those likely to form the basis of imaging techniques and whether the energy probes described are the only ones to which the body is semi-opaque. If readers reach this conclusion affirmatively, then they must also conclude that the present time has its own unique historical place in the evolution of medical imaging.

As readers proceed, they should also enquire what combinations of factors led to each imaging modality appearing when it did? In some cases, the answer lies simply in the discovery of the underlying physical principle itself. In others, however, developments were predicted years before they were achieved. Tomography using x-rays, internal radionuclides or nuclear magnetic resonance all have a requirement for fast digital computing and, although simple analogue reconstruction predated the widespread use of the digital computer, the technique was impracticable. (The reader may like to work out (after Chapter 4) how long it would take to form a computed tomography scan 'by hand'.) Inevitably, some techniques owe their rapid development to parallel military research in wartime (e.g. the development of imaging with ultrasound following sonar research particularly in the Second World War) or to the by-products of nuclear reactor technology (e.g. radionuclide imaging), or to research in high-energy nuclear physics (e.g. particle and photon detector development).

Even if one were to conclude that the fundamental fields of imaging have all been identified, there is no reason for complacency or a sense of completion. Let us immediately recognise an area in which virtually nothing has been achieved in image production. Most of the images we shall encounter demonstrate macroscopic structure with spatial scales of 1 mm or so being described as 'good' resolution. Diagnosticians are accustomed to viewing images of organs or groups of organs, even the whole body, and requesting digital images with pixels of this magnitude. Yet cancer biology is proceeding at the cellular level and it is at this spatial scale that one would like to perform investigative science. It should be appreciated that before abnormal pathology can be viewed by *any* of the existing modalities, one of two events must occur. Either some 10^6 cells must in concert demonstrate a resolvable physical difference from their neighbours, or a smaller number of abnormal cells must give rise to a sufficiently large signal that they swamp the cancelling effect of their 'normal' neighbours within the smallest resolvable cell. An analogy for the latter might be the detection of a single plastic coin in a million metal coins by a measurement technique that cannot register less than a million coins but possesses a sensitivity better than 1 in 10^6.

Two further questions challenge any false complacency. First, we may enquire whether the time will come when all investigative imaging, which currently carries some small risk or is otherwise unpleasant, will be replaced by hazard-free alternative procedures that are more acceptable to the patient? A number of contenders are certainly known to carry far less risk than those based on the use of ionising radiation but currently by no means make x-ray imaging redundant. Secondly, will medical diagnosis ever become a complete science or must it remain partly an art? An enormous number of questions concerning the interpretation of images require to be resolved before the imprecision is removed; the randomness, introduced by the biological component of the process, may make this quite impossible. The importance of this question is enormous, not only for the management of the patient but also regarding the training and financing of clinical staff.

The human body is in a sense remarkably uncooperative as an *active* component of imaging. It emits infrared photons; it generates surface electrical potentials and some acoustic energy in the thorax relating to cardiac blood flow and pulmonary air movement. All these natural emissions have been used in diagnosis. It is, however, the paucity of natural signals that calls for the use of external probes or artificial internal emissions. Indeed, two- or three-dimensional sectional images with good spatial correlation are difficult or impossible to form with these natural emissions, and they are generally disregarded in any description of medical imaging. Also disregarded is the most primitive (but very important) form of imaging, namely visual inspection. Indeed, medical imaging is generally restricted to imply the imaging of *internal* structure, and we shall here adopt this convention. Finally, as is customary, the reader is informed what this book does *not* attempt to do. In keeping with the intention to overview the whole field of medical imaging in one teaching text, it should go without saying that each chapter cannot be regarded as a complete review of its subject. Whole volumes have already appeared on aspects of each imaging modality. I hope, however, to have tempted the student to the delights of the physics of medical imaging and to have honed that natural inquisitiveness which is the basis for all significant progress.

CHAPTER 1

IN THE BEGINNING

S WEBB

In the chapters that follow, an attempt is made to describe the physical principles underlying a number of imaging techniques that prove useful in diagnostic medicine. For brevity, we have concentrated largely on describing state-of-the-art imaging, with a view to looking to the future physical developments and applications. For the student anxious to come to grips with today's technology and perhaps about to embark on a research career, this is in a sense the most realistic approach. It does, however, lead to a false sense that the use of physics in medicine has always been much as it is today and gives no impression of the immense efforts of early workers whose labours underpin present developments. In this chapter we take a short backwards glance to the earliest days of some aspects of medical imaging.

Even a casual glance at review articles in the literature tells us that historical perspective is necessarily distorted by the prejudices of the writer. In a sense, history is best written by those who were involved in its making. During the 75th anniversary celebrations of the *British Journal of Radiology* in 1973, an anniversary issue (*Br. J. Radiol.* **46** 737–931) brought together a number of distinguished people to take stock, and although the brief was wider than to cover imaging, the impact of physics in medical imaging was a strong theme in their reviews. Since that time, of course, several new imaging modalities have burst into hospital practice and, with the excusable preoccupation with new ideas and methods, it is these which occupy most of current research effort, feature most widely in the literature and are looked to for new hope particularly in the diagnosis, staging and management of cancer and other diseases.

Against this background, what features of the landscape do we see in our retrospective glance? Within any one area of imaging, there exists a detailed and tortuous path of development, with just the strongest ideas surviving the passage of time. Numerous reviews chart these developments and are referred to in subsequent chapters. Most of us are, however, fascinated to know what was the *first* reported use of a technique or announcement of a piece of equipment, and might then be content to make the giant leap from this first report to how the situation stands today, with cavalier disregard for what lies between. So, by way of introduction, this is what we shall do here. Even so, a further difficulty arises. In a sense each variation on a theme is new and would certainly be so claimed by its originators, and yet it is not all these novelties that history requires us to remember. The passage of time acts as a filter to perform a natural selection. Perhaps what was regarded as important at its discovery has paled and some apparently unimportant announcements have blossomed beyond expectation. The organised manner in which research is required to be documented also veils the untidy methods by which it is necessarily performed. Most of the imaging modalities that are in common use were subject to a period of laboratory development and it is useful to distinguish clearly between the first reported experimental laboratory equipment and the first truly clinical implementation. Physicists might be tempted to rest content that the potential had been demonstrated in a laboratory, but were they patients they would take a different view! In most of the subjects we shall meet in later chapters, there has been (or still is!) a lengthy intermediate time between these two 'firsts'.

Wilhelm Conrad Röntgen's laboratory discovery of x-rays, when he was Professor of Physics at Wurzburg, is perhaps the only 'first' that can probably be pinned down to the time and day!—the late evening of 8 November 1895. This date was reported in *McClure's Magazine* by the journalist H J W Dam (1896): Röntgen himself never apparently stated a date. The discovery must also rank as one of the fastest ever published—submitted on 28 December 1895 and made known to the world on 5 January 1896. The prospects for x-ray diagnosis were immediately recognised. Röntgen refused, however, to enter into any commercial contract to exploit his discovery. He was of the opinion that his discovery belonged to humanity and should not be the subject of patents, licences and contracts. The result was undoubtedly the wide availability of low-cost x-ray units. A portable set in the USA cost $15 in 1896. Since it is believed that the first x-radiograph taken with clinical intent was on 13 January 1896 by two Birmingham (UK) doctors to show a needle in a woman's hand, the 'clinical first' followed the 'experimental first' with a time lapse also surely the shortest by comparison with any subsequent development. A bromide print of the image was given by Ratcliffe and Hall-Edwards to the woman, who next morning took it to the General and Queen's Hospital where the casualty surgeon J H Clayton removed the needle—the first x-ray guided operation. The single discovery of x-rays has clearly proved so important that it has already been the subject of many reviews (see e.g. Mould 1986, Brailsford 1946, Burrows 1986). It is amusing to note that many newspaper reports of the discovery were anything but enthusiastic. As equipment was readily available to the general public, there were at the time an abundant number of advertisements for x-ray sets. Mould (1986) has gathered together a plethora of these and other photographs of clinical procedures with x-rays. These are with hindsight now known to have been risky for both patient and radiologist, and many early radiation workers became casualties of their tools.

Digital subtraction angiography involves the subtraction of two x-rays, precisely registered, one using contrast material, to eliminate the unwanted clutter of common structures. Historically we have a precedent. Galton (1900) wrote (in the context of photography):

> If a faint transparent positive plate is held face to face with a negative plate, they will neutralise one another and produce a uniform grey. But if the positive is a photograph of individual, A, and the negative a photograph of individual, B, then they will only cancel one another out where they are identical and a representation of their differences will appear on a grey background. Take a negative composite photograph and superimpose it on a positive portrait of one of the constituents of that composite and one should abstract the group peculiarities and leave the individuality.

We shall not attempt to document the first application of x-radiography to each separate body site, but it is worth noting (Wolfe 1974) that Salomon (1913) is reported to have made the first mammogram. Thereafter the technique was almost completely abandoned until the early 1950s.

The announcement of a machine used to perform x-ray computed tomography (CT) in a clinical environment, by Hounsfield at the 1972 British Institute of Radiology annual conference, has been described as the greatest step forward in radiology since Röntgen's discovery. The relevant abstract (Ambrose and Hounsfield 1972) together with the announcement entitled 'X ray diagnosis peers inside the brain' in the *New Scientist* (27 April 1972) can be regarded as the foundation of clinical x-ray CT. The classic papers that subsequently appeared (Hounsfield 1973, Ambrose and Hounsfield 1973) left the scientific community in no doubt as to the importance of this discovery. Hounsfield shared the 1979 Nobel Prize for Physiology and Medicine with Cormack. The Nobel lectures (Hounsfield 1980, Cormack 1980) were delivered on 8 December 1979 in Stockholm. It was made quite clear, however, by Hounsfield that he never claimed to have 'invented CT'. The importance of what was announced in 1972 was the first practical realisation of the technique, which led to the explosion of clinical interest in the subsequent years. Who really did 'invent CT' has been much debated since. The original concept is usually credited to Radon (1917), whilst Oldendorf (1961) is often quoted as having published the first laboratory x-ray CT images of a 'head' phantom. What Oldendorf actually did was to rotate a head phantom (comprising a bed of nails) on a gramophone turntable and provide simultaneous translation by having an HO-gauge railway track on the turntable and the phantom on a flat truck, which was pulled slowly through a beam of x-rays falling on a detector. He showed how the internal structures in the phantom gave rise under such conditions to characteristic signals in the projections as the centre of rotation traversed the phantom relative to the fixed beam and detector. He was well aware of the medical implications of his experiment, but he did not actually generate a CT image. In his paper he referred to the work of Cassen and also Howry, who appear elsewhere in this chapter in another context. It is certainly not difficult to find papers throughout the 1960s describing the potential of reconstruction tomography in medicine, suggesting methods and testing them by both simulation and experiment. Cormack in particular was performing laboratory experiments in CT in 1963 (Cormack 1980). It is perhaps less well known that a CT scanner was built in Russia in 1958. Korenblyum *et al* (1958) published the mathematics of reconstruction from projections together with experimental details and wrote: 'At the present time at Kiev Polytechnic Institute, we are constructing the first experimental apparatus for getting X-ray images of thin sections by the scheme described in this article'. This was an analogue reconstruction method, based on a television detector and a fan-beam source of x-rays. Earlier reports from Russia have also been found (e.g. Tetel'Baum 1957).

The history of the detection of gamma-ray photons emitted from the body after injection of a radionuclide is a fascinating mix of contemporary detector physics and the development of radiopharmaceuticals. Detection techniques are almost as old as the discovery of radioactivity itself. The Crookes' spinthariscope (1903), the Wilson cloud chamber (1895), the gold-leaf electroscope and the Geiger counter (1929) were all used to detect, although not image, radiation. Artificial radionuclides did not arrive until Lawrence invented the cyclotron in 1931. Interestingly $^{99}Tc^m$ was first produced in the 37 inch cyclotron at Berkeley in 1938. Following the first experimental nuclear reactor in 1942, several reactors, notably at Oak Ridge and at the Brookhaven National Laboratory, produced medically useful radionuclides. Nuclear medicine's modern era began with the announcement in the 14 June 1946 issue of *Science* that radioactive isotopes were available for public distribution (Myers and Wagner 1975). A famous one-sentence letter from Sir J D Cockcroft at the Ministry of Supply, AERE, Harwell, to Sir E Mellenby at the MRC, dated 25 November 1946, said: 'I have now heard that the supply of radioactive isotopes to the UK by the US Atomic Energy Project is approved'. In September 1947, UK hospitals were receiving the first shipments from the GLEEP (graphite low-energy experimental pile) reactor at AERE, Harwell, Europe's first nuclear reactor (N G Trott, private communication).

Mallard and Trott (1979) have reviewed the development in the UK of what is today known as nuclear medicine. The *imaging* of radiopharmaceuticals was a logical extension of counting techniques for detecting ionising radiation, which go back to the invention of the Geiger–Müller tube in 1929. The development of Geiger–Müller counting was largely laboratory-based even in the early 1940s (notably in the UK at the National Physical Laboratory), and commercial detectors did not appear until after the Second World War (McAlister 1973). It was not, however, until 1948 that the first point-by-point image (of a thyroid gland) was constructed by Ansell and Rotblat (1948), which might be regarded as the first clinical nuclear medicine scan. The advantages of employing automatic scanning were recognised by several early workers. Cassen *et al* (1950) developed a scintillation (inorganic calcium tungstate) detector and wrote of their desire to mount it in an automatic scanning gantry, which they later achieved (Cassen *et al* 1951). Cassen *et al*'s paper appeared in August 1951, whilst in July 1951 Mayneord *et al* (1951a,b) introduced an automatic gamma-ray scanner based on a Geiger detector. Mayneord *et al* (1955) reported an improved scanner, which used a coincident pair of scintillators and storage tube display.

The concept of the gamma camera might be credited to Copeland and Benjamin (1949), who used a photographic plate in a pinhole camera. Their invention was made in the context of replacing autoradiographs, and long exposure times of the order of days were required. It is interesting to note that they came to criticise their own instrument's usefulness because 'many of the tracers used in biological work have little gamma activity', a situation rather different from what we know today! Anger (1952) first announced an electronic gamma camera with a crystal acting as an image intensifier for a film (also with a pinhole collimator), which used a sodium iodide crystal of size $2 \times 4 \times \frac{5}{16}$ inch^3. This was regarded as a large crystal at the time. The first electronic gamma camera with multiple photomultiplier tubes (PMT) was reported in 1957 (Anger 1957, 1958). It had a 4 inch diameter crystal of thickness 0.25 inch and just seven 1.5 inch diameter photomultiplier tubes. Commercial cameras followed soon afterwards, amongst the first in the UK being the prototype Ekco Electronics camera evaluated by Mallard and Myers (1963a,b) at London's Hammersmith Hospital. In 1968 Anger (1968) was also the first to report how a gamma camera could be used in a rectilinear scanning mode to perform multiplane longitudinal tomography. The new machine made redundant the need to perform several rectilinear scans with different focal-length collimators on single or double detectors.

Single-photon emission computed tomography (SPECT) stands in relation to planar Anger camera imaging as x-ray CT stands to planar x-radiology. Its importance in diagnostic nuclear medicine is now clearly established. Who invented SPECT and when? Once again, the honours are disputable, largely because what was suspected to be possible did not become a clinical reality for some while. It is also possible to identify several 'firsts' since SPECT has been achieved in several widely different ways. Kuhl and Edwards (1963) published the first laboratory single-photon emission tomography (SPET) (note: no 'c') images based on a rotate–translate arrangement and collimated crystal detectors. What we would now call a transverse section tomogram was generated entirely by analogue means without the need for a computer. The angular increment for rotation was a coarse 15°. Kuhl and Edwards (1964) published a photograph of the first SPET scanner (which was also capable of other scanning modes such as rectilinear and cylindrical scanning) and, in their explanation of how the analogue image is built up on film, provided what is possibly the first description of windowed tomography by which potentially overbright values were 'top-cut'. They also refined the image formation method to produce an image on paper tape whose contrast could be adjusted *a posteriori*. One of the first tomographic images visualised a malignant astrocytoma using an intravenous injection of chlormerodrin ^{197}Hg. Interestingly, in an addendum to the 1964 paper, they wrote: 'we have had good results with technetium 99m in pertechnetate form for brain scanning since March 1964'. This represents one of the earliest reports of the use of this isotope, which was to have such an important impact thereafter. The technetium generator was one of the first 'radioactive cows' and was conceived at the Brookhaven National Laboratory by Green, Tucker and Richards around the mid-1950s, being first reported in 1958 (see Ketchum 1986, Tucker *et al* 1958, Tucker 1960, Richards *et al* 1982), although the original identification of technetium as an impurity in the aluminium oxide generator of ^{132}I was reportedly made by a customer for one such generator. Generators were also available in the UK from AERE in the 1950s, supplied by G B Cook, and were in use at the Royal Cancer Hospital (N G Trott, private communication).

Anger himself showed how SPECT could be achieved with a gamma camera as early as 1967, rotating the patient in a chair in front of the stationary camera and coupling the line of scintillation corresponding to a single slice at each orientation to an optical camera (Anger *et al* 1967). Tumours were satisfactorily delineated and the technique was established, but it was far from today's clinical situation. These results were reported on 23 June 1967 at the 14th Annual Meeting of the Society of Nuclear Medicine in Seattle. In 1971 Muehllehner and Wetzel (1971) produced some laboratory images by computer, but the lack of a clinical SPECT system based on a rotating camera was still being lamented as late as 1977 when Jaszczak *et al* (1977) and Keyes *et al* (1977) were reporting clinical results obtained with a home-made camera gantry. The first commercial gamma-camera-based SPECT systems appeared in 1978 and at much the same time single-slice high-resolution SPECT systems were also marketed (Stoddart and Stoddart 1979). Almost a decade earlier, however, SPECT tomograms had been obtained in the clinic by Bowley *et al* (1973) using the Mark 1 Aberdeen Section Scanner, whose principles were largely similar to those of Kuhl and Edwards' laboratory scanner. We see, therefore, that with regard to SPECT imaging there was no clear date separating the impossible from the possible.

Logically complementing imaging with single photons is the detection of the annihilation gammas from positron emitters in order to form images of the distribution of a positron-labelled radiopharmaceutical in the body. The technology to achieve positron emission tomography (PET) is now established commercially but the market is small, although a number of specialist centres have been conducting clinical PET since the early 1960s. Perhaps it is surprising, therefore, to find that the technique of counting gammas from positron annihilation was discussed as early as 1951 by Wrenn *et al* (1951). They were able to take data from a source of ^{64}Cu in a fixed brain enclosed within its skull using thallium-activated sodium iodide detectors. Images as such were not presented, but certainly by 1953 simple scanning arrangements had been engineered for the creation of images. Brownell and Sweet (1953) showed the *in vivo* imaging of a recurrent tumour using ^{74}As. In this paper they write: 'we have been working independently on this [i.e. PET imaging] problem for a period of approximately two years'. It would be reasonable then to assign the beginnings of PET imaging to the year 1951.

The discovery of the phenomenon of nuclear magnetic resonance (NMR) was announced simultaneously and independently in 1946 by groups headed by Bloch and by Purcell, who shared a Nobel Prize. Thereafter, there was a steady development of NMR spectroscopy in chemistry, biology and medicine. NMR imaging followed much later and several 'firsts' are worth recording. In a letter to *Nature* in 1973, Lauterbur (1973) published the first NMR image of a heterogeneous object comprising two tubes of water, but the date of publication is preceded by a patent filed in 1972 by Damadian (1972), who proposed without detail that the body might be scanned for clinical purposes by NMR. The first human image of a live finger was reported by Mansfield and Maudsley (1976) and there followed the first NMR image of a hand (Andrew *et al* 1977) and of a thorax (Damadian *et al* 1977) in 1977. An article in the *New Scientist* in 1978 amusingly entitled 'Britain's brains produce first NMR scans' (Clow and Young 1978) was the first NMR image produced by a truly planar technique. In the same year, the first abdominal NMR scan was reported by Mansfield *et al* (1978). This (1978) was also the year in which the first commercial NMR scanner became available, and the first demonstration of abnormal human pathology was reported in 1980 by Hawkes *et al* (1980). The beginnings of NMR imaging clearly require the specification of a large number of 'firsts'!

It is believed that after x-radiology the use of ultrasound in medical diagnosis is the second most frequent investigative imaging technique. The earliest attempts to make use of ultrasound date from the late 1930s, but these mimicked the transmission method of x-rays and cannot really be recorded as the beginnings of ultrasound imaging as we know it today. Ultrasonic imaging based on the pulse–echo principle, which is also the basis of radar, became possible after the development of fast electronic pulse technology during the Second World War. The use of ultrasound to detect internal defects in metal structures preceded its use in medicine and was embodied in a patent taken out by Firestone in 1940. The first two-dimensional ultrasound scan was obtained using a simple sector scanner and showed echo patterns from a myoblastoma of the leg in a living subject (Wild and Reid 1952). This paper was received for publication on 25 October 1951. Prior to this, ultrasonic echo traces from human tissue had been demonstrated as early as 1950 by Wild (1950), but two-dimensional images had not been constructed. Very shortly afterwards, on 2 June 1952, Howry and Bliss (1952) published the results of their work, which had been in progress since 1947, and their paper included a two-dimensional image of a human wrist. Wild and Reid (1957) went on to develop the first two-dimensional ultrasound scanner and used it to image the structure of the breast and rectum. It was not until 1958 that the prototype of the first commercial two-dimensional ultrasonic scanner was described by Donald *et al* (1958) as a development of a one-dimensional industrial flaw detector made by Kelvin and Hughes. This machine was used to carry out the first investigations of the pregnant abdomen (Donald and Brown 1961). For a more detailed history of ultrasonic imaging, one might consult Hill (1973), White (1976) and Wild (1978). A number of other groups were actively investigating the use of ultrasound in medicine in the early 1950s, including Leksell in Sweden and a group led by Mayneord in what was then the Royal Cancer Hospital in London (now the Royal Marsden Hospital). Wild visited the Royal Cancer Hospital in 1954 and concluded that the group were quite familiar with the high-amplitude echo from the cerebral midline and the connection between its displacement and cerebral disease, which has subsequently become the basis of cerebral encephalography. The early UK work was documented only in the Annual Reports of the British Empire Cancer Campaign but may have provided a basis for the subsequent work of Donald's group in obstetrics.

In the following chapters we shall encounter the attempts that have been made to image a wide variety of different physical properties of biological tissue. Some of these are the bases of well established diagnostic techniques whose origins have been mentioned above. Other imaging methods are less widely applied or are still to reach the clinic. Some are still subject to controversy over their usefulness. For example, we shall find that diaphanography (the measurement of the light transmission of tissue) is receiving renewed interest of late for early diagnosis of breast disease. Commercially available equipment appeared for the first time in the late 1970s and yet Cutler (1929) reported the first attempts at transillumination some 50 years earlier in New York. Perhaps in contrast to the time lapse between the discovery and clinical use of x-rays, this ranks as the longest delay! Thermometric methods for showing the pattern of breast disease were first reported by Lloyd-Williams *et al* (1961). The use of xeroradiographic techniques for mammography were pioneered in the late 1960s and early 1970s when the image quality of the electrostatic technique became comparable with film imaging and the usefulness of the extra information offered was appreciated (Boag 1973). The first medical xeroradiographic image (of a hand) was, however, published in 1907 by Righi (1907), and was reproduced by Kossel (1967). (Righi had been working on the method since 1896.) The process was patented in 1937 by Carlson. This long delay was largely due to the inadequacies of the recording process, which were to be dramatically improved by the development of the Xerox copying process in the 1940s and 1950s. In 1955 Roach and Hilleboe (1955) described their feelings that xeroradiography was a logical replacement for film radiography particularly for mobile work. Their paper begins with an extraordinary justification for the work, namely the preservation of the lives of US casualties in the aftermath of a nuclear attack. They wrote:

In the event of the explosion of an atomic bomb over one of the major cities in the United States, the number of casualties produced and requiring emergency medical care would be tremendous. ... xeroradiography offers a simple, safe and inexpensive medium for the recording of rontgen images. No darkrooms or solutions of any type are needed. No lead lined storage vaults are required and there is no film deterioration problem. No transport of large supplies is involved.

Hills *et al* (1955) were already comparing xeroradiography and screen–film radiography.

Imaging the electrical impedance of the body is a very new technique whose description in Chapter 10 also serves for its history. The first *in vivo* cross-sectional clinical images were recorded early in the 1980s and to date there is no commercially available equipment for applied potential tomography.

In figure 1.1 some of these firsts have been plotted on a non-linear

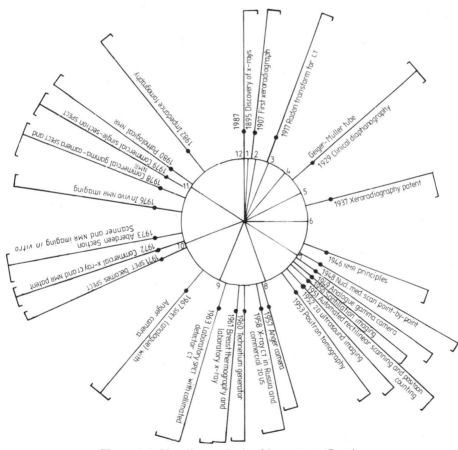

Figure 1.1 Non-linear clock of important 'firsts'.

clock. The period between 1895 and 1987 has been divided up such that the 92 years correspond to 12 h. The 'hour hand' sweeps the fraction of 360°, which is the square of the fraction of the 92 year period elapsed since 1895.

In the introduction it was tentatively suggested that the period between the mid-1940s and the present may have 'completed the set' of all physical probes to which the patient is semi-opaque and which are therefore available as the basis of imaging modalities. This is graphically illustrated in the figure by the crowding of events in the latter half of the timespan. With this brief historical scene setting, let us proceed to examine the physical basis of imaging.

REFERENCES

AMBROSE J and HOUNSFIELD G 1972 Computerised transverse axial tomography *Br. J. Radiol.* **46** 148–9
—— 1973 Computerised transverse axial scanning (tomography). Part 2: Clinical applications *Br. J. Radiol.* **46** 1023–47
ANDREW E R, BOTTOMLEY P A, HINSHAW W S, HOLLAND G N, MOORE W S and SIMAROJ C 1977 NMR images by the multiple sensitive point method: application to larger biological systems *Phys. Med. Biol.* **22** 971–4
ANGER H O 1952 Use of a gamma-ray pinhole camera for in-vivo studies *Nature* **170** 200–1
—— 1957 A new instrument for mapping gamma-ray emitters *Biology and Medicine Quarterly Report* UCRL-3653, January 1957 **38**
—— 1958 Scintillation camera *Rev. Sci. Instrum.* **29** 27–33
—— 1968 Multiplane tomographic gamma camera scanner *Medical Radioisotope Scintigraphy* STI Publ. 193 (Vienna: IAEA) pp203–16
ANGER H O, PRICE D C and YOST P E 1967 Transverse section tomography with the gamma camera *J. Nucl. Med.* **8** 314
ANSELL G and ROTBLAT J 1948 Radioactive iodine as a diagnostic aid for intrathoracic goitre *Br. J. Radiol.* **21** 552–8
BOAG J W 1973 Xeroradiography *Modern Trends in Oncology I* Part 2 *Clinical Progress* ed R W Raven (London: Butterworths)
BOWLEY A R, TAYLOR C G, CAUSER D A, BARBER D C, KEYES W I, UNDRILL P E, CORFIELD J R and MALLARD J R 1973 A radioisotope scanner for rectilinear, arc, transverse section and longitudinal section scanning (ASS—the Aberdeen Section Scanner) *Br. J. Radiol.* **46** 262–71
BRAILSFORD J F 1946 Röntgen's discovery of X-rays *Br. J. Radiol.* **19** 453–61
BROWNELL G L and SWEET W H 1953 Localisation of brain tumours with positron emitters *Nucleonics* **11** 40–5
BURROWS E H 1986 *Pioneers and Early Years: A History of British Radiology* (Alderney: Colophon)
CASSEN B, CURTIS L and REED C W 1950 A sensitive directional gamma ray detector *Nucleonics* **6** 78–80
CASSEN B, CURTIS L, REED C W and LIBBY R 1951 Instrumentation for ^{131}I use in medical studies *Nucleonics* **9** (2) 46–50
CLOW H and YOUNG I R 1978 Britain's brains produce first NMR images *New Scientist* **80** 588

COPELAND D E and BENJAMIN E W 1949 Pinhole camera for gamma ray sources *Nucleonics* **5** 44–9

CORMACK A M 1980 Early two-dimensional reconstruction and recent topics stemming from it (Nobel Prize lecture) *Science* **209** 1482–6

CUTLER M 1929 Transillumination as an aid in the diagnosis of breast lesions *Surg. Gynaecol. Obstet.* **48** 721–9

DAM H J W 1896 The new marvel in photography *McClure's Mag.* **6** 403 *et seq*

DAMADIAN R V 1972 Apparatus and method for detecting cancer in tissue *US Patent* 3789832, filed 17 March 1972

DAMADIAN R, GOLDSMITH M and MINKOFF L 1977 NMR in cancer: Fonar image of the live human body *Physiol. Chem. Phys.* **9** 97–108

DONALD I and BROWN T G 1961 Demonstration of tissue interfaces within the body by ultrasonic echo sounding *Br. J. Radiol.* **34** 539–46

DONALD I, McVICAR J and BROWN T G 1958 Investigation of abdominal masses by pulsed ultrasound *Lancet* **i** 1188–95

GALTON F 1900 Analytic portraiture *Nature* **62** 320

HAWKES R C, HOLLAND G N, MOORE W S and WORTHINGTON B S 1980 NMR tomography of the brain: a preliminary clinical assessment with demonstration of pathology *J. Comput. Assist. Tomogr.* **4** (5) 577–86

HILL C R 1973 Medical ultrasonics: an historical review *Br. J. Radiol.* **47** 899–905

HILLS T H, STANFORD R W and MOORE R D 1955 Xeroradiography — the present medical applications *Br. J. Radiol.* **28** 545–51

HOUNSFIELD G N 1973 Computerised transverse axial scanning (tomography). Part 1: Description of system *Br. J. Radiol.* **46** 1016–22

—— 1980 Computed medical imaging (Nobel Prize lecture) *Science* **210** 22–8

HOWRY D H and BLISS W R 1952 Ultrasonic visualisation of soft tissue structures of the body *J. Lab. Clin. Med.* **40** 579–92

JASZCZAK R J, MURPHY P H, HUARD D and BURDINE J A 1977 Radionuclide emission computed tomography of the head with $^{99}Tc^m$ and a scintillation camera *J. Nucl. Med.* **18** 373–80

KETCHUM L E 1986 Brookhaven, origin of $^{99}Tc^m$ and ^{18}F FDG, opens new frontiers for nuclear medicine *J. Nucl. Med.* **27** (10) 1507–15

KEYES J W, ORLANDEA N, HEETDERKS W J, LEONARD P F and ROGERS W L 1977 The Humongotron—a scintillation camera transaxial tomograph *J. Nucl. Med.* **18** 381–7

KORENBLYUM B I, TETEL'BAUM S I and TYUTIN A A 1958 About one scheme of tomography *Bull. Inst. Higher Educ.– Radiophys.* **1** (3) 151–7 (translated from the Russian by H H Barrett, University of Arizona, Tucson)

KOSSEL F 1967 Physical aspects of xeroradiography as a tool in cancer diagnosis and tumour localisation *Progress in Clinical Cancer* ed I M Ariel, vol 3 (New York: Grune and Stratton) pp176–85

KUHL D E and EDWARDS R Q 1963 Image separation radioisotope scanning *Radiology* **80** 653–62

—— 1964 Cylindrical and section radioisotope scanning of the liver and brain *Radiology* **83** 926–36

LAUTERBUR P C 1973 Image formation by induced local interactions: examples employing nuclear magnetic resonance *Nature* **242** 190–1

LLOYD-WILLIAMS K, LLOYD-WILLIAMS F J and HANDLEY R S 1961 Infra-red thermometry in the diagnosis of breast disease *Lancet* **2** 1378–81

McALISTER J 1973 The development of radioisotope scanning techniques *Br. J. Radiol.* **46** 889–98

MALLARD J R and MYERS M J 1963a The performance of a gamma camera for the visualisation of radioactive isotopes in vivo *Phys. Med. Biol.* **8** 165–82
—— 1963b Clinical applications of a gamma camera *Phys. Med. Biol.* **8** 183–92
MALLARD J R and TROTT N G 1979 Some aspects of the history of nuclear medicine in the United Kingdom *Semin. Nucl. Med.* **9** 203–17
MANSFIELD P and MAUDSLEY A A 1976 Planar and line scan spin imaging by NMR in *Magnetic Resonance and Related Phenomena Proc. 19th Congr. Ampere, Heidelberg* (IUPAP) pp247–52
MANSFIELD P, PYKETT I L, MORRIS P G and COUPLAND R E 1978 Human whole-body line-scan imaging by NMR *Br. J. Radiol.* **51** 921–2
MAYNEORD W V, EVANS H D and NEWBERY S P 1955 An instrument for the formation of visual images of ionising radiations *J. Sci. Instrum.* **32** 45–50
MAYNEORD W V, TURNER R C, NEWBERY S P and HODT H J 1951a A method of making visible the distribution of activity in a source of ionising radiation *Nature* **168** 762–5
—— 1951b A method of making visible the distribution of activity in a source of ionising radiation *Radioisotope Techniques* vol 1 (*Proc. Isotope Techniques Conf. Oxford* July 1951) (London: HMSO)
MOULD R F **1980** *A History of X-Rays and Radium* (Sutton, Surrey: IPC)
MUEHLLEHNER G and WETZEL R A 1971 Section imaging by computer calculation *J. Nucl. Med.* **12** 76–84
MYERS W G and WAGNER H N 1975 Nuclear medicine: how it began *Nuclear Medicine* ed H N Wagner (New York: H P Publishing)
OLDENDORF W H 1961 Isolated flying spot detection of radiodensity discontinuities; displaying the internal structural pattern of a complex object *IRE Trans. Biomed. Electron.* **BME-8** 68–72
RADON J 1917 Uber die Bestimmung von Funktionen durch ihre Integralwerte langs gewisser Mannigfaltigkeiten *Ber. Verh. Sachs. Akad. Wiss. Leipzig Math. Phys.* K1 **69** 262–77
RICHARDS P, TUCKER W D and SHRIVASTAVA S C 1982 Technetium-99m: an historical perspective *Int. J. Appl. Radiat. Isot.* **33** 793–9
RIGHI A 1907 *Die Bewegung der Ionen bei der elektrischen Entladung* (Leipzig: J Barth)
ROACH J F and HILLEBOE H E 1955 Xeroradiography *Am. J. Roentgenol.* **73** 5–9
SALOMON A 1913 Beitrage zue Pathologie und Klinik der Mammacarcinoma *Arch. Klin. Chir.* **101** 573–668
STODDART H F and STODDART H A 1979 A new development in single gamma transaxial tomography: Union Carbide focussed collimator scanner *IEEE Trans. Nucl. Sci.* **NS-26** (2) 2710–12
TETEL'BAUM S I 1957 About a method of obtaining volume images with the help of x-rays *Bull. Kiev Polytechnic Inst.* **22** 154–60 (translated from the Russian by J W Boag, Institute of Cancer Research, London)
TUCKER W D 1960 'Radioisotopic cows' *J. Nucl. Med.* **1** 60
TUCKER W D, GREENE M W, WEISS A J and MURRENHOFF A P 1958 BNL 3746 *American Nuclear Society Annual Meeting, Los Angeles* June 1958 *Trans. Am. Nucl. Soc.* **1** 160
WHITE D N 1976 *Ultrasound in Medical Diagnosis* (Ontario: Ultramedison)
WILD J J 1950 The use of ultrasonic pulses for the measurement of biological tissues and the detection of tissue density changes *Surgery* **27** 183–8
—— 1978 The use of pulse–echo ultrasound for early tumour detection: history and prospects *Ultrasound in Tumour Diagnosis* ed C R Hill, V R McCready and D O Cosgrove (London: Pitman Medical)

WILD J J and REID J M 1952 The application of echo-ranging techniques to the determination of structure of biological tissues *Science* **115** 226–30

—— 1957 Progress in the techniques of soft tissue examination by 15 MC pulsed ultrasound *Ultrasound in Biology and Medicine* ed E Kelly (Washington, DC: Am. Inst. Biol. Sci.) pp30–48

WOLFE J N 1974 Mammography *Radiol. Clin. N. Am.* **12** (1) 189–203

WRENN F R, GOOD M L and HANDLER P 1951 The use of positron emitting radioisotopes in nuclear medicine imaging *Science* **113** 525–7

CHAPTER 2

DIAGNOSTIC RADIOLOGY WITH X-RAYS

D R DANCE

2.1 INTRODUCTION

X-rays have been used to produce medical images ever since their discovery by Wilhelm Röntgen in 1895 (see Chapter 1). In Great Britain, it has been estimated that there are 644 medical and dental radiographic examinations per 1000 population per year (Wall *et al* 1986), so that the technique is of major importance in medical imaging.

In view of this place in history as well as the vast application of x-rays, it is entirely appropriate to begin the discussion of the physics of medical imaging by considering diagnostic radiology with x-rays. In later chapters we shall see how many of the concepts formed for describing radiography with x-rays are also useful for other modalities. In a sense, the language of imaging was framed for x-radiology, including concepts such as image contrast, noise and spatial resolution, and it has subsequently been taken across to describe these other techniques for imaging the human body. In this chapter is covered the essential physics of the design of x-ray imaging equipment. In Chapter 3, some of the practical aspects of using x-ray equipment are further developed.

The radiographic image is formed by the interaction of x-ray photons with a photon detector and is therefore a distribution of those photons which are transmitted through the patient and are recorded by the detector. These photons can either be primary photons, which have passed through the patient without interacting, or secondary photons, which result from an interaction in the patient (figure 2.1). The secondary photons will in general be deflected from their original direction and carry little useful information. The primary photons do carry useful information. They give a measure of the probability that a photon will pass through the patient without interacting and this probability will itself depend upon the sum of the x-ray attenuating properties of all the tissues the photon traverses. The image is therefore a *projection* of the attenuating properties of all the tissues along the paths of the x-rays. It is a two-dimensional projection of the three-dimensional distribution of the x-ray attenuating properties of tissue.

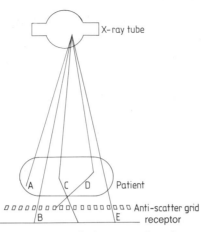

Figure 2.1 The components of the x-ray imaging system and the formation of the radiographic image. B and E represent photons that have passed through the patient without interacting. C and D are scattered photons. D has been stopped by an anti-scatter grid. Photon A has been absorbed.

It is possible to obtain a two-dimensional slice from the three-dimensional distribution of attenuating properties using the techniques of either classical or computed tomography. The reader is referred to Forster (1985 pp130–9) for details of classical tomography and to Chapter 4 for details of computed tomography. This chapter is solely concerned with projection radiography.

2.2 THE IMAGING SYSTEM AND IMAGE FORMATION

The components of a typical x-ray imaging system are shown in figure 2.1. The photons emitted by the x-ray tube enter the patient, where they may be scattered, absorbed or transmitted without interaction. The primary photons recorded by the image receptor form the image, but the scattered photons create a background signal which degrades contrast. In most cases, the majority of the scattered photons can be removed by placing an anti-scatter device between the patient and the image receptor (see §3.7). This device can simply be an air gap or a grid formed from a series of parallel lead strips, which will transmit most of the primary radiation but reject most of the scatter.

The image recorded by the receptor is processed (e.g. an x-ray film is developed) and can then be viewed by the radiologist. It is important here that the radiograph is correctly illuminated and is viewed using a distance and magnification appropriate to the detail in the image and the angular-frequency response of the eye. The interpretation of the radiograph is a very skilled task and involves both the perception of small differences in contrast and detail as well as the recognition of abnormal patterns. The mechanisms of perception are discussed in Chapter 13 and our treatment here is limited to the production of the radiographic image.

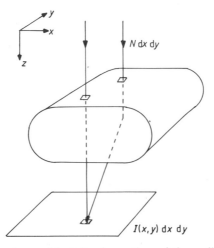

Figure 2.2 Simple model of the formation of the radiographic image showing both a primary and a secondary photon.

It will be useful in what is to follow to have a simple mathematical model of the radiographic imaging process. We start by considering a monochromatic x-ray source that emits photons of energy E and is sufficiently far from the patient that the photon beam can be considered as being parallel (figure 2.2). The incident photon beam is parallel to the z direction and the image is recorded in the xy plane. We assume that each photon interacting in the receptor is locally absorbed and that the response of the receptor is linear, so that the image may be considered as a distribution of absorbed energy. If there are N photons per unit area incident on the patient and $I(x, y)\,dx\,dy$ is the energy absorbed in area $dx\,dy$ of the detector, then

$$I(x, y) = N\varepsilon(E, 0)E \exp\left(-\int \mu(x, y, z)\,dz\right)$$

$$+ \int \varepsilon(E_s, \theta)E_s S(x, y, E_s, \Omega)\,d\Omega\,dE_s \qquad (2.1)$$

$$= \text{primary} + \text{secondary}$$

where the line integral is over all tissues along the path of the primary photons reaching the point (x, y) and $\mu(x, y, z)$ is the linear attenuation coefficient. The scatter distribution function S is defined so that $S(x, y, E, \Omega)\,dE\,d\Omega\,dx\,dy$ gives the number of scattered photons in the energy range E to $E + dE$ and the solid angle range Ω to $\Omega + d\Omega$ which pass through area $dx\,dy$ of the detector. The energy absorption efficiency ε of the receptor is a function of both the photon energy and the angle θ between the photon direction and the z axis. The effects of the anti-scatter device can be easily added to this equation if required.

In most applications the receptor will not have an efficiency close to unity and the path length of the photon through the receptor will have an important effect on efficiency. Scattered photons will be absorbed more efficiently than primary photons, so that inefficient receptors will enhance the effects of scatter on the image.

The scatter function S has a complicated dependence on position and on the distribution of tissues within the patient. For many applications, it is sufficient to treat it as a slowly varying function and to replace the very general integral in equation (2.1) with the value at the centre of the image. As the scatter will decrease away from the centre, this will give a maximal estimate of the contrast-degrading effects of the scatter. Equation (2.1) then simplifies to

$$I(x, y) = N\varepsilon(E, 0)E \exp\left(-\int \mu(x, y, z)\,dz\right) + \bar{S}\bar{\varepsilon}(E)E \qquad (2.2)$$

where

$$\bar{S} = \int S(0, 0, E_s, \Omega)\,d\Omega\,dE_s \qquad (2.3)$$

and

$$\bar{\varepsilon}(E)E = \int \varepsilon(E_s, \theta)E_s S(0, 0, E_s, \Omega)\,d\Omega\,dE_s / \bar{S}. \qquad (2.4)$$

In practice, it is the ratio R of the scattered to primary radiation that is either measured or calculated, and an appropriate form of equation (2.2) is then

$$I(x, y) = N\varepsilon(E, 0)E \exp\left(-\int \mu(x, y, z)\,dz\right)(1 + R). \qquad (2.5)$$

2.3 PHOTON INTERACTIONS

To understand the x-ray imaging system, we need to understand the interactions of photons with matter. It is not our intention here to give a full treatment of these interactions, and the reader is referred to Johns and Cunningham (1983) for a more detailed discussion. There are, however, several aspects of these interactions which it is important to consider.

We start by establishing the appropriate photon energy range for diagnostic radiology. Figure 2.3 shows how the transmission of monoenergetic photons through tissue varies with photon energy and tissue thickness. If the transmission is very low, then very few photons will reach the image receptor and the radiation dose to the tissue will be very high. If the transmission is close to unity, then there will be very little difference in transmission through different types of tissue and the contrast in the image will be poor. The choice of energy will therefore be a compromise between the requirements of low dose and high contrast. This is an example of the general principle of requiring semi-opacity for imaging discussed in the introduction. The photon

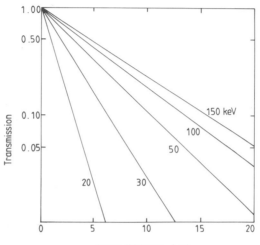

Figure 2.3 Transmission of monoenergetic photons through soft tissue. Curves are shown for photon beams with energies 20, 30, 50, 100 and 150 keV.

energy range used in practice is typically 17–150 keV, with the higher energies being used to image the thicker body sections. In this energy range the important photon interactions are the photoelectric effect and scattering, and figure 2.4 shows the variation of the linear attenuation coefficient with photon energy for both photoelectric absorption and scattering processes. The latter interaction does not result in absorption of the photon and the coefficient for the energy-absorptive component of the scatter, which is due to the transfer of energy to recoil electrons, is also shown. It will be seen that, for soft tissue, the photoelectric cross section is larger than the scatter cross section for energies up to about 25 keV.

Figure 2.5 shows the variation of the linear attenuation coefficients for both bone and soft tissue. The difference between the two coefficients is largely due to differing photoelectric cross sections and densities, and explains why x-rays are so good at imaging broken bones. The difference between the two curves, and hence the contrast between soft tissue and bone, decreases with increasing photon energy.

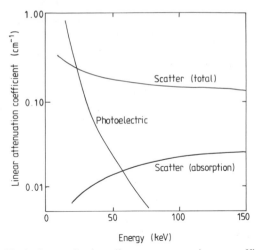

Figure 2.4 Variation of the linear attenuation coefficient with photon energy for soft tissue. The coefficient is shown for photoelectric absorption, scatter and the absorptive part of the scatter process.

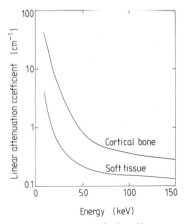

Figure 2.5 Variation with energy of the linear attenuation coefficients for soft tissue and cortical bone.

The photoelectric cross section varies approximately as the fourth power of the atomic number and inversely as the third power of the photon energy, and it shows discontinuities at absorption edges where there is an increase in the cross section due to new processes becoming energetically possible (Johns and Cunningham 1983). The existence of an absorption edge can affect the performance of the image receptor, which can have different efficiency and resolution immediately below and above the edge.

A photoelectric interaction is followed by the ejection of a photoelectron and by one or more characteristic x-rays and Auger electrons. The highest electron energy likely to be produced in this way in diagnostic radiology is 150 keV, and such an electron would have a range in water of only 0.03 cm. For most purposes, therefore, the electrons can be considered as being locally absorbed and will greatly contribute to patient dose and to the energy absorbed in the image receptor. Characteristic x-rays, however, cannot always be considered as being locally absorbed. This is illustrated in table 2.1, which gives the energies, mean free paths and fluorescent yields for selected K-shell characteristic x-rays. (The fluorescent yield gives the probability of a fluorescent x-ray being emitted after the ejection of a photoelectron.) Elements are tabulated to illustrate characteristic x-ray production both in tissue (oxygen and calcium) and in image receptors (silver and gadolinium). It will be seen that the spread of the characteristic x-rays is important for the elements of higher atomic number that are used in image receptors, where it can lead to both a loss of efficiency and to image blurring.

Table 2.1 Properties of K-shell characteristic x-rays.

Element	Mean energy (keV)	Fluorescent yield	Mean free path in water (cm)
O	0.5	0.01	<0.001
Ca	3.7	0.16	0.01
Ag	22.6	0.83	1.8
Gd	43.9	0.93	4.0

The photon scattering cross section varies more slowly with energy than does the photoelectric cross section and is approximately proportional to atomic number. It is therefore less important for providing contrast between tissues with differing average atomic numbers than the photoelectric effect, except at the higher photon energies where the photoelectric cross sections for tissue become small. The energy of the scattered photon is similar to that of the incident photon (for incident photon energies of 25 and 100 keV, the energies of backscattered photons are 22.8 and 71.9 keV, respectively) so that, as illustrated in figure 2.4, the absorptive component of the scatter interaction is small. Because of this and because of the high atomic numbers used, most of the energy deposited in the image receptor is due to photoelectric interactions.

2.4 IMPORTANT PHYSICAL PARAMETERS

The performance of the diagnostic radiology imaging system and the performance of its components can be assessed and specified in terms of just a few physical parameters, and in this section we discuss the most important of these, namely contrast, unsharpness, radiation dose and noise. This discussion will form a basis for the remainder of the chapter, which covers the construction and performance of the various components of the radiographic imaging system.

2.4.1 Contrast and unsharpness

In §2.2 we derived an equation for the radiographic image and we now use this result to obtain an expression for the radiographic contrast. Consider the simple model shown in figure 2.6. The patient is replaced by a uniform block of tissue of thickness t and linear attenuation coefficient μ_1, containing an embedded block of 'target' tissue of thickness x and linear attenuation coefficient μ_2. The 'target' tissue is that volume which it is required to image clearly in the projection radiograph. The contrast C of the 'target' tissue is defined in terms of the image distribution functions I_1 and I_2, which give the energy absorbed per unit area of the receptor outside and inside the (shadow) image of the 'target' tissue, respectively. It is given by

$$C = (I_1 - I_2)/I_1. \tag{2.6}$$

The 'target' is considered to be in the centre of the image and the region with which it is compared is close by, so that the scatter fields in the two regions are very similar. For this to be a reasonable assumption, the lateral extent of the uniform tissue is considered to be large compared with that of the 'target'. The functions I_1 and I_2 can then be obtained from equation (2.2) and are given by

$$I_1 = N\varepsilon(E, 0)E\exp(-\mu_1 t) + \bar{S}\bar{\varepsilon}(E)E \tag{2.7}$$

$$I_2 = N\varepsilon(E, 0)E\exp[-\mu_1(t - x) - \mu_2 x] + \bar{S}\bar{\varepsilon}(E)E. \tag{2.8}$$

So the contrast is given by

$$C = N\varepsilon(E, 0)E\exp(-\mu_1 t)\{1 - \exp[-(\mu_2 - \mu_1)x]\}/I_1 \tag{2.9}$$

which can be simplified using equation (2.5) to

$$C = \{1 - \exp[-(\mu_2 - \mu_1)x]\}/(1 + R). \tag{2.10}$$

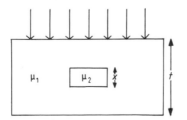

Figure 2.6 Simple model used for the estimation of contrast.

The factors that affect the contrast are therefore seen to be the thickness of the 'target', the difference in linear attenuation coefficients and the scatter-to-primary ratio R. Figure 2.7 shows the dependence of contrast on photon energy for two objects that are important in mammography (the x-ray examination of the breast). It will be seen that the contrast decreases rapidly with increasing photon energy, so that for the best contrast we should use a low photon energy. However, as we have seen already, low energy means high patient dose, and a compromise must be reached between contrast and dose. In deriving equation (2.10), we have assumed that the response of the receptor is linear with the energy absorbed. In practice, however, this is often not the case and the contrast given by equation (2.10) then needs to be modified. In §3.6 the practical implications of equation (2.10) are further developed.

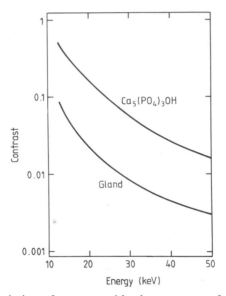

Figure 2.7 Variation of contrast with photon energy for two objects of importance in mammography. The upper curve is for a 100 μm calcification (calcium hydroxyapatite) and the lower curve is for 1 mm of glandular tissue. The contrast is relative to normal breast tissue; contrast degradation due to scatter has been ignored. (Reproduced from Dance and Day (1981) with permission of the Office for Official Publications of the European Communities.)

The unsharpness of the radiological imaging system is another very important parameter and can be expressed in a variety of ways. The most general and useful approach is to use the modulation transfer function (MTF) (see, for example, Barrett and Swindell (1981)), which characterises how well the system can image information at any spatial frequency. It is particularly useful for comparing the contributions to the total unsharpness from the various components in the imaging system. In our case, the contributions to the unsharpness arise from the focal spot of the x-ray tube, which gives a penumbra to the image (geometric unsharpness — see §2.5.3), from the receptor itself and from patient movement during the exposure. In many cases, movement unsharpness can be neglected but it is very important when imaging parts of the body that are in motion, such as the heart and its associated blood vessels. Figures 2.21 and 2.29 give examples of the MTF.

An alternative, and somewhat simpler, approach to unsharpness is to image a bar pattern and to determine the highest bar frequency that can be visualised by the system in line pairs/mm.

2.4.2 Noise and dose

Although the imaging system may have high contrast and good resolution, the radiologist will fail to identify even a large object if the noise level in the image is very high. There are two major contributions to the noise in the radiographic image: statistical fluctuations in the number of x-ray photons detected per unit area (quantum noise) and fluctuations due to the properties of the image receptor and display system. The appearance of the noise in the image will be affected by the spatial-frequency response of the system. A poor high-frequency response will give rise to a somewhat blotchy image and this has led to the use of the terms 'quantum mottle' and 'radiographic mottle'. Because of the frequency variation, it is sometimes convenient to express the noise in terms of its Wiener spectrum or power spectral density function (see Barrett and Swindell (1981) for details).

The noise due to quantum mottle can be reduced by increasing the number of photons used to form the image. This will also increase the dose to the patient and it is instructive to explore the relationship between these two quantities. We use the model shown in figure 2.6 and answer the following question: 'What is the surface dose required to be able to see a contrast C over an area A against a background noise arising purely from quantum mottle'? We first compare the signal we are trying to observe with the background noise to obtain a signal-to-noise ratio (SNR). The signal $\Delta I A$ that we are trying to detect may be found from equations (2.5) and (2.6) by putting $\Delta I = I_1 - I_2$. We then have

$$\text{signal} = \Delta I A = ICA = CAN\varepsilon E \exp(-\mu_1 t) (1 + R). \qquad (2.11)$$

The quantum noise in the image arises from fluctuations in the energy absorbed by the receptor. For simplicity, we assume that each photon that interacts with the receptor is completely absorbed and that the receptor efficiency ε is the same for both primary and secondary photons. The number of photons detected per unit area of the receptor is a Poisson process and the image noise in an area A adjacent to our target area is $E(IA/E)^{1/2}$, or

$$\text{noise} = E[N\varepsilon A \exp(-\mu_1 t)\ (1 + R)]^{1/2}. \tag{2.12}$$

The signal-to-noise ratio is therefore

$$\text{SNR} = C[N\varepsilon A \exp(-\mu_1 t)\ (1 + R)]^{1/2}. \tag{2.13}$$

Substituting equation (2.10) for C gives

$$\text{SNR} = \{1 - \exp[-(\mu_2 - \mu_1)x]\}\ [N\varepsilon A \exp(-\mu_1 t)/(1 + R)]^{1/2}. \tag{2.14}$$

According to Rose (1974), an object becomes detectable when its signal-to-noise ratio exceeds a certain minimum or threshold value. Rose suggested that this threshold was a ratio of 5, but for the moment we shall use the symbol k to denote this quantity. The minimum patient dose will occur at this threshold and, equating equation (2.14) to k and solving, we obtain a value for the number of photons incident on the patient per unit area as

$$N = k^2(1 + R)\exp(\mu_1 t)/[\varepsilon(\Delta\mu\ x)^2 x^2]. \tag{2.15}$$

In deriving this equation we have assumed that the contrast is small, expanded the first exponential in equation (2.14) to the second term and used $\Delta\mu = \mu_2 - \mu_1$. We have also assumed that the object of interest is a cube of side x and have substituted the appropriate value for the area A. The surface dose is then simply obtained as the product of the number of photons per unit area (N), the mass energy absorption coefficient for tissue (μ_{En}/ρ) and the photon energy (E):

$$\text{dose} = (\mu_{En}/\rho)Ek^2(1 + R)\exp(\mu_1 t)/[\varepsilon(\Delta\mu)^2 x^4]. \tag{2.16}$$

An important result follows from this equation: the minimum dose required to visualise an object increases as the inverse fourth power of the size of the object. For fixed dose and contrast, there will be a minimum object size that can be visualised and the low-contrast resolution of the system will vary with object size. This subject is taken up again in some detail in Chapter 13. It should be pointed out, however, that our model is very idealised and the real problem of viewing abnormality against a background with complicated architecture is considerably more difficult. It should also be noted that film-based receptors constrain the dose to within certain limits because a minimum dose is required to achieve any appreciable blackening and the film itself will saturate above a maximum dose. The speed of the image receptor will have a critical effect on the noise in the image and hence on the resolution attainable at low contrasts.

It is instructive to substitute numerical values into equations (2.15) and (2.16) and to calculate the number of incident photons per unit area and the surface dose for imaging 1 mm^3 of tissue with 1% contrast. We use the values $E = 50$ keV, $\varepsilon = 0.3$, $x = 1$ mm, $k = 5$, $\mu_1 = 22.6$ m^{-1}, $(\mu_{En}/\rho) = 0.004$ m^2 kg^{-1}, $\Delta\mu x = 0.03$, $(1 + R) = 3$ and $t = 0.2$ m, which gives $N = 2.6 \times 10^{13}$ photons/m^2 and a surface dose of 0.8 mGy.

The signal-to-noise ratio is a very important imaging parameter and is a quantity that will vary through the various stages of image production as more noise is introduced. For example, the noise level will increase because a receptor does not absorb every incident photon, and the signal-to-noise ratios for the pattern of photons before the detector and for the pattern of photons absorbed by the detector are then related by

$$[\text{SNR(out)}/\text{SNR(in)}]^2 = \varepsilon. \tag{2.17}$$

This ratio forms a very useful parameter, which tells us how the signal-to-noise ratio varies through the system. It is known as the detective quantum efficiency (DQE). The DQE for a complete imaging process is the product of the DQE values for each stage in the imaging process, and study of the individual DQE and their product gives a very clear picture of the transfer of noise through the imaging process.

Table 2.2 gives some typical doses obtained in clinical practice (Shrimpton *et al* 1986, and unpublished data from the Royal Marsden Hospital). It will be seen that they are comparable in magnitude to the above dose estimate and therefore that radiographs are taken fairly close to noise-limited conditions. It is important to reduce the radiation dose to the patient as far as possible because there is a small but significant risk associated with the use of ionising radiation (see Chapter 15).

Table 2.2 Doses for some common radiological examinations.

Examination[a]	Dose (mGy)
CC breast	1.2
AP chest	0.3
AP lumbar spine	9.2
AP pelvis	6.6
AP skull	4.4

[a] CC = cranio-caudad view or projection.
  AP = antero-posterior view or projection.

This risk can be split into genetic and somatic components, which must be considered separately. The genetic injury associated with the former component will depend upon dose to the gonads and the number of children conceived after the irradiation. The genetic dose for a population is usually expressed in terms of the *genetically significant dose* (GSD). This is defined as the dose which, if given to the whole population, would produce the same total genetic injury as the actual doses received by the individuals concerned (UNSCEAR 1972 p134). The GSD to the population of Great Britain arising from diagnostic radiology is 0.118 mGy, which represents 10.6% of the total GSD. Other man-made sources of radiation account for a further 2.4% of the GSD and the remainder arises from naturally occurring background radiation (Darby *et al* 1980, Wall and Shrimpton 1981).

The most important somatic effect of radiation is carcinogenesis. The somatic risk for an individual organ will depend upon the dose to the organ, but any estimate of the risk to the whole body should take into account the different risk factors for different organs. The ICRP (1977) have suggested the use of the quantity *effective dose equivalent*, which is a prescription for calculating the dose which, if given uniformly to the whole body, would produce the same somatic detriment as the actual exposure received by the patient. Individual organ doses will depend not only on the surface dose but also on the depth and position of the organ and the size and quality of the x-ray field. The relationship is complex but can be estimated by Monte-Carlo calculation or by measurement. An alternative approach is to use the total energy imparted to the patient by the radiation field. This is much simpler to estimate and is reasonably well correlated with the effective dose equivalent (Shrimpton 1985).

Table 2.3 gives the effective dose equivalent and selected organ doses for various diagnostic procedures. It is based on the national survey of Shrimpton *et al* (1986). The risks from ionising radiation are further discussed and set in the context of risks from other imaging techniques in Chapter 15.

Table 2.3 Effective dose equivalent (mSv) and organ doses (mGy) (breast, red bone marrow, lung, thyroid, skin, ovaries and testes) for selected radiological examinations.

Examination	Doses per examination							
	Eff. dose equiv.	Breast	RBM	Lung	Thyroid	Skin	Ovary	Testes
Barium meal	3.8	2.2	2.6	8.7	1.1	2.1	3.6	0.3
Barium enema	7.7	0.7	8.2	3.2	0.2	5.1	16.0	3.4
IVU[a]	4.4	0.7	1.9	7.0	0.2	1.9	0.8	0.1
Cholecystography	1.0	0.4	0.8	1.6	0.1	0.8	0.4	0.0

[a] Intravenous urography.

2.5 X-RAY TUBES

2.5.1 Tube construction

The x-ray tube used in diagnostic radiology consists of an oil-filled housing containing an insert (Forster 1985 pp56–68), which is an evacuated envelope of heat-resistant borosilicate glass within which are mounted a filament and an anode (figure 2.8). The filament is heated by passage of an electric current. It produces a narrow beam of electrons, which are accelerated by a potential difference of between 25 and

150 kV and strike the anode. The electrons interact with the material of the anode, slow down and stop. Most of the energy absorbed from the electrons appears in the form of heat, but a small amount (less than 1%) appears in the form of x-rays. Some of these x-rays pass through the exit windows of the insert and housing and through the patient to form the x-ray image. X-rays that are emitted in other directions are absorbed by the housing. The complete tube assembly is mounted on a support and provided with collimation so that the beam direction and the size of the radiation field may be varied as necessary.

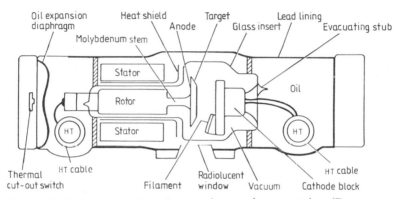

Figure 2.8 The construction of a rotating-anode x-ray tube. (Reproduced with permission of MTP Press Ltd from Forster (1985).)

The design of the filament and the electron optics that guide the electrons to the anode is very important because the unsharpness in the image may be limited by the size of the x-ray source, and the output from the tube is determined by the electron current striking the anode. The filament is constructed from a spiral of tungsten wire (melting point 3410 °C), which is set in a nickel block. This block supports the filament and is shaped to create an electric field that focuses the electrons into a slit beam. The anode has a bevelled edge, which is at a steep angle to the direction of the electron beam. The exit window accepts x-rays that are approximately at right angles to the electron beam so that the x-ray source as viewed from the receptor appears to be approximately square even though the electron beam impinging on the target is slit-shaped (figure 2.9). The choice of the anode angle θ will depend upon the application, with the angle being varied according to the requirements of field and focal spot sizes and tube output. For general-purpose units, an angle of about 17° is appropriate. In many cases the anode disc will have two bevels at different angles and two filaments, so that either a fine or a broad focus may be selected.

It has already been observed that most of the energy in the electron beam is deposited in the target in the form of heat, and one of the problems faced by the tube designer is how to limit the heat deposited in the target area and how to remove it from that area as quickly as possible. The use of a slit source of electrons helps by spreading out the target area and this idea can be extended by using a rotating anode, so that the electron beam impinges on the bevelled edge of a rotating disc and the target area is spread out over the periphery of the disc. A rotation speed of about 3000 RPM and an anode diameter of 10 cm are used in general-purpose units. It is also possible to use a stationary target and these are sometimes to be found in simple, low-power equipment, such as mobile and dental x-ray units.

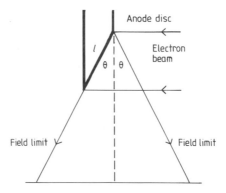

Figure 2.9 The use of a bevelled anode to reduce the effective focal-spot size. The width of the electron beam is $l \cos \theta$ whereas the focal-spot size as viewed from the central axis of the x-ray field is $l \sin \theta$.

The anode is usually constructed from tungsten although molybdenum is used for special applications where a low-energy x-ray beam is required (see §2.5.2). Tungsten has an atomic number of 74, acceptable thermal conductivity and thermal capacity, and a high melting point. The high atomic number is important because the bremsstrahlung yield from the target increases with atomic number and the x-ray spectrum from an element of high atomic number is well suited to imaging the thicker body sections. Improved tube lifetime can be obtained by using a 90/10 tungsten/rhenium alloy. This reduces crazing of the anode surface caused by the continual heating and cooling processes to which it is subjected. It is important that the anode disc has a high thermal capacity. A larger anode disc will give a higher rating and a shorter exposure time, and the greater thermal capacity associated with an increase in anode volume will allow the possibility of a shorter time interval between exposures. For heavier-duty applications, the thermal capacity can be increased by using a molybdenum backing to the anode. Molybdenum has a higher specific heat than tungsten (table 2.4) and the heat capacity for an anode of this type would typically be 250 000 J.

The anode disc is mounted on a thin molybdenum stem. This reduces heat flow backwards and prevents the rotor bearings, which are made from copper, from overheating. The heat loss from the rotating anode is mainly radiative. The requirements of x-ray tubes for CT are mentioned in Chapter 4.

2.5.2 X-ray spectra

The shape of the x-ray spectrum will depend upon the target material, the potential and waveform applied to the tube and the effects of any filters placed in the x-ray beam. Table 2.4 shows the properties of the two most commonly used target materials and figures 2.10, 2.11 and 2.12 show three typical x-ray spectra. The first two spectra are for tubes with tungsten targets and the other spectrum is for a tube with a molybdenum target. Also shown are the spectra after modification by transmission through the body.

Table 2.4 Properties of molybdenum and tungsten.

	Mo	W
Atomic number	42	74
K x-ray energies (keV)	17.4–19.8	58.0–67.7
Relative density	10.2	19.3
Melting point (°C)	2617	3410
Specific heat (J kg^{-1} °C^{-1})	250	125

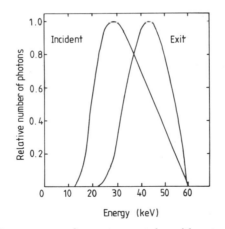

Figure 2.10 X-ray spectra for an x-ray tube with a tungsten target; 60 kV constant potential with 1.5 mm aluminium added. The spectra are shown both before and after attenuation by 9.5 cm soft tissue plus 0.5 cm bone. (The spectra are based on the work of Birch *et al* (1979).)

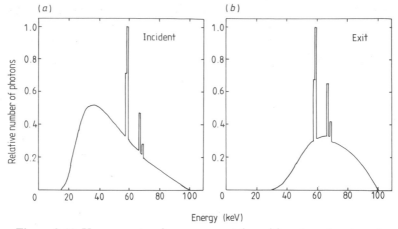

Figure 2.11 X-ray spectra for an x-ray tube with a tungsten target; 100 kV constant potential with 2.5 mm aluminium added. The spectra are shown both before and after attenuation by 18.5 cm soft tissue plus 1.5 cm bone. (The spectra are based on the work of Birch *et al* (1979).)

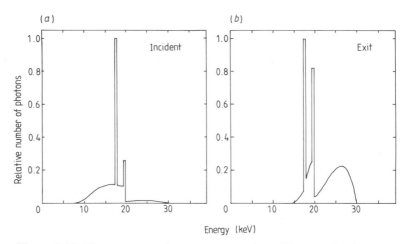

Figure 2.12 X-ray spectra for an x-ray tube with a molybdenum target; 30 kV constant potential with 0.03 mm molybdenum filter. The spectra are shown both before and after attenuation by 5 cm tissue. (The spectra are based on the work of Birch *et al* (1979).)

The tungsten spectra are well suited to imaging the thicker body sections because of the energies of the tungsten characteristic x-rays. Molybdenum has lower-energy x-rays, which are more appropriate for imaging thinner body sections at high contrast. It is used in x-ray units specifically designed for mammography. In figure 2.12 the x-ray spectrum from the molybdenum target has been modified by the use of a molybdenum filter, which heavily attenuates any x-rays that have energies above its K-edge of 20.0 keV. This is standard practice for a mammographic x-ray unit.

There is a large difference between the x-ray spectra before and after passage through the patient. The difference between these spectra is due to the photons that interact in the patient and deliver dose. If the spectrum is too soft, then the lower-energy photons will only contribute to patient dose and not to image contrast, and it is important that they be removed from the x-ray beam before it reaches the patient. This can be achieved by the use of aluminium or copper filters, depending on the tube potential.

The choice of the tube potential and filtration most appropriate to a particular application will be a compromise between contrast and dose, as explained in §2.4. As we have seen already, we cannot reduce the dose to a very low level because of the noise in the image and it is possible to use this fact to predict optimum energies for imaging different body thicknesses. Dance and Day (1981) have modelled the mammographic examination and have investigated how the signal-to-noise ratio in the image varies with photon energy and breast thickness. This is a particularly important investigation to optimise because it is at present the technique of choice for population screening for breast cancer and, even though dose levels are low, there is a small risk of carcinogenesis associated with the examination. (Comments on the relation between the roles of x-ray mammography and other breast imaging modalities are to be found in Chapters 10 and 11. The question of risk is addressed in Chapter 15.) Sample results from Dance and Day (1981) are shown in figure 2.13. Each curve is for fixed breast thickness and gives the signal-to-noise ratio for imaging a 0.1 mm microcalcification (a speck of calcified material that is often found within the breast and which it is important to be able to see on an x-ray) against a uniform background. The breast dose is fixed at 1 mGy. Each curve passes through a maximum and the position of this maximum gives the optimum energy for imaging a microcalcification for a breast of a particular thickness. At this energy, the exposure could be reduced to achieve the desired signal-to-noise ratio at the least possible patient dose. The curves shown in the figure are appropriate to imaging systems where acceptable results can be obtained over a wide range of exposures, so that the dose can be adjusted to give the required noise level without the image being either under- or overexposed. Digital imaging systems satisfy this requirement but the more conventional film-based systems may not. Nevertheless, the curves shown in figure 2.13 provide useful data, which can be of value in clinical practice. They demonstrate clearly how the optimum energy varies with breast thickness. If we look at the mammographic x-ray spectrum shown in figure 2.12, we see that it is most appropriate for imaging the smaller breast. We can also see that it is less appropriate for imaging very large breasts where it would be preferable to use a slightly higher-energy spectrum. Such a spectrum can be obtained by using a different K-edge filter and a tungsten target.

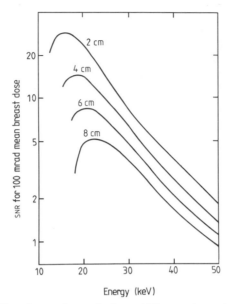

Figure 2.13 Signal-to-noise ratio for a 100 μm microcalcification (calcium hydroxyapatite, $Ca_5(PO_4)_3OH$) embedded in a breast. The SNR has been calculated for a Kodak Min-R screen and a mean breast dose of 1 mGy. Curves are given for breast thicknesses in the range 2–8 cm. (Reproduced from Dance and Day (1981) with permission of the Office for Official Publications of the European Communities.)

2.5.3 Geometric unsharpness

Geometric unsharpness is the unsharpness U_g in the image due to penumbra from the finite size of the x-ray source. Using the geometry and notation shown in figure 2.14, the blurring B in the image is given by

$$B = ad_2/d_1 = a(m - 1) \tag{2.18}$$

where a is the effective size of the focal spot of the x-ray tube and m is the image magnification. It is convenient in practice to correct for the image magnification by dividing by m, so that the geometric unsharpness becomes

$$U_g = a(1 - 1/m). \tag{2.19}$$

In order to assess the importance of the geometric unsharpness, we must combine it with the other components of the image unsharpness (§2.4.1). If we neglect movement unsharpness and if U_r is the unsharpness due to the receptor, then the overall unsharpness U in the image is given by

$$U = (U_r^2 + U_g^2)^{1/2} \qquad (2.20)$$

or

$$U = F[1/m^2 + (1 - 1/m)^2 a^2/F^2]^{1/2} \qquad (2.21)$$

where we have assumed for simplicity that the two types of unsharpness can be added in quadrature and have used the relation

$$U_r = F/m \qquad (2.22)$$

in deriving (2.21). Here F is the intrinsic receptor unsharpness which would obtain if the object had zero thickness and was in contact with the receptor.

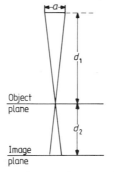

Figure 2.14 Geometry and notation used for the calculation of geometric unsharpness (equation (2.18)) (not to scale).

We can see from equation (2.21) that, if the geometric unsharpness is much smaller than the receptor unsharpness, then the overall unsharpness will be inversely proportional to the magnification, and magnification radiography will reduce the overall image unsharpness. If, however, the receptor unsharpness is much smaller than the geometric unsharpness, then the overall unsharpness will increase with increasing magnification. This is illustrated in figure 2.15, which shows how U/F varies with magnification. Each curve is for a different value of the ratio a/F. A value of less than about 2 for a/F indicates that magnification radiography will reduce unsharpness and a value of more than about 2 indicates that magnification radiography will increase the unsharpness. The following conclusions can be drawn from these curves.

(*a*) For general radiography with $a/F > 2$: the overall unsharpness will increase with increasing magnification; the patient should be as close as possible to the image receptor and the focus–receptor distance should be as large as possible.

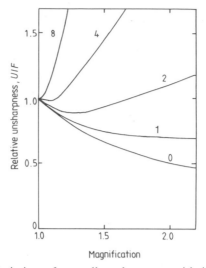

Figure 2.15 Variation of overall unsharpness with image magnification. The unsharpness has been normalised to the receptor unsharpness F. The curves are labelled 0, 1, 2, 4, 8 for focal-spot sizes of 0, F, $2F$, $4F$ and $8F$ respectively.

(*b*) For magnification radiography with $a/F < 2$: the overall unsharpness will decrease with increasing magnification at small to moderate magnifications and will then increase; a significant reduction in unsharpness is only possible if an appropriate magnification is used.

The focal-spot sizes of a general radiography unit with twin foci are typically 0.6 and 1.0 mm, whereas the receptor unsharpness for a screen–film receptor might be 0.1–0.2 mm. The condition $a/F > 2$ holds for such a unit and the overall unsharpness will be limited by the size of the focal spot and the geometry used. Magnification radiography uses a very small focal spot and exposure times will need to be increased to compensate for the reduced tube output. In addition, it is necessary to increase dose or use a noisier receptor to maintain the same density on the x-ray film. The image will also increase in size and may become too large for the receptor! For magnification mammography, a focal-spot size of 0.1 mm and a magnification of 1.5 are often used but the increase in dose and the image size mean that the technique is used to provide an additional view of a suspicious area rather than as a standard part of each examination. Recently, tomographic techniques have been proposed using a microfocal-spot tube, an area detector with small receptor unsharpness and high magnifications. The requirement for a/F to be very much less than 1 for such a system is apparent from figure 2.15.

2.6 IMAGE RECEPTORS

Many different types of image receptor are used in modern diagnostic radiology. They all form an image by the absorption of energy from the x-ray beam but a variety of techniques are used to convert the resulting energy distribution into something that can be visualised by eye. In this section are discussed the various image receptors and the physical factors that determine and limit their performance.

2.6.1 Direct-exposure x-ray film

Direct-exposure x-ray film is only used in special applications in radiology because it has a low absorption efficiency for photons in the diagnostic energy range. However, film *is* used as the final display medium for many types of imaging system and we therefore study its properties in some detail.

2.6.1.1 Construction and image formation. Figure 2.16 shows the construction of a typical direct-exposure (or non-screen) film. The film consists of two photographic emulsions deposited on either side of a transparent polyester or acetate sheet, which is known as the film base. The emulsions are attached to the base by a subbing layer and have a thin surface coating to give protection against abrasion. Each emulsion consists of grains of silver bromide suspended in gelatin and each grain has a diameter of about 1 μm. Two emulsions are used to increase the x-ray absorption efficiency.

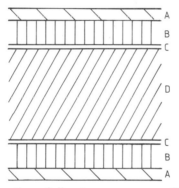

Figure 2.16 Construction of direct-exposure x-ray film: A, protective coating; B, film emulsion (20 μm) of silver halide grains in gelatin; C, subbing layer; D, film base (200 μm). (After Barrett and Swindell (1981) p196.)

The first stage of the image formation is the interaction of the x-ray photons with the atoms within the emulsion. As the silver and bromine atoms within the halide grains have much larger interaction cross sections than do the light elements within the gelatin, most of these interactions will occur within the grains. One or more electrons are released following each interaction and, as these initial electrons slow down, they release further electrons by ionisation. Some of these electrons are eventually captured at trapping centres within the silver bromide grains. This trapping process sensitises the grains and a latent image is built up of such grains. When the image is developed and fixed, the sensitised grains are converted to silver and the unsensitised grains are removed. The photographic process is in fact very complicated and the description given here has been considerably simplified. The reader is referred to Mees and James (1966) for a detailed treatment.

2.6.1.2 The characteristic curve. The blackening on film is usually expressed in terms of the optical density D, which is defined using the transmission of light through the film by

$$D = \log_{10}(I_0/I) \qquad (2.23)$$

where I_0 and I are the intensities of a light beam before and after passage through the film. Our initial aim is to find a relationship between the optical density and the exposure to the film. This is straightforward if we make a number of simplifying assumptions. Our approach is similar to that adopted by Barrett and Swindell (1981). We start by considering the film exposure. Let us suppose that we have a single emulsion containing G grains per unit area and that after irradiation with N photons per unit area there are g sensitised grains per unit area. Let b be the cross-sectional area of a single grain. Assume that a single hit is sufficient to sensitise a grain. The photon absorption efficiency is low, so we need not worry about overlapping grains. The change dg in the number of sensitised grains per unit area caused by a change dN in the number of incident photons per unit area is given by

$$dg = (G - g)b \, dN \qquad (2.24)$$

so that

$$g = G[1 - \exp(-bN)] \qquad (2.25)$$

or

$$g = G[1 - \exp(-kX)] \qquad (2.26)$$

where we have rewritten equation (2.25) in terms of the dose X to the film and k is appropriately defined. Next we relate the number of sensitised grains to the optical density on the developed film. Let us assume that every sensitised grain results in a silver speck of cross-sectional area σ after the film has been developed and let us illuminate the film with light of intensity I_0. Consider a thin slice of the developed emulsion at depth y and of thickness dy and let $I(y)$ be the light intensity transmitted through thickness y of the emulsion. Any light photon that hits a developed grain will be absorbed so that the change dI in the intensity $I(y)$ arising from the slice dy of the emulsion is given by

$$dI = -I(y)(g/t)\sigma \, dy \qquad (2.27)$$

where t is the thickness of the emulsion. This equation can be integrated to give

$$I(t) = I_0 \exp(-g\sigma) \qquad (2.28)$$

or, using equation (2.23) for the optical density,

$$D = 0.434g\sigma. \qquad (2.29)$$

This result is known as Nutting's law. If we substitute $g = G$ then we obtain an expression for the maximum possible density on the film, i.e.

$$D_{max} = 0.434G\sigma. \qquad (2.30)$$

A typical size for a developed grain is 2.5 μm so that a maximum optical density of 3.0 would correspond to 1.1×10^8 grains/cm^2.

We now combine equations (2.26), (2.29) and (2.30) to obtain our relationship between the optical density and the dose to the film, which is

$$D = D_{max}[1 - \exp(-kX)]. \qquad (2.31)$$

This very simple relationship is, of course, a result of our simplifying assumptions. Nevertheless, it provides good insight into the behaviour of the film. Figure 2.17 shows equation (2.31) in graphical form. We have followed standard practice and have plotted the film dose on a logarithmic scale. A baseline optical density has been added to the optical density given by equation (2.31) to allow for the fact that, when an unexposed film is developed, some of the grains are still reduced to metallic silver. The magnitude of this background or 'fog' level will depend upon both the emulsion and the development of the film.

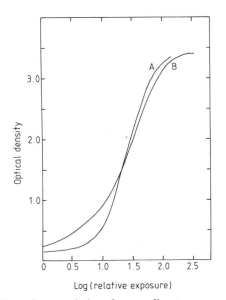

Figure 2.17 Film characteristics for a direct-exposure x-ray film (curve A, equation (2.31)) and for a screen film (curve B). The speed of the direct exposure film has been increased so that the shapes of the two curves can be compared. The two films have the same fog level and maximum density.

Equation (2.31) was derived on the basis that a single hit is sufficient to sensitise a grain. This assumption works well for direct-exposure film but is not appropriate for a film that is exposed by light photons, where it is necessary to develop a multi-hit theory. We have therefore included in the figure a curve for a screen film that has been exposed by light photons.

Curves that relate optical density to film exposure are known as characteristic curves or H and D curves (after Hurter and Driffield). Their important features are as follows.

(*a*) The film characteristic will have a central portion for which the relationship between optical density and the logarithm of dose is approximately linear, so that

$$D = \Gamma \log_{10}(X/X_0). \tag{2.32}$$

For a small contrast $\Delta X/X$, we can expand the logarithm and obtain the associated change in optical density ΔD as

$$\Delta D = 0.434\Gamma\Delta X/X. \tag{2.33}$$

The constant Γ is known as the film gamma and would typically be in the range 2–3. It should be noted that the change in optical density associated with the contrast $\Delta X/X$ remains constant over the portion of the characteristic where equation (2.33) is valid. For the higher values of gamma, some contrast gain will be possible (see also §3.8.2). The film gamma can be increased by increasing the grain size (equation (2.29)).

(*b*) For exposures that result in an optical density which does not lie on the log-linear region of the film characteristic, the response of the film receptor is poor. For high exposures, most of the grains will be sensitised and the slope of the characteristic curve will decrease, resulting in a shoulder and then saturation at $D_{\max}$. For low exposures using light photons, where several hits may be required to sensitise a grain, the characteristic will have a toe region where its gradient is small and the contrast poor. It is important, therefore, that films are exposed so that the regions of interest in the patient produce densities that lie in the log-linear portion of the characteristic. The range of exposures for which this is possible is known as the film latitude. High-contrast films with a high gamma will have a narrow latitude, and low-contrast films with a low gamma will have a wide latitude.

It will be seen from the above that correct exposure of the radiographic film is very important. Over- or underexposure will result in loss of contrast and possibly in the loss of useful diagnostic information. The contrast and latitude of the film must be matched as far as possible to the application. In many cases, it will be difficult to judge the correct exposure from the size of the patient and it is advantageous to use automatic exposure control. This can be achieved by using a suitably positioned ionisation chamber or other detector as a monitor. In §2.7 it will be seen that the recent developments in digital radiology can help to overcome these somewhat hit-and-miss limitations of analogue radiology.

2.6.1.3 Sensitivity and unsharpness. We now look at the suitability of direct-exposure film as an image receptor for use in diagnostic radiology. We have already seen that we can achieve good contrast using film and now consider dose and unsharpness. A discussion of the effects of film granularity on image noise is deferred until §2.6.2.5. The sensitivity of an x-ray film will depend upon the size and packing density of the grains, the emulsion thickness, the x-ray absorption efficiency and the development process. Figure 2.18 shows the energy absorption efficiency of a typical double-emulsion film. It will be seen that, except at the lowest energies, its efficiency is very low and the dose associated with its use will therefore be high. The discontinuity in the efficiency arises from the K absorption edge of silver, which is at 25.5 keV.

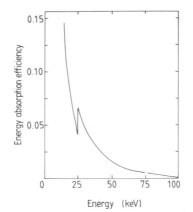

Figure 2.18 Variation of the energy absorption efficiency of a direct-exposure film with photon energy. The film emulsions contain 3 mg cm^{-2} AgBr.

The resolution of the film is limited by the ranges of the secondary particles produced within it. The range of a 50 keV photoelectron in photographic emulsion is 70 μm and the mean free path of a characteristic x-ray from a silver atom is 400 μm. From the dimensions given in figure 2.16 we see that the crossover of electrons from one emulsion to the other is unimportant. Furthermore, the absorption of characteristic x-rays in either emulsion is also unimportant because of their low mean free path. This means that the lateral spread in energy will be approximately limited by the thickness of the emulsion (about 20 μm). The receptor will therefore have excellent resolution and will find clinical application where this feature is more important than the relatively high associated dose. Examples include the radiology of the hands and teeth.

2.6.2 Screen–film combinations

Screen–film receptors are faster than direct-exposure films but do not have such good inherent resolution. They are, however, the receptor of choice for many radiographic examinations where the dose saving is more important than the loss of very fine detail.

2.6.2.1 Construction and image formation. Screen–film receptors produce an image in a four-stage process. An x-ray photon is absorbed by the screen and some of its energy is then re-emitted in the form of light fluorescent photons by the phosphor contained within the screen. The light fluorescent photons expose the emulsion of a film in contact with the screen and the film is then developed and viewed in the normal way. Figure 2.19 shows the construction of a typical fluorescent screen. The phosphor layer consists of active phosphor particles in a binding material. The average size of the phosphor particles is about 10 μm and the phosphor thickness some 70–300 μm with a total coating of typically 50–170 mg cm^{-2} (Barnes 1982, Barrett and Swindell 1981). The packing fraction of the phosphor within the binder is about 50%. The active layer of the screen is supported by a sheet of cardboard or plastic and, depending upon the application, the backing layer between the phosphor and this sheet may be reflective to increase light output. Absorptive dyes may also be used to control the light output.

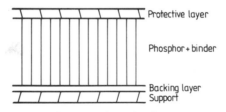

Figure 2.19 Construction of a fluorescent screen.

2.6.2.2 Unsharpness. In §§2.6.1.2 and 2.6.1.3 we saw that the unsharpness for direct-exposure x-ray film is very small. The contribution of the film to the unsharpness of the screen–film combination is also small. The resolution is limited by the construction of the screen and in particular by the lateral spread of the light fluorescent photons as they pass from screen to film. This spread increases with increasing separation between the point of emission of the light and the emulsion; good contact between screen and film is essential. This is achieved by exposing the film in a specially designed light-tight cassette or similar device.

Fluorescent screens can be used in pairs, with a screen on either side of a double-emulsion film, or singly, with the screen placed behind the film but in contact with the emulsion (figure 2.20). More x-ray photons are absorbed in the front half of a screen than in the back half and this configuration for a single screen brings the production point for the fluorescent x-rays as close as possible to the emulsion. Single screens are used for applications where the improved resolution is more important than the dose reduction associated with the use of two screens. Mammography is an important example of the use of a single-screen single-emulsion receptor (Dance and Davis 1983).

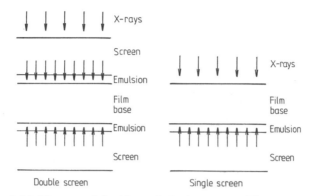

Figure 2.20 The use of double and single screens. The arrows at the top of each figure represent the incident x-rays, which interact in the screens, and the arrows originating in the screens represent the light fluorescent photons, which expose the film emulsion.

The unsharpness of the screen–film receptor can be controlled by varying the thickness of the phosphor and by the use of absorptive dyes that favour those light fluorescent photons produced closest to the emulsion. There is, however, another factor that may limit the resolution of double-screen systems. Light fluorescent photons produced in the screen on one side of a film can expose the emulsion on the other side of the film. This phenomenon is known as crossover.

Blurring is also caused by the interactions of characteristic x-rays with either screen. These may be emitted by a heavy atom in the screen following the photoelectric absorption of the initial x-ray. In such a case, the unsharpness of the receptor may be different below and above the position of the absorption edge of the atom concerned (Arnold and Bjarngard 1979).

Figure 2.21 shows the modulation transfer functions for two screen–film receptors of different speeds (Doi *et al* 1982). The unsharpness becomes worse as the speed of the receptor increases, but nevertheless the resolution is more than adequate for most applications in diagnostic radiology (cf §2.5.3).

2.6.2.3 Sensitivity. The sensitivity of a screen–film receptor will depend upon the x-ray absorption efficiency of the screen(s), the efficiency with which the energy deposited in the screen(s) is converted to light fluorescent photons, the probability of a light fluorescent photon reaching the emulsion, the sensitivity of the emulsion and the development of the film. Absorption of the x-rays by the film emulsion contributes only a few per cent of the total film blackening.

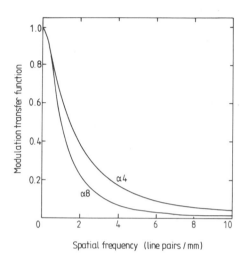

Spatial frequency (line pairs / mm)

Figure 2.21 Modulation transfer functions for 3M Trimax Alpha 4 and Trimax Alpha 8 screens used with OG film. The Alpha 8 screen is approximately twice as fast as the Alpha 4 screen. (The data are taken from Doi *et al* (1982).)

There are several different phosphors that can be used in the construction of fluorescent screens, but for simplicity we limit our discussion to two of the most important: calcium tungstate and terbium-activated rare-earth oxysulphide (X_2O_2S with X being either gadolinium, lanthanum or yttrium) phosphors. Both of these phosphors contain atoms of high atomic number and have a high energy absorption coefficient. Figure 2.22 shows the mass energy absorption coefficients for calcium tungstate and for gadolinium oxysulphide and figure 2.23 shows the energy absorption efficiency for a pair of gadolinium oxysulphide screens. The figures both illustrate the effect of absorption edges on the screen efficiency. The K-edges of gadolinium and of tungsten are at 50.2 and 69.5 keV, respectively, so that the rare-earth screen has the higher absorption efficiency for the mid-energy range. The broken curve in figure 2.23 gives the interaction efficiency whereas the full curve gives the energy absorption efficiency. The difference between the two arises mainly from those characteristic x-rays which are produced in the screen but escape without being absorbed. If the fluorescent yield is high, such photons can carry off a large fraction of the energy of the incident x-ray photon. This fraction will depend upon the photon energy and receptor thickness and can be estimated analytically using the method developed by Dance and Day (1985).

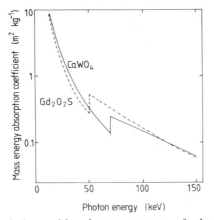

Figure 2.22 Variation with photon energy of the mass energy absorption coefficient for calcium tungstate and gadolinium oxysulphide.

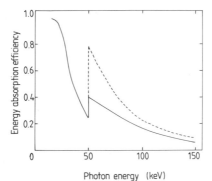

Figure 2.23 Variation with photon energy of the energy absorption efficiency of a pair of gadolinium oxysulphide screens. The phosphor thickness in each screen is 50 mg cm^{-2}. The broken curve has been calculated on the assumption that every photon that interacts in the screen is completely absorbed and the full curve is the true energy absorption efficiency.

Comparison of figure 2.23 with figure 2.18 shows that the energy absorption efficiency of a screen–film combination is much larger than that for a non-screen film.

The efficiency for conversion of energy absorbed in a screen to light fluorescent photons is very different for rare-earth and tungstate phosphors. This efficiency will vary with phosphor thickness and the presence or absence of absorptive dyes, but for an idealised screen it is 3.5% for calcium tungstate and 12%, 15% and 18% for lanthanum, gadolinium and yttrium oxysulphides, respectively (Stevels 1975). The efficiency for this light to reach the emulsion varies with screen construction but a typical value would be about 0.5. We can use this result to calculate the number of light fluorescent photons reaching the emulsion per incident x-ray absorbed. If we take the incident x-ray energy to be 50 keV and use a gadolinium oxysulphide screen that emits green light with a wavelength of say 0.5 μm (2.5 eV), then the number of light photons reaching the emulsion is (x-ray energy/photon energy) × (photon production efficiency) × (photon detection efficiency), namely

$$(50 \times 10^3 \times 0.15 \times 0.5)/2.5 = 1500$$

i.e. 1500 light photons per incident absorbed x-ray photon.

The final stage of the image production process is the exposure and development of the film, and it is important here that the screen and receptor be matched: the light from a tungstate screen is blue whereas the light from a terbium-activated rare-earth screen is green. Conventional screen film is sensitive to blue light and is suitable for tungstate screens, but for rare-earth screens it is necessary to use special film that has molecules of a green-sensitive dye adsorbed on the surface of the silver halide grains.

For most exposures, the density on the film depends only on dose and shows little dependence upon exposure time. However, for very long or for short exposures, it does depend upon dose rate. This phenomenon is known as reciprocity-law failure and it can be important when long exposures are used. This is illustrated in figure 2.24, which is taken from the work of Arnold *et al* (1978) and shows how the exposure for fixed optical density varies with exposure time.

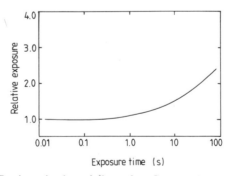

Figure 2.24 Reciprocity-law failure for Cronex 2 x-ray film (blue-sensitive). Relative exposure is the light exposure required for an optical density of 1.0 above fog. (After Arnold *et al* (1978).)

We have seen that there are many factors that influence the sensitivity of a screen–film detector and that there is a compromise between speed and unsharpness. Figure 2.25 quantifies this compromise and shows both the relative sensitivity and modulation transfer function (at a frequency of 2 cycles/mm) for a range of screen–film systems. The data are taken from Doi *et al* (1982) and clearly demonstrate the dose advantage of using a rare-earth screen.

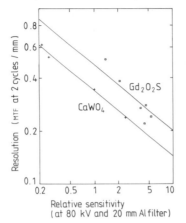

Figure 2.25 Relationship between relative sensitivity and resolution for screen–film systems. The MTF at a frequency of 2 cycles/mm is chosen as a measure of resolution. The sensitivity of the screen–film systems is relative to that of the Par/XRP system measured at 80 kV. (The data are taken from Doi *et al* 1982.)

2.6.2.4 Contrast. Screen–film receptors can either enhance or reduce the contrast in the radiographic image. Two factors are at play. Most obviously, the contrast will depend upon the slope or gamma of the film characteristic curve, which can be selected to give wide latitude and low contrast, or narrow latitude and high contrast, depending upon the requirements of the object. Additionally, the contrast will be affected by the efficiency of the screen, which in general decreases with energy but does contain a large discontinuity at the position of an absorption edge (see previous section). As a consequence, the contrast can either be enhanced or reduced by the energy response of the receptor, depending upon the x-ray spectrum that is transmitted through the patient and the type of screen.

2.6.2.5 Noise. The noise in the image produced by a screen–film receptor arises from five principal sources:

(a) fluctuations in the number of x-ray photons absorbed per unit area of the screen(s) (quantum mottle);

(b) fluctuations in the energy absorbed per interacting photon;

(c) spatial fluctuations in the screen absorption associated with in-homogeneities in the phosphor coating (structure mottle);

(d) fluctuations in the number of light fluorescent photons emitted per unit energy absorbed and

(e) fluctuations in the number of silver halide grains per unit area of the emulsion (film granularity).

The most important of these are the quantum mottle and the film granularity. Structure mottle contributes typically 10% to the total noise and noise source contributions (*b*) and (*d*) may be regarded as making small modifications to the quantum mottle in the image.

In §2.6.1.2 we derived an expression that related optical density to contrast and we can now use this expression to calculate the quantum mottle. We average over area *A* of the image and substitute $(A\varepsilon N)^{1/2}/(A\varepsilon N)$ for $\Delta X/X$ in equation (2.33). Here ε is the probability that an x-ray photon will interact in the screen and *N* is the number of photons incident per unit area of the screen. The resulting expression for ΔD_Q, the noise due to quantum mottle, is

$$\Delta D_Q = 0.434\Gamma(A\varepsilon N)^{1/2}/(A\varepsilon N) = 0.434\Gamma(A\varepsilon N)^{-1/2}. \qquad (2.34)$$

This expression neglects noise sources (*b*) and (*d*). Their inclusion gives (Barnes 1982)

$$\Delta D_Q = 0.434\Gamma(A\varepsilon N)^{-1/2}(1 + \langle\Delta E^2\rangle/\langle E\rangle^2)(1 + \langle\Delta w^2\rangle/\langle w\rangle^2) \qquad (2.35)$$

where $\langle E\rangle$ and $\langle\Delta E^2\rangle$ are the mean and variance of the energy absorbed in the screen per interacting photon and $\langle w\rangle$ and $\langle\Delta w^2\rangle$ are the equivalent quantities for the light photon yield. This expression is for monoenergetic photons and the various quantities must be suitably integrated to calculate the noise from an x-ray spectrum.

The film granularity can also be estimated using the results of §2.6.1.2. We use equation (2.29) to calculate the fluctuation ΔD_G in the average density in area *A* arising from a fluctuation $(gA)^{1/2}$ in the number of developed grains (gA). The result is

$$\Delta D_G = 0.434(gA)^{1/2}\sigma/A \qquad (2.36)$$

or, substituting for *g* from equation (2.29),

$$\Delta D_G = (0.434D\sigma/A)^{1/2}. \qquad (2.37)$$

This expression was first derived by Selwyn in 1935. It neglects fluctuations arising from variation in grain size but may be extended to include these fluctuations by using a multiplicative factor similar to those used in equation (2.35). In practice, however, this refinement is not necessary and equation (2.37) gives reasonable agreement with measurement (Barnes 1982).

Figures 2.26(*a*) and (*b*) show the characteristic curve and the gamma for Kodak X-Omatic RP film and calculations of the quantum mottle and film granularity for this film when used with a DuPont Cronex HiPlus screen. The quantum mottle is proportional to the slope Γ of the film characteristic curve (equation (2.35) and a term which is inversely proportional to the exposure. It therefore shows a maximum value in the log-linear region of the characteristic and approaches zero in the toe and shoulder regions. The film granularity is proportional to the square root of the optical density (equation (2.37)) and therefore increases with increasing density. As a consequence of these effects, quantum mottle is the dominant effect for a correctly exposed film but is overtaken by the film granularity for regions where the film is either under- or overexposed.

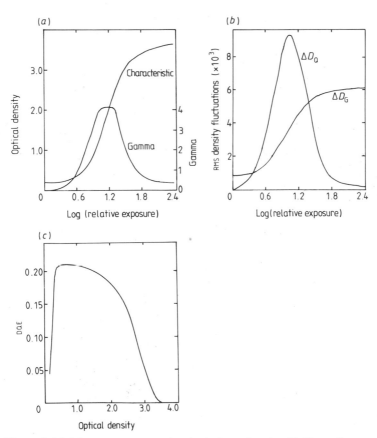

Figure 2.26 Measurements and calculations for the DuPont Cronex HiPlus/Kodak X-Omatic RP screen–film combination. (*a*) Film characteristic and gamma. (*b*) Calculations of the variation of quantum mottle and film granularity with exposure. (*c*) Calculations of the variation of DQE with optical density. Noise estimates are for a circular sampling area of diameter 0.5 mm. (After Barnes (1982).)

A circular sampling area of diameter 0.5 mm was used to calculate the results shown in figure 2.26. For this sampling area and an optical density of say 1.5, the total radiographic mottle is about $0.01D$. If we use a threshold signal-to-noise ratio for visibility of 5 (see §2.4.2), then we might expect to be able to see a contrast corresponding to a density change of $0.05D$, which is a result similar in magnitude to the model we adopted in §2.4.3.

The radiographic noise first increases with density and then decreases, and we must look at the detective quantum efficiency (§2.4.2) to see its significance. Figure 2.26(c) shows the DQE for the Cronex HiPlus screen used with X-Omatic RP film. The DQE is at most 20% and is quite close to this maximum for the density range 0.5 to 2.5. Outside of this range, it falls off rapidly, and this once again illustrates the importance of correct film exposure.

The above treatment has neglected the effects of image unsharpness on the noise. This is important for viewing small objects at low contrast and is best treated by considering the noise power density (or Wiener) spectrum. The reader is referred to Barrett and Swindell (1981) for details.

2.6.3 *Image intensifiers*

The radiographic image intensifier is a high-gain device for imaging x-ray photons. The radiation dose associated with its use for simple procedures is very low but the corresponding unsharpness and noise are inferior to those associated with screen–film systems. The image intensifier can produce single and serial radiographs with low radiation dose and can also operate in a fluoroscopy mode where the x-ray tube runs continuously, but at very low current. It is particularly valuable for the study of processes that involve movement, flow or filling, for intra-operative control during surgery and for the fluoroscopic control of the insertion of cannulae and catheters. It is used for investigations employing radiographic contrast media and is the source of the image for many digital radiology systems.

2.6.3.1 Construction and image formation. Figure 2.27 shows the construction of a typical image intensifier. It consists of an evacuated tube with an intensifying screen at either end, a photocathode and some electron optics (Mistretta 1979, Forster 1985 pp105–7). The entry window of the tube should have a high x-ray transmission. It can be constructed from glass but thin metal windows are often used in modern tubes. For example, the aluminium window used in a certain Siemens image intensifier has a transmission of 92%. After passage through the entry window, the x-rays strike a fluorescent screen deposited on the inner surface of the window. The light fluorescent photons emitted by this screen then hit a photocathode, where they eject photoelectrons. The photoelectrons in turn are accelerated through a potential difference of some 20–30 kV and are guided by electron optics to strike the exit fluorescent screen and produce further fluorescent photons, which can be viewed at the exit window in various ways. The diameter of the input screen will limit the field of view of the intensifier and is typically in the range 12.5–35 cm, although image intensifiers with a 57 cm field of view are now available. The diameter of the output screen is typically 2.5 cm and this demagnification of the image coupled with the electron acceleration means that the image intensifier has a very high photon gain. In fact, the increase in the light output intensity of the image intensifier compared with that from a standard fluorescent screen can be as high as 10 000.

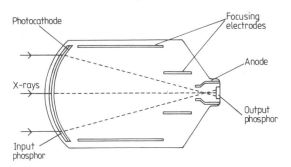

Figure 2.27 Construction of the image intensifier.

The image produced at the exit window is recorded on film or viewed with a television (TV) camera. Direct viewing by eye using appropriate optics is also possible but is rarely used nowadays. In some circumstances, the patient can be screened† or monitored using the image intensifier and the final image is then recorded using a standard screen–film combination.

Modern image intensifiers have caesium iodide input screens. This phosphor has the major advantage that it can be deposited with a packing density of 100%. No binding material is required and a high energy absorption efficiency is possible. Figure 2.28 shows how the mass attenuation coefficient for the caesium iodide screen varies with photon energy. This coefficient has discontinuities at the iodine and caesium K-edges at 33.2 and 36.0 keV but gives rise to an energy absorption efficiency that can be as high as 0.6. The caesium iodide screen has the further advantage that its crystals can be aligned with the direction of the x-ray beam, which reduces the lateral spread of the light fluorescent photons produced within the screen. The spread is also minimised by ensuring close contact between the input screen and the photocathode.

The caesium iodide screen produces about 2000 light fluorescent photons per incident x-ray absorbed (Nudelman *et al* 1982). Their wavelength is matched to the response of the photocathode by the use of sodium doping. The photocathode efficiency is about 0.1, so that 200 photoelectrons are ejected from the photocathode per incident x-ray photon absorbed. These photoelectrons are then focused and accelerated towards the output screen. The focusing electrodes are designed to reduce the image size with as little distortion as possible but, because of the shape of the input window, the images may still suffer from distortion.

†The use of this term should not be confused with screening a population for early disease recognition.

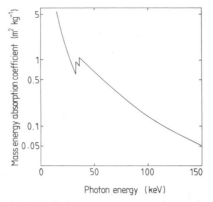

Figure 2.28 Variation of the mass energy absorption coefficient for caesium iodide with photon energy.

The output phosphor should have high resolution but will still contribute significantly to the image unsharpness because of the minification of the image. In addition, there will be loss of contrast or glare due to the scatter and reflection of the fluorescent photons produced in the output phosphor. This glare can be reduced by using a tinted-glass exit window. Direct coupling of the exit window to the TV camera using fibre optics can also be used to reduce glare, and this has the added advantage of a greater photon collection efficiency than a lens system. The unsharpness of the image intensifier will depend upon the degree of minification used. Some tubes are designed to operate at more than one magnification and their response can then be varied depending upon the application, with the unsharpness improving as the input field size is reduced. This is illustrated in figure 2.29, which shows the modulation transfer function for an image intensifier with a large field of view operating at field sizes of 57, 47 and 34 cm. The MTF curves for two screen–film systems have also been included to show the increase of unsharpness associated with the use of the image intensifier.

2.6.3.2 Viewing and recording the image. We have seen in the previous section that the output from the image intensifier may be recorded on film or may be viewed with a TV camera. Both of these methods can in fact be employed for the same image intensifier by using a partially silvered mirror, which will allow an image to reach both film and TV camera.

Still photography uses 70 or 100 mm roll film or 100 mm cut film. The film is often known as a spot film and the image size will depend upon the focal length of the camera. The unsharpness will be slightly worse than that of the output screen but there is a dose saving of typically a factor of 5 compared with a conventional screen–film receptor. The dose saving is limited by the quantum mottle, which is large because of the high photon gain of the system and the film granularity, which becomes more important because of the smaller size of the image. Spot films also have some more practical advantages: they are cheaper than conventional film because they contain less silver, and they are easier to store.

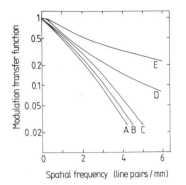

Figure 2.29 Modulation transfer functions for an image intensifier and for two screen–film systems. Curves A, B and C are for the Siemens 57 cm image intensifier at field diameters of 57, 47 and 34 cm, respectively, and were taken from a Siemens data sheet. Curves D and E are for DuPont Cronex Par Speed and Detail screens, respectively, and were taken from the tabulation of Doi *et al* (1982).

A particular advantage of the high photon gain of the image intensifier is the fact that images can be taken with very short exposure times. This is important for imaging in the presence of motion. Figure 2.30 shows the modulation transfer function for images taken with both a 15 cm image intensifier and a screen–film combination and objects moving at speeds of 0.5 and 2.0 cm s^{-1}. There is a factor of 4 difference in the exposure time for the two techniques and the MTF for the image intensifier system is in fact superior to that for the screen–film system.

The high speed of the image intensifier makes ciné photography of the output screen possible. A 35 mm film is used with a frame rate of between 30 and 200 images per second. The dose per frame is about one-twentieth of that from a conventional screen–film combination. The unsharpness is inferior to that obtainable using the larger-format spot films and this is shown in table 2.5, which gives the resolution for various image recording devices attached to an image intensifier.

The TV camera has the advantage over the film camera in that it produces an image that can be viewed in real time on a TV screen. It performs best when it is directly coupled to the output screen of the image intensifier using fibre optics, but this, of course, precludes the use of the film camera to record the image. The image produced by the TV system can be stored on video disc or tape or used as input to a digital radiology system. CT systems have also been constructed based on the use of an image intensifier and digitised TV system (see Chapter 5). The use of a frame store offers the possibility of further dose reduction because the x-ray beam can be switched off while the current image is studied.

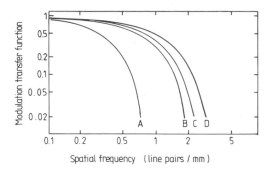

Figure 2.30 Modulation transfer functions for a 15 cm image intensifier used with 70 mm film and for a universal screen. Patient movement speeds are 0.5 and 2.0 cm s^{-1}. The exposure times for the 70 mm film and the screen–film system were 20 and 80 ms, respectively. Curves are as follows: A, screen–film, 2 cm s^{-1}; B, screen–film, 0.5 cm s^{-1}; C, image intensifier, 2 cm s^{-1}; D, image intensifier, 0.5 cm s^{-1}. (After Mistretta (1979).)

Table 2.5 Typical resolutions (line pairs/mm) found in fluorographic systems. (After Theris (1979).)

Image intensifier	Output screen	100 mm film	70 mm film	35 mm ciné	TV monitor
15 cm	5.0	4.2	3.5	2.5	1.5–2.0
25 cm	4.2	3.7	3.1	2.2	1.2–1.5

The choice of TV camera will depend upon the application. There are several types of camera available, but we restrict our discussion here to the vidicon and the plumbicon (Hay 1982). The vidicon camera has an antimony trisulphide photoconductor, which has a significant dark current. Its output current is proportional to the 0.7 power of the input light intensity, so it has a gamma of 0.7 and will reduce contrast. Its redeeming feature is the fact that its response is slow. It will therefore integrate the light output from the intensifier and reduce quantum mottle. It is well suited to imaging stationary organs but would not be suitable for cardiac studies. The plumbicon camera has a lead monoxide photoconductor in the form of a semiconducting p–n junction. It has a very small dark current, a gamma of 1 and negligible lag. It is well suited to imaging moving organs.

The noise level in the video signal produced by the TV camera and associated preamplifier will depend upon the bandwidth used (or the read-out time per pixel). Nudelman *et al* (1982) quote a noise current of about 2.0 nA for a bandwidth of 5 MHz and a signal or operating current of 1.6 μA for a plumbicon camera. This gives an idealised signal-to-noise ratio or dynamic range of 800:1 and would mean that the noise introduced by the TV camera is negligible compared with the noise in the image due to quantum mottle. In practice, the SNR may be less than this and there may be a transition from quantum to video noise domination with increasing dose rate.

The gain of the system and the dose rate and/or x-ray voltage must be adjusted so that the video-tube current is sufficiently high that electronic noise is unimportant and is sufficiently low that the tube does not saturate. The overall gain of the system can be controlled at the design stage and during the exposure (by varying the aperture of the optical coupling between intensifier and TV camera), but the design of a system that can accommodate both fluoroscopic and radiographic dose rates can be difficult.

Although the TV camera can reduce contrast, the TV monitor will enhance the contrast in the image because it has a gamma of about 2.5 (Hay 1982), but the use of the TV system will result in a loss of resolution because of the line-scanned nature of the image. We have already seen this in table 2.5, which shows the resolution achievable for various methods of recording the output from the image intensifier. The resolution of the TV system has to be considered in terms of its vertical and horizontal components. The vertical resolution V (lines per image) can be expressed in terms of the actual number of image scan lines N via the relation

$$V = kN. \tag{2.38}$$

The parameter k is known as the Kell factor and is usually taken to be 0.7 (Moores 1984). For an input diameter of 20 cm and a 625-line TV system, the smallest vertically resolved element would be $200/(625 \times 0.7) = 0.45$ mm. For a 1250-line system the resolution would be 0.22 mm. The horizontal resolution of the system is limited by the bandwidth. The number H of horizontal resolution elements across the picture is given by

$$H = \text{bandwidth}/(\text{frames/s} \times \text{lines/frame}). \tag{2.39}$$

If the bandwidth is selected to make H and V equal and we use 625 lines with 25 interlaced frames per second, we find that we need a bandwidth of 6.8 MHz. (In an interlaced TV the image is scanned in two passes with alternate lines being skipped on each pass.) Doubling the number of lines would increase the bandwidth by a factor of 4 and would also increase the video noise in the image, as noted previously. For a further treatment of TV display, see §14.4.

2.6.4 Xeroradiography

Xeroradiography is a dry non-silver photographic system and produces images on paper which are viewed by reflected light (Boag 1973a, b). It is slower than systems using screen–film receptors but has better resolution and produces images that are edge-enhanced. The principal application of the xeroradiographic technique is in mammography, where the resolution and edge enhancement are of particular value in the perception of microcalcification and small changes in tissue density (Dance and Davis 1983).

2.6.4.1 Construction and image formation. The xeroradiographic image process exploits the photoconductive properties of selenium. The receptor consists of a 125 μm layer of amorphous selenium evenly deposited on a 2 mm thick aluminium backing plate. Prior to radiographic exposure, the surface of the selenium is given a uniform positive charge and this is maintained until the exposure by placing the receptor plate in a light-tight box. (In practice, there is a very small dark current and the surface charge leaks away with a half-discharge period in excess of 1 h.) When the plate is exposed to x-rays, the energy absorbed from the incident photons creates electrons and holes within the bulk of the material. These charge carriers migrate towards the surfaces of the selenium layer under the influence of the internal electric field. The electrons that reach the top surface of the plate form a latent or charge image by subtraction from the original uniform distribution.

The charge image is developed as a two-stage process. The plate is sprayed with an aerosol of fine particles of a blue powder. These particles are charged by friction and are attracted by the charge distribution on the plate so that a powder pattern is built up which is representative of the charge pattern remaining after exposure. The powder image is then transferred to plastic-coated paper by a contact process and is fused into its surface by heating. The image is then ready for viewing in reflected light (figure 2.31). The powdering process is controlled by applying a bias to the plate, and it is possible to produce either positive or negative images. In the positive mode, the most highly charged areas of the plate attract the most powder and are dark blue whereas the discharged areas of the plate are a light blue. The opposite configuration obtains for the negative mode.

2.6.4.2 Sensitivity. The speed of the xeroradiographic receptor depends upon the x-ray absorption efficiency of the selenium layer, the surface charge neutralised per unit energy absorbed and the sensitivity of the powder cloud development process. Figure 2.32 shows the energy absorption efficiency for the selenium plate. The efficiency is high at low photon energies because of the density ($4.25\ \mathrm{g\,cm^{-3}}$) and thickness of the plate and the fairly high atomic number (34) of selenium but falls off rapidly as the photon energy increases. The K-edge and fluorescent yield of selenium (12.66 keV and 0.56, respectively) are fairly low and the energy absorption efficiency does not show the large changes in efficiency associated with the K-edge of materials of higher atomic number (cf figure 2.23).

The efficiency W for conversion of energy absorbed in the selenium to surface charge neutralised depends upon the field strength within the plate. It is given by (Fender 1975)

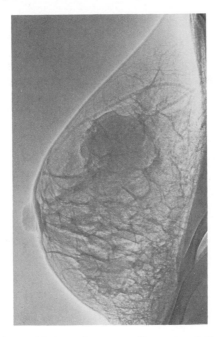

Figure 2.31 A positive-mode xeroradiograph of the female breast (see also figure 11.2(b)).

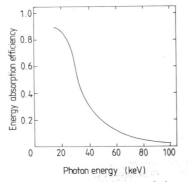

Figure 2.32 Variation with photon energy of the energy absorption efficiency of the selenium plate used in xeroradiography.

$$W = 510/E \qquad (2.40)$$

where W gives the energy (in eV) for one electronic charge neutralised on the surface of the selenium and E is the electric field (in V μm^{-1}). The value of W decreases inversely with the electric field and it is important to have a high initial surface charge. In practice, the initial potential difference across the selenium layer is about 1600 V so that the initial value of W is 44 eV per electronic charge. It is not possible to increase the surface charge to very high values because this would cause the electrical breakdown of the plate. The value of W is also inversely proportional to the surface charge and it is straightforward to show that the surface charge remaining after the exposure decreases exponentially with the radiation dose. This exponential decrease gives the xeroradiographic process a reasonable exposure latitude.

The charge-collection efficiency given by equation (2.40) is quite low and only about 16% of the charge pairs originally created reach the surface of the photoconductor. In addition, it turns out that the optimum quality of the powder image occurs when the plate is discharged to about $1/e$ of its original value (Fatouros 1979). As a consequence, the radiation dose associated with the use of the xeroradiographic receptor is quite high. In fact, for population screening using xeromammography, it is not possible to use an x-ray tube with the molybdenum target–molybdenum filter combination discussed in §2.5.2 because the dose is too high. A tube with a tungsten target is used instead with a 2 mm filter at about 43 kV. Because of the edge-enhanced nature of the powder image, the image quality is still excellent even though the x-ray spectrum is considerably harder. However, the dose is still 2–4 times that for a screen–film combination used with a molybdenum x-ray tube and many centres prefer to image the breast using a screen–film combination.

The sensitivity of the system also depends upon the polarity of the image and is 30% better for negative-mode images.

2.6.4.3 Image properties. The properties of the powder image are best understood by considering the powdering process. The plate is developed in a horizontal position so that the charged powder particles fall towards it under the influence of gravity and the electrostatic field created by its surface charge. The electrostatic field dominates so that the trajectories of the powder particles follow the lines of electric force. Figure 2.33 shows the pattern of the lines of force in the neighbourhood of a step in charge density. Far from the step, the lines of force are perpendicular to the plate and the powder falls vertically so that distant regions on either side of the step have quite similar density. *The technique therefore has poor broad-area contrast* and this is illustrated in figure 2.34, which shows the characteristic curve and gamma for the xeroradiographic process. Close to the charge step, however, the field pattern is complex and this leads to a powder depletion on one side of the step and powder piling on the other. It is for this reason that the xeroradiographic images are edge-enhanced and it is this feature that can make the radiograph easier to interpret than a conventional image and small structures easier to visualise.

The unsharpness in the charge image is determined by the ranges of the secondary particles produced within the plate and the development process. The range of a 50 keV photoelectron in selenium is 20 μm and the mean free path of the 11 keV characteristic x-ray from selenium is 70 μm, so that the unsharpness due to spread of secondary particles is small. The diameter of a powder particle is 4 μm and the toner thickness about 45 μm (Fatouros 1979), so that the total receptor unsharpness is also small.

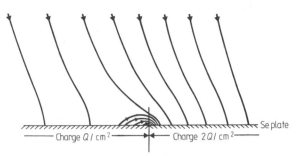

Figure 2.33 Enlarged view of the electric field lines in the vicinity of a charge step on the surface of a selenium plate. There is a grounded electrode (not shown in the figure), which is distant from the plate. (From Dance and Davis (1983) and reproduced with permission from Chapman and Hall (Publishers) Ltd.)

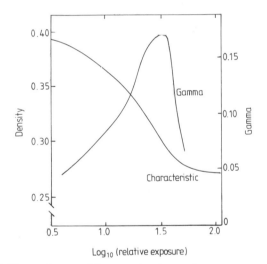

Figure 2.34 Dependence of image density and gamma on exposure for a powder xeroradiograph. (After Shaw (1979).)

The noise in the xeroradiographic image has been discussed by Fatouros (1979), who has shown that a treatment similar to that given in §2.6.2.5 is possible with the film granularity being replaced by fluctuations in the number of toner particles within the sampling area.

2.6.5 *Recent developments*

2.6.5.1 *Ionography.* Conventional x-ray imaging systems use a solid receptor to achieve a high x-ray absorption efficiency. The ionographic

system, however, uses a gas-filled chamber and achieves a high efficiency by using a gas pressure × gas thickness product of some 5–10 atm cm (Johns *et al* 1974). The ionographic image system has quite a good imaging performance but can still be regarded as being at an experimental stage. Our treatment is brief but, for the future, the work of Moores *et al* (1985) suggests that the ionographic receptor may find an application in digital radiology.

Figure 2.35 shows a prototype ionography system used by Boag and coworkers (Boag 1979). The gas-filled chamber is under a pressure of some 5–10 atm, which is contained by a carbon-fibre window. The plates of the chamber are maintained at a high potential difference so that there is a strong electric field in the 1 cm gap between the plates. The interaction of x-ray photons with the gas molecules liberates charge carriers, which are swept towards the chamber plates by the electric field. Either polarity of charge carrier may be collected by stretching an insulating Mylar foil across the appropriate electrode. The charge distribution collected on the foil differs from that collected during xeroradiography in that the image is formed by addition rather than subtraction. The charge pattern may be developed using a powder or a liquid toner or can be read out using a guarded electrometer probe and used as the basis of a digital imaging system (Moores *et al* 1985).

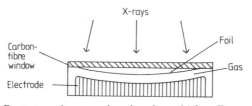

Figure 2.35 Prototype ionography chamber. (After Dance and Boag (1977).)

The efficiency of the ionography chamber will depend upon the choice of gas, its pressure × thickness product and the energy required to produce one ion pair in the gas. The efficiency for collection of the carriers is also very important as the chamber works in proportional mode and the field strength should be selected so that there is little recombination. Avalanche multiplication of the charge carriers by a very high field is undesirable because of the associated increase in quantum mottle. The noble gases krypton and xenon have the best ion-production yield but the Freon CF_3Br has an acceptable yield and is considerably cheaper (Dance and Boag 1977). The energy absorption efficiency of the gas can be comparable to that for a screen–film system but there will be some photon loss due to absorption in the carbon-fibre window. Moores *et al* (1980, 1985) avoided this problem by using a gas at atmospheric pressure. With the Freon CF_3Br gas and a gap of 4 cm between the electrodes, they achieved an efficiency of some 35% for a 40 kV x-ray beam from a tungsten target. The breast dose from their prototype mammographic system was comparable to or less than that from a screen–film combination (Moores *et al* 1980).

The resolution of the ionographic system will depend upon both the lateral spread of ions collected on the Mylar foil and the charge development process. Several factors contribute to the lateral spread of the ions. First, there may be effects due to the fact that the electric field between the chamber plates may not be parallel to the x-ray beam. These can be minimised by using a chamber with spherical electrodes centred on the tube focus and auxiliary electrodes to correct for fringe fields. Secondly, there is a lateral spread in the production points of the ion pairs created by a single x-ray photon. This is due to the ranges of the secondary particles (fast electrons and characteristic x-rays) it produces. This has been discussed by Johns *et al* (1974), who have demonstrated that even for fast electrons with a range of several millimetres the lateral charge distribution reaching the foil is forward-peaked. In fact, for xenon gas at 10 atm cm irradiated by 60 kV x-rays, they claim a resolution of 10 line pairs/mm. This resolution, however, would be expected to decrease with increasing photon energy. The lateral spread of the characteristic x-rays produced within the gas does not affect the resolution because their mean free path is large. They contribute instead to a fog level and decrease the 'useful' energy absorption efficiency of the gas because they are not locally absorbed. Finally, the lateral diffusion of the charge carriers can be important, but is minimised by collecting positive ions rather than electrons.

2.6.5.2 Stimulated luminescence. The receptors we have studied so far fall into two categories: those producing an image that is immediately transferred to another medium and those with a storage capacity themselves. The photostimulable phosphor falls into the latter category. It has recently been developed by the Fuji Photo Film Company and has exciting possibilities as a source of digital images (Sonoda *et al* 1983).

The image receptor is a flexible plate coated with a photostimulable phosphor. Such phosphors can store the energy absorbed from the incident x-ray beam in quasi-stable states and emit this energy in the form of light photons when stimulated by visible or infrared radiation. The requirements for this phosphor are that it should have a high x-ray absorption efficiency and a high light output per unit energy absorbed. The emitted light should be of a suitable frequency for monitoring with a photomultiplier and the response time of the phosphor should be less than 10 μs to allow rapid read-out of the image. Europium-activated barium fluorohalide compounds match these requirements well and are the basis of the stimulated luminescence receptors that are now commercially available.

Figure 2.36 shows the technique used to read out the image plate.

The phosphor is scanned using a laser beam which is deflected by an oscillating mirror so as to pass to and fro across the plate. The plate is advanced after each traverse of the laser beam so that the complete image is read out in a raster pattern. The laser beam has a spot size of 0.1 mm and the image is sampled at 5–10 pixels/mm. The radiation emitted by the plate is collected by a light guide, which is coupled to a photomultplier tube, and the resulting signal is digitised using an 8- or 10-bit analogue-to-digital converter (ADC) and stored.

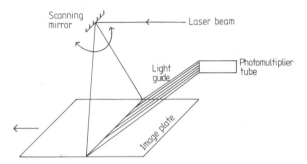

Figure 2.36 System for reading out the photostimulable image plate. (After Sonoda *et al* (1983).)

The Fuji imaging system has a dynamic range of more than 10 000:1 and can accommodate a wide range of radiographic exposures. A detailed study of its imaging characteristics is still awaited but it would appear to have a high photon absorption efficiency (50% for 80 kV radiation) and a resolution limited by the pixel size of 0.1–0.2 mm. Quantum mottle is probably the dominant source of noise in the image.

There are a number of major advantages associated with the direct collection of a digital image and these are discussed in the following section.

2.7 DIGITAL RADIOLOGY

Conventional radiological imaging systems record and display data in an analogue way. They often have very strict exposure requirements because of narrow latitude and offer very little possibility of image processing. Digital radiography systems, however, offer the possibility of imaging at whatever dose level is required and provide an image that may be processed and displayed in a variety of ways. Such systems are more expensive than conventional radiological systems but are finding increasing application as imaging and computer technology develops.

2.7.1 *Systems for digital radiology*

Figure 2.37 shows the components of a typical digital radiographic system. The x-ray tube and image receptor are interfaced to and controlled by a computer and the resulting digitally stored and processed image is displayed on a TV screen, which forms part of the operator's control console or the radiologist's reporting console. Similar consoles may be found as part of CT or NMR imaging systems. The digital image can be stored on magnetic media or optical disc and a film writing device can be used to make a permanent analogue copy of the image.

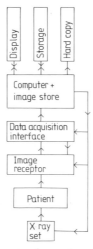

Figure 2.37 Components of a system for digital radiology.

Two classes of image receptor can be used in digital radiology: those recording the whole image at once and those recording only part of the image at a time and obtaining a complete image by scanning both x-ray beam and receptor (scanned projection radiography). The receptors discussed in the previous section all fall into the former category, but of these only the image intensifier, the ionographic chamber and the stimulated luminescence plate are well suited to digital radiology. These three receptors can produce digital images directly, and the production and storage of a conventional image as an intermediate stage is unnecessary. The performance of these image receptors has already been discussed but it is appropriate to note the important differences between them. Image intensifiers will not give the highest spatial resolution or the best contrast but have the advantage that they are fast. The analogue-to-digital conversion of a 512×512 digital fluorographic image can take as little as 0.03 s (Smathers and Brody 1985). Even at an image size of 2048×2048, the image conversion time is only a few seconds. The read-out of images from the stimulated luminescence plate or the ionographic chamber is much slower, but these receptors have the important advantages of better resolution and dynamic range.

It is possible to digitise a film image using a scanning microdensito-meter, but any information that is recorded on the film at very low or high density will be degraded because of the form of the film character-istic. The powder-developed xeroradiograph can also be digitised using a scanning densitometer with reflected light, but this suffers from the disadvantage that the image is already edge-enhanced. There is no commercially available device for directly reading out the charge image on the xeroradiographic plate.

Imaging systems that use a scanning x-ray beam and receptor have the important advantage that they have excellent scatter rejection. A collimator is used before the patient to restrict the primary beam to that area necessary to irradiate the receptor and another collimator is used after the patient to reject scatter. This is illustrated in figure 2.38, which shows the line-scanning system developed by Tesic *et al* (1983) for digital imaging of the chest. The receptor here is a strip of gadolinium oxysulphide screen viewed by a linear array of 1024 photodiodes. Scanned projection radiographs are also produced by CT scanners as an aid to CT slice selection.

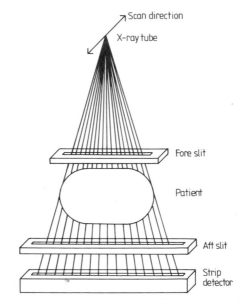

Figure 2.38 Line-scanning system for digital chest radiography.

A major disadvantage of scanning systems is that much of the useful output of the x-ray tube is wasted and long exposure times are necessary. For the method developed by Tesic *et al*, the exposure time was about 10 s, which will decrease the useful working life of the x-ray tube and may create problems with movement. It should be noted that, although the total exposure time is long, the time spent in any region of the image is quite short, so that movement losses are much less important than those for a conventional radiograph with the same exposure time.

The use of a small-area scanning receptor is interesting because of the possibility of altering the dwell time in any area according to the quantum noise level. In this way it is possible to ensure low noise in areas of low transmission through the patient and low dose for areas of high transmission. This technique is known as scanned equalisation radiography and is being used by Plewes (1983) for chest imaging.

In general terms, the digital radiographic system will have inferior resolution to a screen–film receptor but, provided the matrix size and image receptor are matched to the application, this will not be of clinical significance. A 512 × 512 image may be sufficient for digital fluoroscopy whereas a system for digital chest imaging might require a 1024 × 1024 matrix and a pixel size of 0.4 mm. For mammography a 2048 × 2048 system is required for a pixel size of 0.1 mm.

The pixel depth in the image will also depend upon the application. An 8-bit analogue-to-digital converter, giving a precision of 0.4%, will be adequate for noisy images or a large image matrix (smaller pixels are noisier), but 10-bit conversion (0.1% precision) may be necessary for some applications.

The size of the image matrix will also affect the data storage requirements, and image compression may be necessary for storage of large numbers of high-resolution digital images. Modern magnetic-disc technology has the capacity to store images produced over a few days, but it would be necessary to use optical-disc technology if rapid access is required to images obtained over a long period of time. The storage capacity of a single 12 inch optical disc is about 2 Gbyte, corresponding to 1900 1024 × 1024 8-bit uncompressed images. A jukebox-type device can be used to select the optical disc, thereby giving rapid access to any desired image. The possibility of handling all images digitally is attractive and systems for doing this are known as picture archiving and communications systems (PACS).

2.7.2 Advantages and applications

The advantages of digital radiographic systems may be divided into four classes: those associated with the image display; those associated with dose reduction; those associated with image processing; and those associated with image storage and retrieval (which we do not discuss here). We distinguish between mathematical manipulations that affect the analogue display but leave the original digital image unchanged and those which process the digital image before it is displayed. We consider first the improvements in the image display. The function that maps the digital image on to levels of brightness on a TV screen or density on a film can be completely controlled by the user. For example, every film produced from a digitally recorded image can be properly exposed and can have a characteristic that is properly matched to the range of pixel values in the image. Alternatively, the full range of density or brightness can be used to display just part of the range of pixel values, thereby enhancing the contrast in a region of interest (this technique is known as windowing). Algorithms are available for controlling the analogue image so that best use is made of the capabilities of the display system. The histogram equalisation technique maps the digital image on to the display in such a way that there are equal numbers of pixels in each brightness or density level in the analogue image.

The second advantage of digital radiology is the possibility of dose reduction. In conventional radiology, the dose is determined by the sensitivity of the image receptor and the film latitude. In digital radiology, both these constraints can be relaxed. Dose savings can be achieved by adjusting the dose to give the required noise level in the image. Further reductions are possible by using the x-ray spectrum that gives the lowest dose for a given signal-to-noise ratio and by recovering any losses in contrast using the display techniques described above.

The third advantage of digital radiology is the possibility of digital processing of the image. Probably the most important application of such processing is subtraction imaging (Smathers and Brody 1985). We observed in §2.4.2 that radiologists have to identify the abnormal from a complicated background of normal tissue architecture. They may miss a small object that the system has resolved, or a low-contrast object that is visible above the noise level in the image, simply because of the complexity of the surrounding and overlying architecture. Subtraction radiology can remove much of the unwanted background architecture and can thus improve the visualisation of the important features of the radiograph. Computed tomography (see Chapter 4) may be thought of as an example of subtraction radiography because the overlying architecture in the conventional projected image is removed.

Digital subtraction angiography (DSA) is a subtraction technique that is increasingly used for the visualisation of blood vessels following the intravenous or intra-arterial injection of contrast media. An image of the region of interest is obtained before the arrival of the iodinated contrast agent and is used as a mask to subtract from the images showing the passage of the contrast medium through the blood vessels. The signal-to-noise ratio in the subtracted image can be improved by using spatial and temporal filters on the complete image set, and corrections can also be applied for patient movement. For intravenous injection, the contrast medium may be much diluted by the time it reaches the region of interest, so that the high-contrast resolution achievable with digital techniques is very important (a contrast resolution of 1% is achievable using DSA (Ergun and Giordano 1985)). For peripheral imaging, the transit time for passage of the contrast through the region of interest will be several seconds and a moderate frame rate (one per second) is used with a pulsed x-ray beam (Ergun and Giordano 1985). For intra-arterial injection, there is greater contrast in a shorter time interval and this allows the visualisation of smaller vessels. A 1024×1024 matrix together with a short x-ray beam pulse (4–33 ms) and a frame rate of four per second may be appropriate. For cardiac imaging, a frame rate of 30 per second may be necessary with continuous x-ray exposure and a 512×512 image matrix. Figure 2.39 shows a typical DSA image.

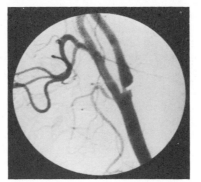

Figure 2.39 Digital subtraction angiogram showing a stenosis at the origin of the internal carotid artery. (Image provided by Philips Medical Systems.)

Dual-energy radiology is another form of subtraction imaging. In this technique two images are obtained using different x-ray spectra. The signal recorded in each pixel gives an indication of the attenuation caused by the overlying tissue and the attenuation itself can be divided into components due to photoelectric absorption and to photon scatter. As these two components have a different energy dependence, it is possible to extract a photoelectric and a scatter component from the two initial images. Equally, since different materials will have different amounts of scatter and photoelectric absorption, it is possible to produce images of the thicknesses of two selected materials, which, in combination, will reproduce the attenuation maps recorded in the two original images (Lehmann *et al* 1981). Suitable selection of these materials can produce images that show or exclude any desired tissue type. For example, it is possible to choose soft tissue and bone (with some nominal composition) as basis materials and to view the soft tissue or bony images separately.

The processing we have discussed so far involves the manipulation of more than one image, but it can also be useful to process a single image. We have already seen in our discussion of xeroradiography that edge enhancement can aid the perception of abnormality, and an edge-enhanced digital image is readily obtained by high-pass filtering. Alternatively, various low-pass filters are available to reduce the noise in the image. Some of these filters (for example the median filter) are able to reduce noise yet retain high-frequency information in the vicinity of an edge. However, the value of applying image processing techniques still needs to be assessed and the ultimate goal of the automatic reading of radiographs remains far in the future.

REFERENCES

ARNOLD B A and BJARNGARD B E 1979 Effect of phosphor K x-rays on MTF of rare earth screens *Med. Phys.* **6** 500–3

ARNOLD B A, EISENBERG H and BJARNGARD B E 1978 Measurements of reciprocity law failure in green-sensitive x-ray films *Radiology* **126** 493–8

BARNES G T 1982 Radiographic mottle: a comprehensive theory *Med. Phys.* **9** 656–67

BARRETT H H and SWINDELL W 1981 *Radiological Imaging* vols I and II (London: Academic Press)

BIRCH R, MARSHALL M and ARDRAN G M 1979 *Catalogue of Spectral Data for Diagnostic X-Rays* SRS-30 (London: Hospital Physicists' Association)

BOAG J W 1973a Xeroradiography *Phys. Med. Biol.* **18** 3–37

—— 1973b Xeroradiography *Modern Trends in Oncology I* Part 2 *Clinical Progress* ed R W Raven (London: Butterworths) pp79–89

—— 1979 Electrostatic imaging in radiology: limitations and prospects *Phil. Trans. R. Soc. Lond.* A **292** 273–83

DANCE D R and BOAG J W 1977 Optimisation of design parameters in ionography *J. Photogr. Sci.* **25** 135–40

DANCE D R and DAVIS R 1983 Physics of mammography *Diagnosis of Breast Disease* ed C A Parsons (London: Chapman and Hall) pp76–100

DANCE D R and DAY G J 1981 Simulation of mammography by Monte Carlo calculation — the dependence of radiation dose, scatter and noise on photon energy *Patient Exposure to Radiation in Medical X-Ray Diagnosis* ed G Drexler, H Eriskat and H Schibilla, EUR7438 (Brussels: CEC) pp227–43

—— 1985 Escape probabilities for fluorescent x-rays *Phys. Med. Biol.* **30** 259–62

DARBY S C, KENDALL G M, RAE S and WALL B F 1980 *The Genetically Significant Dose from Diagnostic Radiology in Great Britain in 1977* NRPB-R106 (Harwell: National Radiological Protection Board)

DOI K, HOLJE G, LOO L-N, CHAN H-P, SANDRIK J M, JENNINGS R J and WAGNER R F 1982 *MTFS and Wiener Spectra of Radiographic Screen–Film Systems* FDA 82–8187 (Rockville: Bureau of Radiological Health)

ERGUN D L and GIORDANO T A 1985 Digital subtraction angiography: system architecture for optimal image quality and future growth *Recent developments in Digital Imaging* ed K Doi, L Lanzi and P-J P Lin (New York: AAPM) pp 351–67

FATOUROS P P 1979 Detail visibility in xeromammography *The Physics of Medical Imaging* ed A G Haus (New York: AAPM) pp 239–87

FENDER W D 1975 Quantification of the xeroradiographic discharge curve *Proc. Soc. Photo-Opt. Instrum.* **70** 364–71

FORSTER E 1985 *Equipment for Diagnostic Radiography* (Lancaster: MTP Press)

HAY G A 1982 Traditional x-ray imaging *Scientific Basis of Medical Imaging* ed P N T Wells (Edinburgh: Churchill Livingstone) pp1–53

ICRP (INTERNATIONAL COMMISSION ON RADIOLOGICAL PROTECTION) 1977 *Recommendations of the International Commission on Radiological Protection* ICRP 26, Ann. ICRP vol 1, no. 3 (Oxford: Pergamon)

JOHNS H E and CUNNINGHAM J R 1983 *The Physics of Radiology* 4th ed (Springfield, IL: C C Thomas)

JOHNS H E, FENSTER A, PLEWES D, BOAG J W and JEFFREY P N 1974 Gas ionisation methods of electrostatic image formation in radiography *Br. J. Radiol.* **47** 519–29

LEHMANN L A, ALVAREZ R E, MACOVSKI A, BRODY W R, PELC N J, RIEDERER S J and HALL A L 1981 Generalised image combinations in dual kV digital radiography *Med. Phys.* **8** 659–67

Mees C E K and James T H 1966 *The Theory of the Photographic Process* 3rd edn (New York: Macmillan)

Mistretta C A 1979 X-ray image intensifiers *The Physics of Medical Imaging* ed A G Haus (New York: AAPM) pp182–205

Moores B M 1984 Physical aspects of digital fluorography *Digital Radiology—Physical and Clinical Aspects* ed R M Harrison and I Isherwood, IPSM1 (London: Hospital Physicists' Association) pp45–57

Moores B M, Dovas T, Pullan B and Booler R 1985 A prototype digital ionographic imaging system *Phys. Med. Biol.* **30** 11–20

Moores B M, Ramsden J A and Asbury D L 1980 An atmospheric pressure ionographic system suitable for mammography *Phys. Med. Biol.* **25** 893–902

Nudelman S, Roehrig H and Capp M P 1982 A study of photoelectronic–digital radiology—Part III: Image acquisition components and system design *Proc. IEEE* **70** 715–27

Plewes D 1983 A scanning system for chest radiography with regional exposure control: theoretical considerations *Med. Phys.* **10** 646–54

Rose A 1973 *Vision: Human and Electronic* (New York: Plenum) pp21–3

Shaw R 1979 Fundamental image quality parameters for xeroradiography *The Physics of Medical Imaging* ed A G Haus (New York: AAPM) pp231–8

Shrimpton P C 1985 Energy imparted as a measure of radiological hazard to patients from X-ray examinations *Med. Biol. Eng. Comput.* **23** Suppl. 2, 1135–6

Shrimpton P C, Wall B F, Jones D G, Fisher E S, Hillier M C, Kendall G M and Harrison R M 1986 *A National Survey of Doses to Patients Undergoing a Selection of Routine X-Ray Examinations in English Hospitals* NRPB-R200 (Harwell: National Radiological Protection Board)

Smathers R L and Brody W R 1985 Digital radiology: current and future trends *Br. J. Radiol.* **58** 285–307

Sonoda M, Takano M, Miyahara J and Kato H 1983 Computed radiography utilising scanning laser stimulated luminescence *Radiology* **148** 833–8

Stevels A L N 1975 New phosphors for x-ray screens *Medicamundi* **20** 12–22

Tesic M M, Mattson R A, Barnes G T, Sones R A and Stickney J B 1983 Digital radiology of the chest: design features and considerations for a prototype unit *Radiology* **148** 259–64

Theris N A 1979 Imaging system requirements and film for fluorography *The Physics of Medical Imaging* ed A G Haus (New York: AAPM) pp216–27

UNSCEAR (United Nations Scientific Committee on the Effects of Atomic Radiation) 1972 *Ionising Radiation: Levels and Effects* vol 1 *Levels* (New York: United Nations)

Wall B F, Hillier M C and Kendall G M 1986 *An Update on the Frequency of Medical and Dental X-Ray Examinations in Great Britain—1983* NRPB-R201 (Harwell: National Radiological Protection Board)

Wall B F and Shrimpton P C 1981 Methods and results of the recent assessment of the GSD from diagnostic radiology in Great Britain *Patient Exposure to Radiation in Medical X-Ray Diagnosis* ed G Drexler, H Eriskat and H Schibilla, EUR7438 (Brussels: CEC) pp129–41

CHAPTER 3

QUALITY ASSURANCE AND IMAGE IMPROVEMENT IN DIAGNOSTIC RADIOLOGY WITH X-RAYS

S H EVANS

3.1 INTRODUCTION TO QUALITY ASSURANCE

In Chapter 2 we have discussed at length the physical principles of diagnostic radiology with x-rays. Much of the discussion has centred on what may be done at the design stage to produce a useful imaging tool. In this chapter we consider the practical performance of the x-ray imaging system. In particular, we first highlight some of the tools in the kit of the physical scientist whose task it is to ensure that what has been well designed, works well in practice. Later in the chapter we shall show how improved images of greater diagnostic usefulness may be obtained. Quality assurance in diagnostic radiology is a means of maintaining standards in imaging and working towards minimising patient and staff doses. To accomplish these objectives, a number of physical parameters that affect the performance of the x-ray imaging system need to be measured. The characteristics of these parameters can vary with time; hence the tests need to be made at regular intervals. It is therefore necessary to learn how an image is influenced by these parameters and how the characteristics of these parameters can be measured using suitable test objects.

3.2 BASIC QUALITY-ASSURANCE TESTS FOR X-RAY SETS

Details of the tests to be made as part of a routine quality-assurance programme are presented. The important characteristics of the test objects are identified and, where appropriate, comparisons between film-based test objects and electronic test objects are made.

3.2.1 Tube potential

The excitation potential across an x-ray tube affects both the number of photons transmitted through the patient and the film contrast. Radiological examinations require specific excitation potentials depending upon which organs are being examined. For instance, high potentials ($120–140\ kV_{peak}$, hereafter written as 'kilovoltage') are often used for chest examinations and low potentials are used in mammography (typically $30\ kV_{peak}$), where maximum contrast of the soft tissues is needed (see §3.5.2).

The voltage shown on the x-ray control unit normally indicates the peak value of the potential applied across the tube. The peak potential can be measured to within $1\ kV$ using either a penetrameter or an electronic kilovoltmeter. The most widely used version of the penetrameter is described by Ardran and Crooks (1967). A penetrameter usually consists of a copper step wedge, which is placed on top of a fast screen. Adjacent to this wedge lies a slow screen, which produces a reference density for the exposure. The relative speeds of the two screens are chosen so that the film density under the slow screen will lie within the range of the densities under the fast screen. To achieve accurate measurement of the peak potential, the penetrameter is covered with a 1.5 mm thick copper plate, which preferentially absorbs the lower-energy photons. Additionally, this plate simulates the attenuation of the beam when a patient is examined and gives a more realistic assessment of the beam quality. Details of the penetrameter are shown in figure 3.1. Attenuation coefficients vary with photon energy, and the film density observed under the slow screen will roughly equal the density underneath one region of the copper step wedge. The potential can be found if the penetrameter is calibrated to give the relationship between the kilovoltage and the step number that produces the same density as the reference density. The penetrameter should be calibrated regularly (annually or every two years) by a primary-standards laboratory such as (in the UK) the National Physical Laboratory at Teddington, so that the results can be interpreted with confidence. A radiograph of a GEC penetrameter is shown in figure 3.2, and the arrow indicates where the reference density equals the step density, yielding a measurement of the kilovoltage.

The second type of potential-measuring device uses solid-state electronics. This equipment usually employs two copper plates of different thicknesses, which attenuate the x-ray beam by different amounts. Photodiodes convert the x-rays that pass through the copper plates into two low voltages. The ratio of the magnitudes of these voltages is proportional to the kilovoltage of the x-ray beam. Some kilovoltmeters are calibrated to display these ratios as peak potentials, which provides a fast and easy method for determining potentials. With some instruments the kilovoltage waveform can be displayed on a storage oscilloscope. This display shows any irregularities in the waveform and provides the most accurate and detailed analysis of the tube potential. Electronic kilovoltmeters also, of course, require regular calibration against a primary standard.

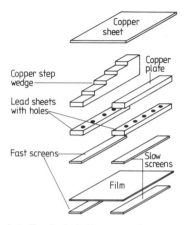

Figure 3.1 Exploded diagram of penetrameter.

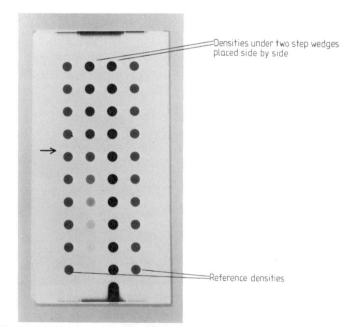

Figure 3.2 Radiograph of GEC penetrameter. The right-hand step wedge is used for the mammographic energy range and the left-hand side is used for excitation potentials above 50 kV. The above radiograph was produced at 80 kV.

Measurements should be made at least every 10 kV throughout the normal operating range of the x-ray system. There is no national standard for acceptance limits but it is usual to consider that a variation in measured kilovoltage of up to 5 kV is acceptable. If the measured kilovoltage is greater than 5 kV and less than 10 kV from the nominal value, the service engineer should be informed. If the measured kilovoltage is more than 10 kV different from the nominal kilovoltage, the equipment should not be used until seen by the service engineer.

3.2.2 Timing

Exposure times should be checked to ensure that this contribution to determining the correct film density is accurate. This can be achieved by exposing a spinning top on a film cassette or by using an electronic detector. The spinning top consists of a brass disc connected to a small motor, which rotates the disc at a fixed rate. The disc has at least one slit cut out of it and, by radiographing the spinning top, an image of the movement of this slit will be formed. In figure 3.3 a radiograph of a spinning top is shown. This top has three slits cut into the brass disc so that the image of at least one of the slits will not be superimposed by the image of the motor. The maximum time that can be measured with this instrument is 0.3 s. The angle of the arc produced on the film by the slit will correspond to the duration of the x-ray exposure. The disc is

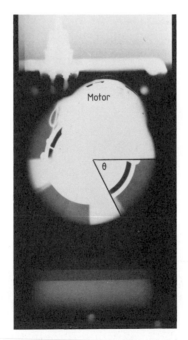

Figure 3.3 Radiograph of spinning top. Time is given in seconds as $t = \theta/360$.

usually made to spin at a constant one revolution per second and therefore, by dividing the angle of the arc in degrees by 360, the exposure time can be calculated. In addition to the measurement of the exposure time, the time taken for the x-ray beam to reach maximum potential can be subjectively studied by observing the density of the exposed arc at the start of the exposure. This will give some indication of the minimum exposure time that can be set to give adequate results.

A simpler spinning-top method can be used for single-phase machines. With a mains frequency of 50 Hz, a 1 s exposure will produce 50 discrete exposures for a half-wave-rectified unit and 100 discrete exposures for a full-wave-rectified unit. If a brass disc with a small hole in it is made to spin manually on top of a cassette just before the exposure, a number of spots proportional to the exposure time will be formed on the film. The exposure time can therefore be calculated by dividing the number of spots by 50 for half-wave-rectified units and by 100 for full-wave-rectified units.

An electronic timer employs a photodiode that switches on an electronic timing circuit during the exposure. Most electronic timers can be used to give a direct display of the exposure time for multiphase units and can be operated to display the number of pulses for single-phase units. These units provide an immediate measurement of the exposure time and they are not limited, as in the case of the spinning top, to a maximum of a 1 s exposure. The electronic timers do not, however, give any indication of the time taken for the x-ray beam to reach maximum voltage.

Exposure times should ideally be measured throughout the normal working range of the unit being investigated. This is not always possible using the spinning top and also, because of the relative ease of use of the electronic timer, the spinning-top device is not recommended for routine quality assurance.

It is recommended that exposure times should be within 0.01 s or 5% (whichever is the greater) of the stated value.

3.2.3 Output

The excitation potential across an x-ray tube can be modified by the tube current (mA) and it is therefore important to check the potentials at different tube currents. It is not always possible to measure voltages at all the tube currents used for clinical examinations owing to the high tube-current demands of many electronic kilovoltmeters and penetrameters. Output measurements, however, can be compared at various exposure settings to provide a consistency check on the exposure parameters. Outputs are measured with suitably calibrated ionisation chambers and should be made at 75 cm, which is the usual patient–focal-spot distance. It is recommended that outputs are measured at various tube currents for the most commonly used kilovoltage and also at various kV × s for the most commonly used tube current. Additionally, it is important to check the variation of output with exposure time at one particular kilovoltage and tube current to ensure that there is no significant effect of voltage risetime. Output values should not differ by more than about 20% at the same value of tube current × exposure time (mA s) (as tube current and exposure time are varied separately) and kilovoltage, otherwise the film density might be unacceptable. Finally, successive output measurements at the same exposure values should be made to provide a check on the consistency of the x-ray system. The consistency of the unit should be better than 5%. A summary of the various parameters studied is shown in figure 3.4.

(1) Variable tube current (mA) $\left\{ \begin{array}{l} \text{Fixed tube potential (kV)} \\ \text{Fixed exposure time (s)} \end{array} \right.$

(2) Variable exposure time (s) $\left\{ \begin{array}{l} \text{Fixed tube potential (kV)} \\ \text{Fixed tube current (mA)} \end{array} \right.$

(3) Variable tube potential (kV) $\left\{ \begin{array}{l} \text{Fixed exposure time (s)} \\ \text{Fixed tube current (mA)} \end{array} \right.$

Figure 3.4 Output measurements.

3.2.4 Half-value layer

In an unfiltered x-ray beam, the lower-energy component does not present any diagnostic information because it is all absorbed by the patient (see §2.5.2). This situation results in an unacceptable dose. These lower energies can be preferentially filtered out by inserting an aluminium filter of suitable thickness at the exit port of the x-ray tube. The minimum thickness that has to be used varies with the maximum operating potential of the x-ray unit but is typically 2.5 mm for the majority of units operating above 100 kV. By measuring the transmission of the beam through a number of aluminium filters of different thicknesses, the amount of filtration in the beam can be found without having to take apart the x-ray set. The amount of filtration required to reduce the transmission to half its original value is known as the half-value layer (HVL) of the beam. The total filtration can be found from graphs relating the HVL to this quantity (HPA 1977).

For accurate measurements of the half-value layer, a geometry must be used which reduces the scatter component in the measuring chamber. This can be achieved by collimating the x-ray beam and by placing the measuring chamber at least 75 cm from the floor. Additionally, a monitoring chamber should be inserted between the x-ray tube and added filters to provide a means of normalising the output to take account of variations in the unfiltered beam intensity. A suitable arrangement is shown in figure 3.5.

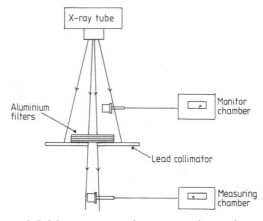

Figure 3.5 Measurement of HVL: experimental set-up.

Once the half-value layer is determined, the effective linear attenuation coefficient μ_{eff} can be calculated using

$$I = I_0 \exp(-\mu_{\text{eff}}\text{HVL}) \tag{3.1}$$

where I_0 is the intensity of the beam at the monitoring chamber and I is the intensity of the filtered beam ($I = 0.5I_0$). Hence

$$\mu_{\text{eff}} = \ln(2)/\text{HVL} = 0.693/\text{HVL}. \tag{3.2}$$

The effective energy of the beam, which is a useful parameter to characterise the beam, is that energy whose narrow-beam monoenergetic linear attenuation coefficient is equal to μ_{eff}.

3.2.5 Focal-spot size

The finite area of the focal spot causes geometric blurring (see §2.5.3). This blurring can therefore be kept to a minimum by employing a tube with a small focal spot. The focal-spot size is related to the size of the tube filament and there is a limit to which the filament size can be reduced because of the heat production. For the same output requirements, the smaller the filament, the greater the target temperature. Most modern x-ray units have two focal-spot sizes. These are referred to as the broad and fine foci. They are usually of the order of 1.2 and 0.6 mm square. The fine focus is used when special requirements of sharpness are needed.

The effective size of the focal spot can be measured by taking an image of the focal spot using a pinhole camera. A pinhole camera for this purpose consists of a small hole typically 75 μm in diameter drilled through a small gold/platinum plate. The hole is usually fluted at 8° to allow for divergence of the x-ray beam. The pinhole camera should be inserted into a lead plate to absorb the x-rays that do not pass through the pinhole. To reduce blurring of the image, a non-screen film should be used. Because exposure requirements can be large, however, a film–screen combination can be used if necessary. A typical image of a focal spot produced using a pinhole camera is shown in figure 3.6. The length and width of the image are l and w respectively.

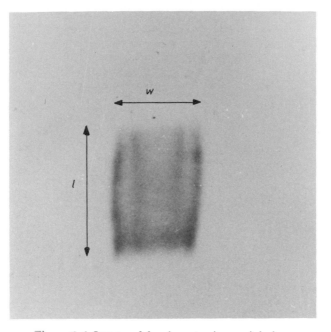

Figure 3.6 Image of focal spot using a pinhole.

The pinhole projects the image of a focal spot onto a film and, if the x-ray tube is horizontal when the pinhole is directly below the focal spot and in a plane parallel to the axis of the x-ray tube it produces an image of the 'effective focal spot'. Lateral displacement of the pinhole will produce a different size image. A Perspex plate embedded with a series of lead cylinders provides a simple device for estimating the position of the focal spot. If this plate is inserted into the light-beam diaphragm and imaged on a film, the cylinders off-axis will appear distorted. The position of the cylinder that shows no distortion indicates the position at which the pinhole should be placed.

It is desirable to obtain a magnified image of the focal spot to reduce errors in measurement. To achieve adequate magnification, the pinhole should be placed near to the focal spot with the film at a distance no less than 1 m away. The exposure geometry is shown in figure 3.7. The magnification is given by $M = d_2/d_1$, where d_2 is the pinhole-to-film

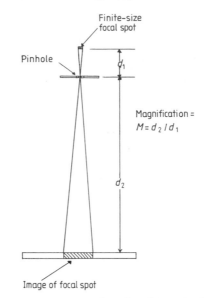

Figure 3.7 Exposure geometry for the determination of focal-spot sizes using a pinhole.

distance and d_1 is that between pinhole and focal spot. Not all x-ray units have the position of their focal spot identified and the distance d_1 can be found by measuring the magnification produced by a 1 cm lead strip placed at the same position as the pinhole. The new geometry is shown in figure 3.8 and should not be confused with the pinhole geometry. Here the magnification is given by $X = (d_1 + d_2)/d_1$ and therefore $d_1 = d_2/(X - 1)$.

The effective width of the focal spot is given by

$$W = (d_1/d_2)w = w/(X - 1).$$

The length of the focal spot is modified by multiplying the measured length by 0.7 (see Robinson and Grimshaw (1975) for a discussion of the value) to account for intensity fall-off of the x-ray beam along the axis of the x-ray tube. The factor adopted is somewhat approximate and should ideally be chosen depending on the target angle. However, to a good approximation the effective focal-spot length is given by $L = 0.7l/(X - 1)$. The effective focal-spot dimension is given by $(WL)^{1/2}$.

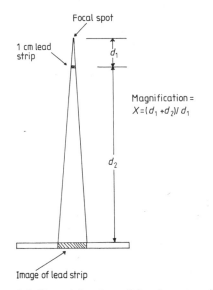

Focal spot

1 cm lead strip

d_1

Magnification =
$X = (d_1 + d_2)/d_1$

d_2

Image of lead strip

Figure 3.8 Determination of focal-spot position.

The IEC (Geldner and Schnitger 1982) recommended tolerances for focal-spot sizes are quite large, being 50% for focal spots below 1 mm^2 and between 30% and 40% for focal spots greater than 1 mm^2. More stringent tolerances have been suggested (Robinson and Grimshaw 1975).

From the above discussion on the use of the pinhole camera, it may be appreciated that this method of measuring the focal-spot size is not without difficulties and is time-consuming. A simpler method is to determine the focal-spot size indirectly by measuring the sharpness of an image produced by the focus. This can be achieved using a test object containing a series of line pairs of diminishing separation. Such an object should consist of line pairs running parallel and perpendicular to each other so that the width and length of the focal spot can be determined. These objects can be purchased complete with holder (normally a Perspex cylinder) and calibration chart, which relates the effective focal-spot size to the minimum size of line pairs that can be resolved at a particular magnification. An image of a line-pairs test object is shown in figure 3.9. The line pairs do not require the same magnification requirements as the pinhole camera and can be placed further from the focal spot. Off-axis distortions are therefore reduced and the line pairs do not need to be positioned exactly in line with the focal spot. Because of the simplicity of this test, it can be carried out as part of a routine quality-assurance programme. The resolution test will not, however, provide visual identification of any defects in the target, such as pitting of the anode surface, and the pinhole camera should be used if any adverse affects are suspected.

(a)   (b)

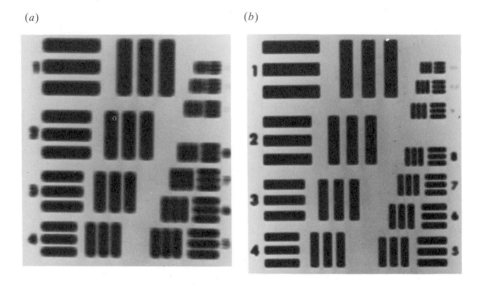

Figure 3.9 Radiograph of line-pairs test object: (*a*), broad focus 1.3×1.5 mm^2; (*b*), fine focus 0.6×0.6 mm^2.

3.2.6 Alignment and perpendicularity of the light-beam diaphragm

To collimate the x-ray field to the correct size and position, a light beam is projected through the x-ray collimators to coincide with the x-ray field (see figure 3.10). The box housing the light source and collimators is referred to as the light-beam diaphragm and is situated immediately below the x-ray tube. In addition to defining the x-ray field, the light source passes through light-opaque cross-wires, which indicate the centre of the beam. Field alignment is checked by adjusting the light field to lie between two copper-wire rectangles. By exposing these frames to the x-ray field, the position of the field relative to the light beam can be assessed. The centre of the field can be investigated at the same time by placing a screw and washer, vertically displaced by 15 cm, in the centre of the field. If the centre of the light field coincides with the x-ray field, then the image of the screw should be seen in the centre of the image of the washer. The focus–film distance for the above tests should be 1 m. The light field should be within 1 cm of the x-ray field in all four directions and the screw should not be visible outside the washer. Figure

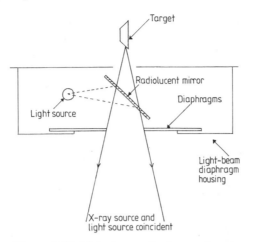

Figure 3.10 Light-beam diaphragm (LBD).

3.11 shows the correspondence between the light-beam diaphragm and the x-ray field.

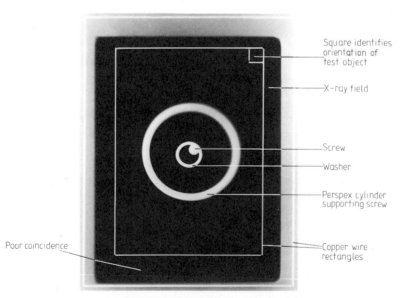

Figure 3.11 Radiograph of LBD test object.

3.3 SPECIFIC QUALITY-ASSURANCE TESTS

In addition to the basic quality-assurance tests, the following tests should be made as part of the routine programme where appropriate.

3.3.1 Tomographic tests

A (classical) tomographic x-ray system produces a two-dimensional image in which a particular plane at some desired depth within a patient is in focus whilst the images of other displaced structures, not in this plane, are superimposed in a highly blurred way. The position of this plane is determined by the position of the patient's couch relative to the x-ray tube and cassette holder. Classical tomography is achieved by moving the x-ray tube and cassette holder relative to the stationary patient in a precise geometry. Modern units will indicate the height of the in-focus plane. A simple test to check the accuracy of the height positioning is to image the test object shown in figure 3.12. This consists of a series of lead numbers embedded in a Perspex cylinder. The numbers are arranged concentrically at 1 mm height intervals and the height of the tomographic cut should be adjusted to 0.5 cm above the base of this cylinder. If the height positioning is accurate, the number 5 should be most clearly visible, and the appearance of the other numbers should be progressively blurred (figure 3.13).

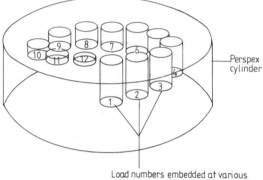

Figure 3.12 Tomographic test object to determine the height of cut.

Figure 3.13 Radiograph of tomographic test object used to determine the height of cut. Numbers 3 and 4 show most sharply indicating that the height of the cut is 1 to 2 mm lower than indicated on the x-ray unit.

Another important parameter to check is the coincidence of tube and cassette movement. This can be achieved by imaging a 1 mm diameter hole in a lead sheet placed at the height of the tomographic cut. If the coincidence is poor, the image of the hole will appear distorted.

The angle of rotation of the x-ray tube can be measured by exposing a film inclined at about 15° to the vertical to a narrow beam. The film should be supported on the table top and the height of the tomographic cut adjusted to the centre of the film. The exposed film should look like figure 3.14. The angle of rotation of the x-ray tube is given by $180 - \theta$.

Figure 3.14 Film exposed at an angle of about 15° to the vertical. Angle of rotation of x-ray tube is $180 - \theta$.

3.3.2 Basic image-intensifier tests

Image intensifiers (see §2.6.3) normally have under-couch x-ray tubes. It is therefore necessary to support the penetrameter (or kilovoltmeter), the focal-spot test object and the timer upside down. The position of the test objects can be viewed on the TV monitor so that correct positioning is ascertained.

The coincidence of the x-ray field size and the region viewed on the monitor can be checked by exposing a film at the position of the input phosphor. It is important that the maximum x-ray field size is not larger than the area of the input phosphor since photons outside the input phosphor will not contribute to the image and the patient will receive an unnecessary dose.

To monitor the performance of the TV monitor, it is first necessary to place a 1 mm copper plate between the x-ray tube and intensifier. This hardens the beam and allows the test phantoms to be assessed under clinical exposure conditions. Distortion of the TV screen can be checked by measuring the diameter both horizontally and vertically of the image formed by exposing a circular copper wire. Resolution of the monitor can be measured by exposing a line-pairs test object at the input phosphor. Contrast can be assessed by exposing a Perspex or aluminium step wedge, although more critical test objects can be purchased commercially.

3.3.3 Mammographic tests

All the basic quality-assurance parameters should be measured on a mammography x-ray unit. It will be necessary, however, to use a penetrameter that is not filtered by a copper sheet (which will absorb too much radiation) and is capable of measuring low kilovoltages (down to 20 or 25 kV). Exposure requirements at low kilovoltages are very large (e.g. 5000 mA s at 25 kV) and can prohibit frequent testing. Recently solid-state detectors have been developed that reduce the exposure requirements and provide an immediate display of the kilovoltage.

Success of the mammographic diagnosis depends upon the perception of low-contrast structures and the resolution of microcalcifications within the image (see §2.6.4). It is therefore important to monitor the contrast and resolution of the system under clinical conditions. Breast phantoms that allow these two parameters to be assessed are therefore required. A simple test phantom can be constructed from D-shaped Perspex slabs. Film contrast can be measured by exposing a Perspex step wedge on the slabs. The sharpness of the imaging system can be quantified by exposing a line-pairs test object on the slabs. Many commercial test objects are available which attempt to provide a more detailed analysis of the performance of the system.

3.4 DATA COLLECTION AND PRESENTATION OF THE RESULTS

To make full use of the quality-assurance data, the recording, analysis and application of the results must be well organised. Use of a microcomputer to assist with the above problems has proved successful (Evans 1985). The results in figure 3.15 were first manually entered into the computer and any result outside specification was identified by the computer and brought to the attention of the user. Duplicate hard copies of the results are automatically provided. One copy is given to the x-ray department and any comments are discussed.

3.5 SUMMARY OF QUALITY ASSURANCE

Quality assurance is used in diagnostic radiology to maintain standards in imaging and to minimise patient and staff doses. There are six basic

```
THE ROYAL MARSDEN DIAGNOSTIC QUALITY CONTROL SERVICE          21/08/87
------------------------------------------------------------------------

RESULTS FOR THE RMH                    EXAMPLE              SUMMER 87


GENERATOR TYPE & No. - ------------
TUBE TYPE & No.      - ------------
FILTRATION (mmAl)    - INHERENT =1.0    ADDED = 1.5   TOTAL = 2.5


      :-----------------------::-----:-----:-----:-----:-----:-----:-----:-----:
  K   :NOMINAL KV              ::  50 :  60 :  70 :  80 :  90 : 100 : 110 : 120 :
  V   :-----------------------::-----:-----:-----:-----:-----:-----:-----:-----:
  P   :MEASURED KV             ::* 57 :  65 :  71 :  81 :  93 : 103 : 110 : 120 :
      :-----------------------::-----:-----:-----:-----:-----:-----:-----:-----:


       :-----------------------::----------------------::----------------------::-------------:
       :    VARIABLE mA        ::    VARIABLE TIME     ::    VARIABLE KV       ::CONSISTENCY:
       :-----------------------::----------------------::----------------------::-------------:
       : 80KVp  0.1sec         :: 80KVp  200mA         :: 0.1sec  200mA        ::    80KVp    :
       :-----------------------::----------------------::----------------------::    200mA    :
 O     : mA  :  O  :O/mAs::  sec :  O  :O/mAs:: KV  :  O  :O/mAs::    0.1sec    :
 U     :-----:-----:-----::-----:-----:-----::-----:-----:-----::------:------:
 T     : 100 : 1.28: 0.12:: 0.01: 0.25: 0.12:: 60  : 1.25: 0.06::No. :  O  :
 P     :-----:-----:-----::-----:-----:-----::-----:-----:-----::------:------:
 U     : 200 : 2.06: 0.10:: 0.05: 1.08: 0.10:: 80  : 1.97: 0.09:: 1  : 1.85:
 T     :-----:-----:-----::-----:-----:-----::-----:-----:-----::------:------:
 (O)   : 320 : 3.88: 0.12:: 0.5 :10.00: 0.10:: 100 : 3.24: 0.16:: 2  : 2.03:
 (mGy) :-----:-----:-----::-----:-----:-----::-----:-----:-----::------:------:
 @75cm : 400 : 4.77: 0.11:: 1.0 :20.00: 0.10:: 120 : 4.73: 0.23:: 3  : 1.97:
       :-----:-----:-----::-----:-----:-----::-----:-----:-----::------:------:
                                                                             *


       :--------------::-------------------------------::----------------------------------:
       :CHECKS        ::     TIMING (sec)              ::     FOCAL SPOT (mm)              :
       :--------------::-----:-----:-----:-----::------:------------------:-------:--------:
       :NOMINAL       ::0.05 :0.5  :2.0  :5.0  ::broad:1.2 X 1.2:fine :0.6 X 0.6:
       :--------------::-----:-----:-----:-----::------:------------------:-------:--------:
       :MEASURED      ::0.053:0.476:1.981:4.890::broad:1.5 X 1.5:fine :0.6 X 0.6:
       :--------------::-----:-----:-----:-----::------:------------------:-------:--------:



HALF-VALUE LAYER AT 81kV  =   2.7 mmAl
FIELD COLLIMATION         - Good
PERPENDICULARITY          - Good


COMMENTS :-  * Indicates result outside specification
Low kV value high; Radiographer alerted.
Output consistency poor.

                        SURVEYED BY :S. E.
                                PHYSICS DEPT.
```

Figure 3.15 Hard copy of quality-assurance results produced by microcomputer.

parameters to test as part of a routine quality-assurance programme.

   (*a*) Potential (kV) can be measured using either a penetrameter or an electronic kilovoltmeter. The kilovoltages should be within ±5 kV of their nominal value. If the measured kilovoltage is more than 10 kV out of specification, the x-ray unit should be immediately serviced.

   (*b*) Exposure times (s) can be measured using either a spinning top or an electronic timer. The electronic timer is recommended. Exposure times should be within 0.01 s or 5%, whichever is the greater.

   (*c*) Outputs (mGy) are measured using an ionisation chamber. Output measurements are useful for checking the overall performance of the x-ray unit. Output values for the same exposures (mA s) and potentials (kV) should not differ by more than 20–40% at different tube currents or exposure times.

(*d*) Focal-spot size (mm^2) is measured using a pinhole camera or line-pairs test object. The line pairs provide a more convenient method of assessment for routine quality assurance but the pinhole method produces an image of the focal spot and should be used if the focal spot is suspected of being damaged. The measured focal spot should be within 50% of the nominal value if the nominal value is less than or equal to 1 mm^2 and within 30–40% if the nominal value is greater than 1 mm^2.

(*e*) Half-value layers (mm Al) are measured by placing varying amounts of aluminium filters in the x-ray beam. The HVL is the amount of beam filtration required to reduce the transmitted beam to half its original intensity. It can be used to calculate the effective energy of the beam and to find the total filtration of the x-ray unit. For most units, the total filtration should be equivalent to at least 2.5 mm Al.

(*f*) Alignment and perpendicularity of the light-beam diaphragm is assessed using a wire frame and vertically displaced screw and washer. The x-ray field should be within 1 cm of the light-beam field and the screw should not be visible outside the washer.

In addition to the basic quality-assurance tests, additional checks should be made on the height of the tomographic cut and coincidence of the x-ray tube movement and cassette; image intensifier monitors should be checked for resolution, contrast and distortion; and resolution and contrast should be measured with suitable breast phantoms on mammography units.

3.6 IMPROVEMENT IN RADIOGRAPHIC QUALITY

In the previous sections of this chapter we have seen how a quality-assurance programme helps to maintain imaging standards in the x-ray department, concentrating particularly on x-ray production. From time to time, however, new equipment is purchased, new film–screen combinations are developed and different imaging techniques are employed.

To get the maximum performance out of the imaging system, it is necessary to understand the physics of the imaging process to be able to assess and advise on the most appropriate instruments and techniques for the investigations being undertaken. In Chapter 2 the physics of diagnostic radiology with x-rays was extensively discussed and the following section further considers some of the most important processes that influence the quality of the image. In particular, we concentrate here on the practical measures that might be employed to improve image quality once the performance of the x-ray production equipment has been assured by the methods outlined earlier in the chapter.

3.6.1 Contrast and the effect of scatter in the image

One of the most important factors that allows different features to be distinguished in an image is the difference in density or contrast between a feature and its background. Clearly, the greater the contrast, the easier it is to observe any feature in the image. Other important physical factors were discussed in §2.4.

In §2.4.1 a model for radiographic contrast was developed, where it was shown (equation (2.6)) that contrast is related to the transmitted intensity I_2 of photons through an object and its background transmitted intensity I_1 by

$$C = (I_1 - I_2)/I_1. \tag{3.3}$$

In this equation I_1 and I_2 both include the same scatter component and are defined in the sense of equations (2.7) and (2.8). As shown in §2.2, scattered radiation has a complicated spatial dependence, but for simplicity may be considered as a slowly varying function of position. With this simplification, scattered radiation is considered to fall indiscriminately on the image receptor, causing a uniform increase in density of the image, and under these conditions equation (3.3) reduced to the simple form (2.10). Adding a subscript 's' to C to indicate that equation (3.3) includes scatter, we have

$$C_s = (I_1 - I_2)/I_1. \tag{3.4}$$

If, however, scatter were ignored, the corresponding result would be

$$C_0 = (I_1 - I_2)/(I_1 - \bar{\varepsilon}E\bar{S}) \tag{3.5}$$

where the scatter component $\bar{\varepsilon}E\bar{S}$ (see §2.2) has been subtracted from both I_1 and I_2. Hence we find

$$C_s = C_0[(I_1 - \bar{\varepsilon}ES)/I_1].$$

The factor in square brackets is simply $(1 + R)^{-1}$, where R is the scatter-to-primary ratio. Hence, we see that scatter degrades contrast via

$$C_s = C_0/(1 + R). \tag{3.6}$$

This has isolated one contrast-degrading feature from equation (2.10). The ability to distinguish low-contrast objects is of particular importance in mammography and the following discussions are particularly relevant to this area. We shall for the moment concentrate on the degrading effects of scattered radiation.

3.6.2 Magnitude of scatter

3.6.2.1 Dependence on field size. Scatter obviously increases with an increase in the x-ray field size due to the presence of more scattering centres. For small field sizes, the ratio R increases roughly linearly with the field size. For large field sizes (typically above 8 cm in mammography), the ratio R increases slowly because x-rays are absorbed within the irradiated volume and not transmitted to areas on the film well away from the primary originating paths.

3.6.2.2 Dependence on thickness of the irradiated volume. Scatter increases in rough proportion to the thickness of the irradiated volume due again to the increase in scattering centres.

3.6.2.3 Dependence on energy. The magnitude of scatter production decreases with increasing photon energy because Compton scattering decreases with increasing energy. However, the effect of scattering becomes more evident at higher kilovoltages because a greater fraction of the scattered radiation is in the forward direction and penetrates through to the film at higher energies. We have already seen in Chapter 2 that the variation of contrast with energy depends on factors other than just the scatter-to-primary ratio (see equation (2.10)), specifically the difference in linear attenuation coefficient of two tissues at a particular energy. This difference decreases with increasing energy for biological tissues and contrast is generally low at high energies (see figure 2.7).

3.6.2.4 Modification of effect by image receptors. The ratio R is modified by the conversion efficiency of the image receptors at different energies to give

$$R_R = \text{scattered energy absorbed/primary energy absorbed}$$

$$= Re_s/e_p \tag{3.7}$$

where e_s and e_p are the conversion efficiencies at the energies of the scattered photons and primary (transmitted) photons, respectively.

The fractional amount of energy deposited in the detector normally decreases with increasing photon energy and increases with the path length traversed by the photon through the detector. Both these qualities result in the ratio R_R being greater than the ratio R. First, the scattered radiation will mainly be reduced in energy and, secondly, since the scattered photons will not traverse the detector at normal incidence, their path lengths through the detector will be greater than those of the primary photons.

3.7 SCATTER REMOVAL

To improve image quality, the scatter reaching the detector must be reduced. Some of the techniques used for reducing the scatter contribution and improving the image quality are now discussed.

3.7.1 Grids

Grids are used to reduce the amount of scatter reaching the detector. Grids have directional properties; they consist typically of extremely thin lead strips interspaced by either cotton fibre (low energy), plastic or aluminium (high energy). Because the scattered photons will not meet the grid at normal incidence, they will be largely absorbed by the lead strips, whereas most of the primary photons will pass through the grid. Because some of the primary and most of the secondary radiation is absorbed by both the lead strips and the interspace medium, the exposure requirements are increased.

3.7.1.1 Grid parameters. If D is the separation between the lead strips and h is the height of the strips, then the grid ratio is defined as $h/D = R_G$. This ratio determines the relative number of scatter-to-primary photons that reach the detector. The higher the ratio R_G, the smaller the scatter-to-primary ratio R. Typical ratios R_G lie between 5 and 15. The grid ratio can be improved by increasing h but this will also increase the exposure requirements since there is more interspace material. A small interspace (D) will also reduce the ratio R, but fabrication difficulties limit this distance to about 150 μm.

The selectivity is defined as the ratio of primary-to-scatter transmission, i.e.

$$\text{selectivity} = \frac{\text{transmitted primary radiation}}{\text{transmitted scatter radiation}}$$

$$= T_p/T_s \tag{3.8}$$

and should be high for an efficient grid.

The contrast improvement factor K is given by the ratio of the contrast with the grid to the contrast without the grid. From equation (3.6) we obtain

$$K = (1 + R)/(1 + RT_s/T_p). \tag{3.9}$$

Typical values of K lie between 1.5 and 3.5.

The exposure increase with the grid is expressed in terms of an exposure factor or Bucky factor F and is given by

$$F = \frac{\text{exposure required with grid}}{\text{exposure required without grid}} \quad \genfrac{}{}{0pt}{}{\text{to produce}}{\genfrac{}{}{0pt}{}{\text{same film}}{\text{blackening}}}$$

$$= (1 + R)/T_p(1 + RT_s/T_p) \tag{3.10}$$

$$= K/T_p.$$

Typical values of F lie between 2 and 8.

3.7.1.2 Types of grid. There are two types of grid, which are now discussed.

Stationary grids Stationary grids are effective in removing scatter but the grid lines will be observed on the image unless ultra-fine lead strips are used. There are two types of stationary grid: (i) parallel, unfocused grids; and (ii) focused grids, which can only be used at a fixed focus-to-grid distance.

Moving grids The grid lines will be blurred out if the grid is moved during an exposure. The mechanism used to move the grid is referred to as a Bucky or a Potter–Bucky mechanism. Uniform movement of the grid is rarely used. There are three main types of grid movement: reciprocating movement, where the grid moves fast in one direction and slow in another; oscillating movement, where the grid vibrates back and forth across the film during exposure; and catapult movement, where the grid is accelerated quickly and decelerated in an exponential fashion.

3.7.2 Air gaps

The air-gap technique takes advantage of the inverse square law to reduce the ratio R. If d_1 is the distance between the patient and cassette and d_2 is the distance between the focus and cassette, then

$$\text{scatter} \propto 1/d_1^2 \qquad \text{and} \qquad \text{primary} \propto 1/d_2^2. \qquad (3.11)$$

Therefore

$$R \propto d_2^2/d_1^2. \qquad (3.12)$$

The larger the separation between the patient and the film, the smaller the ratio R.

As well as lowering the ratio R, an air gap will also produce a magnification effect. Exposure requirements are therefore increased to achieve the same optical density on the film. As was discussed in §2.5.3, magnification radiography can be associated with decreased unsharpness.

3.7.3 Scanning slit

A scanning slit employs two narrow transverse slits each about 1 mm wide. These slits are positioned above and below the patient and move in coincidence with a thin fan beam along the body of the patient. The moving fan beam is produced by an oscillating x-ray tube. Scatter produced within the patient is rejected by the cassette collimator. The scanning-slit method is illustrated in figure 3.16. This technique provides an excellent method for scatter rejection. Lower patient doses are achieved compared with grids since no primary transmission is absorbed by the collimators. Unfortunately, scan times are slow, which could result in blurred radiographs through patient movement, and high tube loadings are required, which limit the application of this technique.

3.8 CONTRAST ENHANCEMENT

Not all imaging techniques employing good scatter rejection provide enough radiographic contrast to allow a diagnosis to be made. It is sometimes necessary to improve the image quality by introducing new techniques. Some of the ways in which contrast can be improved are now discussed.

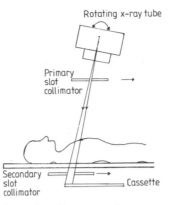

Figure 3.16 Scanning-slit system for scatter reduction.

3.8.1 Contrast media

Some anatomical features can be so similar in density and atomic number to their surroundings that they cannot be radiographically distinguished from their background. It is sometimes possible, however, to introduce a contrast medium into the patient which has a different attenuation coefficient to normal tissue and hence enhances the image of the feature that has taken it up. The medium needs to be either of lower density than normal tissue, e.g. air, which is used for imaging brain ventricles, or of higher density, e.g. barium compounds for examination of the digestive tract and iodine compounds for kidney and ureter investigations. The contrast medium therefore needs to be of a suitable atomic number Z and, because it is introduced into the body, it must be non-toxic and possess a viscosity suitable for either injection or ingestion. It must also have a suitable persistence.

Examinations requiring contrast media involve an element of risk. They are less desirable because contrast media make the procedures more invasive. Fortunately, digital radiography employing contrast enhancement has the potential to reduce the need for some of these examinations.

3.8.2 Film gradients

The relationship between film density and exposure is shown in figure 3.17. A graph of the film's density against the logarithm of the x-ray exposure is called its characteristic curve (see also §2.6.1.2). From the graph of the characteristic curve, it will be noted that the film does not have such a good response to changes in exposure at regions of high density (shoulder region) and low density (toe region). Since contrast is proportional to the change in density with exposure, it follows that the contrast is reduced in the toe and shoulder regions. It is therefore desirable to expose the film so that the density range within the image lies on the straight-line portion of the characteristic curve. It is apparent that the contrast can be improved using a film with a steep gradient. Unfortunately, exposure values become more critical with the steepness of the gradient, because the latitude of exposure decreases as the gradient increases.

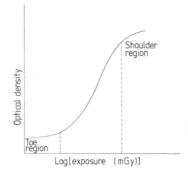

Figure 3.17 Typical characteristic curve for x-ray film.

The amount of radiation required to produce the required blackening on the film varies with the kilovoltage and focus–film distance, and also between patients. The correct exposure can be achieved automatically using radiolucent ionisation chambers placed between the grid and film. Exposure is terminated when a predetermined amount of radiation reaches the chamber. These systems are sometimes referred to as iontomats and it is common to have three chambers, with each unit covering different regions of the patient.

If the range of densities within an image is large, it can sometimes be beneficial to use a film that does not have a steep gradient so that none of the features within the image lie on the toe or shoulder region.

3.9   SUMMARY OF METHODS OF IMAGE ENHANCEMENT

Image quality includes the requirement for good radiographic contrast. Scatter degrades the contrast and must be reduced. Scatter increases with the size of the area imaged and the thickness of the irradiated volume. The scatter-to-primary ratio R at the detector increases with increasing photon energy and is increased by the receptor.

Grids are most commonly used for reducing scatter. The disadvantage of using grids is that exposure requirements are increased. Other methods for reducing scatter include air gaps and scanning slits.

Contrast can be artificially enhanced by introducing contrast media into the patient. Some examinations using contrast media carry high risks. Digital radiography can be used to increase the contrast within an image and reduce the need for the use of contrast media.

Contrast can also be increased using films with steeper characteristic curves. The characteristic curve for a film shows reduced sensitivity at high densities (shoulder region) and low densities (toe region). To obtain maximum contrast, the film must be exposed between the toe and shoulder regions. The correct film density can be achieved using iontomats, which are radiolucent ionisation chambers. These chambers terminate the exposure when a preselected amount of radiation falls on the chambers.

In this chapter we have highlighted the importance of ensuring that equipment is functioning optimally and have described techniques for evaluating performance. Even when equipment has been well designed and is functioning optimally, the physics of the x-ray scattering process conspires to reduce the diagnostic usefulness of the images. A number of ways of overcoming this problem have been described, most of which are in routine use in x-ray departments. Chapters 2 and 3 have essentially concentrated, however, on projection radiography. In the next two chapters we shall meet the revolution in x-ray imaging known as computed tomography.

REFERENCES

ARDRAN G M and CROOKS H E 1967 *Diagnostic X-Ray Beam Quality Report* AERE-M 1909

EVANS S H 1985 Management of quality assurance in diagnostic radiology by microcomputer *J. Soc. Radiol. Prot.* **5** (3) 117–19

GELDNER E and SCHNIIGER H 1982 IEC Publication 336/1981 — Revised edition of the standard for focal spots of x-ray tubes *Electromedica* **1/82** 6–10

HPA (HOSPITAL PHYSICISTS' ASSOCIATION) 1977 *The Physics of Radiodiagnosis* Scientific Report Series 6 (London: HPA)

ROBINSON A and GRIMSHAW G M 1975 Measurement of the focal spot size of diagnostic x-ray tubes — a comparison of pinhole and resolution methods *Br. J. Radiol.* **48** 572–80

CHAPTER 4

X-RAY TRANSMISSION COMPUTED TOMOGRAPHY

W SWINDELL AND S WEBB

4.1 THE NEED FOR SECTIONAL IMAGES

When we look at a chest x-ray (see figure 4.1), certain anatomical features are immediately apparent. The ribs, for example, show up as a light structure because they attenuate the x-ray beam more strongly than the surrounding soft tissue, so the film receives less exposure in the shadow of the bone. Correspondingly, the air-filled lungs show up as darker regions.

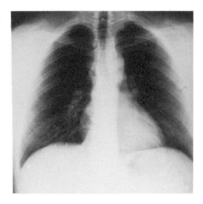

Figure 4.1 Typical chest x-radiograph.

A simple calculation illustrates the type of structure that one could expect to see with this sort of conventional transmission radiograph. The linear attenuation coefficients for air, bone, muscle and blood are

$$\mu_{\text{air}} = 0$$

$$\mu_{\text{bone}} = 0.48 \text{ cm}^{-1}$$

$$\mu_{\text{muscle}} = 0.180 \text{ cm}^{-1}$$

$$\mu_{\text{blood}} = 0.178 \text{ cm}^{-1}$$

for the energy spectrum of a typical diagnostic x-ray beam. Thus, for a slab of soft tissue with a 1 cm cavity in it, the results of table 4.1 follow at once using Beer's expression for the attenuation of the primary beam, namely

$$I(x) = I_0 \exp(-\mu x). \tag{4.1}$$

Table 4.1 Contrast in a transmission radiograph.

Material in cavity	$I(x)/I_0$ ($x = 1$ cm)	Difference (%) with respect to muscle
Air	1.0	+20
Blood	0.837	+ 0.2
Muscle	0.835	0
Bone	0.619	−26

X-ray films usually allow contrasts of the order of 2% to be seen easily, so a 1 cm thick rib or a 1 cm diameter air-filled trachea can be visualised. However, the blood in the blood vessels and other soft-tissue details, such as details of the heart anatomy, cannot be seen on a conventional radiograph. In fact, to make the blood vessels visible, the blood has to be infiltrated with a liquid contrast medium containing iodine compounds; the iodine temporarily increases the linear attenuation coefficient of the fluid medium to the point where visual contrast is generated (see §3.8.1, where contrast media are discussed in detail). Consideration of photon scatter further degrades contrast (see §§2.4.1 and 3.6.1).

Another problem with the conventional radiograph is the loss of depth information. The three-dimensional structure of the body has been collapsed, or projected, onto a two-dimensional film and, while this is not always a problem, sometimes other techniques such as stereoscopic pairs of radiographs or conventional tomography (see §3.3.1) are needed to retrieve the depth information. Some early history of conventional tomography is provided by MacDonald (1981), and different geometries and equipment are reviewed by Coulam *et al* (1981).

It is apparent that conventional x-radiographs are inadequate in these two respects, namely the inability to distinguish soft tissue and the inability to resolve spatially structures along the direction of x-ray propagation.

4.2 THE PRINCIPLES OF SECTIONAL IMAGING

With computed tomography, a planar slice of the body is defined and x-rays are passed through it only in directions that are contained within, and are parallel to, the plane of the slice (see figure 4.2). No part of the body that is outside of the slice is interrogated by the x-ray beam, and this eliminates the problem of 'depth scrambling'. The CT image is as though the slice (which is usually a few millimetres thick) had been physically removed from the body and then radiographed by passing x-rays through it in a direction perpendicular to its plane. The resulting images show the human anatomy in section with a spatial resolution of about 1 mm and a density (linear attenuation coefficient) discrimination of better than 1% (see figure 4.3). This chapter is about the method of converting the x-ray measurements of figure 4.2 into the image shown in figure 4.3.

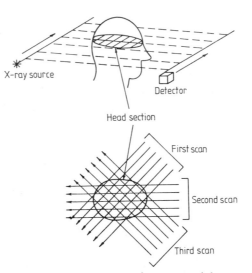

Figure 4.2 Simple scanning system for transaxial tomography. A pencil beam of x-rays passes through the object and is detected on the far side. The source–detector assembly is scanned sideways to generate one projection. This is repeated at many viewing angles and the required set of projection data is obtained. (Reproduced from Barrett and Swindell (1981).)

There are many good reviews on the subject of computed tomography: see, for example, Brooks and Di Chiro (1975, 1976), Kak (1979) and Zonneveld (1979). The commercial development of x-ray computed tomography has been described as the most important breakthrough in diagnostic radiology since the development of the first planar radiograph.

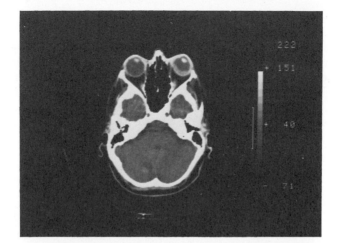

(a)

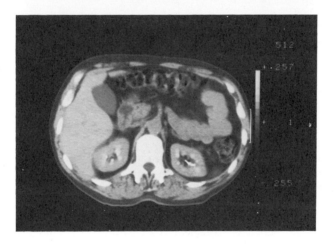

(b)

Figure 4.3 (*a*) CT image of a head taken at eye level. (*b*) Abdominal section through the kidneys.

4.2.1 Scanner configurations

As far as the patient is concerned, the CT scanner is a machine with a large hole in it. The body or the head is placed inside the hole in order to have the pictures taken (see figure 4.4). The covers of the machine hide a complicated piece of machinery, which has evolved through several versions since its inception (Hounsfield 1973). Here follows a short description of this development.

A finely collimated source defines a pencil beam of x-rays, which is then measured by a well collimated detector. This source–detector combination measures parallel projections, one sample at a time, by stepping linearly across the patient. After each projection, the gantry rotates to a new position for the next projection (see figure 4.5). Since there is only one detector, calibration is easy and there is no problem with having to balance multiple detectors; also costs are minimised. The scatter rejection of this first-generation system is higher than that of any other generation because of the two-dimensional collimation at both source and detector. The system is slow, however, with typical acquisition times of 4 min per section, even for relatively low-resolution images.

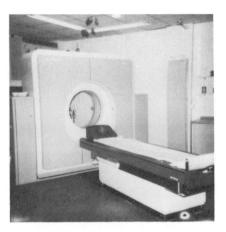

Figure 4.4 Typical CT scanner.

Data gathering was speeded up considerably in the second generation. Here a single source illuminated a bank of detectors with a narrow (~10°) fan beam of x-rays (see figure 4.6). This assembly traverses the patient and measures N parallel projections simultaneously (N is the number of detectors). The gantry angle increments by an angle equal to the fan angle between consecutive traverses. These machines can complete the data gathering in about 20 s. If the patient can suspend breathing for this period, the images will not be degraded by motion blur, which would otherwise be present in chest and abdominal images.

In third-generation systems, the fan beam is enlarged to cover the whole field of view (see figure 4.7). Consequently, the gantry needs only to rotate, which it can do without stopping, and the data gathering can be done in 4–5 s. It is relatively easy for a patient to remain still for this length of time. Detector balancing is critical for this geometry if circular 'ring' artefacts are to be avoided. Xenon detectors are often chosen because of their stable nature of operation.

Fourth-generation systems use a stationary ring of typically 1000 detectors and only the source rotates (see figure 4.8). Scan speeds remain fast and the ring artefact is overcome. Since every detector is, at some time during the scan, sampling the unattenuated x-ray beam,

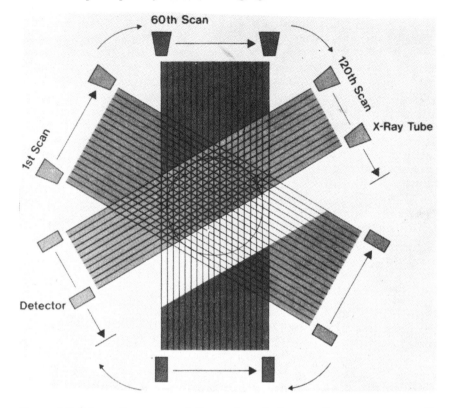

Figure 4.5 Schematic representation of a first-generation CT scanner. It utilises a single pencil beam and single detector for each scan slice. The x-ray source and detector are passed transversely across the object being scanned, with incremental rotations of the system at the end of each transverse motion. (Reproduced from Maravilla and Pastel (1978).)

calibration in 'real time' can be performed.

In the race for speed, the next clinically useful break point comes at around 0.1 s for the data acquisition time. This permits cardiac motion to be frozen. This will allow clearer images not only of the heart but also of organs that are well perfused with blood, such as the liver, and which pulsate in synchrony to the heart beat. Mechanical movement is ruled out and multiple stationary sources are prohibitively cumbersome and expensive. The fifth-generation device (Peschmann *et al* 1985) has no moving parts. A target of the x-ray tube follows the shape of a circular arc of approximately 210°. The patient is placed at the centre of this arc and the effective source of x-rays is made to move by scanning the electron beams around the circumference of the target (see figure 4.9). Scan times can thus be reduced to a few milliseconds.

Special x-ray sources and detectors have been designed and manufactured for use in CT scanners. Each generation imposes its own special requirements. There are also special requirements imposed on the power supplied for the x-ray tubes, especially with regard to stability. A good review of these problems is provided by Webb (1987). The properties of some photon scintillation detectors are shown later in table 6.1.

In addition to the gantry, which houses the scanning mechanism, x-ray sources and detectors, there are other essential components to a CT system. These include the computer, which controls the hardware and processes the data, and the operator's viewing console. These parts of the system will not be covered in this book.

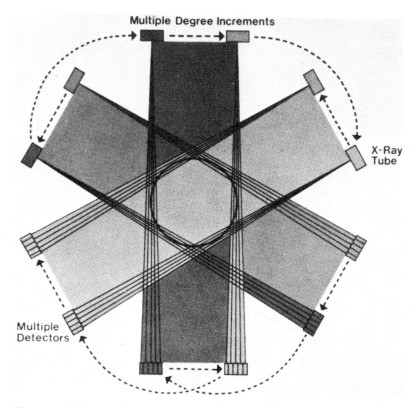

Figure 4.6 Schematic representation of a second-generation CT scanner. A narrow-angle fan beam of x-rays and multiple detectors record several pencil beams simultaneously. As the diverging pencil beams pass through the patient at different angles, this enables the gantry to rotate in increments of several degrees and results in markedly decreased scan times of 20 s or less. (Reproduced from Maravilla and Pastel (1978).)

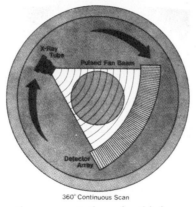

360° Continuous Scan

Figure 4.7 Schematic representation of a third-generation CT scanner in which a wide-angle fan beam of x-rays encompasses the entire scanned object. Several hundred measurements are recorded with each pulse of the x-ray tube. (Reproduced from Maravilla and Pastel (1978).)

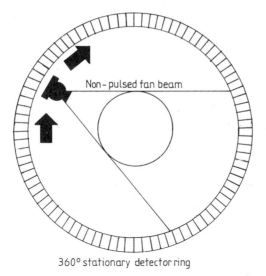

Non-pulsed fan beam

360° stationary detector ring

Figure 4.8 Schematic representation of a fourth-generation CT scanner. There is a rotating x-ray source and a continuous 360° ring of detectors, which are stationary. Leading and trailing edges of the fan beam pass outside the patient and are used to calibrate the detectors. (Reproduced from Maravilla and Pastel (1978).)

4.2.2 Line integrals

The data needed to reconstruct the image are transmission measurements through the patient. Assuming, for simplicity, that we have (i) a very narrow pencil beam of x-rays, (ii) monochromatic radiation and (iii) no scattered radiation reaching the detector, then the transmitted intensity is given by

$$I_\phi(x') = I_\phi^0(x') \exp\left(- \int_{AB} \mu[x, y] \, dy'\right) \tag{4.2}$$

where $\mu[x, y]$ is the two-dimensional distribution of the linear attenuation coefficient, ϕ and x' define the position of the measurement and $I_\phi^0(x')$ is the unattenuated intensity (see figure 4.10). The $x'y'$ frame rotates with the x-ray source position such that the source is on the y' axis. Equation (4.2) is simply an extension of Beer's law (equation (4.1)) to take the spatial variation of μ into account. Frequently $\mu[x, y]$ is simply referred to, somewhat loosely, as the density distribution, and we shall adopt that practice here. In this context, 'density' refers to electron density (electrons/cm^3), which for most practical CT scanners is a parameter found to be related to attenuation coefficient by a series of linear relationships (see e.g. Parker *et al* 1980).

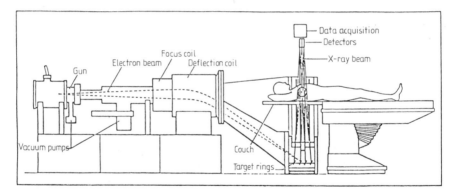

Figure 4.9 Imatron CT-100 ciné CT scanner; longitudinal view. Note the use of four target rings for multislice examination. (Courtesy of Imatron.)

A single projection of the object $\lambda_\phi(x')$ is defined as

$$\lambda_\phi(x') = -\ln[I_\phi(x')/I_\phi^0(x')]$$
$$= \int_{-\infty}^{\infty} \int_{-\infty}^{\infty} \mu[x, y]\delta(x \cos \phi + y \sin \phi - x') \, dx \, dy \tag{4.3}$$

where, now, the Dirac delta function δ picks out the path of the line integral, since the equation of AB is $x' = x \cos \phi + y \sin \phi$.

Equation (4.3) expresses the linear relationship between the object function $\mu[x, y]$ and the measured projection data λ_ϕ. The problem of reconstructing is precisely that of inverting equation (4.3), i.e. recovering $\mu[x, y]$ from a set of projections $\lambda_\phi(x')$.

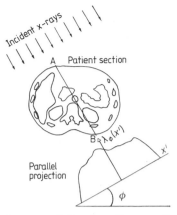

Figure 4.10 Projections are defined as the negative logarithm of the fractional x-ray transmittance of the object, $\lambda_\phi(x') = -\ln[I_\phi(x')/I^0_\phi(x')]$. ϕ is the angle at which the projection data are recorded.

4.2.3 Projection sets

The quantity $\lambda_\phi(x')$ in equation (4.3) may be interpreted as the one-dimensional function λ_ϕ of a single variable x' with ϕ as a parameter, and with the arrangement of figure 4.10 this $\lambda_\phi(x')$ is referred to as a parallel projection. To gather this sort of data, a single source and detector are translated across the object at an angle ϕ_1, producing $\lambda_{\phi_1}(x')$. The gantry is then rotated to ϕ_2 and $\lambda_{\phi_2}(x')$ is obtained, and so on for many other angles. As we mentioned in the previous section, the inefficiencies of this first-generation scanning are no longer tolerated in commercial systems, and the projection data are measured using a fan beam. In this case, the distance x' is measured in a curvilinear fashion around the detector from the centre of the array and ϕ is the angle of the central axis of the projection (see figure 4.11). In what follows we analyse the case for parallel projections simply because it is the easier case to study. The added complexity of fan beam geometry obscures the basic solution method, while adding but little to the intellectual content (see also end of §4.3.1).

In practice, the x-ray source and x-ray detector are of finite size. The projection data are better described as volume integrals over long, thin 'tubes' rather than as line integrals. One effect of this is to average over any detail within the object that is small compared to the lateral dimensions of the tube. The highest spatial frequencies that would be present in a 'perfect' projection are thus not measurable and the object appears to be 'band-limited' because of this low-pass filtering by the measuring system. This has important consequences, which will be discussed later (§4.3).

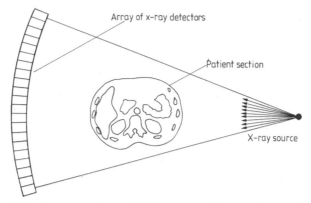

Figure 4.11 For second- and higher-generation systems, data are collected using the fan-beam geometry as shown here.

4.2.4 Information content in projections and the central section theorem

Up to this point, we have assumed that equation (4.3) has a solution. We shall now show that a complete set of projection data do indeed have enough information contained to permit a solution. In doing so, we shall point the way to the method of solution that is most commonly used in x-ray CT scanners.

First we specify the notation. The Fourier transform of the density distribution $\mu[x, y]$ is $M[\zeta, \eta]$. The square brackets serve to remind us that the coordinates are Cartesian. In polar coordinates, the corresponding quantities are $\mu^{\mathrm{p}}(r, \theta)$ and $M^{\mathrm{p}}(\rho, \phi)$. ($M$ is upper-case Greek 'mu'.) The various quantities defined in the $x'y'$ frame are $\mu'[x', y']$, $M'[\zeta', \eta']$, etc. It is not necessary to use the prime on $\lambda_\phi(\quad)$, etc, since the different functional form of λ for each ϕ value is implicitly denoted by the subscript ϕ.

The angular orientation of the $[x, y]$ reference frame is arbitrary, so without loss of generality we can discuss the projection at $\phi = 0$. From equation (4.3) we have

$$\lambda_0(x') = \int_{-\infty}^{\infty} \int_{-\infty}^{\infty} \mu[x, y]\delta(x - x')\,\mathrm{d}x\,\mathrm{d}y. \qquad (4.4)$$

The integration over x is trivial, i.e.

$$\lambda_0(x) = \int_{-\infty}^{\infty} \mu[x, y]\,\mathrm{d}y \qquad (4.5)$$

which is an obvious result anyway.

The next step is to take the one-dimensional Fourier transform of both sides of equation (4.5). Readers unfamiliar with the basic concepts of the Fourier transform may care to study the appendix to Chapter 12 (§12.9) at this point. Writing the transformed quantity as $\Lambda_0(\zeta)$, we have

$$\Lambda_0(\zeta) \equiv \int_{-\infty}^{+\infty} \lambda_0(x) \exp(-2\pi i \zeta x) \, dx$$

$$= \int_{-\infty}^{\infty} \int_{-\infty}^{\infty} \mu[x, y] \exp[-2\pi i(\zeta x + \eta y)] \, dy \, dx \Bigg|_{\eta=0}. \quad (4.6)$$

An extra term $\exp(-2\pi i \eta y)$ has been slipped into the Fourier kernel on the right-hand side (RHS), but the requirement that the integral be evaluated for $\eta = 0$ makes this a null operation. However, in this form the RHS of equation (4.6) is recognisable as the two-dimensional Fourier transform $M[\zeta, \eta]$ evaluated at $\eta = 0$, so equation (4.6) can be rewritten as

$$\Lambda_0(\zeta) = M[\zeta, 0]. \quad (4.7)$$

Because the Cartesian ζ axis (i.e. $\eta = 0$) coincides with the polar coordinate ρ at the same orientation, equation (4.7) can be written as

$$\Lambda_0(\zeta) = M^P(\rho, 0). \quad (4.8)$$

This is an important result. In words, it says that if we take the one-dimensional Fourier transform of the projection λ_0, the result Λ_0 is also the value of the two-dimensional transform of μ along a particular line. This line is the central section that is oriented along the direction $\phi = 0$. Now we can restore the arbitrary angular origin of the reference frames and state the general result, namely

$$\Lambda_\phi(\zeta') = M'[\zeta', \eta']$$

$$= M^P(\rho, \phi). \quad (4.9)$$

This important result is known as the central section or central slice theorem. To illustrate the theorem, consider a general, bounded object. This object can always be synthesised from a linear superposition of all of its two-dimensional spatial frequency components. Now, consider just one of those cosinusoidal frequency components (see figure 4.12(*a*)). Only when the projection direction is parallel to the wave crests does the projection differ from zero. However, for that particular direction, the full cosine distribution is projected onto the x' axis. The Fourier transform of this one component is shown in figure 4.12(*b*). The original object is a superposition of many component waves of various phases, periods and directions, and it follows that only those waves that are parallel to the first one will have their transforms located on the ξ' axis, and that these are the only waves that will change the form of $\lambda_\phi(x')$. Thus, the transform of the slice is identical to the corresponding section (or slice) of the two-dimensional transform.

4.2.5 Reconstruction by two-dimensional Fourier methods

It now follows that a complete set of projections contains the information that is needed to reconstruct an image of μ. This can be seen by considering a large number of projections at evenly spaced angles ϕ_n.

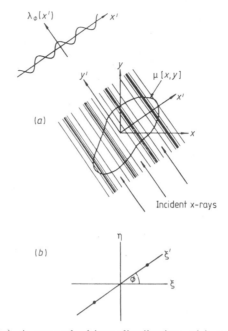

Figure 4.12 (*a*) A general object distribution $\mu(r)$ can be decomposed into Fourier components of the form $\sin(2\pi\rho r)$ or $\cos(2\pi\rho r)$. One of the latter is depicted here. There is only one direction ϕ for which the projection of this component is non-zero, and at this particular ϕ the component is fully mapped onto the projection. (*b*) The Fourier transform of this component is a pair of δ functions (shown here by dots) located on the ξ' axis. (Reproduced from Barrett and Swindell (1981).)

The value of $M^p(\rho, \phi)$ can then be determined along the radial spokes of the same orientations. If M^p is thus defined on a sufficiently well sampled basis (more about this later—§4.5.1), then $\mu[x, y]$ can be obtained by a straightforward two-dimensional transformation of M, which can be obtained from M^p by means of interpolation from the polar to the Cartesian coordinate systems.

It is worth noting that the projections must be taken over a full 180° rotation without any large gap in angle. If there are large gaps, there will be corresponding sectors in Fourier space that will be void of data. The object μ cannot faithfully be constructed from its transform M if this latter is incompletely defined. We shall see in Chapter 6 that certain classes of positron emission tomography scanners suffer the problem of limited-angle projection data.

The solution method just outlined is not a very practicable one for a number of reasons, but the discussion demonstrates that, in principle, an object can be reconstructed from a sufficiently complete set of its projections. The commonly used 'filtered backprojection' method is described in §4.3.

4.2.6 Displaying the image

The reconstruction of μ is usually made on a rectangular array, where each element or pixel has a value μ_i ascribed to it $(1 \leqslant i \leqslant I)$. Before these data are displayed on a video screen, it is conventional to rescale them in terms of a 'CT number', which is defined as

$$\text{CT number} = \frac{\mu_{\text{tissue}} - \mu_{\text{water}}}{\mu_{\text{water}}} \times 1000. \qquad (4.10)$$

Thus the CT number of any particular tissue is the fractional difference of its linear attenuation coefficient relative to water, measured in units of 0.001, i.e. tenths of a per cent. The CT numbers of different soft tissues are relatively close to each other and relatively close to zero. However, provided that the projection data are recorded with sufficient accuracy, then different soft tissues can be differentiated with a high degree of statistical confidence. Similar tissues, which could not be resolved on conventional transmission x-radiographs, can be seen on CT reconstructions. Small differences in CT number can be amplified visually by increasing the contrast of the display. The output brightness on the screen is related to the CT number by means of a level and a width control (see figure 4.13). These windowing controls can be varied by the operator while looking at the image, so the small range of CT numbers corresponding to the soft tissues within the body can be selected to drive the screen from black to white.

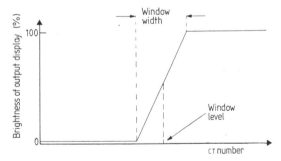

Figure 4.13 The 'windowing' facility allows the display brightness range to be fully modulated by any desired range of CT values as determined by adjusting the window 'level' and 'width'.

4.3 FOURIER-BASED SOLUTIONS: THE METHOD OF CONVOLUTION AND BACKPROJECTION

The mathematics of transmission computed tomography, or the theory of reconstruction from projections, has itself acquired the status of attracting attention as an independent research area. The literature is enormous and there already exist many excellent books reviewing the field. Some, such as Herman (1980), review the subject from the viewpoint of the theoretically minded physicist or engineer, whereas others, for example Natterer (1986), are really only accessible to persons who first and foremost regard themselves as mathematicians. Optionally, for a practical discussion, the book by Barrett and Swindell (1981) may be consulted. Against the backcloth of this formidable weight of literature, the purpose of this chapter is to provide a simplified view of theory applicable to the most elementary scanning geometry in order to come to grips with some basic principles. From this beginning, it should be possible to go on to view the analogous developments in single-photon emission computed tomography (Chapter 6) as well as reconstruction techniques using ultrasound (Chapter 7) and nuclear magnetic resonance (Chapter 8). The serious research student will find the treatment in this chapter over-simple and could do no better than branch out starting with one of the above books.

The theory of reconstruction from projections pre-dates the construction of any practical scanner for computed tomography. It is generally accepted that the problem was first analysed by Radon (1917) some 70 years or so ago. The theory has been 'rediscovered' in several distinct fields of physics including radioastronomy (Bracewell and Riddle 1967) and electron microscopy (Gilbert 1972). An account of a method and the first system for reconstructing x-ray medical images probably originated from Russia (Tetel'baum 1956, 1957, Korenblyum *et al* 1958) (see Chapter 1), although it is the work of Hounsfield (1973) that led to the first commercially developed system. This was a head scanner marketed by EMI. Many different techniques for solving the reconstruction problem have been devised and, in turn, each of these has received a great deal of attention regarding subtle but important additional features that improve image quality. It would be true to say that, just as scanner hardware has developed rapidly, so parallel developments in reconstruction theory have kept pace to service new designs and to predict optimal scanning conditions and the design criteria for new scanners.

Reconstruction techniques can be largely classified into two distinct groups, namely the convolution and backprojection methods (or equivalent Fourier techniques) and iterative methods. For a long time, there was much debate as to the relative superiority of one algorithm or another, and, in particular, whether one of the two classes was in some way superior. Today, this debate has largely subsided, with the inevitable conclusion that each method has its advantages, it being important to tailor the reconstruction technique to the scanner design and (in a wider context) to the physics of the imaging modality. For example, iterative techniques have found some important applications in emission CT where photon statistics are poorer.

Next we derive the algorithm that is most commonly used in CT scanners. It is the method of 'filtered backprojection' or 'convolution and backprojection'. A formal statement of the two-dimensional inverse polar Fourier transform yielding $\mu[x, y]$ is given by

$\mu^p(r, \theta) = \mu[x, y]$

$$= \int_0^\pi \int_{-\infty}^\infty M^p(\rho, \phi) \exp[2\pi i \rho(x \cos \phi + y \sin \phi)]|\rho| \, d\rho \, d\phi \quad (4.11)$$

where $x(= r \cos \theta)$ and $y(= r \sin \theta)$ denote the general object point.

If equation (4.11) is broken into two parts, the method of solution becomes immediately apparent (indeed, these two equations give the method its name):

$$\mu[x, y] = \int_0^\pi \lambda^\dagger_\phi(x') \, d\phi \bigg|_{x'=x \cos \phi + y \sin \phi} \quad (4.12)$$

where

$$\lambda^\dagger_\phi(x') = \int_{-\infty}^\infty M^p(\rho, \phi)|\rho|\exp(2\pi i \rho x') \, d\rho. \quad (4.13)$$

Consider equation (4.13), which defines an intermediate quantity $\lambda^\dagger$. For reasons that will become obvious, $\lambda^\dagger$ is called the filtered projection. The first point to notice is that equation (4.13) is the one-dimensional Fourier transform of the product of M^p and $|\rho|$. As such, it should also be possible to write it as the convolution of the Fourier transforms of M^p and $|\rho|$ (see the appendix of Chapter 12, §12.9.2). Taking M^p first, its transform is known from the central slice theorem. It is just the projection data, $\lambda_\phi(x')$, i.e.

$$\lambda_\phi(x') = \int_{-\infty}^\infty \Lambda_\phi(\zeta') \exp(2\pi i \zeta' x') \, d\zeta' \quad (4.14)$$

where, from (4.9),

$$\Lambda_\phi(\zeta') = M^p(\rho, \phi).$$

Now consider $|\rho|$. This is not a sufficiently well behaved function for its transform to exist. In practice, however, we have seen in §4.2.3 that $M(\rho, \phi)$ is band-limited by the measuring system, so if the maximum frequency component of $M(\rho, \phi)$ is ρ_{max} then $|\rho|$ can be similarly truncated. Thus we need the transform $p(x')$ of $P(\rho)$, where

$$\begin{aligned} P(\rho) &= 0 & |\rho| \geqslant \rho_{max} \\ P(\rho) &= |\rho| & |\rho| < \rho_{max} \end{aligned} \quad (4.15)$$

i.e.

$$p(x') = \int_0^{\rho_{max}} \rho \exp(2\pi i \rho x') \, d\rho - \int_{-\rho_{max}}^0 \rho \exp(2\pi i \rho x') \, d\rho. \quad (4.16)$$

Equation (4.16) is straightforward to evaluate (see appendix, §4.6), with the result

$$p(x') = \rho^2_{max}[2 \, \text{sinc}(2\rho_{max}x') - \text{sinc}^2(\rho_{max}x')] \quad (4.17)$$

which is perfectly well behaved.

Using the convolution theorem, equation (4.13) can now be written as

$$\lambda^\dagger_\phi(x') = \int_{-\infty}^\infty \lambda_\phi(x)p(x' - x) \, dx \quad (4.18)$$

or in the conventional shorthand notation (∗ denoting convolution)

$$\lambda^\dagger_\phi(x') = \lambda_\phi(x') * p(x').$$ (4.19)

The dagger (†) indicates a *filtered* projection because the original projection is convolved with $p(x')$, which constitutes a filtering operation.

Now we look at equation (4.12). This represents the process of *backprojection* in which a given filtered projection $\lambda^\dagger_\phi$ is distributed over the $[x, y]$ space. For any point x, y and projection angle ϕ, there is a value for x' given by

$$x' = x \cos \phi + y \sin \phi.$$

This is the equation of a straight line (parallel to the y' axis), so the resulting distribution has no variation along the y' direction. A simple analogy is to think of dragging a rake, with a tooth profile given by $\lambda^\dagger_\phi(x')$, though gravel in the y' direction. The one-dimensional tooth profile is transferred to the two-dimensional bed of gravel. Backprojection is not the inverse of projection. If it were, the reconstruction-from-projection problem would be trivial! It is very important to be clear that pure backprojection of unfiltered projections will not suffice as a reconstruction technique. Equation (4.12) also contains an integration over ϕ, which represents the summation of the backprojections of each filtered projection, each along its own particular direction. It is like raking the gravel from each projection direction with a different tooth profile for each filtered projection. The analogy breaks down, however, since each raking operation would destroy the previous distribution rather than adding to it, as required by the integration process.

The total solution is now expressed by equation (4.19) and equation (4.12). In words, each projection $\lambda_\phi(x')$ is convolved (filtered) with $p(x')$ (equation (4.17)). The filtered projections are each backprojected into $[x, y]$ and the individual backprojections (for each projection angle) are summed to create the image $\mu[x, y]$.

4.3.1 A practical implementation

In practice, the data are discretely sampled values of $\lambda_\phi(x')$. Thus the continuous convolution of equation (4.18) must be replaced by a discrete summation, as must also the angular integration of equation (4.12). We deal first with the convolution. The Whittaker–Shannon sampling theorem states that a band-limited function with maximum frequency component ρ_{max} can be completely represented by, and reconstructed from, a set of uniform samples spaced s apart, where $s \leq (2\rho_{max})^{-1}$. This requirement corresponds to adjacent samples being taken approximately $w/2$ apart, where w is the width of a detector. Provided that the data are band-limited in this manner, the continuous convolution can be replaced by a discrete convolution. Grossly misleading results can occur if the sampling is too wide to satisfy this Nyquist condition.

From equation (4.17) and using $s = (2\rho_{max})^{-1}$, it is seen that

$$
\begin{aligned}
p(ms) &= 0 & m \text{ even, } m \neq 0 \\
p(ms) &= -(\pi ms)^{-2} & m \text{ odd} \\
p(ms) &= (2s)^{-2} & m = 0
\end{aligned}
\tag{4.20}
$$

where ms denotes the positions along x' at which the discrete filter is defined. The projection data λ_ϕ are sampled at the same intervals, so that equation (4.18) can be replaced by its discrete counterpart

$$
\lambda_\phi^+(ms) = \frac{1}{4s} \lambda_\phi(ms) - \frac{1}{\pi^2 s} \sum_{\substack{n \\ (m-n)\text{odd}}} \frac{\lambda_\phi(ns)}{(m-n)^2}
\tag{4.21}
$$

where m and n are integers. Figure 4.14 shows the continuous and sampled versions of $p(x')$.

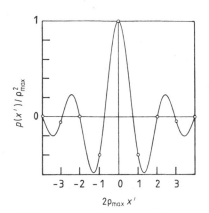

Figure 4.14 The full curve shows continuous form of the Ramachandran–Lakshminarayanan filter. The open circles show the points at which the filter is sampled for digital filtering methods. (Reproduced from Barrett and Swindell (1981).)

Equation (4.21) is the result obtained in a quite different way in the classic paper by Ramachandran and Lakshminarayanan (1971) and is the discrete version of the result obtained by Bracewell and Riddle (1967). Note that, although the Fourier transform has featured in its derivation, the reconstruction technique is entirely a real-space operation. The convolution function (the transform of the bounded $|\rho|$) is the same for all the projections and can therefore be computed, stored and re-used for each projection. Equations (4.12) and (4.21) show that the contributions to the reconstruction $\mu[x, y]$ can be computed from the projections one by one, as they arrive from the scanner, and once 'used' the projection may be discarded from computer memory. This is a distinct advantage over the two-dimensional Fourier transform method (§4.2.5) of recovering $\mu[x, y]$, when all the transformed projections are in use simultaneously. One can even view the reconstruction 'taking shape' as the number of contributing projections increases.

The discrete backprojection is shown in figure 4.15. It is necessary to assign and then sum to each element in the image array μ_i the appropriate value of $\lambda^\dagger_\phi$. This can be done on a nearest-neighbour basis, but it is better to interpolate between the two nearest sampled values of $\lambda^\dagger_\phi$. Formally, the process is described by

$$\mu_i = \sum_{n=1}^{N} \lambda^\dagger_{\phi_n}(m^*s) \tag{4.22}$$

where subscript n denotes the nth projection and m^* denotes an interpolated point within m for which the interpolated value of $\lambda^\dagger$ is calculated. The backprojection through m^* passes through the centre of the ith pixel.

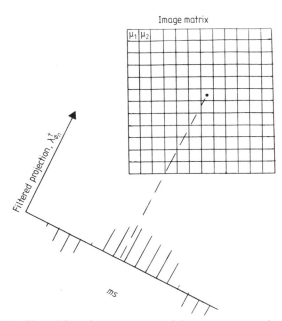

Figure 4.15 The object is reconstructed into an array of pixels μ_1, μ_2, ... by backprojecting each filtered projection onto the array and summing the results for each projection angle.

The total process is not so daunting as it seems. For parallel projection data, the whole process can be coded into less than 25 lines of FORTRAN code (see figure 4.16).

```
      DIMENSION P(65),PSTAR(65),F(4225)
      DATA W/.3333333/, M/50/, F/4225*0./
C  FOLLOWING STEPS ARE DONE FOR EACH PROJECTION
      DO 50 K=1,M
C  READ ONE SET PROJECTION DATA AND ANGLE
      READ (1,100) P,PHI
100   FORMAT (66F6.2)
      SINE = SIN(PHI)
      COSINE = COS(PHI)
C  CALCULATE FILTERED PROJECTION PSTAR
      DO 30 I=1,65
      Q = P(I)*2.467401
      JC = 1 + MOD(I,2)
      DO 20 J=JC,65,2
20    Q = Q - P(J)/(I-J)**2
30    PSTAR(I) = Q/(3.141593*M*W)
C  BACK PROJECT FILTERED PROJECTION ONTO IMAGE ARRAY
      DO 50 J=1,65
      IMIN = J*65-32-INT(SQRT(1024.-(33-J)**2)
      IMAX = (2*J-1)*65 - IMIN + 1
      X = 33 + (33-J)*SINE + (IMIN-J*65+31)*COSINE
      DO 50 I=IMIN,IMAX
      X = X + COSINE
      IX = X
50    F(I) = F(I) + PSTAR(IX) + (X-IX)*(PSTAR(IX+1)-PSTAR(IX))
C  DENSITY VALUES ARE NOW STORED IN F ARRAY READY FOR PRINTOUT
      STOP
      END
```

Figure 4.16 Despite the apparent complexity, the reconstruction process of filtering the projections and backprojecting into the image array can be coded into just a few lines of FORTRAN. This code is for parallel-beam reconstruction. (After Brooks and Di Chiro (1975).)

Returning briefly to the filtering operation, it is sometimes advantageous to reduce the emphasis given to the higher-frequency components in the image for the purpose of reducing the effects of noise. One widely used filter due to Shepp and Logan (1974) replaces $|\rho|$ with

$$|(2\rho_{max}/\pi)\,\sin(\pi\rho/2\rho_{max})|$$

(see figure 4.17). The digital filter has the form

$$p(ms) = -2(\pi s)^{-2}(4m^2 - 1)^{-1} \qquad m = 0, \pm1, \pm2, \ldots . \quad (4.23)$$

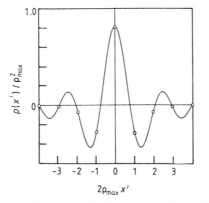

Figure 4.17 The continuous and discrete versions (full curve and open circles, respectively) of the Shepp and Logan filter. (Reproduced from Barrett and Swindell (1981).)

Another widely used filter, the Hanning window, uses an apodising factor $A(\rho)$ given by

$$A(\rho) = \alpha + (1 - \alpha) \cos(\pi\rho/\rho_{max})$$

which multiplies into $|\rho|$. The quantity α is a variable parameter that gives $A(0) = 1$ for all α and $A(\rho_{max}) = 2\alpha - 1$ varying from -1 for $\alpha = 0$ to 1 for $\alpha = 1$. In practical terms, when $A(\rho)$ is included on the RHS of equation (4.13), this becomes equivalent to convolving the projection data $\lambda_\phi(x')$ with a modified kernel $p(x')$ whose form is different from that in equation (4.17) (or discretely, equation (4.20)). A fuller discussion of windowing is given in the section on single-photon emission computed tomography (§6.7).

In this section, the filtered backprojection method has been described. Since both the filtering and the backprojection are linear and shift-invariant operations, it does not matter which is performed first. When backprojection is performed first, however, the filtering becomes a two-dimensional operation. Backprojecting unfiltered projections would yield a result $\mu_B[x, y]$, where

$$\mu_B[x, y] = \int_0^\pi \lambda_\phi(x') \, d\phi \Bigg|_{x' = x \cos \phi + y \sin \phi} . \qquad (4.24)$$

Rewriting equation (4.14) for the central slice theorem, we have

$$\lambda_\phi(x') = \int_{-\infty}^\infty \left(\frac{M_P(\rho, \phi)}{|\rho|} \right) |\rho| \exp(2\pi i \rho x') \, d\rho. \qquad (4.25)$$

Comparing the pairs of equations (4.24) and (4.25) with (4.12) and (4.13), it is quite clear that $\mu_B[x, y]$ is related to $\mu[x, y]$ by a function that compensates for the denominator in the integral (4.25). Deconvolution of this function from $\mu_B[x, y]$ to yield $\mu[x, y]$ is possible, but is a very clumsy way of tackling the reconstruction problem. If it is also remembered that filtering can take place in real space (by convolution) or Fourier space, it is clear that there are many equivalent ways of actually performing the reconstruction process (Barrett and Swindell 1981).

After the first generation of transmission CT scanners, the technique of rotate–translate scanning was largely abandoned in favour of faster scanning techniques involving fan-beam geometry. Viewed at the primitive level, however, these scanning geometries merely in-fill Fourier space in different ways and a reconstruction of some kind will always result. Indeed, it is perfectly possible to imagine merging projection data for the same object taken in quite different geometries. Once this is realised, it is soon apparent that the multitude of reconstruction methods that exist are in a sense mere conveniences for coping with less-simple geometry. The methods do, however, possess some elegance and many of the derivations are quite tricky! Without wishing to be overdismissive of a very important practical subject, we shall make no further mention of the mathematics of more complex geometries for reconstructing two-dimensional tomograms from one-dimensional projections. Similar reconstruction techniques have been used for x-ray transmission cone-beam tomography, whereby the x-rays are collimated to a cone and impinge on an *area* detector (see e.g. Feldkamp *et al* 1984, Webb *et al* 1987). There are also other quite ingenious methods for obtaining transmission tomograms that rely on very different mathematics: for example, circular tomography (Smith and Kruger 1987), whereby every point within the patient is projected onto a circle on the face of an image intensifier by the circular motion of the focal spot of a custom-designed x-ray tube. Selected planes are brought into focus by optically tracking an annular viewing field across the image intensifier with the diameter of the annulus defining the plane of interest. Circular tomography is somewhat intermediate between classical tomography, which requires no reconstruction mathematics, and x-ray CT, since circular tomography demands that the recorded data be decoded before they are able to be interpreted.

4.4 ITERATIVE METHODS OF RECONSTRUCTION

In the early days of computed tomography, iterative methods were popular. Various techniques with names such as ART (algebraic reconstruction technique), SIRT (simultaneous iterative reconstruction technique) and ILST (iterative reconstruction technique) were proposed and implemented. Such methods are no longer used for x-ray CT but still find application where the data sets are very noisy or incomplete, as they often are in emission computed tomography (see Chapter 6).

The principle of the method is described in figure 4.18. The image (the estimate of the object) is composed of I two-dimensional square pixels with densities μ_i, $1 \le i \le I$. The projections $\hat{\lambda}(\phi, x')$ that would occur if this were the real object are readily calculated using

$$\hat{\lambda}(\phi, x') = \sum_{i=1}^{I} \alpha_i(\phi, x')\mu_i \tag{4.26}$$

where $\alpha_i(\phi, x')$ is the average path length traversed by the (ϕ, x') projections through the ith cell. These coefficients need only be calculated once; they can then be stored for future use. For a typical data set, equation (4.26) represents 10^5 simultaneous equations. The solution method is to adjust the values of the μ_i iteratively until the computed projections $\hat{\lambda}$ most closely resemble the measured projections λ. These final values μ_i are then taken to be the solution, i.e. the image. Equation (4.26) is not soluble using direct matrix inversion for a variety of reasons that relate not only to the size (α is typically a $10^5 \times 10^5$ square matrix, albeit a very sparse one) but also to the conditioning of the data.

Because of measurement noise, and the approximations engendered by the model, there will not be a single exact solution. The arguments are very similar to those in Chapter 12 explaining why image deconvolution is difficult. Furthermore, there are usually far more equations than there are unknowns, so a multiplicity of solutions may exist. Part of the difficulty of implementing the solution is in deciding upon the correct criteria for testing the convergence of the intermediate steps and knowing when to stop. This is but one example of a whole class of *ill-posed* problems, which in recent years has necessitated the development of a new branch of mathematics bearing this same name.

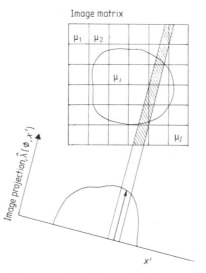

Figure 4.18 In iterative reconstruction methods, a matrix of *I* cells represents the object. The line integrals for the projection data are then replaced by a set of linear equations (4.26). (Reproduced from Barrett and Swindell 1981)

The many iterative algorithms differ in the manner in which the corrections are calculated and reapplied during each iteration. They may be applied additively or multiplicatively; they may be applied immediately after being calculated; optionally, they may be stored and applied only at the end of each round of iteration. The order in which the projection data are taken into consideration may differ as well.

The simple example shown in figure 4.19 illustrates additive immediate correction. Four three-point projections are taken through a nine-point object O, giving rise to projection data sets P_1 through P_4. Taken in order, these are used successively to calculate estimates E_1 through E_4 of the original object.

The initial estimate is obtained by allocating, with equal likelihood, the projection data P_1 into the rows of E_1. Subsequent corrections are made by calculating the difference between the projection of the previous estimate and the true projection data and equally distributing the difference over the elements in the appropriate row of the new estimate. For example, the difference between the projection of the first column of E_1 shown in parentheses (15), and the true measured value (16) is 1. In creating the first column of E_2, one-third of this difference ($\frac{1}{3}$) is added to each element of the first column of E_1. The first iteration is completed with the calculation of E_4. That the process converges in this numerical example is demonstrated by calculating the root-mean-square (RMS) deviation of elements of E_1 through E_4 from the true values in O. As the figure shows, these RMS errors decrease monotonically.

Further discussion of iterative algorithms is detailed in the book by Herman (1980) and the review article by Webb (1981).

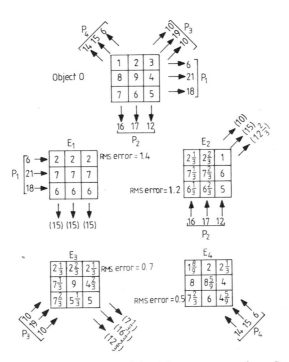

Figure 4.19 A simple example of iterative reconstruction. See text for explanation.

4.5 OTHER CONSIDERATIONS

4.5.1 Angular-sampling requirements

Assuming a point-like source of x-rays, the effect of having a rectangular detector profile of width w in the direction of the projection is to modulate the spectrum of the projection with a $\sin(\pi\rho w)/\pi\rho w$ apodisation. The first zero of this function is at $\rho = 1/w$. If we equate this to the maximum frequency component, i.e. $\rho_{max} = 1/w$, then the sampling interval s along the projection must be $s \leqslant w/2$, as required by the sampling theorem (see also §12.6). Frequencies higher than ρ_{max} will, of course, persist in the sidelobes of the apodising function but at greatly reduced amplitudes, so the sampling requirement is only approximately fulfilled. Additional high-frequency attenuation will take place, however, owing to the finite source size (and possibly patient motion), and this $w/2$ criterion is found to be an acceptable compromise between generating aliasing artefacts and processing massive amounts of data. The question regarding the number N_ϕ of angular samples remains. The number N_ϕ is taken to be the number of projections in the range $0 \leqslant \phi < 180°$.

If the final image is to have equal resolution in all directions, then the highest spatial-frequency components must be equally sampled in the radial and azimuthal directions in the neighbourhood of $\rho = \rho_{max}$ in the (ρ, ϕ) Fourier space.

For an object space of diameter D and projection data that are sampled with an interval d, the number of samples per projection is $N_s = D/d$ and the radial sampling interval in Fourier space is thus $2\rho_{max}d/D$. The azimuthal interval at $\rho = \rho_{max}$ is $\rho_{max}\Delta\phi$, where $\Delta\phi = \pi/N_\phi$. Equating these sampling intervals yields the result

$$N_\phi = (\pi/2)N_s. \tag{4.27}$$

In practice, projections are usually taken over 360° to reduce partial-volume and other artefacts, so $2N_\phi$ projections are usually taken. Equivalently, the angular increment in projection angle is

$$\Delta\phi = 2/N_s \tag{4.28}$$

for a uniformly sampled image data set.

4.5.2 Dose considerations

The projection data are subject to measurement noise. If a particular measurement were repeated many times, yielding an average measured value of n detected x-ray photons, then the random noise associated with a single reading will be $\sqrt{n}$. These fluctuations result from the Poisson statistics of the photon beam, and cannot be eliminated. These measurement fluctuations propagate through the reconstruction algorithms, with the result that a perfectly uniform object of density μ will appear to have a mottled appearance. A signal-to-noise ratio (SNR) can be defined as

$$\text{SNR} = \mu/\Delta\mu \tag{4.29}$$

where $\Delta\mu$ is the RMS fluctuation in the reconstructed value of μ about its mean value.

On the assumption that this photon noise is the only source of noise in the image, several authors (see for example Barrett *et al* 1976) have determined an expression relating the x-ray dose U delivered to the centre of a cylindrical object to the spatial resolution ε and the signal-to-noise ratio SNR. The expression has the form

$$\eta U = k_1 (\text{SNR})^2 / \varepsilon^3 b \qquad (4.30)$$

where b is the thickness of the slice, η is the detective quantum efficiency of the detector, and k_1 is a constant that depends on beam energy, the diameter of the object and the precise manner in which ε is defined. The points to note are that the dose depends on the second power of the signal-to-noise ratio and, to all intents and purposes, on the inverse fourth power of the resolution. This latter claim is made because in any reasonable system the slice width will be scaled in proportion to the resolution required: thickness $b = k_2 \varepsilon$ where k_2 is typically 2–5, i.e. the reconstruction voxel is a rather skinny, rectangular, parallelepiped. If k_2 becomes too large, partial-volume effects, as described in the next section, will become obtrusive.

Figure 4.20 shows how the quantities are related for a typical beam energy and object size. It is seen that dose levels in the range of 0.01 to 0.1 Gy are delivered to produce images with millimetre resolution at approximately 1% density discrimination. These are the dose levels that are actually found in practice, and one therefore presumes that commercial CT scanners are operating at, or close to, this photon-limited situation. The 16-fold increase in dose that would be needed for even a modest two-fold increase in resolution would seem to negate the possibility of improving the spatial resolution. Fortuitously, submillimetre resolution of bony structures can be obtained without invoking these unacceptably high dose levels. This is because the bone/tissue interface has a density ratio of almost 2:1 and the SNR can be traded off to improve the spatial resolution without escalating the dose. Of course, it requires finer sampling of the projection data. A new research tool has recently been announced—Superscope (Flannery *et al* 1987a,b) to perform tomographic microscopy with a resolution 1000 times that of medical x-ray CT. In view of the dose–resolution constraint expressed by equation (4.30), the enormous intensity of the Brookhaven National

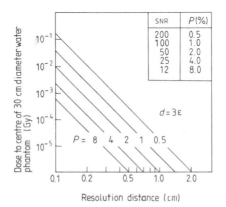

Figure 4.20 Nomogram relating dose U, density discrimination (SNR) and resolution ε for a typical diagnostic CT scanner.

Laboratory synchrotron source was harnessed. The instrument has been used to visualise the internal structure of insects.

4.5.3 Partial-volume artefacts

Because the x-ray beam diverges in a direction perpendicular to the slice, a projection measured in one direction may be slightly different from the projection taken along the same path but in the opposite direction. This provides one reason for requiring a full 360° scan of the patient. The inconsistencies in the data can be compensated by combining data from opposite directions.

A different but related partial-volume effect arises from the observation that the anatomical structures do not in general intersect the section at right angles. A long, thin voxel could well have one end in soft tissue and the other end in bone. As a result the reconstructed μ would have an intermediate value that did not correspond to any real tissue at all. This is the main reason for scaling d with ε and not letting k_2 get too large (see previous section).

4.5.4 Beam-hardening artefacts

As the x-ray beam passes through tissue, the lower-energy components are attenuated more rapidly than the higher-energy components. The average photon energy increases; the beam becomes *harder*. As a result, the exponential law of attenuation no longer holds.

With no absorber in the beam, the detector output is

$$I_0 = k_3 \int_0^{E_{max}} S(E) \, dE \qquad (4.31)$$

where the source spectrum $S(E)$ is defined such that $S(E)\,dE$ is the energy fluence in the energy range E to $E + dE$.

With an object in the beam, Beer's law must be weighted by the energy spectrum and integrated, to give

$$I = k_3 \int_0^{E_{max}} S(E) \exp\left(- \int_{AB} \mu_E[x, y] \, dy'\right) dE \qquad (4.32)$$

and the projection λ is thus

$$\lambda = - \ln\left(\frac{\int_0^{E_{max}} S(E) \exp(- \int_{AB} \mu_E[x, y] \, dy') \, dE}{\int_0^{E_{max}} S(E) \, dE}\right). \qquad (4.33)$$

It is this *non*-linear relationship between λ and μ that causes the problem. The principal effects of this artefact show up as a false reduction in density in the centre of a uniform object and the creation of false detail in the neighbourhood of bone/soft tissue interfaces. These artefacts are particularly troublesome in the skull. There are several ways that the problems can be overcome, and at this point the reader is referred to the appropriate literature.

We thus complete our, albeit brief, presentation of the important concepts of x-ray CT. The reconstruction theory itself has a certain elegant simplicity using simple assumptions concerning the physics of the data assembly. We have seen how imaging hardware is constructed to realise the theory and how in a relatively brief timespan this equipment has been optimised. In the next chapter we shall complement this discussion by describing how, in addition to its role as a diagnostic device, the CT scanner has found a fundamental place in aiding the planning of radiotherapy treatment.

4.6 APPENDIX

4.6.1 Evaluation of the Fourier transform $p(x')$ of the band-limited function $|\rho|$

We start from equation (4.16), which is integrated as follows:

$$p(x') = \int_0^{\rho_{max}} \rho \exp(2\pi i \rho x') \, d\rho - \int_{-\rho_{max}}^0 \rho \exp(2\pi i \rho x') \, d\rho$$

$$= \left[\frac{\rho \exp(2\pi i \rho x')}{2\pi i x'} \right]_0^{\rho_{max}} - \int_0^{\rho_{max}} \frac{\exp(2\pi i \rho x') \, d\rho'}{2\pi i x'}$$

$$- \left[\frac{\rho \exp(2\pi i \rho x')}{2\pi i x'} \right]_{-\rho_{max}}^0 + \int_{-\rho_{max}}^0 \frac{\exp(2\pi i \rho x') \, d\rho'}{2\pi i x'}$$

$$= \frac{\rho_{max} \sin(2\pi \rho_{max} x')}{\pi x'} - \left[\frac{\exp(2\pi i \rho x')}{(2\pi i x')^2} \right]_0^{\rho_{max}} + \left[\frac{\exp(2\pi i \rho x')}{(2\pi i x')^2} \right]_{-\rho_{max}}^0$$

$$= 2\rho_{max}^2 \operatorname{sinc}(2\rho_{max} x') + \frac{\cos(2\pi \rho_{max} x') - 1}{2\pi^2 x'^2}$$

$$= 2\rho_{max}^2 \operatorname{sinc}(2\rho_{max} x') - \rho_{max}^2 \operatorname{sinc}^2(\rho_{max} x').$$

We thus obtain the result given by equation (4.17), i.e.

$$p(x') = \rho_{max}^2 [2 \operatorname{sinc}(2\rho_{max} x') - \operatorname{sinc}^2(\rho_{max} x')].$$

REFERENCES

Barrett H H, Gordon S K and Hershel R S 1976 Statistical limitations in transaxial tomography *Comput. Biol. Med.* **6** 307

Barrett H H and Swindell W 1981 *Radiological Imaging: The Theory of Image Formation, Detection and Processing* vol II (New York: Academic Press) p384

Bracewell R N and Riddle A C 1967 Inversion of fan-beam scans in radio astronomy *Astrophys. J.* **150** 427–34

BROOKS R A and DI CHIRO G 1975 Theory of image reconstruction in computed tomography *Radiology* **117** 561–72
—— 1976 Principles of computer assisted tomography (CAT) in radiographic and radioisotope imaging *Phys. Med. Biol.* **21** (5) 689–732
COULAM C M, ERICKSON J J and GIBBS S J 1981 Image and equipment considerations in conventional tomography *The Physical Basis of Medical Imaging* ed C M Coulam, J J Erickson, F D Rollo and A E James (New York: Appleton-Century-Crofts)
FELDKAMP L A, DAVIS L C and KRESS J W 1984 Practical cone-beam algorithm *J. Opt. Sci. Am.* A **1** (6) 612–19
FLANNERY B P, DECKMAN H W, ROBERGE W G and D'AMICO K L 1987a Superscope *Sci. Am.* July pp21–2
—— 1987b Three dimensional X-ray microtomography *Science* **237** 1439–44
GILBERT P 1972 Iterative methods for the three-dimensional reconstruction of an object from projections *J. Theor. Biol.* **36** 105–17
HERMAN G T 1980 *Image Reconstruction from Projections: The Fundamentals of Computed Tomography* (New York: Academic Press)
HOUNSFIELD G N 1973 Computerised transverse axial scanning (tomography). Part 1: Description of system *Br. J. Radiol.* **46** 1016–22
KAK C K 1979 Computerised tomography with x-ray, emission and ultrasound sources *Proc. IEEE* **67** (9) 1245–72
KORENBLYUM B I, TETEL'BAUM S I and TYUTIN A A 1958 About one scheme of tomography *Bull. Inst. Higher Educ. – Radiophys.* **1** 151–7 (translated from the Russian by H H Barrett, University of Arizona, Tucson)
MACDONALD J S 1981 Computed tomography in a clinical setting *Computerized Axial Tomography in Oncology* ed J E Husband and P A Hobday (Edinburgh: Churchill Livingstone) p5
MARAVILLA K R and PASTEL M S 1978 Technical aspects of CT scanning *Comput. Tomogr.* **2** 137–44
NATTERER S 1986 *The Mathematics of Computerised Tomography* (New York: Wiley)
PARKER R P, CONTIER DE FREITAS L, CASSELL K J, WEBB S and HOBDAY P A 1980 A method of implementing inhomogeneity corrections in radiotherapy treatment planning *J. Eur. Radiother.* **1** (2) 93–100
PESCHMANN K R, NAPEL S, COUCH J L, RAND R E, ALEI R, ACKELSBERG S M, GOULD R and BOYD D P 1985 High speed computer tomography: systems and performance *Appl. Opt.* **24** 4052–60
RADON J 1917 Uber die Bestimmung von Funktionen durch ihre Integralwerte langs gewisser Mannigfaltigkeiten *Ber. Verh. Sachs. Akad. Wiss. Leipzig Math. Phys. Kl.* **69** 262–77
RAMACHANDRAN G N and LAKSHMINARAYANAN A V 1971 Three-dimensional reconstruction from radiographs and electron micrographs: applications of convolutions instead of Fourier transforms *Proc. Natl Acad. Sci. USA* **68** 2236–40
SHEPP L A and LOGAN B F 1974 The Fourier reconstruction of a head section *IEEE Trans. Nucl. Sci.* **NS-21** 21–43
SMITH S W and KRUGER R A 1987 Fast circular tomography device for cardiac imaging: image deflection mechanism and evaluation *IEEE Trans. Med. Imaging* **MI-6** (2) 169–73
TETEL'BAUM S I 1956 About the problem of improvement of images obtained with the help of optical and analog instruments *Bull. Kiev Polytechnic Inst.* **21** 222

—— 1957 About a method of obtaining volume images with the help of x-rays *Bull. Kiev Polytechnic Inst.* **22** 154–60 (translated from the Russian by J W Boag, Institute of Cancer Research, London)

WEBB S 1981 Reconstruction of cross-sections from transmission projections: a short simplified review and comparison of performance of algorithms *Physical Aspects of Medical Imaging* ed M Moores, R P Parker and B Pullan (New York: Wiley)

—— 1987 A review of physical aspects of x-ray transmission computed tomography *IEE Proc.* A **134** (2) 126–35

WEBB S, SUTCLIFFE J, BURKINSHAW L and HORSMAN A 1987 Tomographic reconstruction from experimentally obtained cone-beam projections *IEEE Trans. Med. Imaging* **MI-6** (1) 67–73

ZONNEVELD F W 1979 *Computed Tomography* (Eindhoven: Philips Medical Systems)

CHAPTER 5

CLINICAL APPLICATIONS OF X-RAY COMPUTED TOMOGRAPHY IN RADIOTHERAPY PLANNING

H J DOBBS AND S WEBB

5.1 X-RAY COMPUTED TOMOGRAPHY SCANNERS AND THEIR ROLE IN PLANNING

Cancer is a major cause of mortality in the Western world, ranking second only to cardiovascular disease. Overall, 30% of patients with cancer are cured and return to a normal life. Chemotherapy has had a dramatic improvement on the survival rate of patients treated for some of the less common cancers, such as testicular and paediatric tumours and the lymphomas, which present with disseminated disease. However, the large majority of patients who are cured have localised disease at presentation and are treated by surgery or radiotherapy or by a combination of the two modalities.

The success of radiotherapy is dependent upon ensuring that a tumour and its microscopic extensions are treated accurately to a tumoricidal dose. It is therefore essential to define the precise location and limits of a tumour with clinical examination and by using the optimum imaging methods available for a particular tumour site. The normal organs that surround a tumour limit the radiation dose that can be given because of their inherent and organ-dependent radiosensitivity. If this radiation tolerance is not taken into account when planning treatment, permanent damage to normal tissues would result.

Conventionally, localisation of a tumour and adjacent sensitive organs within the body has been carried out using orthogonal radiographs with contrast media where appropriate, e.g. introduced into the bladder via a catheter. All available patient data, i.e. clinical, surgical and radiological, are taken into account when determining the volume for treatment, in addition to a knowledge of the natural history of the tumour and likely routes of microscopic spread. Radiotherapy plans are produced in the cross-sectional plane with the use of a planning computer (figure 5.1). The limitations of the conventional process are due to the inability of plane radiography to visualise the tumour and the difficulty in transcribing data into the transverse plane required for dosimetry.

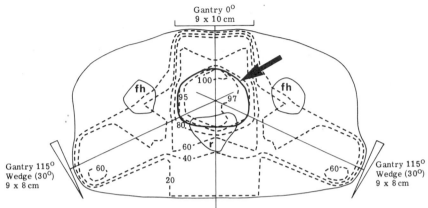

Figure 5.1 Conventional radiotherapy plan for treatment of a prostatic tumour. Arrow indicates target volume. fh, femoral head; r, rectum. (From Dobbs and Barrett (1985).)

Computed tomography (CT) has made an important contribution to the local staging of primary tumours. For instance, it delineates extra-vesical extension of bladder tumours (Hodson *et al* 1979), extracapsular spread of prostatic tumours (Dobbs and Husband 1985) and mediastinal extension of bronchial tumours (Baron *et al* 1982). CT images are ideal for planning radiotherapy treatment because they are obtained in the transverse plane, provide detailed visualisation of the tumour and adjacent organs and also depict the complete body contour necessary for dosimetry (figure 5.2). CT examinations must, of course, be made under geometrical conditions identical to those during radiotherapy in order to ensure accurate reproduction of subsequent treatments. This is acheived with the use of lasers for alignment of the patient and reference marks on the skin such as a permanent ink tattoo. The subject of the reproducibility of patient position is, however, a matter of ongoing investigation.

With CT scanning incorporated into the procedure leading towards therapy, a new planning sequence has been developed. Initial simulation is unnecessary and invasive techniques such as cystography can be avoided. Following a CT 'therapy scan' (meaning scan used for therapy planning), the images are interpreted by a diagnostic radiologist and localisation of the tumour and treatment volume are performed in three dimensions by a radiotherapist. CT data can be transferred directly to a radiotherapy planning computer, which provides an accurate approach to dosimetric calculations. Where treatment of thoracic tumours involves entry of radiation beams via the lungs, heterogeneity corrections can be made on a pixel-by-pixel basis, hence allowing for the difference between normal aerated lungs and those affected by consolidation or pleural effusion.

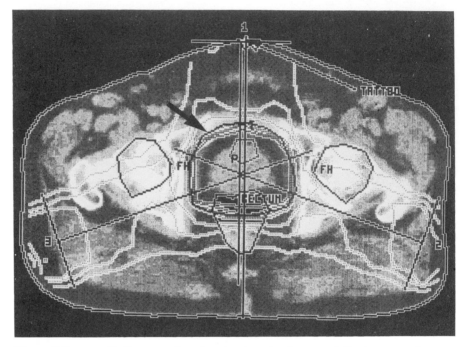

Figure 5.2 CT radiotherapy plan for treatment of a prostatic tumour. Arrow indicates target volume.

Several prospective studies have been carried out comparing the localisation of tumours by conventional means and by CT, with a reported range of between 20% and 45% alteration in treatment plan following CT examination (table 5.1). The study from the Royal Marsden Hospital (Dobbs *et al* 1983) included 320 patients and showed that 33% of patients had to have an alteration made in the conventional plan following information gained by CT to ensure that the tumour was adequately treated and that a geographical miss was avoided (figure 5.3). It is interesting to note that, when analysis was made of the changes in treatment volume, 34% resulted in a reduction in volume after CT, leading to the possibility of an increase in tumour dose and the potential for improvement in local control.

Table 5.1 Alterations in treatment plans following comparison of conventional and CT localisation of tumour volume.

Authors	Number of patients	Alteration in plan (%)
Munzenrider *et al* (1977)	75	45
Ragan and Perez (1978)	45	20
Goitein *et al* (1979)	77	36
Dobbs *et al* (1983)	320	33

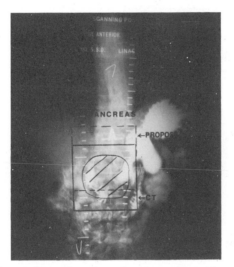

Figure 5.3 Comparison of proposed target volume from conventional localisation and that obtained using CT for a pancreatic tumour (shown with oblique hatched lines).

The use of CT for radiotherapy planning is dependent on the anatomical site of the tumour (Dobbs and Parker 1984). Table 5.2 illustrates those tumours where CT contributes new diagnostic information, more accurate details of adjacent normal organs or improved heterogeneity corrections. CT is of more value where treatment is given to a small volume and where greater precision is needed than where radiation is given to a larger volume in order to cover the risk of microscopic spread over a wide area.

For patients with bladder tumours, the most valuable information demonstrated by CT is soft-tissue extravesical disease (figure 5.4). CT findings have been correlated with pathological specimens and show an accuracy of around 80% for CT. For prostatic tumours, CT, often with

Table 5.2 Tumour sites suitable for CT planning.

Head and neck	Antrum
	Orbit
	Parotid
Chest	Bronchus
	Oesophagus
	Mediastinum, e.g. thymus
Abdomen	Pancreas
	Retroperitoneal mass
	Kidney
Pelvis	Bladder
	Prostate
	Rectum
Miscellaneous	Soft-tissue sarcomas

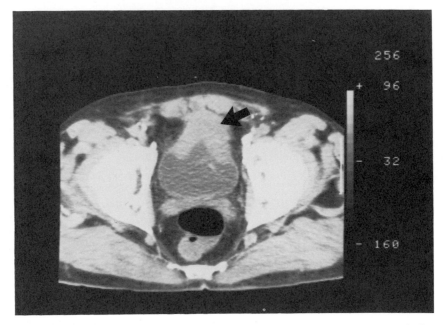

Figure 5.4 CT section showing extensive bladder tumour with extravesical spread anteriorly (arrowed).

the use of coronal and sagittal reconstructions, detects unsuspected involvement of the seminal vesicles. The planning of bronchial tumours is improved by the use of CT in demonstrating local extension of tumour into the mediastinum and occult mediastinal lymphadenopathy. It cannot, however, detect disease in normal-size nodes and gives poor evaluation of hilar lymph nodes.

There are, however, limitations of CT in some anatomical sites. For instance, irradiation of patients with primary breast cancer involves the abduction of the patient's arm, with the elbow flexed, a movement not feasible within the limited aperture of most CT scanners (see §5.2). Patients with cerebral tumours wear a Perspex immobilisation shell during treatment, and hence a CT scan must be taken with the patient wearing the shell fixed to the scanner table. The scans must be taken in the treatment plane, which involves a steep gantry angle, not reproducible by the gantry of many CT scanners. This currently limits the use of CT planning for cerebral tumours in many centres and conventional simulation is therefore still employed.

Therapy CT scans have also provided new information about dynamic aspects of some internal body organs. For instance, it has been noted that, if a patient has a bladder full of urine, the small bowel is displaced superiorly out of the pelvis when compared with an empty bladder (figure 5.5). This can be used to advantage when irradiating pelvic organs, such as the prostate or cervix, where small-bowel tolerance to radiation is a dose-limiting factor. Whole pelvic nodal treatment may be given to a patient with a full bladder following oral fluids to displace some small bowel out of the treatment volume. Conversely, for treatment of a bladder tumour, where the smallest bladder volume is desirable, a patient is asked to empty his or her bladder prior to irradiation. When a comparison was made between the volume of an empty bladder using a conventional planning cystogram and that after voluntary micturition, as used for CT planning, it was found that there was a large discrepancy (figure 5.6). Measurements of bladder volume were made in the small series of patients using CT images and it was found that there was a variation in residual urine of between 103 and 260 ml. In contrast, urinary catheterisation for localisation falsely reduces the bladder volume by allowing urinary drainage and installation of only around 30 ml of contrast. This discrepancy could lead to undetected displacement of the tumour outside the target volume by a distended bladder if allowances are not made. It is recognised that irradiation of the bladder may lead to radiation cystitis and increases in the frequency of micturition, which may also affect the volume of residual urine within the bladder.

There are no reports yet of randomised prospective studies comparing local control or survival rates of patients irradiated following the use of CT versus conventional planning. Recently Rothwell *et al* (1985) reported a retrospective analysis of survival of patients with bladder cancer in relation to coverage of tumour by the 90% isodose as defined by CT. It showed that patients treated for tumours of which a part received less than 90% of the prescribed dose fared significantly worse than those receiving 90% or more of the radiation dose to the entire tumour. If the survival of the underdosed population could be improved

(a)

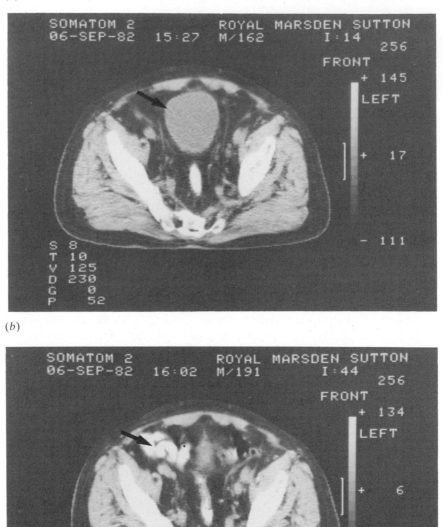

(b)

Figure 5.5 Comparison of pelvic CT scans taken at the same anatomic level: (a) with bladder full of urine and (b) after micturition. Note presence of small bowel (arrowed on (b)) when bladder is not distended as arrowed in (a).

by CT planning so that it was equal to the rest of the group, Rothwell *et al* (1985) predicted a 9% increase in survival of the whole group at 3.5 years. This is comparable to the 3.5% improvement in 5 years survival from improved planning predicted by Goitein (1979) using a mathematical model. It indicated that a substantial 30% increase in the accuracy of planning radiation therapy translates into only a modest expectation of improvement in local control or survival.

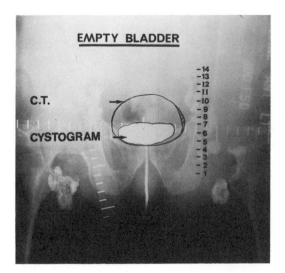

Figure 5.6 Comparison of bladder size using a cystogram for localisation and CT scans taken after voluntary micturition, showing that residual urine remains in the bladder.

However, the advantages in terms of better staging of the primary tumour, improved visualisation of anatomical detail in the transverse plane and the ease of computerised dosimetric calculation mean that the use of CT in radiotherapy treatment planning is of recognised value. It remains essential to study prospectively, and with well defined endpoints, each new major technological advance in this field. Such studies are necessary to see whether more sophisticated developments, such as three-dimensional CT planning, will contribute more, in terms of either improved local control of tumours or reduced morbidity for patients undergoing radiation treatment.

5.2 NON-STANDARD COMPUTED TOMOGRAPHY SCANNERS

The first part of this chapter has discussed how CT images may be used to assist the planning of radiotherapy and illustrated how the role of x-ray CT is much wider than providing for diagnosis and staging. The machines utilised to produce such images are referred to as commercial diagnostic CT scanners, since they generate images that have the required resolution and contrast for diagnosis. It is important, however, to recognise that there are applications in radiotherapy for which diagnostic-quality images are not required and for which a number of special-purpose CT scanners have been constructed during the 1980s. In view of their different purposes, it is not surprising that these CT scanners differ in their technological bases. It is possible that their principles of construction may be considered too transitory to be included in a teaching text, but the outline principles are presented if only to draw attention to what are clearly important developments in the history of CT and which may become developing areas of interest.

A CT scanner was constructed based around a radiotherapy simulator by Harrison and Farmer (1978) and was used for planning the treatment of radiotherapy of the thorax (Kotre *et al* 1984). The projection data (see Chapter 4) were obtained by digitising the lines of a TV camera viewing the transmitted x-ray intensity on a fluorescent screen. The digital values required correction for the non-linear response characteristics of the screen and the TV camera. An attenuating compensator was also included in the x-ray collimator to overcome saturation of the detecting system. The orientation of the gantry and the data acquisition were controlled by a small minicomputer. Typically, each projection was digitised into 128 samples, projections were recorded over 180° and 128 × 128 images were reconstructed from some 100 projections via convolution and backprojection techniques. Values for spatial and density resolution have not been published, but typical images appear suitable for radiotherapy planning. There have been several independent CT systems, developed around a simulator, based on the same principles, notably those by Arnot *et al* (1984), Redpath and Wright (1985) and Redpath (1988). In view of the small size of image intensifying systems, these were based on sets of half-projections obtained with a fan beam with one edge of the fan along the line from the x-ray source to the centre of rotation and the detector offset. Wright *et al* (1984) and Kijewski *et al* (1984) have performed similar work.

An alternative method of forming projection data is to utilise a position-sensitive linear detector to construct a single-slice machine. This was the approach taken by Webb *et al* (1984) and Leach *et al* (1985), who constructed a special-purpose detector using plastic scintillator viewed by two photomultiplier tubes and a moving collimator to give position sensitivity. In their system, a wide fan beam completely spanned the patient and the rotation was through 360°. Projection data comprised 331 pixels per projection sampled at 2 mm intervals and of the order of 100 projections were recorded. The gantry and detector movements together with the data acquisition were controlled by a PDP11/05 minicomputer. The collimation of the detector led to a unique geometry of data collection, for which a special mathematical algorithm was necessary (Herman 1982) in order to rearrange the data for reconstruction by convolution and backprojection techniques. Images were reconstructed with 256 × 256 pixels and 6 mm spatial resolution.

Figure 5.7 shows typical images from this system in which the external contour, internal lung boundaries and other anatomical structures are clearly visible. The Royal Marsden Hospital CT Simulator has provided the basis for an extensive investigation into the accuracy and

(a)

(b)

(c)

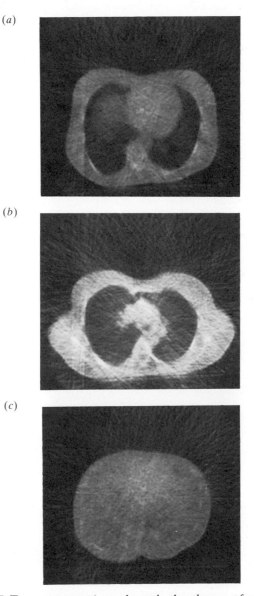

Figure 5.7 Transverse sections through the thorax of a female patient obtained with the Royal Marsden Hospital CT Simulator: (a) is a slice at the midplane of the breast, (b) is some 7 cm more superior, and (c) is some 7 cm more inferior.

homogeneity of delivering dose to the breast as part of the post-operative conservative management of breast cancer (Webb *et al* 1987). Green *et al* (1987) have shown images from a CT Simulator which used a silicon diode strip detector.

Moving on from CT at energies of the order of 100 keV, other special-purpose CT scanners have been constructed for imaging with megavoltage radiation. Simpson *et al* (1982) utilised a one-dimensional plastic scintillator detector to image transmitted x-rays from a linear accelerator and Swindell *et al* (1983) constructed a bismuth germanate detector for attachment to a linear accelerator. Fan-beam data were captured over 220° rotation at every 2°. The spatial resolution of the latter system was 3 mm and the density resolution was 1%. A typical image from this system is shown in figure 5.8. The *raison d'être* of this development was to provide a basis for incorporating heterogeneity corrections into treatment planning and also to assist with assessing the accuracy of patient positioning at the time of treatment. Lewis and Swindell (1987) have recently upgraded this system to provide 1–2 mm spatial resolution and a larger (about 38 cm) field of view, achieving this by adjusting the magnification and arranging for the bank of detectors to be offset, relative to the line from the source to the centre of rotation, by 0.25 of each detector width ('interdigitation'). The importance of such developments in the precision of radiotherapy has yet to be established but is clearly the logical complement to obtaining good spatial accuracy in diagnosis and in planning treatment.

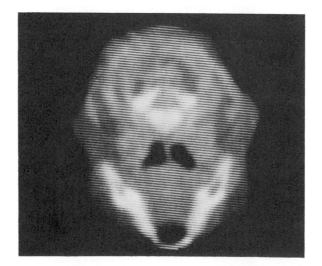

Figure 5.8 Oblique view through the head from a megavoltage CT system. The image shows the mandible, top vertebra at the base of the skull and the interfaces between muscle and fat. (From Swindell *et al* (1983).)

There have been proposals for performing proton CT (Hanson *et al* 1982). A system was developed at the Clinton P Anderson Meson Facility at Los Alamos and used to image human autopsy specimens. A stable finely focused beam of protons of energy between 224 and 236 MeV was swept across the specimen, which was rotated within the field of view of the swept beam. The exit radiation was detected by a combination of multiwire proportional chamber (MWPC) position-sensitive detectors and a range telescope comprising 32 counters of NE102 plastic scintillator. Reconstruction was by convolution and backprojection and 3 mm spatial resolution was achieved. The main disadvantage of such a system is the need for a high-energy beam of protons. However, proton CT does not suffer from streak artefacts and the CT data are immediately applicable to dose planning for charged-particle radiotherapy. Proton CT of small animals has also been reported recently using lower-energy (25–70 MeV) beams, and figure 5.9 shows a proton CT image of a mouse from such a system.

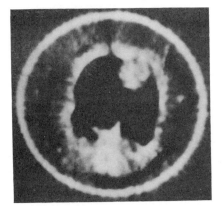

Figure 5.9 Proton CT image of the lower lung region of a mouse. (From Ito and Koyama-Ito (1985).)

Recently it has been shown that elastic scatter tomography can be achieved, yielding cross-sectional data on the tissue angular differential cross section for elastic scatter. The technique (Harding *et al* 1985) uses a first-generation scanning modality with an x-ray tube and scintillation detectors. By using several detectors placed at an angle to the forward direction of the x-ray pencil, a number of images may be computed corresponding to different orders of scatter. It is too early to assess to what biological property of tissue the images might correspond, and also the present equipment is only at the laboratory stage. Other tomographic imaging methods based on the use of Compton-scattered photons without reconstruction techniques have been reported. Further considerations of the ways in which the physics of CT may progress have been included in the review by Webb (1987).

REFERENCES

ARNOT R N, WILLETS R J, BATTEN J R and ORR J S 1984 X-ray image intensifier and TV camera for imaging transverse sections in humans *Br. J. Radiol.* **57** 47–56

BARON R L, LEVITT R G, SAEL S S, WHITE M J, ROPER C L and MARBARGER J P 1982 Computed tomography in the preoperative evaluation of bronchogenic carcinoma *Radiology* **145** 727–32

DOBBS H J and BARRETT A 1985 *Practical Radiotherapy Planning: Royal Marsden Hospital Practice* (London: Edward Arnold)

DOBBS H J and HUSBAND J E 1985 The role of CT in the staging and radiotherapy planning of prostatic tumours *Br. J. Radiol.* **58** 429–36

DOBBS H J and PARKER R P 1984 The respective roles of the simulator and computed tomography in radiotherapy planning; a review *Clin. Radiol.* **35** 433–9

DOBBS H J, PARKER R P, HODSON N J, HOBDAY P and HUSBAND J E 1983 The use of CT in radiotherapy treatment planning *Radiother. Oncol.* **1** 133–41

GOITEIN M 1979 The utility of computed tomography in radiation therapy: an estimate of outcome *Int. J. Radiat. Oncol. Biol. Phys.* **5** 1799–807

GOITEIN M, WITTENBERG J, MENCLIONDO M, DOUCETTE J, FRIEDBERG C, FERRUCCI J, GUNDERSON L, LINDGOOD R, SHIPLEY W and FINEBERG H 1979 The value of CT scanning in radiation therapy treatment planning: a prospective study *Int. J. Radiat. Oncol. Biol. Phys.* **5** 1787–98

GREEN T, SKRETTING A, LARSSEN H, MYRVOLL T, LEVERNES S, OLSEN J B and SIMONSEN B 1987 Development of the radiotherapy simulator at the CART Demonstrator in Oslo *Proc. 9th Int. Conf. on Computers in Radiation Therapy (Scheveningen) 1987* ed I A D BRUINVIS, P H VAN DER GIESSEN, H J VAN KLEFFENS and F W WITTKÄMPER (Amsterdam: North-Holland/Elsevier)

HANSON K M, BRADBURY J N, KOEPPE R A, MAKEC R J, MACHEN D R, MORGADO R, PACIOTTI M A, SANDFORD S A and STEWARD V W 1982 Proton computed tomography of human specimens *Phys. Med. Biol.* **27** 25–36

HARDING G, KOSANETZKY J and NEITZEL U 1985 Elastic scatter computerized axial tomography *Med. Biol. Eng. Comput.* **23** Suppl. 1, 238–9

HARRISON R M and FARMER F T 1978 The determination of anatomical cross sections using a radiotherapy simulator *Br. J. Radiol* **51** 448–53

HERMAN G T 1982 Reconstruction algorithms for non-standard CT scanner designs *J. Med. Syst.* **6** 555–68

HODSON N J, HUSBAND J E and MACDONALD J S 1979 The role of computed tomography in the management of bladder cancer *Clin. Radiol.* **30** 388–95

ITO A and KOYAMA-ITO H 1985 Development of a proton computed tomography system *Med. Biol. Eng. Comput.* **23** Suppl. 1, 240–1

KIJEWSKI M F, JUDY P and SVENSSON G K 1984 Image quality of an analog radiation therapy simulator-based tomographic scanner *Med. Phys.* **11** 502–7

KOTRE C J, HARRISON R M and ROSS W M 1984 A simulator based CT system for radiotherapy treatment planning *Br. J. Radiol.* **57** 631–5

LEACH M O, WEBB S and BENTLEY R E 1985 An x-ray detector system and modified simulator providing CT images for radiotherapy dosimetry planning *Phys. Med. Biol.* **30** 303–11

LEWIS D and SWINDELL W 1987 A MV CT scanner for radiotherapy verification *Proc. 9th Int. Conf. on Computers in Radiotherapy, Holland* June 1987 in *The Use of Computers in Radiation Therapy* ed I A D Bruinvis *et al* (Amsterdam: North-Holland/Elsevier)

Munzenrider J E, Pilepich M, Rene-Ferrero J B, Tchakarova I and Carter B L 1977 Use of body scanner in radiotherapy treatment planning *Cancer* **40** 170–9

Ragan D P and Perez C A 1978 Efficacy of CT assisted two dimensional treatment planning: analysis of 45 patients *Am. J. Roentgenol.* **131** 75–9

Redpath A T 1988 Clinical use of the Edinburgh simulator computed tomography system *Proc. Hospital Physicists' Association Conf.* (*Brit. J. Radiol.* **61** 561)

Redpath A T and Wright D H 1985 The use of an image processor system in radiotherapy simulation *Br. J. Radiol.* **58** 1081–9

Rothwell R I, Ash D V and Thorogood J 1985 An analysis of the contribution of computed tomography to the treatment outcome in bladder cancer *Clin. Radiol.* **36** 369–72

Simpson R G, Chen C T, Grubbs E A and Swindell W 1982 A 4 MV CT scanner for radiation therapy: the prototype system *Med. Phys.* **9** 574–9

Swindell W, Simpson R G, Oleson J R, Chen C T and Grubb E A 1983 Computed tomography with a linear accelerator with radiotherapy applications *Med. Phys.* **10** 416–20

Webb S 1987 A review of physical aspects of x-ray transmission computed tomography *IEE Proc.* A **134** (2) 126–35

Webb S, Leach M O, Bentley R E, Maureemootoo K, Yarnold J R, Toms M A, Gardiner J and Parton D 1987 Clinical dosimetry for radiotherapy to the breast based on imaging with the prototype Royal Marsden Hospital CT Simulator *Phys. Med. Biol.* **32** 835–45 (and corrigendum *Phys. Med. Biol.* **32** 1516–17)

Webb S, Leach M O and Herman G T 1984 Reconstructions from a non-standard CT scanner *IEEE Trans. Med. Imaging* **MI-3** 193–6

Wright D H, Redpath A T, Jarvis J H G and Harris J R 1984 Image processing techniques applied to fluoroscopic X-ray pictures obtained from a radiotherapy simulator *Proc. 8th Int. Conf. on the Use of Computers in Radiation Therapy (Toronto) 1984* pp343–9

CHAPTER 6

THE PHYSICS OF RADIOISOTOPE IMAGING

R J OTT, M A FLOWER, J W BABICH AND P K MARSDEN

6.1 INTRODUCTION

From methods of imaging human anatomy with x-rays emanating external to the body and passing through it, attention is now turned to imaging physiological function using radiation emanating from inside the human body. The use of radioisotopes in tracer quantities for the clinical diagnosis of human disease has grown rapidly in the last two decades and has become a recognised medical speciality known as nuclear medicine (Wagner 1975, Maisey 1980, Sharp *et al* 1985). Radioisotope imaging involves a range of techniques used in the production of images of the distribution of radionuclide-labelled agents in the body (figure 6.1). These agents, known as radiopharmaceuticals, are designed to show physiological function of individual organs in the body. The distribution of these agents within the body is determined by route of administration and by such factors as blood flow, blood volume and a variety of metabolic processes. The techniques discussed in this chapter contrast with most other medical imaging techniques, which essentially provide anatomical details of the body organs. For a comparison of imaging methods, see also Chapter 15.

The first use of a radioisotope, ^{131}I, to investigate thyroid disease was carried out in the late 1930s. Early developments in imaging equipment include the production of the rectilinear scanner and the scintillation camera during the 1950s. Both these devices became widely available in the mid-1960s. Since then, the Anger camera or gamma camera has become the instrument of choice for radioisotope imaging. Within this chapter, we will explain the function and development of the modern gamma camera, including the most recent improvements brought about by the use of advanced microprocessor technology.

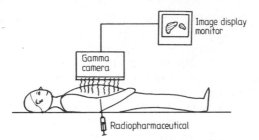

Figure 6.1 Schematic of radioisotope imaging technique.

A range of clinical information is obtainable from radioisotope images. Planar static imaging, known as planar scintigraphy, is still the most common form of clinical image used in nuclear medicine. These single-view images consist of two-dimensional representations of projections of three-dimensional distributions of activity in the detector field of view. Unlike x-ray imaging, where both the emission and detection position of each x-ray is known, only the γ-ray detection position is determined from a radioisotope source. To produce an image it is, hence, necessary to provide some form of collimation, which defines the photon direction. This can take the form of mechanical (i.e. lead) collimation or electronic collimation.

Temporal changes in the spatial distribution of radiopharmaceuticals can be obtained by taking multiple images over periods of time that may vary from milliseconds to hundreds of seconds. This form of imaging, known as dynamic scintigraphy, is fundamental to the use of radioisotopes in showing the basic function of the organ/system being examined.

Since the planar image contains information from a three-dimensional object, it is often difficult to determine clearly the function of tissue deep in the body. Tomographic studies obtained by taking multiview acquisitions of the object overcome most problems caused by superposition of information in a single planar view. The technique of emission computed tomography (ECT) has parallels with x-ray computed tomography (CT) but also some important differences. X-ray CT is based on the determination of photon attenuation in body tissues, whereas in ECT there is a basic need to correct for the effect of photon attenuation whilst determining the distribution of radioactivity within the body. In addition, limited count rates in radioisotope studies show up as a degradation of image quality in comparison with x-ray CT. However, as ECT is not an anatomical imaging technique by nature, it is not sensible to compare image quality directly with those obtained from anatomical modalities such as x-ray, ultrasound and nuclear magnetic resonance.

6.2 RADIATION DETECTORS

The radioactive decay of radioisotopes leads to the emission of α-, β-, γ-

and x-radiation, depending on the radionuclide involved (see §6.4.3). The range of α- and β-particles in tissue is too small to allow *in vivo* imaging using detectors external to the body, but x- and γ-rays penetrate tissue quite successfully. Figure 6.2 shows that at low energies, below 100 keV, x- and γ-rays are greatly attenuated by body tissues owing to photoelectric absorption. At energies above 100 keV, photon interactions in tissue are dominated by Compton scattering (see Chapter 2). Increased absorption due to pair production does not become important until well above 1 MeV. Hence, radioisotope imaging is restricted to the use of radionuclides emitting photons with energies greater than about 50 keV. We will now consider several radiation detectors that have been used for localising the emission of x- and γ-rays from the body.

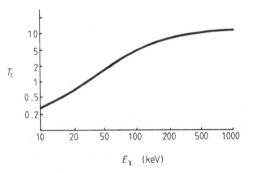

Figure 6.2 Variation of relative γ-ray transmission coefficient (T_c) in tissue versus photon energy (E_γ).

6.2.1 Gas detectors

The traditional gas-filled ionisation chamber (Attix and Roesch 1966) is a very sensitive and accurate detector of ionising radiation (figure 6.3). It is, however, relatively transparent to x- or γ-rays unless high-atomic-number (Z) materials are added to the detector. This may take the form of high-Z, high-pressure gases such as xenon or the addition of metal converters within the chamber gas. At high voltages, the ionisation chamber becomes known as a proportional counter, in which the signal produced by the detection of ionising radiation undergoes gas amplification ($\times 10^8$) and produces a large signal that is proportional to the applied chamber voltage. The use of multiwire proportional chambers as radioisotope imaging detectors is discussed in §6.3.7. At still higher applied voltages, the Geiger counter has been used as a γ-ray detector, particularly in radiation protection, but it has little application in radioisotope imaging.

In general, simple gas detectors have not found much use in radioisotope imaging because of poor sensitivity and time-resolution characteristics.

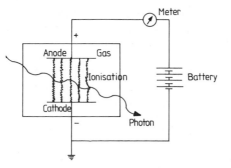

Figure 6.3 A schematic of a simple ionisation chamber.

6.2.2 Scintillation detectors

Materials that emit visible or near-visible light when energy is absorbed from ionising radiation have been utilised both to count and to image radioisotopes (Birks 1964). Table 6.1 (see §6.3.5) shows a range of inorganic scintillators having a high Z and hence good photon stopping power. If the light emission characteristics of the scintillator match the spectral sensitivity of a photomultiplier (PM) and the scintillator is transparent to its own light, ionising radiation detectors using scintillator–photomultiplier combinations will have high sensitivity. Light emission by most inorganic scintillators is proportional to energy deposited in the material. Hence, it is possible not only to detect photons using a scintillation counter but also to determine the energy of the photon detected. Typically, energy resolution is about 10–15% at 100–200 keV (figure 6.4). This enables the scintillation counter to discriminate approximately between γ-rays emitted from the body unscattered and those which have scattered and lost energy in the process. Limitations in the application of scintillators as imaging detectors are mainly due to physical size.

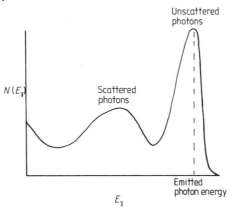

Figure 6.4 The pulse-height spectrum from a γ-ray detected by an organic scintillation counter.

Small-diameter (10 cm), thick (10 cm) single crystals are available but larger-diameter (40–50 cm) single crystals of thicknesses greater than 1–1.5 cm are difficult to manufacture. Figure 6.5 shows the photon detection efficiency of NaI(Tl) crystals as a function of photon energy. When combined with figure 6.2, we can see that there is a window between 50 and 500 keV (figure 6.6), in which scintillation counters can be used in radioisotope imaging detectors.

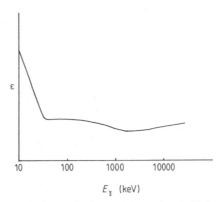

E_γ (keV)

Figure 6.5 The variation of photon detection efficiency (ε) versus photon energy (E_γ).

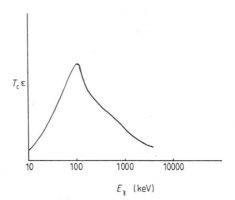

E_γ (keV)

Figure 6.6 The variation of $T_c\varepsilon$ versus E_γ.

6.2.3 Semiconductor detectors

In the last 10 years, the development of low-cost silicon detectors has provided a potentially useful alternative to scintillators. It is possible to make small-area silicon detectors 5×5 cm^2 and about 1 mm thick, and to obtain a spatial resolution down to 10 μm using either drift-chamber or strip-detector technology (Gerber *et al* 1977, Gatti *et al* 1985). This, allied with the excellent energy resolution (a few kiloelectronvolts at room temperature), is still outweighed by the low Z of silicon. The semiconductor may, however, have useful applications in autoradiography as a replacement for film emulsion, and in small-area radioisotope telescopes.

Germanium, with a much higher Z, would appear to have some advantages over silicon, except that high energy resolution (better than 1 keV) (figure 6.7) is obtained only at liquid-nitrogen temperatures. Again, only small-area, thin sections of germanium are readily available. Developmental gamma cameras have been produced that show the potential advantages of the semiconductor over scintillators, but the cost of the raw material and cryogenics is still prohibitive. It is likely that small germanium detectors with high resolution may be available in the near future.

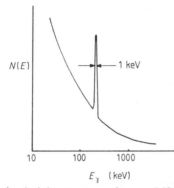

Figure 6.7 The pulse-height spectrum from a 140 keV γ-ray detected by a semiconductor detector.

6.2.4 Summary

From the above discussion, it is clear that scintillation materials provide the greatest potential for radioisotope imaging, although developments in both hybrid proportional chambers and semiconductors may be competitive in the next decade. It is worth noting here the application of film emulsion for the detection of β emission and x-rays from radioisotopes used in autoradiography (the radioisotope imaging of tissue sections) (Lear 1986). Both gas detectors and semiconductors may have a role in autoradiography in the near future.

6.3 RADIOISOTOPE IMAGING EQUIPMENT

6.3.1 History

Early radioisotope images were produced by the use of a scintillation detector connected to a focused lead collimator. The collimator allowed only those x-ray or γ photons emitted from a small region in the object near the collimator focus to be detected efficiently by the scintillator. By scanning this detector in a rectilinear motion across the body, it was possible to build up an image of the radiopharmaceutical distribution. In the 1950s, rectilinear scanners were developed to include one or two detectors with large-volume NaI(Tl) crystals and heavy focused lead collimators, which could image the distribution of isotopes such as ^{18}F, ^{51}Cr, ^{131}I, ^{198}Au and ^{59}Fe (Beck 1964). Although detecting high-energy photons with reasonable sensitivity, they made poor use of the patient dose, spending very little time imaging any particular part of the body. More recently, multiple crystal/collimator combinations have been used as special-purpose imagers for ECT and as linear scanners for whole-body imaging.

Before and during the development of large-area crystals, a few attempts were made to produce a large-area stationary imaging device in which many individual detectors were arranged in a fixed array. One such system was manufactured by Baird Atomic Inc. in the mid-1960s (Bender and Blau 1963). This device comprised a large number of thick NaI(Tl) crystals arranged in a rectangular array. These were coupled by plastic lightpipes to two separate banks of PM tubes that uniquely identified the scintillator position.

However, the development of single large-area crystals of NaI(Tl) led to a radical change in radioisotope imaging with the invention of the gamma camera (Anger 1958, 1964). A single lead collimator with a large area to match the crystal and containing many parallel holes with axes orthogonal to the collimator surface was used to define the photon paths and give 1:1 imaging of organs. By the use of analogue electronics, with either capacitive or, later, resistive coupling to several photomultipliers, attached to the rear surface of the scintillator, it was possible to determine the position of the photon interaction in the scintillator. Crystal/collimator combinations up to 50 cm diameter allow large-organ and whole-body images to be obtained and, for the first time, imaging showing dynamic information relating to fast temporal changes in the radiopharmaceutical distribution was possible. Since the spatial resolution of a gamma camera was about 1–2 cm at depth and better close to the collimator surface, this enabled improved high-resolution images to be made.

These large-area, thin NaI(Tl) crystals have a high sensitivity at low photon energies (about 100–200 keV) but become rapidly less sensitive above 300 keV. The trend to even thinner crystals (about 6 mm) to achieve higher spatial resolution has exacerbated this low-photon-energy use, and only suits the use of newer radioisotopes such as ^{99}Tc^m, ^{123}I, ^{111}In and ^{201}Tl.

6.3.2 Rectilinear scanners

Historically (see also Chapter 1), the first moving-detector radioisotope imaging systems were based on the electromechanical rectilinear scanner

(Cassen *et al* 1951, Mayneord and Newbery 1952, Mallard and Peachey 1959) with one or two detector heads (figure 6.8). These automatic scanning devices allow accurate transverse and longitudinal positioning of a scintillator crystal, which is equipped with a focused lead collimator to define the small area being imaged. The scintillator is usually a large-volume NaI(Tl) crystal with dimensions up to 12.5 cm thick by 12.5 cm diameter. The collimator is a large, truncated lead cone with multiple holes in which the diameter of each hole tapers towards the focal point (figure 6.9). This design maximises sensitivity for a given spatial resolution at the focal plane. Moving away from the focus along the collimator axis, the sensitivity decreases and the spatial resolution deteriorates so that information out of the focal plane is reduced in amplitude and blurred. The septal thickness is determined, obviously, by the energy of the photons being imaged, and the sensitivity and resolution will be much affected by this parameter also. Focal lengths (the distance between focal point and collimator outer face) usually vary from 10 to 20 cm. The hole cross section may be round, square, hexagonal or triangular, the shape chosen being determined by cost and sensitivity factors.

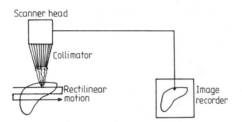

Figure 6.8 Schematic of a rectilinear scanner.

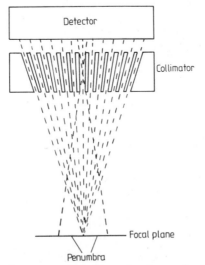

Figure 6.9 A cross section through a focused collimator.

The large-volume crystal and thick collimators allow the rectilinear scanner successfully to image high-energy photons from ^{52}Fe (511 keV) and ^{131}I (364 keV) as well as isotopes more appropriate for gamma camera imaging, ^{99}Tc^m (140 keV).

Images are formed by attaching to the scanner a mechanical arm fitted with some form of recording stylus such as a tapping pen with a colour ribbon or a photorecording system. In either case, the tapping rate or the intensity of light is proportional to the instantaneous counting rate, respectively. Hence an image can be formed on paper or film (figure 6.10). If a colour ribbon is used on the tapper unit, colour coding can also be used to represent the variable counting rates.

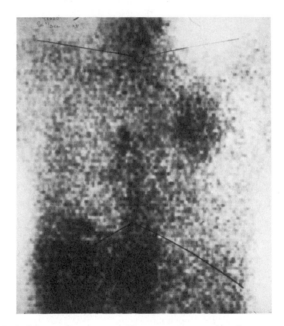

Figure 6.10 Image from a rectilinear scanner showing cancer in the upper left lung using ^{67}Ga citrate.

Care must be taken in setting up the scanning parameters. One or more single-channel analysers can be adjusted to allow the appropriate photopeak(s) to be selected so that the effects of scattered photons on the image can be minimised. The scan speed and the line spacing between each scanning line are also very important parameters, since the information density (ID) in the recorded image is given by

$$\text{ID} = \frac{\text{count rate}}{\text{scan speed} \times \text{line spacing}} \tag{6.1}$$

in units of counts per unit area. An important final procedure is the necessity to determine, prior to scanning, the maximum count rate expected in the scan to allow the image to fit the dynamic range of the recording system and so prevent saturation.

The major surviving role of the rectilinear scanner is high-energy or small-part imaging (e.g. thyroid imaging) or quantitative scanning especially of high-energy photon emitters using a double-headed scanner. In the latter case, two long-focal-length collimators are used; their heads are spaced as shown in figure 6.11 and they have their axes carefully aligned. With careful design of the collimators, the combined sensitivity and spatial response now become very uniform with depth through the object, allowing isosensitive, isoresolution scanning (see also §6.6.4). Thus, deep-seated lesions can be visualised as well as surface lesions, and quantification can be performed if careful calibration is carried out. It is possible to obtain a spatial resolution of about 1 cm at a depth of 10 cm, but the sensitivity will be low because of the short time spent imaging each point in the image. Hence, rectilinear scans are, in general, inferior to scintigrams from a modern gamma camera.

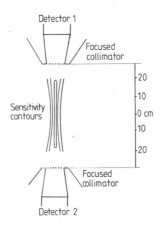

Figure 6.11 Schematic showing how a double-headed scanner can be set up for isosensitive scanning.

The rectilinear scanner has now become the 'steam train' of radioisotope imaging, and those in existence now are usually old commercial or home-made scanners kept going by enthusiasts (Flower *et al* 1986).

6.3.3 Linear scanners

A direct extension of the rectilinear scanner is the development of linear scanners. In this case, the rectilinear motion is replaced by a single longitudinal scan with a laterally extended scanning head. In one design (figure 6.12) the scanning head contains 10 NaI(Tl) crystals, each with a collimator focused in the direction of the scan, providing a detector surface area of $60 \times 12.5 \, cm^2$. The use of two heads provides anterior and posterior images of the whole body. Scan speeds vary between 2.5 and $20 \, cm \, min^{-1}$, taking 10–80 min for a whole-body scan. Images are stored on either Polaroid or x-ray film via a console containing a cathode ray tube. The collimators have a focal length of about 12 cm and a focal-plane spatial resolution of about 8 mm. These short focal-length collimators were optimised for skeletal imaging (figure 6.13), but are not useful for isosensitive scanning.

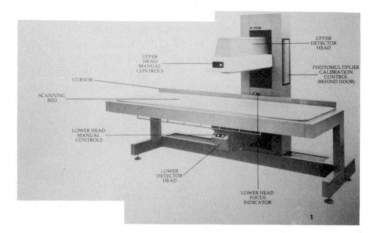

Figure 6.12 The Cleon Linear Bone Scanner.

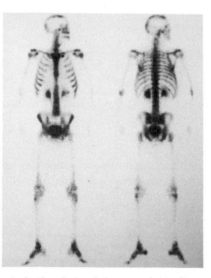

Figure 6.13 Whole-body skeletal images (anterior, posterior) from the Cleon Linear Bone Scanner.

An alternative design (figure 6.14) contains one or two scanning heads consisting of single slabs of NaI(Tl), $50 \times 3.2\ \text{cm}^2$ in area and 2 cm thick, with a single lead collimator specially designed to give a spatial resolution of 0.7–1.2 cm at 8 cm in tissue and a depth of focus in air of 18–21 cm.

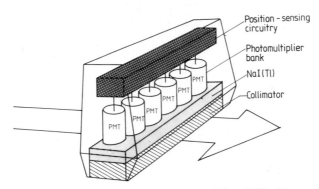

Figure 6.14 A schematic of one detector of the CGR Linear Scanner.

These special-purpose imaging systems are no longer commercially available, being of little use other than for skeletal imaging. Even for this application, they have now been replaced by scanning gamma cameras.

6.3.4 Multicrystal single-photon tomographic scanners

Early transaxial emission tomographic scans were obtained from detectors containing scintillating crystals, each with a focused collimator (Kuhl and Edwards 1963). The crystal/collimator systems were translated and rotated to provide sufficient views of the object for tomographic imaging. A typical example of this type of detector, built by Aberdeen University (the Aberdeen Section Scanner, ASS), consisted of two small-area detectors equipped with long-focal-length collimators (Bowley *et al* 1973). The detector heads were scanned in tandem at one azimuthal angle to the object (figure 6.15) and then rotated to a new azimuthal angle, where scanning was repeated. The system was designed to produce single transaxial sections with high sensitivity and depth-independent resolution.

A more modern version of this system is the Aberdeen Section Scanner Mark II (Evans *et al* 1986). In this case, 24 detectors are arranged along the sides of a square, six detectors on each side, each with a 20 cm focal-length collimator (figure 6.16). The detectors are moved tangentially up to 64 mm and rotated through an angular range of 95°. The spatial resolution achieved in the single-section image plane is 9 mm and the plane thickness is 14 mm. The large increase in scintillation detector volume gives a greatly increased sensitivity ($300\ \mathrm{cps\,kBq^{-1}\,ml}$ within a single slice).

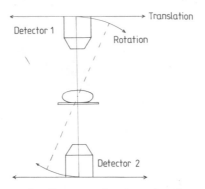

Figure 6.15 Schematic diagram showing the detector motion for transaxial scanning used in the Aberdeen Section Scanner Mark I.

Figure 6.16 The Aberdeen Section Scanner Mark II with the covers removed. (Courtesy of N Evans.)

Few of these multicrystal single-photon tomographic scanners (MSPTS) have been produced commercially, mainly because of their limitation as a special-purpose single-slice imager. One example is the Tomomatic 64 (marketed by Medimatic), similar to the Kuhl Mark IV detector (Kuhl *et al* 1976) but specifically designed for regional cerebral blood-flow (CBF) studies using ^{133}Xe. The Tomomatic 64 (Stokely *et al* 1980) has four banks of 16 detectors and 25 photomultipliers (figure 6.17). Each

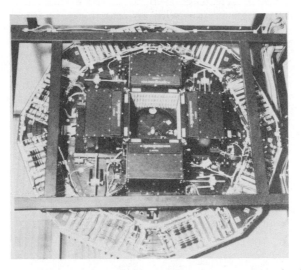

Figure 6.17 The Tomomatic 64 with the covers removed, showing the detector arrays in the centre, with the electronic printed circuit cards on the periphery. (From Stokely *et al* (1980).)

bank of detectors has the photomultipliers arranged in three rows (8 + 9 + 8), so that each scintillator is viewed by three photomultipliers, one in each row. The detectors are hence position-sensitive in the axial direction and, with the use of multislice collimators, can produce three sections simultaneously (figure 6.18).

Further examples of this type of scanner are the Cleon systems of Union Carbide Imaging Systems Inc. (Stoddart and Stoddart 1979), which joined the emission tomographic scene in the late 1970s, vanished in 1980 and have recently risen from the dead (after several false rebirths, Moore *et al* 1984) under the guise of the Novo Tomograph 810, and even more recently the Multi-X 810 Radionuclide Brain Imager marketed by Strichman Medical Equipment Inc. Two scanners were originally produced: the Cleon 710 Brain Imager and the 711 Body Imager. The major features of these two scanners were the use of large-volume, large-area scintillators fitted with short-focal-length collimators. The brain imager had 12 detectors, each mounted on a separate scanning frame at 30° intervals around a central axis (figure 6.19). Each of the large-area detectors required a 9 cm diameter photomultiplier and a 15 cm focal-length collimator, which subtended a 30° angle at the focal plane. Each detector underwent a rectilinear scan within the plane of interest so that the focal point scanned half the field of view. All detectors moved either clockwise or anticlockwise when scanning tangentially but opposing pairs of detectors moved radially together, either towards or away from the central axis, at the end of each tangential scan. The Cleon 710 achieved a transaxial resolution of about 1 cm in a single section with an effective slice thickness of about 2 cm and a sensitivity of approximately 420 cps kBq^{-1} ml (Flower *et al* 1979).

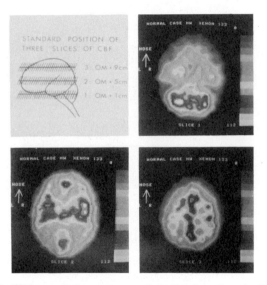

Figure 6.18 ^{133}Xe perfusion images of the brain using the Tomomatic 64. Slices are relative to the orbito-meatal (OM) line. (From Lassen (1985).)

In conclusion, MSPTS are designed to provide single-section images with higher sensitivity and spatial resolution than a rotating gamma camera. However, since the gamma camera, as we shall see (§6.7), can be used for planar, dynamic and multiplane tomographic scintigraphy, this versatility has overshadowed the development of the MSPTS and their contribution to clinical nuclear medicine is consequently very limited. Nevertheless, a number of new SPECT imagers are currently at the development stage, including SPRINT from the University of Michigan, MUMPI from the University of Missouri and ASPECT from Digital Scintigraphics. The latter two imagers utilise an annular single-crystal camera (Heller and Goodwin 1987).

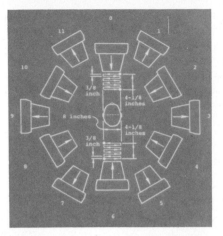

Figure 6.19 Schematic diagram showing the detector movement of the Cleon 710 Brain Imager.

6.3.5 Multicrystal positron emission tomography systems

In positron emission tomography (PET), the decay of the radionuclide produces two approximately antiparallel γ-rays, each of 511 keV. The most efficient method of imaging the *in vivo* distribution of the positron-emitting radiopharmaceutical is to surround the patient with a large number of individual scintillation detectors, with each detector in electronic coincidence with those on the opposite side of the patient. Multicrystal scanners (Nahmias *et al* 1984, Litton *et al* 1984) provide the state-of-the-art technology for PET, although the high cost of such systems has greatly limited their use in clinical nuclear medicine.

Figure 6.20 shows the configurations of typical single-slice multicrystal PET devices. Each crystal is individually coupled to a photomultiplier tube (PMT) and lead septa minimise random and scattered events as described in §6.7.7. Because a large number of applications of PET are in brain metabolism and neurology, most systems are specifically designed for brain imaging and have a correspondingly small ring diameter. Imaging of the torso is more difficult because the number of unscattered photons obtained is typically 2–8 times less than that from the brain for a given activity. Also the larger diameter means that more crystals are required to maintain the same sensitivity as a brain scanner. Despite this, many such systems exist, in particular for dynamic imaging of the heart (Schelbert *et al* 1980).

The most important feature of any PET system is the scintillation material used. Table 6.1 lists the relevant properties of three commonly used scintillators (Cho and Farukhi 1977, Laval *et al* 1983). Bismuth germanate (BGO), because of its high density and atomic number, providing a sensitivity approximately three times that of sodium iodide (NaI), is now used in nearly all multicrystal PET systems. Being non-hygroscopic, BGO crystals can also be packed much closer together than can NaI crystals. Barium fluoride (BaF_2) has a lower sensitivity but, owing to its very fast decay time, it is used in time-of-flight systems (see below).

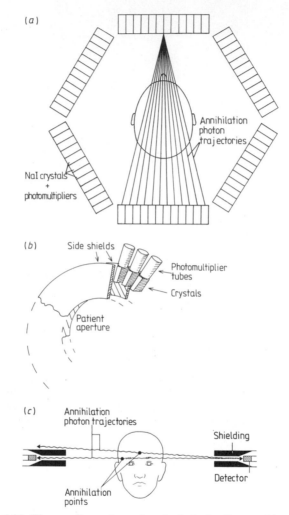

Figure 6.20 The configuration of typical single-slice multicrystal PET cameras: (*a*) plan of a hexagonal system; (*b*) ring system; (*c*) side view of either system.

The following description of the NeuroECAT III (CTI 831/08–12.5) brain scanner (Hoffman *et al* 1983b, 1987) gives a summary of the properties of a state-of-the-art commercially available system. The NeuroECAT III consists of eight 62 cm diameter rings of 320 BGO crystals. Each crystal has dimensions of 5.6 mm (tangential) by 30 mm (radial) by 12.5 mm (axial). Blocks of 32 crystals are read out by four PMTS (see figure 6.21(c)). Fifteen slices 6.5 mm apart can be imaged simultaneously by making use of coincidences between adjacent rings (i.e. eight direct and seven cross slices). The spatial resolution in a reconstructed image is 6 mm FWHM (full width at half-maximum) with a slice thickness of 6.3 mm FWHM. The sensitivity for a 20 cm diameter cylinder uniformly filled with activity is 250 cps kBq^{-1} ml for a direct slice and 430 cps kBq^{-1} ml for a cross slice, and the temporal resolution is approximately 6 ns. Figure 6.21 illustrates the larger whole-body scanner (931 series), which involves the same technology as the smaller brain scanner described above.

Table 6.1 Physical properties of inorganic scintillators.

Scintillator	Density (g cm^{-3})	Effective Z	Relative light yield	Decay constant (ns)	Wavelength of emission (nm)
Sodium iodide (NaI)	3.67	50	100	230	410
Bismuth germanate (BGO)	7.13	74	12	300	480
Barium fluoride (BaF$_2$)	4.89	54	5	0.7	195, 220
			15	620	310

The NEUROECAT III demonstrates most of the current trends in PET instrumentation. It is a multislice machine with comparable resolution in three orthogonal directions. This allows whole organs such as the brain to be imaged without having to move the patient, and also permits the display of sagittal and coronal as well as transaxial sections. The high spatial resolution obtained by using small crystals enables the quantification of activity in small (<1 ml) structures without overlapping into neighbouring regions. However, a consequence of the use of very small crystals is the need for some sort of semi-analogue coding scheme, as it is not possible to pack in a single PMT for every crystal, and there are many different examples of such schemes (Derenzo 1984, Casey and Nutt 1986). We shall return to a discussion of ring PET systems in §6.7.7.

Although not as common as BGO-based devices, PET scanners that utilise photon time-of-flight (TOF) information have been developed at several centres and are available commercially (Mullani *et al* 1984). Such systems make use of the very fast temporal resolution obtainable with BaF$_2$ to localise the annihilation position to within 5–10 cm. Whilst this is not sufficient to enhance the spatial resolution of the system greatly, it significantly reduces statistical noise in reconstructed images and thus makes up for the lower sensitivity obtained with BaF$_2$

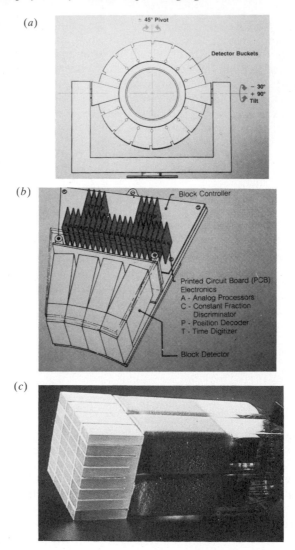

Figure 6.21 (*a*) Schematic of the 931 series PET scanner; (*b*) schematic of one detector bucket; (*c*) photograph of one detector. (Reproduced with permission from Siemens Gammasonics, The Netherlands.)

(Budinger 1983). The main advantages of TOF systems at present appear to be the ability to perform very-high-count-rate dynamic studies without correspondingly high random rates, and the ability to perform emission and transmission scans simultaneously (because the source used in the transmission scan is separated from the object being imaged by more than 10 cm). Time-of-flight systems are more complex and require more data-processing facilities than BGO systems. Both types of scanner are still undergoing rapid development and it is not yet clear (barring the discovery of a new scintillator that combines the high efficiency of BGO with the short resolving time of BaF_2) which of the two approaches will eventually see the most widespread use.

6.3.6 *Gamma camera*

The modern gamma camera (Short 1984) (figure 6.22) is made up of a multihole collimator, a large-area NaI(Tl) crystal, a light guide for optically coupling a hexagonal array of photomultipliers to the crystal and analogue electronics for position encoding and pulse-height analysis. The whole is contained in a lead shield of sufficient thickness to minimise background from radiation sources outside the field of view of the camera.

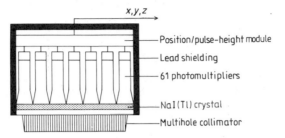

Figure 6.22 The basic components of a gamma camera.

The basic operation of the camera and the importance of each of the component parts is discussed in detail below.

The *collimator* acts to select the direction of the photons incident on the camera. In the case of a parallel-hole collimator (figure 6.23), only photons incident normal to the collimator surface will reach the scintillator. The collimator also defines the geometrical field of view (FOV) of the camera and essentially determines both the spatial resolution and sensitivity of the system. A range of collimators is required to image different photon energies and to achieve sufficient compromise between spatial resolution and sensitivity (table 6.2). As well as parallel-hole collimators, it is useful to have a pinhole collimator for imaging small superficial organs, and diverging/converging-hole collimators for whole-body or medium-sized organ imaging.

For a *parallel-hole collimator* it is possible to express the spatial resolution and geometrical efficiency in terms of the collimator dimensions. If L is the hole length, d the hole diameter and z the source-to-collimator distance, then the collimator spatial resolution (R_c) is given by

$$R_c \simeq d(L + z)/L. \qquad (6.2)$$

Hence, spatial resolution is improved by increasing the hole length or by increasing the number of holes per unit area (provided the septal thickness is adequate), so that a larger number of smaller-diameter holes can be fitted into the same overall area. In addition, and importantly,

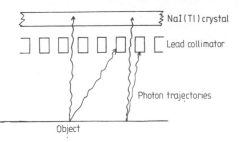

Figure 6.23 A schematic of a parallel-hole collimator showing how only normally incident photons are detected by the camera.

Table 6.2 Examples of parallel-hole collimator types for use with a 400 mm diameter gamma camera.

Description	Hole size[a] (mm)	Number of holes	Septal thickness (mm)
Low-energy, high-resolution (LEHR)	1.8	30 000	0.3
Low-energy, general-purpose (LEGP)	2.5	18 000	0.3
Low-energy, high-sensitivity (LEHS)	3.4	9000	0.3
Medium-energy, high-sensitivity (MEHS)	3.4	6000	1.4

[a] Diameter of round holes, or distance across flats of hexagon.

the spatial resolution can be improved by minimising the distance between the source and collimator surface. Figure 6.24 shows how the spatial resolution of a gamma camera fitted with a parallel-hole collimator falls off with increasing source-to-collimator distance.

The geometrical efficiency g of the collimator is given by

$$g \sim [Kd^2/L(d + t)]^2 \qquad (6.3)$$

where t is the thickness of the lead septa between the holes and K is a constant dependent upon hole shape (e.g. $K = 0.26$ for hexagonal holes in a hexagonal array). Note that g is independent of source–collimator distance for a point source imaged in air since the inverse square-law relationship is compensated by the detector area exposed. Performance characteristics of some commercially available parallel-hole collimators are shown in table 6.3.

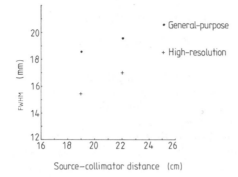

Figure 6.24 The variation of spatial resolution with source-to-collimator distance for a parallel-hole collimator.

Table 6.3 Performance characteristics of some typical commercially manufactured parallel-hole collimators[a].

Type[b]	Maximum energy (keV)	g (%)	R_c (mm) at 10 cm
LEHR	150	0.0184	7.4
LEGP	150	0.0268	9.1
LEHS	150	0.0574	13.2
MEHS	400	0.0172	13.4

[a] Data from Sorenson and Phelps (1980).
[b] See table 6.2.

An approximate relationship between R_c and g exists if z is much greater than L and if d is much greater than t (i.e. for source locations many multiples of the collimator hole length from the surface and for holes whose diameter is much greater than the septal thickness). This becomes (from equations (6.2) and (6.3))

$$g \propto R_c^2. \qquad (6.4)$$

This shows that collimator spatial resolution can be improved only at the expense of reduced collimator geometric efficiency for a given septal thickness. It is possible to calculate a value for the required septal thickness in terms of the length of the septa (and hole), the hole diameter and the linear attenuation coefficient μ of lead at the appropriate photon energy (Sorenson and Phelps 1980). If w is the shortest path in lead traversed by photons passing between adjacent holes, then

$$t \sim 2dw/(L - w). \tag{6.5}$$

If 5% septal penetration is allowed, then $\mu w \gtrsim 3$ and

$$t \sim 6d/\mu[L - (3/\mu)]. \tag{6.6}$$

Typical values for septal thickness vary from 0.3 mm for photon energy less than 150 keV to 4.7 mm for roughly 400 keV photon energy.

Figure 6.25 shows schematically how a *pinhole collimator* can be used to create an image of a small object. The collimator consists of a single small aperture (3–5 mm diameter) at the end of a conical lead shield containing sufficient attenuating material to minimise photon penetration for energies up to about 500 keV. The collimator allows an inverted image to be projected onto the scintillator and, by suitable positioning of the object, significant image magnification can be achieved, although at the expense of some distortion particularly at the image edge. The spatial resolution and sensitivity of a camera fitted with a pinhole collimator is determined by the pinhole diameter (d_p), the scintillator diameter (D), the collimator length (L) and the source-to-collimator aperture distance (z) (figure 6.26). For instance, if D is constant and L and z are increased, the spatial resolution improves, but the sensitivity is decreased. If D is increased, then z can be reduced for a given object size, and thus sensitivity and resolution can be improved at the expense of image distortion. A pinhole collimator is most suitable for small-organ imaging where the object can be placed close to the collimator, e.g. thyroid gland, skeletal joints.

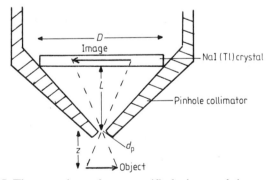

Figure 6.25 The creation of a magnified, inverted image using a pinhole collimator.

A *converging multihole collimator* provides the best combination of high resolution and sensitivity (figure 6.27) at the expense of reduced field of view (Murphy *et al* 1975) and, as with the pinhole collimator, some image distortion. A *diverging multihole collimator* (Muehllehner 1969) will provide a large field of view, especially when attached to a small-area gamma camera (figure 6.28). However, both spatial resolution and sensitivity are reduced and the change of magnification with depth causes distortion. Other special-purpose collimators have been developed for SPECT with a gamma camera (see §6.7.2).

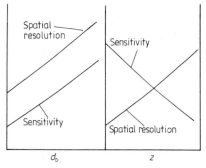

Figure 6.26 The variation of spatial resolution and sensitivity with the parameters d_p and z for a pinhole collimator.

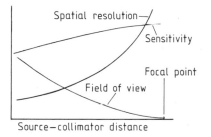

Figure 6.27 The variation of spatial resolution, sensitivity and field of view with source-to-collimator distance for a converging collimator.

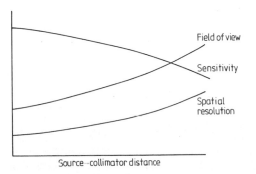

Figure 6.28 The variation of spatial resolution, sensitivity and field of view with source-to-collimator distance for a diverging collimator.

The majority of gamma cameras use a large-area (up to 50 cm diameter) thin (6–12 mm) single *scintillation crystal* of thallium-activated sodium iodide, NaI(Tl). This scintillator emits blue-green light close to 415 nm and the spectral output matches well the light-photon response characteristics of standard bialkali photomultipliers (figure 6.29). The crystal has a high atomic number and density, and the linear attenuation coefficient at 150 keV is 2.22 cm^{-1}. Hence 90% of 150 keV photons are absorbed in about 10 mm. The light-emission decay constant of 230 ns means that event rates of many tens of thousands of counts per second (kcps) can be attained without serious degradation of performance. The light output is the highest of all the readily usable inorganic scintillators (see table 6.1) and the material is highly transparent to its own light emission. Although hygroscopic and, hence, requiring hermetic encapsulation, NaI(Tl) is unsurpassed as a scintillator at photon energies close to 100 keV. The energy resolution (figure 6.30) of a thin NaI(Tl) crystal is typically 10–12% at 150 keV.

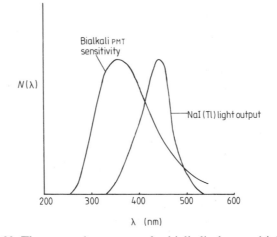

Figure 6.29 The spectral response of a bialkali photomultiplier compared with the light output from a NaI(Tl) scintillator.

Owing to the high refractive index of NaI(Tl) (1.85) it is usually necessary to provide a *light guide* to interface the scintillator to the photomultiplier. This minimises light losses in the transport of light to the photomultiplier—the guides are made of a transparent plastic with a refractive index close to 1.85 and are carefully shaped to match the shape of the photomultiplier cathode. The light guide also helps to minimise the variation of light collection efficiency across the face of the scintillator. More recently, the light guide has been dispensed with in favour of microprocessor-based correction procedures (see §6.5).

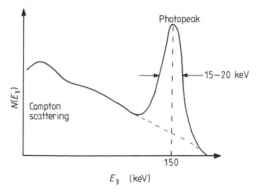

Figure 6.30 The energy resolution of a thin NaI(Tl) crystal at 150 keV.

The optimal arrangement of *photomultiplier tubes* (with circular or hexagonal cross sections) closely packed onto the surface of a circular scintillation crystal is a hexagonal array (figure 6.31) containing 7, 19, 37, 61, etc, photomultipliers. In the 1970s, 50 mm diameter photomultipliers were available, limiting the number used on a 40 cm diameter camera to 37. From about 1980 onwards, gamma cameras utilising smaller-diameter (25–30 mm) photomultipliers enabled cameras with 61 or 75 tubes (by adding extra tubes around the edges) to become available.

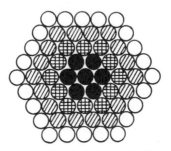

Figure 6.31 The arrangement of photomultiplier tube arrays on the crystal of a gamma camera containing 7, 19, 37 and 61 photomultiplier tubes.

The spectral response of the photocathode of the photomultipliers is matched to the light output spectrum of the scintillator by use of bialkali material (such as SbK_2Cs). The tubes are chosen from batches having closely matched gain response characteristics so that the application of high voltage and the gain adjustment to provide uniform response over the scintillator surface area is simplified.

A capacitor network or, more recently, a resistive-coupled network (figure 6.32), is used to provide *positional information* from the analogue outputs of the photomultiplier tubes. The relative intensity of these signals determines the x, y position of the scintillation event and provides four signals (x^+, x^-, y^+, y^-) to produce an image on a cathode ray oscilloscope (CRO) and/or on a storage oscilloscope. The total intensity of the signal, z (note, *not* a position coordinate), is given by

$$z = x^+ + x^- + y^+ + y^-$$ (6.7)

and the x and y positions by

$$x = k(x^+ - x^-)/z$$ (6.8)

$$y = k(y^+ - y^-)/z$$ (6.9)

where k is a constant. This is sometimes referred to as 'Anger logic'.

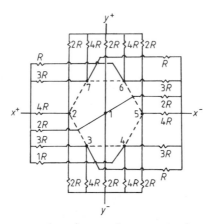

Figure 6.32 An example of a resistor network used to provide positional information from the analogue outputs of the gamma-camera photomultipliers.

The z signal is sent to a *single-channel pulse-height analyser* (SCA), which uses two discriminator levels to determine whether the z signal corresponds to that expected from the gamma photon detected. Figure 6.33 shows how the SCA is used to choose the 'photopeak' (or total-absorption peak) of the radionuclide. The lower-level discriminator is set to reject low-energy scattered photons, which would degrade the spatial accuracy of the image if allowed through the analyser. Modern gamma-camera systems are provided with two or three SCA to allow multiple photopeaks to be imaged simultaneously.

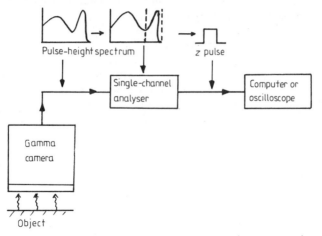

Pulse-height spectrum

z pulse

Single-channel
analyser

Computer or
oscilloscope

Gamma
camera

Object

Figure 6.33 The use of a single-channel analyser to select the photopeak window.

At high count rates, the analogue signal circuitry can become over-loaded due to pulse pile-up of scintillation signals from the detector. In addition, the system will begin to lose counts because of the inherent dead-time generated in the camera electronics.

The true count rate (N) of the system is related to the observed count rate (n) by

$$N = n/(1 - n\tau) \tag{6.10}$$

where τ is the electronic dead-time, typically about 4 μs for a modern camera.

The scintillation crystal and electronics are surrounded by a large *lead shield* to minimise the detection of unwanted radiation from outside the collimator field of view. With the development of rotating gantries for gamma cameras to allow ECT to be performed, this shielding has been severely reduced to minimise the weight of the system undergoing rotation. Many modern cameras have shielding that is sufficient only to minimise penetration of low-energy photons (less than 250 keV) and this, together with the use of thin crystals, enables only low-energy photon emitters ($^{99}Tc^m$, ^{111}In, ^{123}I, ^{201}Tl) to be used.

The intrinsic spatial resolution (R_i) of the gamma camera, as determined by the position encoder, is typically 3–4 mm, but when combined with the geometrical spatial resolution of the collimator (R_c) (equation (6.2)), this gives a *system spatial resolution* (R_s) of

$$R_s = (R_i^2 + R_c^2)^{1/2}. \tag{6.11}$$

Figure 6.34 shows how the value of R_s varies with distance from the collimator surface for a parallel-hole collimator and shows how the system resolution is dominated by the collimator at all distances. Hence there is little to be gained from improving the intrinsic spatial resolution of a gamma camera beyond present values.

A modern gamma camera is equipped with an *image display facility* to provide image hard copy on Polaroid and/or x-ray film. This is achieved by using an optical camera to view the image produced on the CRO. Generally, an image-formatting system will be provided to allow multiple images to be recorded on a single x-ray film. The quality of these 'analogue images' is highly dependent on matching the CRO output intensity to the film response and requires the information density (counts per unit area) to be sufficient to expose the film adequately. The dynamic range of information available often cannot be stored on a single film exposure and computer-stored images are required to optimise image quality and minimise information loss (see §6.5).

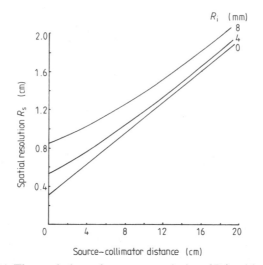

Figure 6.34 The variation of system resolution (R_s) with source-to-collimator distance for a range of intrinsic spatial resolution values (R_i) for a typical parallel-hole collimator.

Gamma-camera developments during the 1970s were mainly aimed at improving intrinsic spatial resolution (White 1979). This was achieved by the following processes.

(*a*) Bialkali photomultipliers were used instead of conventional tubes to give greater photoelectric conversion efficiency and improve the spectral matching.

(*b*) Capacitors were replaced by precision resistors in the encoding matrix to improve stability. Later, some manufacturers replaced the position encoding matrix with delay lines.

(*c*) Preamplifiers were placed directly on the photomultiplier base to reduce noise amplification.

(*d*) The photomultiplier diameter was reduced and the number of photomultiplier tubes was increased accordingly.

Other modifications have included upgrading the lightpipe to use variable-density material, which achieved better uniformity at the expense of intrinsic spatial resolution. In addition, a reduction in collimator septal thickness was achieved by the use of laminates of corrugated lead strips, leading to a significant improvement in system spatial resolution.

More recently, in the 1980s, most development has been aimed at improved uniformity and stability to meet the demands of emission tomography (see §6.7). Non-uniformity is caused by variations in point source sensitivity caused by variable energy response across the camera field of view, spatial distortion (non-linearity) due to inherent errors in the determination of x and y, and variations caused by the Earth's magnetic field. Modern methods of correcting for non-uniformity fall into two categories. Initially automatic photomultiplier tuning is used to correct for gain variations with time. This is achieved by either using light-emitting diodes to monitor photomultiplier gain or using multiple, small-energy windows to monitor photopeak variations. In either case, detected variations are corrected by automatically adjusting the photomultiplier high voltages. Once the photomultiplier gains are stabilised by an auto-tune method, it is possible to apply corrections to reduce spatial and energy response non-linearities. The former is achieved by generating a distortion map using a linearity phantom, and correcting the measured x, y positions accordingly. The distortion map is usually generated at the factory or before the camera is installed. The latter is performed by an energy-correction map generated by the user, using an appropriate uniform radiation source. Again, energy corrections are applied to the z signal before the image is produced. Each of these corrections is applied via on-line microprocessors (see §6.5).

6.3.7 *Multiwire proportional chamber detectors*

One area in which gas-filled detectors may find widespread use in the future is *low-cost positron emission tomography*. Although multiwire proportional chambers (MWPC) have a relatively low efficiency at 511 keV, they can be built in sufficiently large areas (i.e. more than 30×30 cm^2) to obtain adequate sensitivity for PET. The intrinsic resolution is in general very good, i.e. approximately 5 mm or less, and the resolving time can be approximately 10 ns. Imaging is performed by placing a detector on either side of an object, as in figure 6.35, and then rotating these azimuthally in order to acquire the complete set of data necessary to reconstruct a 3D image. Unfortunately, this configuration is more susceptible to random and scattered events than the conventional PET (see §6.3.5) geometry. Combined with the low sensitivity, this makes quantitative measurements difficult to make.

The detectors used vary between different research groups, but all make use of lead converters to convert incident photons to electrons, which are subsequently detected in the usual way by causing ionisation of the chamber gas. The addition of lead converters to an MWPC increases the detection efficiency from a fraction of 1% to approximately 10%. The two most successful designs to date are those developed at the British SERC Rutherford Appleton Laboratory (RAL) (Bateman *et al* 1984, Ott *et al* 1986) and at CERN (Jeavons *et al* 1983, Townsend *et al* 1984). The former design uses a stack of interleaved wire planes and thin lead foil converters, as in figure 6.36. The latest version of this system has a spatial resolution of 6 mm and a sensitivity of 490 cps kBq^{-1} ml for a 20 cm cylinder of activity 12 cm high. The sensitivity per slice is roughly an order of magnitude lower than that of a conventional PET device; future versions of the RAL detector, utilising structured converters, should obtain a spatial resolution of 3.5 mm and a five-fold increase in sensitivity (Marsden *et al* 1986).

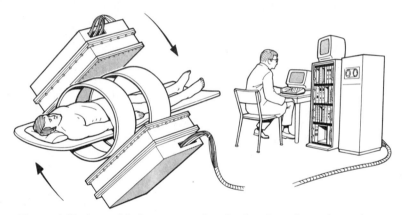

Figure 6.35 A multiwire proportional chamber (MWPC) positron camera.

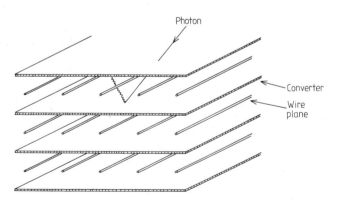

Figure 6.36 A stack of multiwire proportional chambers showing the interleaved wire-plane anodes and thin lead-foil cathode converters.

The CERN design uses two 5 mm converters drilled with a fine matrix of holes (0.8 mm in diameter on a 1 mm pitch), as in figure 6.37. Photoelectrons produced in the lead escape into the holes of the converter. A $10 \, \mathrm{kV \, cm^{-1}}$ electric field then pulls them out of the converter onto a wire plane, where they are detected. For a $^{22}\mathrm{Na}$ line source in a plastic holder, a line-spread function with a FWHM of 2 mm is obtained. The temporal resolution and sensitivity are similar to that of the RAL system. Owing to the poor sensitivity of MWPC systems, the high spatial resolution is presently only realised in practice for small organs (e.g. the thyroid gland) with a high specific activity. Images of larger organs require smoothing in order to reduce statistical fluctuations. In a typical study, about two million coincidences are recorded in 15 min.

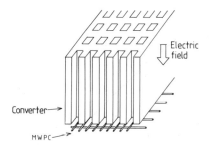

Figure 6.37 A lead channel-plate photon converter for use with a multiwire proportional chamber positron camera.

Because the coincidence events collected by MWPC positron cameras (and indeed any other positron camera using large-area detectors) are not confined to a set of planes through the object, as in ring-system PET, image reconstruction cannot be performed by the standard 2D filtered backprojection method (see Chapter 4). Instead, the acquired data are first backprojected into a 3D array, which is subsequently filtered in 3D (Townsend *et al* 1983). Three-dimensional reconstruction methods such as this may become more widespread in the future as a means of increasing the effective sensitivity of multicrystal ring PET scanners by allowing coincidences between non-adjacent rings.

A second application of MWPC technology to radioisotope imaging is the development of an *MWPC gamma camera* for imaging low-energy photons (less than 100 keV) (Lacy *et al* 1984). The detector (figure 6.38) consists of a drift region and a detection region encased in an aluminium pressure vessel, with a thin entrance window. Photons entering the window interact with the high-pressure gas (90% xenon, 10% methane at 3–5 atm) and the resulting ionisations are 'drifted' into the detection region by an electric field of $1 \, \mathrm{kV \, cm^{-1}}$. The detection region contains three parallel wire planes, the two outer ones at ground potential (cathodes) and the inner one at high positive potential (anode). The gas avalanche is detected by the wire planes, and the two cathodes, containing wires oriented orthogonal to each other, provide the spatial information of the detection position. The detector sensitive area is 25 cm in diameter and the detection efficiency reaches a maximum of about 50–60% at roughly 40 keV, falling to 10–15% at 100 keV. The intrinsic energy resolution at 60 keV is 33% and the count-rate capacity is 850 kcps for a 50% dead-time loss. The intrinsic spatial resolution is about 2.5 mm full width at half-maximum (FWHM) and about 5 mm full width at tenth maximum (FWTM) but, of course, when a collimator is used to provide image capability, the system resolution is between 20 and 25 mm at 10 cm from the object—highlighting once again the disastrous effect of the collimator on system performance in single-photon imaging. Ultimately, the detector has very limited application to medicine, as few useful imaging radioisotopes emit photons in the energy window 50–100 keV. ^{133}Xe and ^{201}Tl are potentially suitable and ^{178}Ta from a ^{178}W/^{178}Ta generator has some applications to first-pass cardiac imaging.

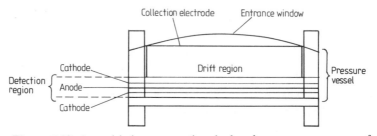

Figure 6.38 A multiwire proportional chamber gamma camera for imaging with γ-ray energies below 100 keV. (After Lacy *et al* (1984).)

A third important application of MWPC technology is the development of *quantitative autoradiography* using agents labelled with β emitters, such as ^{14}C, ^{32}P, ^{35}S, ^{3}H as well as the x-ray emitter, ^{125}I (Bateman *et al* 1985). Autoradiography can be seen as radioisotope imaging of either tissue samples, to determine radiopharmaceutical distribution at cellular levels, or of electrophoresis plates, for determining molecular weights, studying protein-binding properties and DNA sequencing. The detector is described as a multistep avalanche (MSA) MWPC and consists of three distinct sections—an avalanche gap (close to the sample plate), a transfer gap and an MWPC section (figure 6.39). Electrons emitted from the sample plate produce a gas avalanche in the first gap. The process in this section is aided by ultraviolet fluorescence produced in argon gas, which ionises acetone molecules in the gas mixture. Further ionisation and pulse amplification take place in the transfer region. In the MWPC section the ionisation produces an anode-plane pulse that induces pulses in two adjacent cathode planes, which provide the coordinates of the ionisation event.

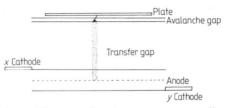
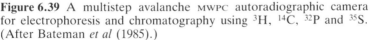

Figure 6.39 A multistep avalanche MWPC autoradiographic camera for electrophoresis and chromatography using ^{3}H, ^{14}C, ^{32}P and ^{35}S. (After Bateman *et al* (1985).)

Using ^{3}H-labelled agents, a spatial resolution of 0.4 mm has been achieved, sufficient to resolve DNA sequence bands in electrophoresis plates. The notable advantage of the MSA MWPC detector over film autoradiography is the high sensitivity (about 19% for ^{3}H), giving exposure times of 1000 times less than conventional film techniques (a few hours instead of a few weeks). However, as the radiation energy detected increases, the spatial resolution deteriorates due to the increase in the electron range. The spatial resolution is 0.8 mm for ^{14}C and more than 1 mm for ^{32}P and ^{35}S (figure 6.40), which are isotopes in common use for conventional autoradiography. Clearly, these detectors can give no hope of measurements at the cellular level (see introduction to book) where better than 10 μm (0.01 mm) resolution is necessary.

Figure 6.40 An electrophoresis plate of a DNA sequence using ^{35}S. (Courtesy of J E Bateman.)

One further application of gas detectors to radioisotope imaging is a *hybrid system* using an MWPC and a scintillation crystal, which promises to provide a detector with some of the best properties of both (Anderson *et al* 1983). Most inorganic scintillators have an optical decay constant in the range 200–5000 ns. Barium fluoride (BaF_2) has two light emission components. Eighty per cent of the light is emitted with a decay constant of 620 ns at a wavelength of about 310 nm and 20% is emitted with a decay constant of 0.7 ns at wavelengths of 195 and

220 nm. The material has a density of 4.89 g cm^{-3} (compared with 3.67 g cm^{-3} for NaI (see table 6.1)) but the light output is only 5% that of NaI. However, BaF$_2$ would be a useful scintillator if the fast-light component (195 and 220 nm) could be detected. Photocathodes of the majority of photomultiplier tubes have little sensitivity at 200 nm, but tetrakis(dimethylamino)ethylene (TMAE) is sensitive at this wavelength. A detector incorporating BaF$_2$, a photocathode of TMAE and a multiwire proportional chamber (figure 6.41) has been evaluated to determine its application in medicine. Gamma-ray photons are detected efficiently by the BaF$_2$, but only a small percentage of the light output is converted by the photocathode. However, preliminary tests have shown the advantage of using a high-Z material, which gives high photon detection sensitivity and acceptable pulse-height resolution. Further, the high event-rate capacity and excellent intrinsic spatial resolution of the MWPC detector gives a combination that may have potential uses in radioisotope imaging.

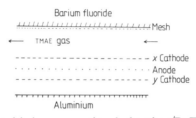

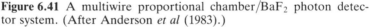

Figure 6.41 A multiwire proportional chamber/BaF$_2$ photon detector system. (After Anderson *et al* (1983).)

6.3.8 Semiconductor detectors

Although inorganic scintillating materials have several properties that make them excellent γ-ray detectors, the physical process of scintillation leads to a relatively poor energy response. This is because it takes tens of electron volts of energy to cause the appropriate excitation process in the scintillator. When the statistical variation of the conversion efficiency (10–30%) of a photomultiplier is convolved with that of the light output characteristics of the scintillator, this results in energy resolutions of 10–15% for 100–200 keV photons detected by NaI(Tl).

If we consider a silicon or germanium semiconductor, an electron–hole (e–h) pair takes only a few electron volts to produce, and, if the material is fully depleted, it functions like a solid-state ionisation chamber, giving a very high detection efficiency for each e–h pair. Hence it is possible to obtain energy resolutions of about 600 eV for high-purity germanium at liquid-nitrogen temperature and a few kiloelectronvolts for silicon at room temperature.

The high atomic number (Z) and excellent energy resolution of germanium attracted the attention of several groups of workers in the 1970s, who constructed small-scale gamma cameras using lithium-drifted germanium crystals. The cryogenic problems and the cost of the material have seriously inhibited further development but interest has been renewed recently with the production of a Compton scatter camera, which, although still using NaI(Tl) as the basic detector, allows replacement of the lead collimator by a semiconductor that effectively acts as a coded-aperture or multi-pinhole system.

More recently, the development of silicon wafers with spatial read-out using either drift or strip techniques has motivated work on devices for autoradiography and the development of a radioisotope Compton telescope. Some of these applications are discussed below to highlight a rapidly developing area of interest in radioisotope imaging.

A *germanium gamma camera* was produced in the UK in the early 1970s (McCready *et al* 1971). The detector used orthogonal electrical contact strips on a single slice of lithium-drifted germanium to form a p–i–n structure (figure 6.42). The detector was $44 \times 44 \, \text{mm}^2$ in area and 10 mm thick. Contacts were 1.9 mm wide with 1.1 mm separations. The energy resolution was a few kiloelectronvolts. The detector was used with a specially designed lead collimator. An intrinsic spatial resolution of 3 mm was measured, but this naturally deteriorated with the use of the collimator in the same way as we have seen for a scintillation gamma camera with a collimator.

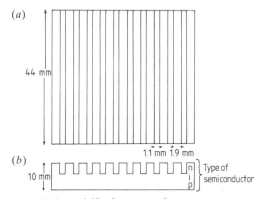

Figure 6.42 A lithium-drifted germanium gamma camera using a p–i–n structure: (*a*), plan view; (*b*), side view. (After McCready *et al* (1971).)

A similar detector developed in the mid-1970s (Gerber *et al* 1977) used a charge-splitting resistor network to determine the spatial position of the photon conversion point (figure 6.43). This detector was $30 \times 30 \, \text{mm}^2$ in area, 5 mm thick and had 14 electrode strips on each side spaced on 3 mm centres. The detector had a 5.5 keV energy resolution and 1.7 mm intrinsic spatial resolution.

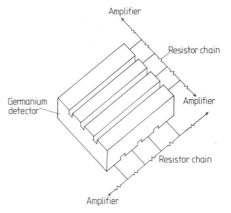

Figure 6.43 The charge-splitting system used to read out a semiconductor gamma camera. (After Gerber *et al* (1977).)

The clear advantages of these types of gamma camera have never been fully realised because of overall cost and specifically the difficulty of making large-area single Ge crystals. In addition, the need to use a collimator to image with single-photon emitters always greatly reduces the sensitivity and spatial resolution of any gamma camera. Several attempts to overcome this have been made using 'coded-aperture' systems. Perhaps the most interesting of these is the *Compton gamma camera*. The idea was also first proposed in the mid-1970s (Everett *et al* 1977), in which a segmented semiconductor (silicon) was suggested as the scatterer in coincidence with a conventional gamma camera (minus collimator) as the absorber (figure 6.44). The Compton scatter of photons is governed by the relationship

$$\cos \theta = (E - k\delta E)/(E - \delta E) \tag{6.12}$$

where $k = 1 + m_0c^2/E$, E is the photon energy, m_0c^2 is the rest mass of the electron, δE is the energy loss in the scatter process and θ is the scatter angle of the photon. If the scatter process can be defined by the measurement of $(x, y, z)_s$, $(x, y, z)_a$ and the two energies $(E - \delta E)$ and δE, then it is possible to determine θ. When each event is backprojected, the incident photon vector is found to be contained in a conical surface. The distribution of the photon source can be obtained by appropriate iterative reconstruction techniques.

The most recent developments of Compton scatter detectors are the use of small germanium scatterers to provide high-sensitivity whole-body measurements (Singh and Doria 1983), and silicon drift or strip detectors to provide the scatterer in a system designed for animal/tumour imaging. The former contains a 33×33 array of high-purity germanium detectors coupled to an uncollimated conventional gamma camera (figure 6.45). Each superconducting element is 5×5 mm^2 in area and 6 mm thick. It is predicted that a spatial resolution of 12 mm at 10 cm from the detector is achievable, and the sensitivity should be 15 times higher than that of a conventional gamma camera/collimator system.

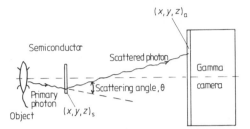

Figure 6.44 A schematic diagram of the semiconductor/scintillation counter Compton camera, showing how an image can be reconstructed from a knowledge of the coordinates of the scattered photon in the semiconductor scatterer $(x,y,z)_s$ and the gamma-camera absorber $(x,y,z)_a$.

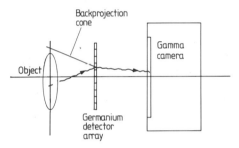

Figure 6.45 A Compton camera system using an array of high-purity germanium detectors and an uncollimated conventional gamma camera.

The recent development in spatially resolving silicon detectors raises the possibility of a small-scale Compton camera in which the scatter position is known to better than 1 mm and the energy loss in the scatter process to a few kiloelectronvolts. The spatial accuracy in the image is a function of the accuracy of both the energy and spatial measurement in the scatterer and absorber. Although the spatial resolution of the scintillation camera is no better than 3 mm (at best), the use of appropriate geometry can provide a spatial resolution of better than 1 mm in an object close to the scatterer. If scatter in the object is ignored, the energy of the photon entering the scattering material is known and, hence, if the energy loss due to scatter is accurately measured, the poor energy resolution in the NaI is of little importance. It should be possible to image small objects (a few square centimetres) with a spatial resolution of better than 1 mm, ideal for measurements of the distribution of radiolabelled agents (antibodies, etc) in small human tumours and animal models. The sensitivity will be 10 times that of a conventional gamma camera, but the system is likely to be limited to coincidence count rates of a few thousand counts per second.

One further and important application of semiconductors is the use of silicon drift or strip detectors to autoradiography. The *drift detector* (figure 6.46) incorporates one or more anode strip conductors and a large number of cathode strips, each separated from its neighbours by about 50 μm. An electron entering the silicon wafer is absorbed close to a single cathode strip and produces many e–h pairs. The holes drift laterally to the nearest cathode, producing a pulse that may be used to start a time digitiser. The electrons drift along the silicon to the nearest anode and produce a second pulse, which can be used to stop the digitiser. If the electron drift time is sufficiently slow (determined by the anode–cathode voltage drop), the spatial position of the detection point can be determined accurately (to within about 5–10 μm). Alternatively, using a simple *strip detector*, where the conducting strips are spaced 50 μm apart, each strip can be read out using amplifier discriminator circuits and spatial accuracy of 50–100 μm achieved (figure 6.47). The advantage of the drift system, apart from the obviously improved resolution, is the low read-out cost, although the long drift time prohibits high-count-rate applications (not a problem in autoradiography). This technology is still in its infancy and neither technique has yet been applied successfully to autoradiography. However, the obvious advantage of sensitivity (10^3 times that of film), which would enable an enormous reduction in the exposure time needed for an autoradiograph, will surely spur on efforts in this area of development.

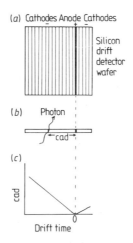

Figure 6.46 A silicon drift detector: (*a*) plan view; (*b*) side view; (*c*) cathode–anode distance, cad, versus drift time characteristic.

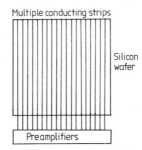

Figure 6.47 A silicon strip detector.

6.3.9 Summary

We can conclude from this section that the detection of γ-photons has been dominated for the last 30 years by the use of inorganic scintillators (mostly NaI(Tl)) and photomultipliers. The scintillator is an efficient γ-ray absorber and produces plenty of light photons, but the isotropy of light emission and the poor photoelectron conversion efficiency of photocathodes significantly degrades the signal. In addition, the extended temporal nature of the light emission is often a problem in high-count-rate imaging. The photomultiplier is clearly the weak link of scintillator-based detectors, particularly as its poor analogue stability seriously inhibits the uniformity of a conventional gamma camera and hence requires sophisticated correction processes to prevent artefacts in the images. One might speculate that the recent advances in gas and semiconductor detectors will slowly result in the replacement of the scintillator–photomultiplier combination in radioisotope imaging within the next decade.

6.4 RADIONUCLIDES FOR IMAGING

One of the primary advantages associated with the use of radionuclides in medicine is the large signal (in this case the emitted radiation) obtained from the relatively small mass of radionuclide employed for a given study. Nuclear medicine takes advantage of this physical characteristic by using various radioisotope-tagged compounds (radio-pharmaceuticals) in order to 'trace' various functions of the body (McAfee and Thakur 1977). The minute mass of radiolabelled material allows for non-invasive observation without disturbance of the system under study through pharmacological or toxicological effect. For most nuclear medicine studies, the mass of tracer used is in the range of nanograms (see table 6.4), and no other physical technique could be employed to measure mass at these levels. Therefore, the sensitive measurement of biochemical and physiological processes through the use of radioactivity and its detection comprise the fundamental basis of nuclear medicine and is the key to its future growth.

Table 6.4 Approximate mass (g) of 37 GBq (1 Ci) of radionuclide for a given half-life and atomic weight.

$T_{1/2}$	Atomic weight of atom (amu)		
	18	99	201
15 s	2.4×10^{-11}	1.3×10^{-10}	2.7×10^{-10}
15 min	1.4×10^{-9}	7.7×10^{-9}	1.6×10^{-8}
6 h	3.5×10^{-8}	1.9×10^{-7}	3.8×10^{-7}
8 d	1.1×10^{-6}	6.3×10^{-6}	1.2×10^{-5}
15 a	7.7×10^{-4}	4.2×10^{-3}	8.3×10^{-3}

6.4.1 Radioactive decay

The radioactivity Q of a number (N) of nuclei is given by

$$Q = -\lambda N = dN/dt \qquad (6.13)$$

where λ is defined as the decay constant for the radioisotope. We can see that the rate of decay of nuclei depends only upon λ and the number of nuclei, N. The solution to equation (6.13) is

$$N = N_0 \exp(-\lambda t) \qquad (6.14)$$

where N_0 is the number of nuclei at some reference time $t = 0$. If $T_{1/2}$ is the time for half the nuclei to decay, the so-called 'half-life', then

$$T_{1/2} = (\ln 2)/\lambda. \qquad (6.15)$$

An alternative form of equation (6.14) is

$$N = N_0(\tfrac{1}{2})^m \qquad (6.16)$$

where m is the number of half-lives since the reference time $t = 0$.

Using these simple equations, it is possible to calculate the radioactivity of any mass of nuclei at any time subsequent to a measurement at a reference-zero time.

6.4.2 The production of radionuclides

The fundamental property of all radioactive elements is the imbalance of the proton-to-neutron ratio of the nucleus. A proper balance of protons and neutrons is essential for maintaining a stable atomic nucleus. The balance must be maintained to overcome electrostatic repulsion of the charged protons, and figure 6.48 shows how the neutron-to-proton (n/p) ratio changes with increasing mass. There are four ways by which radionuclides are produced:

 (*a*) neutron capture (also known as neutron activation);
 (*b*) nuclear fission;
 (*c*) charged-particle bombardment; and
 (*d*) radionuclide generator.

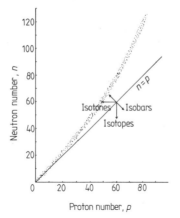

Figure 6.48 Graph of neutron number n versus proton number p, showing the increase in the n/p ratio with atomic number.

Each method affords useful isotopes for nuclear medicine imaging. A description of the methods and examples of some of the useful radioisotopes produced will now be presented.

Neutron capture is the absorption of a neutron by an atomic nucleus, and the production of a new radionuclide via reactions such as

$$\text{n} + {}^{98}\text{Mo} \rightarrow {}^{99}\text{Mo} + \gamma \tag{6.17}$$

$$\text{n} + {}^{32}\text{S} \rightarrow {}^{32}\text{P} + \text{p}. \tag{6.18}$$

To produce radioactive elements through neutron capture, neutrons must have a mean energy of 0.03–100 eV. These 'thermal' neutrons are best-suited for interaction with and absorption into the atomic nucleus. The most efficient means of producing radioisotopes by this method is through the use of a nuclear reactor. To produce a radioactive species, a sample of a target element is placed in a field of thermal neutrons. The reaction yield depends on the flux density of incident particles, ϕ ($\text{cm}^{-2}\text{s}^{-1}$), the number of accessible target nuclei (n_t) and the likelihood or cross section of the reaction, σ (barn). The yield (N_y) is given by

$$N_y = \frac{n_t\phi\sigma}{\lambda}[1 - \exp(-\lambda t)]. \tag{6.19}$$

The radionuclide produced via the (n,γ) reaction is an isotope of the target material, i.e. the two nuclei have the same number of protons. This means that radionuclides produced via the (n,γ) reaction are not carrier-free (compare with nuclear fission), and thus the ratio of radioactive atoms to stable atoms and the specific activity are relatively low. Separation of the radionuclide from other target radionuclides is possible by physical and chemical techniques. Useful tracers produced by neutron absorption are shown in table 6.5.

Nuclear fission is the process whereby heavy nuclei (${}^{235}\text{U}$, ${}^{239}\text{Pu}$, ${}^{237}\text{U}$,

^{232}Th) irradiated with thermal neutrons are rendered unstable due to absorption of these neutrons. Consequently, these unstable nuclei undergo 'fission', the breaking up of the heavy nuclei into two lighter nuclei of approximately similar atomic weight, for example

$$^{235}_{92}U + ^{1}_{0}n \rightarrow ^{236}_{92}U \rightarrow ^{99}_{42}Mo + ^{133}_{50}Sn + 4^{1}_{0}n. \qquad (6.20)$$

As seen from equation (6.20), this reaction produces four more neutrons, which may be absorbed by other heavy nuclei, and the fission process can continue until the nuclear fuel is exhausted. Interaction such as that in equation (6.20) must, of course, conserve Z and A.

Table 6.5 Radionuclides produced by neutron absorption.

Isotope	Gamma-ray energy (keV)	Half-life	Absorption cross section (barn)
^{51}Cr	320	27.7 d	17 [a]
^{59}Fe	1099	44.5 d	1.1 [a]
^{99}Mo	740	66.02 h	0.13
^{131}I [b]	364	8.05 d	0.2

[a] From BRH (1970).
[b] ^{130}Te$(n,\gamma)^{131}$Te $\rightarrow$ ^{131}I.

Nuclides produced by fission of heavy nuclei must undergo extensive purification in order to harvest one particular radionuclide from the mixture of fission products. The fission process affords high specific activity due to the absence of carrier material (non-radioactive isotope of the same element). However, fission products are usually rich in neutrons and therefore decay principally via β^- emission, a physical characteristic that is undesirable for medical imaging, but of interest in therapy. Useful nuclides produced by nuclear fission are shown in table 6.6.

Table 6.6 Radionuclides produced by nuclear fission.

Isotope	Gamma-ray energy (keV)	Half-life	Fission yield [a] (%)
^{99}Mo	740	66.02 h	6.1
^{131}I	364	8.05 d	2.9
^{133}Xe	81	5.27 d	6.5
^{137}Cs	662	30 a	5.9

[a] From BRH (1970).

Charged-particle bombardment is the process of production of radionuclides through the interaction of charged particles ($H^{\pm}$, D^{+}, $^{3}He^{2+}$, $^{4}He^{2+}$) with the nuclei of stable atoms. The particles must have enough kinetic energy to overcome the electrostatic repulsion of the positively charged nucleus. Two basic types of accelerator are used for this purpose, the linear accelerator and the cyclotron. In both systems, charged particles are accelerated over a finite distance by the application of alternating electromagnetic potentials (figure 6.49). In both types of machine, particles can usually be accelerated to various energies. Examples of typical reactions in a target are

$$p + {}^{68}Zn \rightarrow {}^{67}Ga + 2n \tag{6.21}$$

$$\alpha + {}^{16}O \rightarrow {}^{18}F + p + n. \tag{6.22}$$

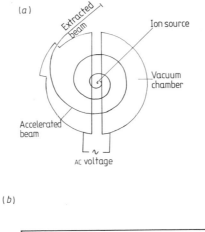

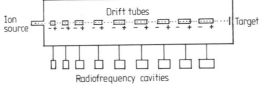

Figure 6.49 Schematic diagrams of (*a*) a cyclotron and (*b*) a linear accelerator for radionuclide production.

For the production of medically useful radionuclides, particle energies per nucleon in the range 1–100 MeV are commonly used. One major advantage of producing isotopes through charged-particle bombardment is that the desired isotope is almost always of different atomic number to the target material. This theoretically allows for the production of radionuclides with very high specific activity and minimal radionuclide impurity. However, the actual activity and purity obtained is related to the isotopic and nuclidic purity of the target material, the cross section of the desired reaction and the cross section of any secondary reaction.

Charged-particle reactions yield radionuclides that are predominantly neutron-deficient and therefore decay by β^+ emission or electron capture. The latter radioisotopes are particularly useful for clinical imaging due to the lack of particulate emission. Examples of accelerator-produced radionuclides routinely used in nuclear medicine are shown in table 6.7.

Table 6.7 Radionuclides produced by charged-particle bombardment.

Isotope	Principal gamma-ray energy (keV)	Half-life	Reaction
^{11}C	511 (β^+)	20.4 min	$^{14}N(p,\alpha)^{11}C$
^{13}N	511 (β^+)	9.96 min	$^{13}C(p,n)^{13}N$
^{15}O	511 (β^+)	2.07 min	$^{15}N(p,n)^{15}O$
^{18}F	511 (β^+)	109.7 min	$^{18}O(p,n)^{18}F$
^{67}Ga	93	78.3 h	$^{68}Zn(p,2n)^{67}Ga$
	184		
	300		
^{111}In	171	67.9 h	$^{112}Cd(p,2n)^{111}In$
	245		
^{123}I	159	13 h	$^{124}Te(p,2n)^{123}I$
			$^{127}I(p,5n)^{123}Xe \rightarrow ^{123}I$
^{201}Tl	68–80.3	73 h	$^{203}Tl(p,3n)^{201}Pb \rightarrow ^{201}Tl$

Radioactive decay can lead to the generation of either a stable or a radioactive nuclide. In either case, the new nuclide may have the same or different atomic number depending on the type of decay (see next section). Radioactive decay leading to the production of a radioactive daughter with a different Z allows for the possibility of simple chemical separation of the parent–daughter combination. If the daughter radionuclide has good physical characteristics compatible with medical imaging and the parent has a sufficiently long half-life to allow for production, processing and shipment, then remote parent–daughter separation means a potentially convenient source of a medically useful short-lived radionuclide. This type of radionuclide production system is known as a *radionuclide generator*.

A radionuclide generator is a means of having 'on tap' a short-lived radionuclide. It is technically achieved by the chemical separation of the daughter radionuclide from the parent. This can be accomplished through the use of chromatographic techniques, distillation or phase partitioning. However, chromatographic techniques have been the most widely explored and are the current state-of-the-art technology (Yano 1975) for the majority of generator systems in use today (figure 6.50).

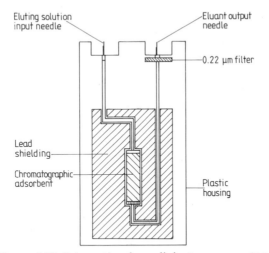

Figure 6.50 Schematic of a radioisotope generator.

The equations governing generator systems stem from the formula

$$A_2 = \frac{\lambda_2}{\lambda_2 - \lambda_1} A_1^0 [\exp(-\lambda_1 t) - \exp(-\lambda_2 t)] \tag{6.23}$$

where A_1^0 is the parent activity at time $t = 0$, t is the time since the last elution of the generator, A_2 is the activity of the daughter product ($A_2^0 = 0$), and λ_1 and λ_2 are the decay constants of parent and daughter radioisotopes, respectively.

For the special case of secular equilibrium, defined by $\lambda_2 \gg \lambda_1$, we have

$$A_2 = A_1^0 [\exp(-\lambda_1 t) - \exp(-\lambda_2 t)]. \tag{6.24}$$

If t is much less than the half-life of the parent, $\ln(2)/\lambda_1$, and greater than approximately seven times the daughter half-life, $\ln(2)/\lambda_2$, then

$$A_2 \simeq A_1^0. \tag{6.25}$$

This is the equilibrium condition. The growth of the daughter here is given by

$$A_2 = A_1^0 [1 - \exp(-\lambda_2 t)]. \tag{6.26}$$

For transient equilibrium, defined by $\lambda_2 > \lambda_1$ but λ_2 not *very* much greater than λ_1, we have

$$A_2 = \lambda_2 A_1^0 / (\lambda_2 - \lambda_1). \tag{6.27}$$

Figure 6.51 shows the growth of $^{99}\text{Tc}^\text{m}$ activity in a $^{99}\text{Mo} \rightarrow {}^{99}\text{Tc}^\text{m}$ generator.

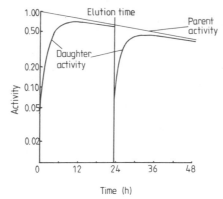

Time (h)

Figure 6.51 The build-up of activity of the daughter product with time for a typical (^{99}Mo/^{99}Tc^m) radioisotope generator.

The most widely used generator-produced radionuclide in nuclear medicine is ^{99}Tc^m. The parent, ^{99}Mo, has a half-life of about 66 h, can be produced through neutron activation or fission, can be chemically adsorbed onto an Al$_2$O$_3$ (alumina) column and decays to ^{99}Tc^m (85%) and ^{99}Tc (15%). ^{99}Tc^m has a half-life of 6.02 h, decays to ^{99}Tc by isomeric transition and emits a 140 keV γ-ray (98%) with no associated particulate radiations:

$$^{99}\text{Mo} \underset{\beta^-}{\rightarrow} {}^{99}\text{Tc}^m \underset{\text{IT}}{\rightarrow} {}^{99}\text{Tc} + \gamma. \tag{6.28}$$

In the early days of their development (see Chapter 1), technetium generators were sometimes referred to as 'radioactive cows'. The ^{99}Tc^m is 'milked' from the chromatographic column of alumina by passing a solution of isotonic saline through the column (0.9% NaCl). This saline solution and the solid phase of Al$_2$O$_3$ allow for efficient separation of ^{99}Tc^m from the ^{99}Mo with only minute amounts of ^{99}Mo breakthrough (less than 0.1%). The eluted ^{99}Tc^m can be chemically manipulated so that it binds to a variety of compounds, which will then determine its fate *in vivo* (see §6.9). Other generator systems exist producing radionuclides useful for gamma-camera imaging as well as for PET and examples of these are given in table 6.8.

6.4.3 Types of radioactive decay

All radionuclides used in nuclear medicine are produced via one of the four ways described above. Each of these radionuclides has a unique process by which it decays. The *decay scheme* describes the type of decay, the energy associated with it and the probability for each type of decay. These decay schemes can be very complex since many of the radionuclides decay via multiple nuclear processes (figure 6.52).

Table 6.8 Generator-produced radionuclides.

Parent P	Parent half-life	Mode of decay P→D	Daughter D	Mode of decay of D	Daughter half-life	Gamma-ray energy from daughter (keV)
^{99}Mo	2.7 d	β^-	^{99}Tc^m	IT	6 h	140
^{82}Sr	25 d	EC	^{82}Rb	EC β^+	1.3 min	777 511
^{68}Ge	280 d	EC	^{68}Ga	EC β^+	68 min	511
^{52}Fe	8.2 h	EC β^+	^{52}Mn^m	EC β^+ IT	21 min	511
^{81}Rb	4.7 h	EC	^{81}Kr^m	IT	13 s	190
^{62}Zn	9.1 h	EC β^+	^{62}Cu	EC β^+	9.8 min	511
^{178}W	21.5 d	EC	^{178}Ta	EC	9.5 min	93

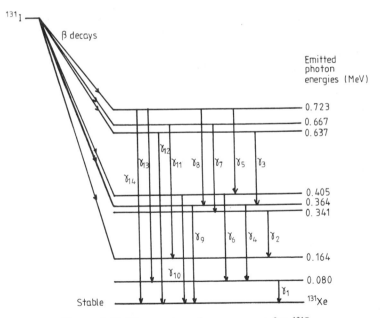

Figure 6.52 The nuclear decay process for ^{131}I.

Alpha decay is the process of spontaneous emission of an α-particle (a helium nucleus) in the decay of heavy radioisotopes, with a discrete energy in the range 4–8 MeV. If α decay leaves the nucleus in an excited state, the de-excitation will be via the emission of γ-radiation. Most of the energy released in the transition is distributed between a daughter nucleus (as recoil energy) and the α-particle (as kinetic energy). In the α-decay process, the parent nucleus loses four units of mass and two units of charge. As an example we show

$$^{226}_{88}\text{Ra} \rightarrow {}^{222}_{86}\text{Rn} + \alpha. \qquad (6.29)$$

Although α-emitting nuclides have no use in medical imaging (since the α-particles travel virtually no distance in tissue), there has been a renewed interest in their use for targeted (i.e. highly localised) therapy.

As discussed previously, many radionuclides are unstable due to the neutron/proton imbalance within the nucleus. The decay of neutron-rich radionuclides involves the ejection of a β^--particle (e^-), resulting in the conversion of a neutron into a proton. Decay via β^- *emission* results in the atomic number of the atom changing but the atomic weight remaining the same. The energy of the emitted β^--particles is not discrete but a continuum (i.e. varies from zero to a maximum, E_m) and, since the total energy lost by the nucleus during disintegration must be discrete, an additional process must be responsible. Energy conservation of β decay is maintained by the emission of a third particle—the neutrino (v). The neutrino has no measurable mass nor charge, and interacts weakly with matter. An example of β^- decay is

$$^{99}\text{Mo} \rightarrow {}^{99}\text{Tc}^m + e^- + \bar{v}. \qquad (6.30)$$

Beta decay may be accompanied by γ-ray emission if the daughter nuclide is produced in an excited state. After β^- decay, the atomic number of the daughter nuclide is one more than the parent nuclide, but the atomic mass remains the same.

Nuclei that are rich in protons or are neutron-deficient may decay by *positron emission* from the nucleus. This decay is also accompanied by the emission of an uncharged particle, the antineutrino ($\bar{v}$). After positron decay, the daughter nuclide has an atomic number that is one less than that of the parent, but again the atomic weight is the same. The range of a positron (e^+) is short (of the order of 1 mm in tissue) and, when the particle comes to rest, it combines with an atomic electron from a nearby atom, and is annihilated. Annihilation (the transformation of these two particles into pure energy) gives rise to two photons both of energy 511 keV emitted approximately antiparallel to each other. These photons are referred to as annihilation radiation. Positron emission only takes place when the energy difference between the parent and daughter nuclides is larger than 1.02 MeV. An example of positron decay is

$$^{68}\text{Ga} \rightarrow {}^{68}\text{Zn} + e^+ + v. \qquad (6.31)$$

An alternative to positron emission for nuclides with a proton-rich nucleus is *electron capture*. Electron capture involves the absorption within the nucleus of an atomic electron, transforming a proton into a neutron. For this process to occur, the energy difference between the parent and the daughter nuclides can be small, unlike positron emission. Usually the K-shell electrons are captured because of their closeness to the nucleus. The vacancy created in the inner electron orbitals is filled by electrons from the outer orbitals. The difference in energy between these electron shells appears as an x-ray that is characteristic of the daughter radionuclide. The probability of electron capture increases with increasing atomic number because electron shells in these nuclei are closer to the nucleus. An example of electron capture is

$$^{123}\text{I} \rightarrow {}^{123}\text{Te} + \gamma. \tag{6.32}$$

A nucleus produced in a radioactive decay can remain in an excited state for some time. Such states are referred to as *isomeric states*, and decay to the ground state can take from fractions of a second to many years. A transition from the isomeric or excited state to the ground state is accompanied by γ *emission*. When the isomeric state is long-lived, this state is often referred to as a metastable state. $^{99}\text{Tc}^{\text{m}}$ is the most common example of a metastable isotope encountered in nuclear medicine (see equation (6.28)).

Internal conversion is the process that can occur during γ-ray emission when a photon emitted from a nucleus may knock out one of the atomic electrons from the atom. This particularly affects K-shell electrons, as they are the nearest to the nucleus. The ejected electron is referred to as the *conversion electron* and will have a kinetic energy equal to the energy of the γ-ray minus the electron binding energy. The probability of internal conversion is highest for low-energy photon emission. Again, vacancies in the inner orbitals are filled by electrons from the outer shells, leading to the emission of *characteristic x-rays*. Furthermore, characteristic x-rays produced during internal conversion may themselves knock out other outer orbital electrons provided the x-rays have an energy greater than the binding energy of the electron with which they interact. This emitted electron is then referred to as an *Auger electron*. Again, vacancies in the electron shells due to Auger emission are filled by other electrons in outer orbitals leading to further x-ray emission.

6.4.4 Choice of radioisotope for imaging

The physical characteristics of radionuclides that are desirable for nuclear medicine imaging include:

(*a*) a suitable physical half-life;
(*b*) decay via photon emission;
(*c*) associated photon energy high enough to penetrate the body tissue with minimal tissue attenuation; but
(*d*) low enough for minimal thickness of collimator septa; and
(*e*) absence of particulate emission.

The effective half-life T_{E} of a radiopharmaceutical is a combination of the physical half-life T_P and the biological half-life T_B, i.e.

$$\frac{1}{T_E} = \frac{1}{T_B} + \frac{1}{T_P}. \tag{6.33}$$

Close matching of the effective half-life with the duration of the study is an important dosimetric as well as practical consideration in terms of availability and radiopharmaceutical synthesis.

The photon energy is critical, for various reasons. The photon must be able to escape from the body efficiently and it is desirable that the photopeak should be easily separated from any scattered radiation. These two characteristics favour high-energy photons. However, at very high energies, detection efficiency using a conventional gamma camera is poor (see figure 6.5) and the increased septal thickness required for collimators decreases the sensitivity further. In addition, high-energy photons are difficult to shield and present practical problems for staff handling the isotope.

The radionuclide that fulfils most of the above criteria is technetium-99m ($^{99}Tc^m$), which is used in more than 90 % of all nuclear medicine studies. It has a physical half-life of 6.02 h, is produced via decay of a long-lived ($T_{1/2} = 66$ h) parent ^{99}Mo, and decays via isomeric transition to ^{99}Tc emitting a 140 keV γ-ray. The short half-life and absence of $\beta^{\pm}$ emission results in a low radiation dose to the patient. The 140 keV γ emission allows for 50% penetration of tissue at a thickness of 4.6 cm but is easily collimated by lead. Most importantly, the radioisotope can be produced from a generator lasting the best part of a week, supplying imaging agents 'on tap'.

Other radioisotopes in common use in nuclear medicine include ^{123}I, ^{111}In, ^{67}Ga, ^{201}Tl and $^{81}Kr^m$. ^{123}I has proved a valuable replacement for ^{131}I as it decays via electron capture emitting a γ-ray of energy 159 keV and has a 13 h half-life. It is easily bonded to proteins and pharmaceuticals that can be iodinated. However, like most of the other radioisotopes in this list, it is cyclotron-produced (table 6.7) and presently still very expensive in a form free of other iodine isotopes.

The radionuclides ^{111}In and ^{67}Ga are very similar chemically and both decay via electron capture (table 6.9). Most recently, there has been an increased interest in their use as antibody labels via bifunctional chelates. ^{111}In is the superior imaging isotope, emitting acceptable photon energies for gamma-camera studies, but again it is expensive as it is produced by charged-particle bombardment (table 6.7). ^{67}Ga has long been used as a tumour localising agent in the form of gallium citrate and has also proved useful in the same form in the detection of abcesses. ^{201}Tl is utilised by the cardiac muscle in a similar fashion to potassium and is in widespread use for imaging myocardial perfusion. However, the photon emissions used in myocardial imaging are the 80 keV x-rays, which are close in energy to lead x-rays produced by the collimator, and this, together with the long half-life (73 h), makes the isotope a poor imaging agent. Hence there is continued search for a $^{99}Tc^m$-labelled myocardial perfusion agent.

Table 6.9 Main gamma emissions of ^{67}Ga and ^{111}In.

Radionuclide	Gamma-ray emission	Gamma-ray energy (keV)	Mean number per disintegration
^{67}Ga	γ_2	93	0.38
	γ_3	185	0.24
	γ_5	300	0.16
	γ_6	394	0.043
^{111}In	γ_1	171	0.90
	γ_2	245	0.94

Radioisotopes emitting positrons (table 6.7) have been used extensively for physiological research, in the main, rather than clinical nuclear medicine because of the need for an on-site or nearby cyclotron to produce them in view of their relatively short half-lives. The radionuclides ^{15}O, ^{13}N, ^{11}C and ^{18}F have many applications in the field of functional imaging but have had little impact, as yet, in routine nuclear medicine because of lack of availability. ^{68}Ga and ^{82}Rb, however, are two generator-produced positron-emitting nuclides that could provide invaluable radiopharmaceuticals for clinical PET. In particular, ^{68}Ga can be used to label many agents in a manner similar to ^{99}Tc^m, and ^{82}Rb is a greatly superior myocardial perfusion agent to ^{201}Tl. These radioisotopes, coupled with the development of low-cost PET cameras (§6.3.7), may bring the much-needed advantages of high sensitivity and spatial resolution to clinical nuclear medicine.

6.5 THE ROLE OF COMPUTERS IN RADIOISOTOPE IMAGING

With the development of minicomputers in the mid-1960s, it became possible to store large quantities of data and to perform complex data manipulation well beyond the range of previous human experience. However, not until the mid-1970s were computers attached to a gamma camera (Lieberman 1977), partly because of the complexity of interface electronics and partly through lack of suitable computer software. In the 1980s it has become possible to have enormously powerful computers on a single microchip and the cost of memory, data storage and input/output facilities has fallen dramatically. Thus, technology has provided a revolution in the application of computers to medical imaging and, in particular, to nuclear medicine (Todd-Pokropek 1982, Harbert and Da Rocha 1984a), where the need for tomographic and dynamic information is important. There are multiple applications for computers in radioisotope imaging, including

(*a*) data acquisition,
(*b*) on-line data correction,

(*c*) data processing,
(*d*) image display and manipulation,
(*e*) data storage and
(*f*) system control,

and we will discuss each of these in some detail.

6.5.1 Data acquisition

Data from radioisotope imaging devices, such as the gamma camera or scanner, can be acquired and stored on a computer in either list or frame mode. In *list mode*, information is recorded as serial spatial and temporal data, which can later be either reordered into dynamic data files (frame mode) or in some special applications directly reconstructed into tomographic images. For example, in a high-data-rate acquisition with a gamma camera, such as a first-pass cardiac study, we can store not only the spatial information (x, y) for each detected event but also a timing parameter (t), which can be used in subsequent reordering of the data into a series of time-dependent frames. Another example is the storage of two sets of spatial information $(x_1, y_1, z_1; x_2, y_2, z_2)$ in a positron camera system, where the reconstruction process will produce a backprojected image directly without the need for framing the raw data.

More commonly, however, in *frame mode* the spatial values (x, y) are stored directly into a 2D matrix (or frame) in the computer using analogue-to-digital converters (ADC). The analogue information from a gamma camera is digitised in such a way that the storage location for the detected event is determined (figure 6.53). The data are then stored in the computer memory if a suitable z pulse is available.

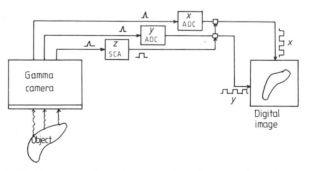

Figure 6.53 The use of analogue-to-digital converters to read x,y positional information from a gamma camera into a computer matrix.

A single frame or image is usually stored in a matrix where the pixel size (picture element size) matches or is somewhat smaller than the desired image resolution (see Chapters 4 and 12 on sampling). For instance, if a 40 cm diameter gamma camera is used to image the body, a 64 × 64 matrix will give a linear sampling size of 0.625 cm. This is adequate for data storage from a detector with a spatial resolution of 15 mm (2–3 times this value). For higher resolution (about 5–6 mm), the image should be stored in a 128 × 128 or 256 × 256 array. For small organs, such as the heart or thyroid, a 32 × 32 matrix is often used if the number of image counts is low. In determining the size of the array, it should be remembered that, for a multiframe data acquisition, a large memory is needed for storage. For example, a typical tomographic acquisition may have 64 frames each containing 64 × 64 pixels, needing more than 0.5 Mbytes of computer memory in a standard 16-bit nuclear medicine computer.

The data acquisition hardware must be supplemented by sophisticated acquisition software that will allow a range of parameters to be varied to suit the study being performed. An example of the types of acquisition procedures provided by a commercially available nuclear medicine computer system is shown in table 6.10.

Table 6.10 Types of acquisition procedures for a gamma camera/computer system.

Type	Example
Static	Single or dual isotope
Dynamic	List or frame mode
Tomographic	Single or dual isotope

6.5.2 On-line data correction

With the advent of cheap, flexible and powerful microprocessors, it has become possible to correct acquired data on-line to remove some of the inadequacies of gamma-camera performance. In particular, the continued use of photomultipliers (PM) requires careful and rapid automatic monitoring and subsequent correction to 'iron out' errors in image production caused by the instability of these devices. The most obvious of these are the necessary corrections for variations in pulse-height gain of the PM with time or temperature and the differences between even well matched PM in pulse-height response to scintillation light.

Hence, the most important feature is the ability to monitor PM pulse-height drift and to correct for this drift by altering the applied high voltage on a timescale much shorter than that over which the image is acquired. Once photomultiplier gains have been stabilised, it is possible to correct for spatial non-linearities by imaging specially designed linearity phantoms (see §6.8.4) and determining a correction matrix, which is then applied to each acquired event. A final example is the use of a pre-acquired uniform flood image to correct for image non-uniformities caused by collimator, crystal and light-guide imperfections.

This uniformity correction process may be performed during or after data acquisition.

6.5.3 Data processing

An example of the types of processing procedure provided by a commercially available nuclear medicine computer system is shown in table 6.11. This area of computer application can be subdivided into several categories. The first of these is the production of new images by applying complex *data reconstruction* techniques such as filtered back-projection (FBP) (see Chapter 4), as in SPECT or PET. The second is the processing of images either to remove statistical noise or to enhance a particular image feature such as the edge of a structure. These are usually known respectively as *smoothing* and *filtering*. Thirdly, both single-frame and multiframe acquisitions can be processed using appropriate software to provide *numerical* or *curve data*, which can add to the functional diagnostic features of the images themselves.

Table 6.11 Types of processing procedures for a gamma camera/computer system.

Type	Example
Data reconstruction	Filtered backprojection
Image enhancement	Smoothing or filtering
Image analysis	Regions of interest and time–activity curves
Image manipulation	Rotation, minification, subtraction

If we consider each of these categories of data processing in terms of the computer needs, we will see that there are requirements for high-speed complex mathematical processes that stretch even the modern mini/microprocessor to its limits.

The processes involved in FBP require the repeated use of algorithms to apply 2D Fourier transforms and various other mathematical manipulations to large data files. A successful *reconstruction program* will provide multislice tomographic images in a matter of 10–20 min. Often the requirement for operator interaction can extend the image production process more than the computer processing. The production of 3D volume images requires reconstruction times of several hours with conventional computer techniques and the process becomes inappropriate for routine clinical use (see Chapter 14). Only by the use of parallel or array processors can, for instance, a 4 Mbyte 3D image be reconstructed in minutes and even then only if the system is configured so that the time for data transfer to and from the processor unit is minimised.

If we now consider the requirements for image processing to remove statistical noise or to enhance image features, the computational requirements are directly related again to the volume of data being handled. The simplest *image filtering techniques* in one or two dimensions require little sophisticated software or hardware if matrices in real space are used. Spatial smoothing with a nine-point filter is performed in real space by convolving the information stored in the pixel to be smoothed with weighted values of its nearest eight neighbours (figure 6.54(*a*)). The filter is applied to each pixel in the image in turn. If a very smooth image is required, the filter elements are made more equal and vice versa.

A 1D filter can be used for temporal smoothing of three adjacent frames (figure 6.54(*b*)). The relative values of the filter elements again decide the smoothness.

The nine-point 2D and three-point 1D filters can, of course, be extended to cover a larger number of frames or pixels, such as 25-point 2D or five-point 1D, for instance, at the expense of computing time.

Edge enhancement, the opposite of smoothing, can be achieved by having negative values in the filtering matrix (figure 6.54(*c*)). This filter is useful for delineating organ boundaries in images.

More complex image filters can be carried out in Fourier (or frequency) space using a range of algorithms to remove or damp out unwanted spatial frequencies from an image. Whereas real-space filters discussed above are applied by convolution, frequency-space filters require simple multiplication (Brigham 1974) (see also Chapter 12). Two commonly used examples in nuclear medicine are the Hanning filter

$$A(f) = \begin{cases} 0.5[1 + \cos(\pi|f|/F)] & |f| \leq F \\ 0 & |f| > F \end{cases} \tag{6.34}$$

and the Butterworth filter

$$A(f) = [1 + (|f|/Q)^p]^{-1/2} \tag{6.35}$$

where $A(f)$ is the amplitude of the filter at a spatial frequency f. The values of the parameters F, Q and p are selected by the operator: F is called the cut-off frequency, Q is a frequency that controls roll-off and p is the power factor, which together with Q determines the Butterworth filter shape. If F or Q is high, the filter is 'sharp', allowing high-frequency noise and sharp edges in the image. For low values of F or Q, the image becomes smooth. Similar filters are used to temper reconstruction processes (see §6.7.2).

The application of frequency-space filters requires the image data to be transformed to and from Fourier space before and after the filtering operation—fast Fourier transform algorithms are necessary for these operations (see Chapter 12).

Examples of the application of both real- and Fourier-space filters are shown in figure 6.55. In most cases, although the intrinsic spatial resolution of an imaging system may be adequate, poor image statistics require the image to be smoothed and the spatial resolution is necessarily degraded at the expense of noise removal. More sophisticated image processing techniques, for instance, to remove the point-source response function from 2D and 3D images (Webb *et al* 1985b, Yanch *et al* 1987) are discussed elsewhere (Chapter 12).

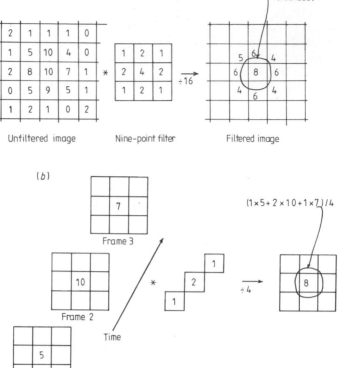

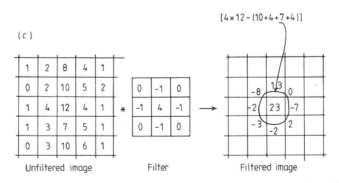

Figure 6.54 (*a*) An illustration of how a nine-point smoothing filter is applied to image data. (*b*) An illustration of how a one-dimensional three-point temporal smoothing filter is applied to sequential frames. (*c*) An illustration of an edge-enhancement filter applied to image data.

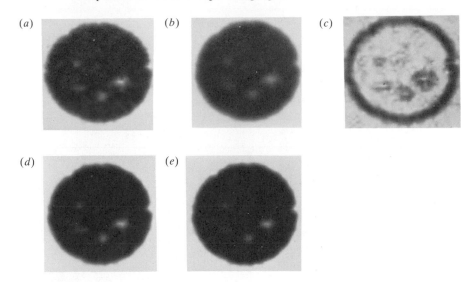

Figure 6.55 The effects of spatial filtering on a radionuclide image. (*a*) The original noisy SPECT image of a phantom containing six 'cold' spheres. (*b*)–(*e*) The results of performing the following filtering on image (*a*): (*b*) a nine-point smoothing filter; (*c*) an edge-enhancement filter; (*d*) a Butterworth filter with $Q = 0.5$ cm^{-1}, $p = 12$; and (*e*) a Hanning filter, with $F = 0.5$ cm^{-1}.

Although frequency-space operations are time-consuming for large data sets, even they are quick in comparison with real-space operations that filter images with large filter kernels. Real-space operations of this kind have been carried out to produce images, the significance of whose pixel values is precisely known in relation to their neighbours (Webb 1987). In these non-linear image processing methods, the value of a pixel is compared with the local average of pixels in neighbouring regions and may be replaced by such an average if its departure from the mean is not significant by some pre-set criterion. It can be shown that images have improved perceptual properties (see also Chapter 13).

The third category of data-processing features to be found on a nuclear medicine image-processing system involves the use of regions of interest (ROI) to produce *numerical and curve data*. If a sequence of frames is acquired during a dynamic study, compound images can be formed by addition (reframing) of groups of frames. If ROI are drawn around the images of organs, background, etc (figure 6.56(*a*)), the ROI counts in each frame can be used to produce 'activity–time' curves, which provide temporal information about the radiopharmaceutical distribution (figure 6.56(*b*)). Curve subtraction and fitting options provide for accurate parametric determination as used in both kidney and heart studies (see §§6.6 and 6.9). Parametric images are formed by putting in each pixel the value of a parameter determined from the temporal variation of that pixel (figure 6.57). Hence, several parametric images can compress a large amount of information into a relatively acceptable format.

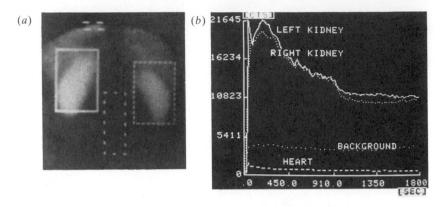

Figure 6.56 (*a*) Regions of interest (ROI) positioned on a summed image of a dynamic renal study. (*b*) Time–activity curves produced from the ROI shown in (*a*).

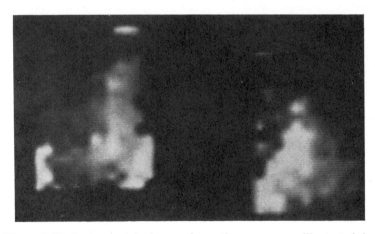

Figure 6.57 A parametric image from the renogram illustrated in figure 6.56 showing the distribution of the time to peak for the two kidneys.

6.5.4 Image display and manipulation

The end-product of data acquisition, correction and processing is the image, which must be displayed in a manner that best highlights the diagnostic information required. The most immediate advantage of digital over analogue image display is the facility to alter the display (or grey) levels to optimise the viewing of the full dynamic range of information in the image. Figure 6.58 shows how an image displayed with all its grey levels can be enhanced by windowing to pick out different grey-level regions. This helps to highlight high- and low-contrast areas separately when both may be of diagnostic value. Similar windowing procedures were discussed in the context of x-ray CT in Chapter 4 (§4.2.6).

(a)   (b)   (c)   (d)

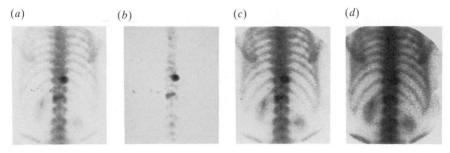

Figure 6.58 The effects of grey-level windowing on a digital image of part of the skeleton: (*a*) 0–100%; (*b*) 10–90%; (*c*) 0–50%; (*d*) 0–25%.

Further enhancements can be achieved by careful selection of colour scales. If adjacent grey levels are now represented by high-contrast colours, a contouring effect can be achieved, which will both highlight areas of low/high uptake and may be used to give a first-order (roughly 10%) quantification of an image directly from the display.

Important display features allowing optimal image viewing are overall image size and the number of pixels used for display. Although a 64 × 64 pixel image (with 6 mm pixel size) adequately describes a typical SPECT image (where spatial resolution is 15–20 mm), the viewer will find the pixel edges disturbing and this may detract from the diagnostic value (figure 6.59(*a*)). Use of 128 × 128 or 256 × 256 arrays provides image quality certainly qualitatively as good as the best analogue images produced in radioisotope imaging (figures 6.59(*b*) and (*c*)).

The ability to rotate images, to provide multi-image display and to add textual information is now an accepted and important feature of image display. In particular, the viewing of dynamic or tomographic images is enhanced by a 'snaking' display (figure 6.60), or by a ciné display, where images are shown sequentially at varying display rates to simulate temporal changes or spatial rotation, respectively.

6.5.5 *Data storage*

The large volumes of data produced during dynamic and tomographic acquisitions require modest computer memory for short-term storage, but considerable hard-disc space for semipermanent storage. In addition, processing will produce further large data files, requiring more computer memory and disc space. A standard SPECT study produces a 0.5 Mbyte acquisition data file and probably similar-sized processed multislice data files. A multiframe dynamic acquisition can be even more demanding of storage space.

(a)   (b)   (c)

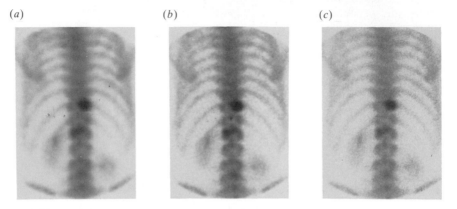

Figure 6.59 The effects of number of pixels on the digital image shown in figure 6.58: (a) 64 × 64; (b) 128 × 128; (c) 256 × 256.

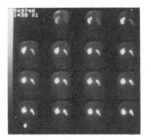

Figure 6.60 A snaking display of multiple images, which are part of a renogram study.

A typical minicomputer used for nuclear medicine imaging will have a minimum 0.5 Mbyte memory and a hard-disc system of between 10 and 100 Mbytes. For a permanent archive of data, floppy discs with a capacity of 0.5–1.0 Mbytes, are somewhat limited and usually magnetic tape is the medium of choice with a storage capacity of 10–20 Mbytes or more. Floppy discs are more appropriate for speedy movement of data files from one computer to another, whereas magnetic tapes are for low-speed, permanent back-up of data.

No doubt in the future video tapes and optical discs will play a more important role in the data transfer/storage process of medical imaging (see also Chapter 14).

6.5.6 *System control*

The availability of low-cost, powerful microprocessors has increased the role of computers in system control. In radioisotope imaging, this falls into several categories, as now discussed.

Gamma-camera gantry control has been necessary for the provision of multiview data acquisition. The gantry control provides rotation needed for transaxial SPECT imaging, translation for planar scanning, and translation or rotation as necessary for longitudinal tomography. In all cases, positional information from the gantry must be monitored and stored by the computer. This information is used to control the camera position during data acquisition and to provide correct spatial and angular information required for data processing.

Patient couch control is necessary for some scanning camera systems and is particularly important for tomographic imaging. If circular gamma-camera orbits are used in SPECT, accurate patient positioning using the couch is necessary to optimise the multiview information and to ensure that at no point does the region of interest move out of the field of view (figure 6.61(*a*)). Elliptical gamma-camera orbits (figure 6.61(*b*)) allow the camera head to be placed optimally close to the patient at all angles and hence maximise spatial resolution. This process is best achieved by computer control of the patient couch during acquisition, so that the lateral and vertical couch motion is closely synchronised with the gantry rotation.

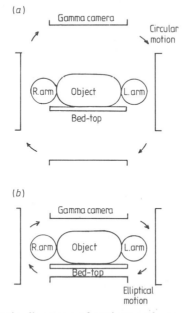

Figure 6.61 Schematic diagrams of patient and camera positions for a tomographic study with (*a*) a circularly orbiting camera and (*b*) an elliptically orbiting camera.

Gamma-camera control to provide a totally 'digital' gamma-camera system is now the main thrust of most commercial manufacturers. Here, several microprocessors are used to provide acquisition set-up, image display, data processing, gantry/couch control and data storage. The analogue data from the gamma-camera photomultipliers are digitised right at the camera head to minimise noise effects. No analogue images are available, but are replaced by high-resolution (256×256 or 512×512) digital images. This total control package has several other advantages, including the provision of multiple correction procedures for a range of radioisotope/collimator/energy window parameter combinations, and the optimisation of image display and processing features by choice of dedicated processors that communicate via a common database.

6.5.7 Summary

The use of computers has both improved image quality and enabled the extraction of quantitative functional information from radionuclide imaging. Further aspects are taken up in Chapter 14. Computers are now an essential part of clinical diagnostic nuclear medicine.

6.6 STATIC AND DYNAMIC PLANAR SCINTIGRAPHY

Conventionally, the most common radioisotope image used in clinical diagnosis is the planar scintigram. With the increased use of on-line computers, dynamic scintigraphy is now playing an ever-increasing role, particularly in the evaluation of heart, lung and kidney function. In this section we discuss the basic principles of planar scintigraphy and give simple examples of its use in nuclear medicine. Specific clinical applications are described in §6.9.

6.6.1 Static planar scintigraphy—basic needs

The requirements for the production of the simplest form of analogue radioisotope image include the following: a modern large-field-of-view gamma camera mounted on a flexible electromechanical gantry; a range of collimators to cover all clinical needs; and a high-quality analogue image production/storage system. This equipment, allied to a range of readily available radiopharmaceuticals, will provide the basic analogue radioisotope image—'the x-ray film' of nuclear medicine. The properties of the gamma camera and collimators have been described in §6.3.6, so here we will comment only on the additional features mentioned above.

The *electromechanical gantry* must be capable of rapid and accurate movements to allow the gamma camera to be raised, lowered or tilted in any direction. This will allow images of any part of the body to be obtained without physical discomfort to the patient. Additionally, aspects of safety must be included in gantry design to make them 'fail-safe'—they are, after all, supporting a weight of about 500 kg. Examples of commercially available gantries are shown in figure 6.62, indicating the enormous range of ideas on how to make the system both easily movable during setting-up and yet stable during imaging.

(a)

(b)

(c)

Figure 6.62 A selection of electromechanical gantries used with gamma cameras: (*a*) Siemens Pho/Gamma IV with collimator changer; (*b*) Nuclear Enterprises whole-body imaging facility; (*c*) GE 400AT with counterbalanced gantry capable of 360° rotation for tomography.

The production of high-quality *analogue images* has been, until recently, the central feature of the scintigraphic system. Most modern gamma cameras will be supplied with a formatting photographic system enabling many images to be produced on a single x-ray film (4–16 typically). Polaroid film is still used as an instant record or back-up, but on the whole provides a poorer dynamic range for information storage when compared with x-ray film (figure 6.63). The most difficult part of analogue imaging is the production of good-quality hard copy. The problem lies in selecting correct intensity settings, so that the wide count-density range in the image is satisfactorily transferred to film. It is

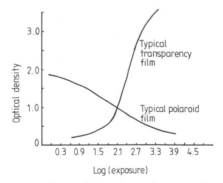

Figure 6.63 A comparison of the dynamic range for an x-ray film and a polaroid film.

here that analogue imaging (figure 6.64) must give way to new digital systems that allow post-imaging grey-scale manipulation to optimise the information on the film and to take advantage of the full dynamic range available (figure 6.58). Here the role of the computer can be seen to make even planar scintigraphy more quantitative, whereas in the past the modality has tended to be qualitative and subjective.

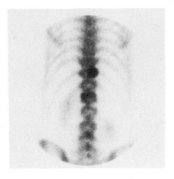

Figure 6.64 An analogue image of the study already shown in the digital images in figure 6.58. Note the lack of contrast in some regions when compared with the digital images where the full dynamic range becomes available.

The availability of suitable radioisotopes is an important feature of all nuclear medicine imaging, and has been considered previously in §6.4. In addition, the role of the pharmaceutical will be highlighted in §6.9, in the discussion of specific clinical applications.

6.6.2 Static planar scintigraphy—important parameters

The optimisation of image quality in static scintigraphy depends on several parameters, and errors in any of these can seriously degrade the information content and hence the diagnostic value of the image. For instance, the choice of a *collimator* and *photopeak window* to match the γ-ray emission from the radioisotope used are obviously important in image production. The field of view of the collimator must also match the size of the object to be studied and the collimator efficiency should be chosen to provide a high enough count rate (possibly at the cost of spatial resolution) to give satisfactory information density in the image.

The *time after injection* can be an important factor if the radiopharmaceutical is required to clear from some tissues before the target organ can be satisfactorily imaged. For example, a 20 min wait time is sufficient for liver/colloid imaging, whereas 2–3 h is necessary for a polyphosphate to clear the soft tissues sufficiently to allow skeletal imaging.

In order to obtain sufficient information about the physical location of abnormalities, it is usually necessary to take *several views* of the object from different angles. For brain scintigraphy, a minimum of four views (anterior, posterior, left and right lateral) are required and are obtained by placing the head as close to the collimator as possible, with the brain well inside the usable field of view. Hence, immobilisation and accurate *patient positioning* are important considerations. The *duration time* of the study is usually determined by a compromise between the need to keep the patient immobile and gathering enough counts to give a satisfactory image.

The film used to record an analogue image has a characteristic relationship between the count density and the film exposure (the *characteristic curve* of the film (see Chapter 2)), which determines the minimum total counts required per study. This total count will also depend on the size of the object and the distribution of the radiopharmaceutical. Typically a single-view image may take 5 min and contain 100 000–500 000 counts, but variations from organ to organ and with radiopharmaceutical can extend outside this range.

Clearly an improper choice of *exposure time* or count density can negate the value of the analogue image, problems which, as we have seen in §6.5, are overcome by the use of a computer to create digital images.

6.6.3 Dual-isotope imaging

As we have said before, one of the major limitations of planar scintigraphy as a diagnostic technique is the lack of specificity of radiopharmaceuticals, which leads to poor definition of the function of the target organ. However, the availability of more than one radiopharmaceutical to image the different functions of any region of interest (ROI) means that we may improve image definition by the use of dual-isotope imaging. In its simplest form, images are obtained of the distribution of two radiopharmaceuticals labelled with different radioisotopes and by a comparison or subtraction of the two images a clearer image of target-organ function is obtained. The philosophy is not unlike that underlying digital subtraction techniques in x-radiology (Chapter 2). There are several clinical examples of the use of this technique, which will be discussed later. However, we use one here to indicate the potential value and highlight the problems inherent in the image comparison/subtraction process. The example chosen is the use of ^{67}Ga citrate and ^{99}Tc^m colloid to image liver function and enhance the detection of space-occupying disease. The former agent is metabolised by both normal and abnormal liver, whereas the latter is not taken up by tumours in the liver. Hence, a subtraction of two images, taken simultaneously to minimise movement artefact, should show the liver tumours more clearly than either agent on its own. Two or more single-channel analysers are required to produce separate images and normalisation is performed using count densities obtained from regions of normal uptake in the liver. If A and B are the counts per pixel in the ^{67}Ga and ^{99}Tc^m images, respectively, then the composite pixel count C is given by

$$C = A - kB \qquad (6.36)$$

where A is made up of the pixel count due to normal liver (A_{NT}), the pixel count due to tumour (A_T), and the pixel count due to over/underlying tissues (A_B), and B is similarly made up of B_{NT}, B_T and B_B. If we assume $B_T = 0$ (i.e. no colloid uptake in tumour) and k is chosen so that

$$A_{NT} + A_B \simeq k(B_{NT} + B_B) \qquad (6.37)$$

then

$$C \simeq A_T. \qquad (6.38)$$

Unfortunately, in nearly all cases, the assumptions made above are incorrect both physiologically and physically. It is not surprising that the normal tissue/liver distributions of the agents are different. In this case, images are taken 48 h or more after the ^{67}Ga citrate administration but only 15–20 min after the ^{99}Tc^m colloid administration. Also the physical attenuation of the photons emitted from two isotopes is quite different and this means that k is depth-dependent. Figure 6.65 shows that, in the case of ^{131}I and ^{99}Tc^m dual-isotope imaging, subtraction can only be correct at one depth in the body, implying over/under subtraction of data obtained from other depths. This can lead to the production of serious image artefacts (figure 6.66), which completely negates the objective of the technique.

6.6.4 Quantification in planar scintigraphy

Single planar views of an object represent a 2D superposition of 3D information, and the image of the target organ may be obscured and confused by under/overlying tissue activity. Furthermore, the spatial resolution of a gamma camera gets poorer rapidly with distance from the collimator, and hence any region on a planar image is a complicated function of the object radioactivity distribution. The counts in an image from any ROI will be depth-dependent (figure 6.65) and any quantitative information will be totally dominated by this effect. As attenuation correction of a single view is impossible (unless the depth of the organ is known), quantification is only possible if attenuation is negligible or small. An example of this is the measurement of the uptake of $^{99}Tc^m$ pertechnetate in the thyroid, where the organ is covered by only a few centimetres of tissue and the attenuation correction is small (only 0.8–0.9 photon attenuation).

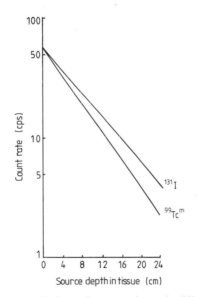

Figure 6.65 The transmission of γ-rays through different thicknesses of tissue for photons from ^{131}I and $^{99}Tc^m$.

In the case where relative organ measurements are needed, such as divided kidney function, depth corrections are possible by measuring the position of the organs in the body using lateral views. Minimisation of correction errors can be done by ensuring that the organs are as close to the collimator face as possible—hence, kidney images are made using the posterior view. It is obvious (figure 6.67(a)) that a lateral image can be helpful in determining the depth correction required for such small organs, but for a large organ, such as the liver, this is of little help. In

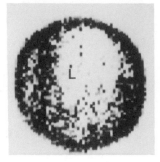

Figure 6.66 The production of image artefacts by subtraction in the dual-isotope technique. The image shown is the result of a subtraction of ^{131}I antibody and ^{99}Tc^m-HSA images and shows a ring due to ^{131}I· scatter plus a band of apparent activity corresponding to the arm/body interface. L is the lung of the patient.

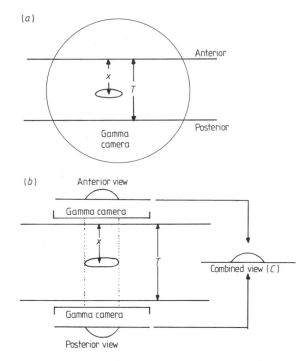

Figure 6.67 (*a*) The use of the lateral view for depth corrections to reduce the effect of photon attenuation. Here x is organ depth and T is patient thickness. The depth corrections are proportional to $\exp(\mu x)$ for anterior image and to $\exp[\mu(T - x)]$ for posterior image. (*b*) An illustration of how the combination of two opposed gamma-camera views reduces the effect of photon attenuation. Here x and T are the same as in (*a*). If the geometric mean is used in the combined view, then $C = \{kA \exp(-\mu x) \; kA \exp[-\mu(T - x)]\}^{1/2}$, i.e. $C = kA \exp(-\mu T/2)$.

the latter case, if it is possible to obtain anterior and posterior images without moving the patient (e.g. using a rotating gantry), then attenuation problems may be minimised and we can make crude quantitative measurements of organ function.

If we consider a point source of activity A at a depth x below the anterior surface of the body, which is of thickness T (figure 6.67(b)), then the counts recorded in the anterior and posterior views will be given by

$$C_A = kA \exp(-\mu x) \tag{6.39}$$

and

$$C_P = kA \exp[-\mu(T-x)] \tag{6.40}$$

respectively, where k represents a calibration factor or sensitivity figure for the gamma camera, and μ represents the linear attenuation coefficient for photons in the body. If the anterior and posterior images are summed pixel by pixel, forming an arithmetic mean (AM) image, then the counts recorded in the combined image will be given by

$$C_{AM} = \tfrac{1}{2}kA\{\exp(-\mu x) + \exp[-\mu(T-x)]\} \tag{6.41}$$

which is less dependent upon source depth than C_A or C_P, but does not, of course, totally remove the dependence on x (figure 6.68(a)). If, however, the geometric mean (GM) is used when combining the images,

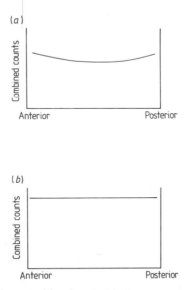

Figure 6.68 (a) The combined counts from opposed views using arithmetic mean, as a function of depth in tissue for a point source. The curve shown is proportional to $\exp(-\mu x) + \exp[-\mu(T-x)]$. ($b$) The combined counts from opposed views using the geometric mean, as a function of depth in tissue. The curve is proportional to $\exp(-\mu T/2)$.

then

$$C_{GM} = (C_A C_P)^{1/2} = kA \exp(-\mu T/2) \qquad (6.42)$$

which is totally independent of source depth (figure 6.68(*b*)). It is important, however, to remember that this result only holds for the situation where the attenuation coefficient is held constant. The technique of combining opposed views to achieve a depth-independent response is known as isosensitive or quantitative scanning. It was used originally with double-headed rectilinear scanners (Hisada *et al* 1967) but has since been used with the gamma camera (Graham and Neil 1974, Fleming 1979). A further property of an image formed by combining opposing views is that the spatial resolution in the image can be made to be depth-independent (figure 6.69). This is an important feature, which is utilised to good advantage in SPECT (§6.7). The geometric mean is also dependent only on the total path length T for a uniform source of activity. For all other source distributions, say including a tumour at depth x, the geometric mean is *not* independent of depth x.

The limitations here, of course, are that first μ is not constant throughout the body and, secondly, the superposition of information may still obscure the target organ or region. For more accurate quantitation, the removal of counts from over- and underlying structures via ECT is required.

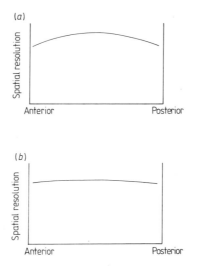

Figure 6.69 (*a*) The effect of using the arithmetic mean to combine opposing views on the spatial resolution as a function of depth in tissue. (*b*) The effect of using the geometric mean to combine opposing views on the spatial resolution as a function of depth in tissue.

6.6.5 High-energy photon emitters

Even though more than 90% of radioisotope images are now obtained using $^{99}Tc^m$-labelled radiopharmaceuticals, there is still a number of very important applications using radioisotopes emitting higher-energy photons. However, the modern gamma camera, which has been optimised for $^{99}Tc^m$ imaging, is often ill-equipped to cope with them. The most important of these radioisotopes are ^{131}I, ^{67}Ga and ^{111}In. Although the major role of ^{131}I now is as a therapy agent, iodine labelling of proteins is becoming increasingly important, with an enormous potential for nuclear medicine. This will be discussed in §6.9. In addition, techniques for chelating ^{67}Ga and ^{111}In to antibodies have been developed and these complexes show promise as specific disease-localising agents. These applications, as well as the more conventional uses of these radioisotopes, indicate a continuing demand for images obtained with medium- and high-energy photons.

An examination of the physical properties of these radioisotopes reveals potential problems for the production of images with a gamma camera. In particular, the energies of the photons emitted (see tables 6.5 and 6.7 in §6.4.2) are, in general, well above the optimal sensitivity range of the modern camera (100–200 keV). Hence, detection efficiency is low (about 25% at 300 keV (figure 6.5)) and is made worse by the difficulty of collimating photons of energy greater than 200 keV. The penetration capability of photons of energy 360 keV can be estimated from the half-value layer (for description, see Chapter 3), which for lead is 0.24 cm, compared with 0.04 cm at 140 keV. Taking the criterion of 5% for the level of acceptable penetration (Anger 1964), the septa of a collimator suitable for ^{131}I imaging should be 1.0 cm thick. This is quite impracticable and, even if the collimator depth is increased, photons can still readily penetrate the 0.3–0.4 cm septa actually used. The result of this septal penetration is a considerable degradation of spatial resolution (2–3 cm typically), with associated image blurring. Finally, on top of these intrinsic problems, the decay characteristics of these radioisotopes (especially their long half-life and, in the case of ^{131}I, the existence of β emission) are such that the radiation dose per unit injected activity is much higher than for equivalent levels of $^{99}Tc^m$. This usually limits the administered activity levels to less than 100 MBq, and the count rates obtained consequently lead to poor image quality.

Hence, the information content of images made with high-energy photons is both statistically poor and lacks spatial definition. Attempts to improve their diagnostic quality using ECT have proved difficult (§6.7).

6.6.6 Dynamic planar scintigraphy—general principles

The technique of dynamic scintigraphy has blossomed with the development of low-cost, high-power minicomputers. Prior to these, the production of a sequence of images spaced throughout a 20–30 min study could supply useful qualitative information about organ function on this timescale. Now, with a modern nuclear medicine computer, quantitative dynamic information is readily obtainable to determine the function of organs such as the heart, the lungs and the kidneys.

The basic technique involves the acquisition of data in frame mode (stored directly into x, y matrices) or list mode (stored sequentially as x, y, t values and framed off-line) to describe the temporal changes in radiopharmaceutical distribution *in vivo*.

For cardiac imaging, fast dynamic acquisition is required to produce up to 32 images of the heart during a single heart beat. By using the signal from an electrocardiogram (ECG) to add information from multiple beats to each of these images, high-count images can be obtained of the heart at different times during the cardiac cycle. For renal function measurements, slow dynamic acquisition is sufficient, as the renal uptake/excretion phase takes typically 20 min. In general, external respiratory gating is not used, as the movement artefact from respiration is small.

Acquired data are processed to produce 'activity–time' curves describing organ uptake/excretion rates. Images of extracted parameters (parametric images) can be formed to condense the information from the large amount of acquired data. The two examples mentioned above will be described in more detail now to highlight the parameters determined and the processes used.

6.6.7 Dynamic cardiac imaging

Cardiac function measurements can be obtained from either a multigated acquisition (MUGA) or from a first-pass study (Harbert and Da Rocha 1984b, Berger *et al* 1979). *MUGA studies* require ECG triggering, where the R-wave signal (figure 6.71), indicating the beginning of the systole phase (contraction of the left ventricle of the heart), is used to align the acquisition data from different cardiac cycles. Hence, a MUGA study will supply information about the cardiac function averaged over many heart beats during a period when the radiopharmaceutical (usually $^{99}\text{Tc}^{\text{m}}$-labelled red blood cells) is uniformly distributed in the blood pool. *First-pass studies* require rapid acquisition of data from the left and right ventricles during the initial period when the injected activity first passes through the heart. The total passage of the bolus injection through the heart requires several heart beats and individual measurements are made of function during each beat.

As the two processes are entirely different, they are considered separately.

6.6.7.1 Multigated acquisition studies.
If we consider the function of the heart as a two-stroke, four-chamber pump (figure 6.70), then in the diastole phase the two atria, right and left (RA and LA), contract and pump blood into the two ventricles (RV and LV), whilst in the systole phase the ventricles contract and pump blood to the lungs (from RV)

and body (from LV) and the atria expand taking blood from the body (to RA) and the lungs (to LA). The most important measurement obtained from a MUGA study is left-ventricular ejection fraction (LVEF), which determines the amount of blood entering the body arterial systems from the left ventricle averaged over many heart beats. The ECG signal (figure 6.71) can supply a clear trigger from end diastole, and this is used to align the acquired data into a series of up to 32 images or frames during each heart beat. The camera must be carefully positioned to optimise the image of the left ventricle and minimise information coming from other chambers (figure 6.72). This is usually made easier by the fact that the left ventricle has the largest muscle to cope with the task of pumping blood around the whole body.

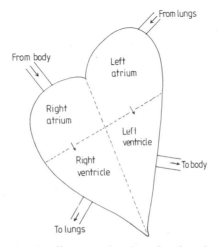

Figure 6.70 A simple diagram showing the function of the heart chambers.

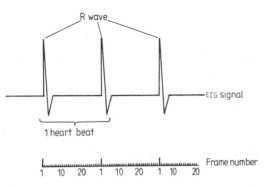

Figure 6.71 The use of an ECG signal to provide gating for the multiple-frame acquisition in a MUGA study.

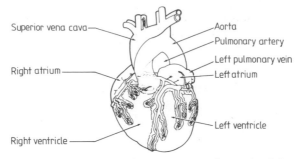

Figure 6.72 Schematic of the heart as seen from the left anterior oblique view used for a MUGA study. (Courtesy of IGE Medical Systems Ltd, Radlett.)

High count rates (up to 40 kcps) may prevent data from being stored immediately into frames, in which case a list-mode acquisition is used. Each image frame is typically a matrix containing 32×32 pixels (12 bits per word) although 64×64 (eight bits) or 128×128 (four bits) are possible.

Acquisition continues until an acceptable count density is reached in each frame (usually in excess of 100 heart beats). The data are then preprocessed, often by smoothing the images spatially and temporally. After this, ROI are either manually positioned or automatically determined to represent the left ventricle in each frame (figure 6.73(*a, b*)).

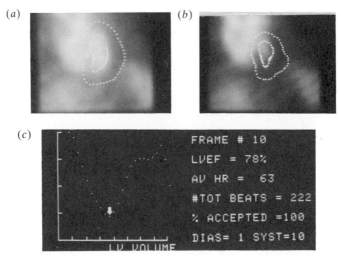

Figure 6.73 (*a, b*) Regions of interest used to outline the left ventricle (inner contours) of the heart at (*a*) end diastole and (*b*) end systole. (*c*) The variation of left ventricular volume as a function of time and its use to determine left-ventricular ejection fraction (LVEF).

Curves are then generated of LV volume versus time during each heart beat (figure 6.73(c)) and the LVEF is calculated as

$$\text{LVEF} = \frac{\text{end-diastole volume} - \text{end-systole volume}}{\text{end-diastole volume}} \times 100\%. \quad (6.43)$$

Other parameters can be obtained from a MUGA study and, in particular, parametric images of the amplitude and phase of the ejection from the LV can be made by fitting the LV volume curve with a single harmonic. These parametric images may show malfunction in the cardiac cycle caused by muscle, valve or septal problems.

In some cases, abnormal cardiac function is only seen if the heart is stressed and, hence, a repeat measurement of LVEF is made after stressing the patient (carefully!).

6.6.7.2 First-pass studies. Such studies always require list-mode acquisition since very high initial count rates are needed to give acceptable statistics. The passage of the injected activity is tracked through the right and left ventricles (RV and LV) during the first few heart beats. Data are acquired with the camera positioned to optimise the view of both ventricles simultaneously as well as the superior vena cava (SVC) (figure 6.74). The data are reordered into frames using the timing information and then the frames are summed to provide a composite image from which the positions of SVC, RV and LV can be determined (figure 6.75(a)). Functional curves of the passage of the radiopharmaceutical through these regions are then derived from the dynamic data (figure 6.75(b)). The SVC curve determines the input function of the activity entering the heart and this should be less than 2.5 s wide to produce acceptable results (i.e. the injection must be a small bolus injected quickly). Frame markers are used to produce curves of the function of the two ventricles during the first pass of the radioactivity before recirculation spreads out the temporal information obtained. By careful positioning of frame markers, the LVEF for each heart beat can be obtained (figure 6.76) free from effects of background, cardiac arythmia, etc. In comparison with a MUGA study, this process is difficult to do well and statistically the data are usually of poor quality.

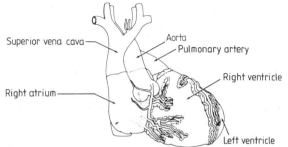

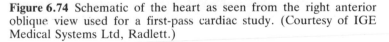

Figure 6.74 Schematic of the heart as seen from the right anterior oblique view used for a first-pass cardiac study. (Courtesy of IGE Medical Systems Ltd, Radlett.)

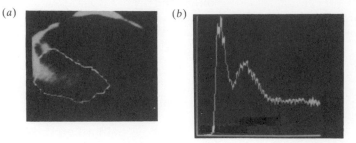

Figure 6.75 (*a*) Summed image from a first-pass study showing the contribution of the superior vena cava, the right ventricle and the left ventricle, and a region of interest used to produce (*b*) the time–activity curve for the ventricles.

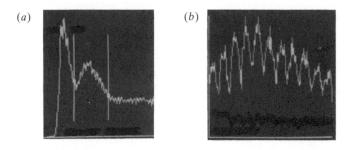

Figure 6.76 (*a*) Frame markers used to define the left-ventricular portion of the study shown in figure 6.75 and (*b*) a left-ventricular volume versus time curve for this study.

6.6.8 *Dynamic renal imaging*

An example of a 'slow' dynamic study is the use of multiple temporal images of the kidney obtained with an appropriate radiopharmaceutical to produce renograms—time–activity curves representing kidney function (Britton and Brown 1971). The kidney extracts urine (containing low-molecular-weight waste products) from the blood in the renal arteries and excretes it via ureters to the bladder. The urine is then passed out of the body from the bladder during micturition.

The three major phases of kidney function (figure 6.77) seen in a renogram are perfusion (0–30 s after injection), when the kidney is perfused by the radiopharmaceutical; filtration (1–5 min after injection), during which the radiopharmaceutical is filtered from the blood supply; and excretion (more than 5 min after injection), when the radiopharmaceutical in the urine is passed to the bladder.

Perfusion is an important indication of the function of a transplanted kidney, when a sequential improvement of perfusion after the transplant can provide evidence of success—poor perfusion often indicates tissue rejection. The glomerular filtration rate (GFR) in absolute terms or in relative terms (when the two kidneys of a patient are compared) can be a measure of the intrinsic health of the kidneys. Abnormal function in the excretion phase can indicate problems in the collection system, obstruction of the ureters or reflux (return of urine from bladder to kidney during micturition), any of which may lead to loss of kidney function, pain, infection and ultimately the need for transplantation.

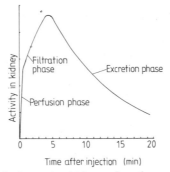

Figure 6.77 Typical dynamic kidney function curve showing the three major phases of a renogram study.

The *perfusion study* requires a rapid injection (bolus) of a suitable radiopharmaceutical and the kidney is imaged anteriorly providing up to 90 one-second frames each containing 64×64 pixels. ROI placed over the kidney, iliac artery and a suitable background area (figure 6.78) provide kidney and arterial function curves (figure 6.79) which can be corrected for background counts. The ratio of the rate of kidney perfusion to arterial activity gives a *perfusion index* (PI) (Hilson *et al* 1978), and sequential measurements of the PI can indicate success or failure of a transplant.

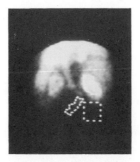

Figure 6.78 Summed perfusion images and regions of interest to outline the kidney, the iliac artery and the background.

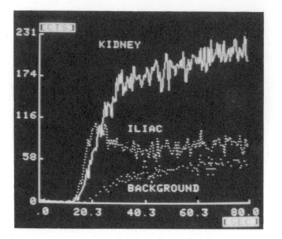

Figure 6.79 Time–activity curves for the regions of interest shown in figure 6.78.

A *renogram study* again requires a bolus of injected activity to be imaged, but this time both kidneys and the LV of the heart should be in the camera field of view. The posterior view is selected in order to minimise attenuation of photons coming from the kidneys. Typically 60–90 twenty-second frames each of 64×64 pixels are acquired. The total acquisition time is 20–30 min and up to 720 kbytes of computer memory are required. An additional study of the micturition phase may be required, again with a number of twenty-second frames. ROI analysis of the activity in both kidneys, the LV and suitable background region(s) is carried out (figure 6.56(*a*)). The LV activity–time curve (figure 6.56(*b*)) represents the spreading of the bolus by recirculation and provides the input function to the kidneys. This function must be removed from the kidney curves before estimations of GFR and excretion rates are made. The removal of the impulse curve is usually performed by deconvolution (Diffey *et al* 1976) or an appropriate parametric analysis such as fitted retention and excretion equations (FREE) (Hyde *et al* 1988). Abnormalities such as renal obstruction, dilated pelvis or ureters can often be diagnosed from the curves. More accurate information is gained from parametric images used to separate out the function of the uptake and excretion parts of the kidney. As with cardiac studies, these parametric images contain a wealth of information, which can often add considerably to diagnosis.

6.6.9 *Other dynamic studies*

Other applications of dynamic scintigraphy include studies of the brain, lungs, liver, hepatobiliary, oesophageal and vascular systems, and such studies can supply invaluable information in relation to abnormal function of these organs. Some of these applications will be discussed in §6.9.

6.7 EMISSION COMPUTED TOMOGRAPHY

In the previous section, it was shown that planar scintigraphy, which involves the production of a 2D projection of a 3D object distribution, is seriously affected by the superposition of non-target activity, which restricts the measurement of organ function and prohibits accurate quantification of that function. Emission computed tomography (ECT) is a technique whereby multi-cross-sectional images of tissue function can be produced, thus removing the effect of overlying and underlying radioactivity (Larsson 1980, Williams 1985, Croft 1986, Ott 1986).

The technique of ECT is usually considered as two separate modalities. *Single-photon emission computed tomography* (SPECT) involves the use of radioisotopes such as ^{99}Tc^m, where a single γ-ray is emitted per nuclear disintegration. *Positron emission tomography* (PET) makes use of radioisotopes such as ^{68}Ga, where two γ-rays, each of 511 keV, are emitted simultaneously when a positron from a nuclear disintegration annihilates in tissue (see §6.4.3). The technique of ECT can also be classified into two general types: limited-angle or transaxial. In *limited-angle* (or *longitudinal*) *ECT*, photons are detected within a limited angular range from several sections of the body simultaneously. In this case, the image planes that are reconstructed are parallel to the face of the detector(s) (figure 6.80). In *transaxial* (or *transverse-section*) *ECT*, the detectors move around, or surround, the body to achieve complete 360° angular sampling of photons from single (or multiple) sections of the body. Here the reconstructed image planes are perpendicular to the face of the detector(s) (figure 6.81). In both cases, when 3D data sets are complete, the data may be redisplayed in other orthogonal planes.

In the following sections we will consider the mechanisms involved in both limited-angle and transaxial SPECT and PET. The advantage of ECT over planar scintigraphy can be seen in the improvement of contrast between regions of different function, better spatial localisation, improved detection of abnormal function and, importantly, greatly improved quantification. In certain ECT systems, these advantages are obtained at the expense of spatial resolution.

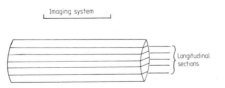

Figure 6.80 Schematic showing the direction of the imaging planes for longitudinal tomography.

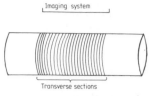

Figure 6.81 Schematic showing the imaging planes for transaxial tomography.

The basic reconstruction technique used in ECT is similar to that of x-ray transmission computed tomography (TCT) (Chapter 4). However, there are several important differences between ECT and TCT. In comparison with TCT, where the *in vivo* distribution of x-ray linear attenuation coefficients μ is determined, the reconstruction of ECT images is more complicated, since an attempt is made to determine the distribution of activity A in the presence of unknown μ. ECT images physiological and metabolic processes with a spatial resolution (ranging from about 5 mm for PET to 15–20 mm for SPECT) much poorer than the anatomical resolution of TCT (about 1 mm). The radiation dose associated with ECT is distributed throughout much of the body for the effective lifetime of the radiopharmaceutical, whereas in TCT the radiation dose is limited to the duration of the x-ray exposure and to the section of the body irradiated. Hence, if identical radiation doses were given to a patient, the photon utilisation in ECT is typically 10^4 times lower when compared with TCT.

One of the techniques used for ECT image reconstruction (filtered backprojection (FBP)) has already been described in Chapter 4. Hence, only a brief reference will be made to the technique of FBP in this chapter. However, the alternative method of iterative reconstruction, which has some advantages for ECT, will be described in more detail in §6.7.3.

6.7.1 Limited-angle emission tomography

Several different types of equipment have been developed for limited-angle or longitudinal-section emission tomography. The basic principle of these systems is the same: the radioactive distribution is viewed from several different angles, so that information about the depth of the radioactive sources can be obtained. The simplest form of longitudinal tomography is that obtained by a rectilinear scanner with a highly focused collimator (figure 6.82). This system produces an image that represents the radioactive distribution of the plane 'in focus' but also contains additional blurred information from regions above and below the focal plane. A more sophisticated system is Anger's multiplane tomographic scanner (Anger 1969), which was later marketed by Searle as the Pho/Con Imager (figure 6.83). This device combined the tomographic features of large-detector scanners with the Anger camera electronics, so that six longitudinal tomograms were produced from a single scan of the object. In order to bring each plane into focus, the images are repositioned electronically. Figure 6.84 shows multiple longitudinal sections through the skeleton produced by a Pho/Con Imager.

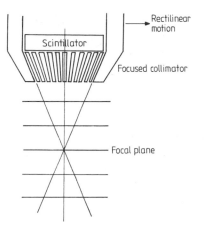

Figure 6.82 A rectilinear scanner used for focal-plane imaging.

The stereoscopic information required for limited-angle ECT can also be acquired using a gamma camera with special collimators (moving or stationary) (Freedman 1970, Muehllehner 1971, Rogers *et al* 1972, Vogel *et al* 1978). Various moving (usually rotating) collimators have been designed. These include slant hole, pinhole, diverging and converging multihole collimators. The stationary collimators fall into two categories: coded apertures (such as the Fresnel zone-plate, stochastic multi-pinhole and the random pinhole), which produce scrambled images that need to be decoded; and segmented apertures (such as the multiple pinhole and the quadrant slant hole), which produce separated images, each providing a different angular view of the object. The reconstruction (decoding) processes for coded-aperture imaging are fairly complicated and time-consuming. Segmented-aperture reconstruction is simpler.

Since insufficient angular sampling of the 3D object is obtained in limited-angle ECT, the tomographic images include blurred information from over- and underlying planes of activity. Methods of deblurring limited-angle tomograms have been developed using either deconvolution or iterative techniques (e.g. Webb *et al* 1978), but the depth resolution (effective slice thickness) is still poor compared with the in-slice resolution and usually gets worse with increasing depth. There are several other problems associated with limited-angle tomography: noise propagation occurs during the decoding process; compensation for attenuation can be complex; the decoding process is computationally costly; and distortions can occur, especially with pinhole systems. As a result of these problems, the tomographic images are essentially qualitative unless orthogonal views are used (Bizais *et al* 1983), and the technique of limited-angle tomography has essentially been replaced by transaxial systems, which, although not without their own problems, do offer the potential for quantitative imaging (in which the number of counts per pixel in a tomogram represents the radioactive concentration in the corresponding volume elements of the object). However, it should be remembered that, for the stationary-detector, limited-angle devices, dynamic ECT is possible, and in general the image times for limited-angle ECT are less than for transaxial ECT.

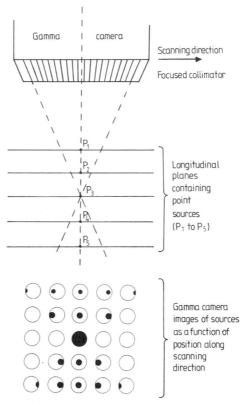

Figure 6.83 Multiplane longitudinal tomography using a Pho/Con Imager. Images of activity at five different planes in front of the focused collimator move at different rates across the detector. By compensating for these differences in motion for each plane, several longitudinal tomograms (i.e. in-focus images) are produced. (After Anger (1969).)

6.7.2 Single-photon emission computed tomography (SPECT) with a rotating gamma camera

If one or more gamma cameras are attached to a computer-controlled gantry (figure 6.85), which allows the detector(s) to be rotated around a

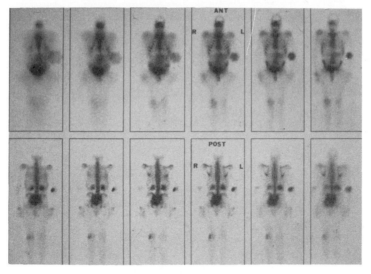

Figure 6.84 Twelve longitudinal tomograms of the skeleton recorded using a Pho/Con Imager.

supine patient, multiple views (or 2D projections) of the 3D radiopharmaceutical distribution can be acquired and stored on the computer. A gamma camera coupled to a parallel-hole collimator provides a set of 2D images consisting of multiple profiles, each profile representing a 1D projection of the radioactivity in a single slice of the patient (figure 6.86). Hence the 3D object is divided up into multiple 2D sections, and each section is represented by a set of discrete 1D profiles. Each point on the profile would represent the linear sum (in the absence of attenuation) of the activity elements along the line of view of the detector as determined by the collimator. If a large number of linear and angular data samples are taken, then it is possible to reconstruct cross-sectional images that represent the radiopharmaceutical distribution in the body.

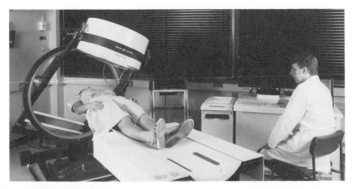

Figure 6.85 A rotating gamma camera/computer system for trans-axial emission tomography.

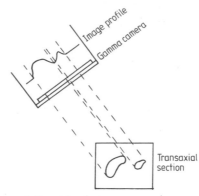

Figure 6.86 The relationship between an image profile and the projection of radioactivity in a single transaxial slice for a camera with a parallel-hole collimator.

Typically, a gamma-camera SPECT acquisition may consist of 64 planar views (or projections), each containing 64×64 image pixels and acquired at 64 discrete angles covering 360° around the patient. Unfiltered backprojection (BP) techniques can provide a blurred image, crudely representative of the object but containing spoke or star artefacts (see Chapter 4) (figure 6.87(*a*)). By the use of filtered backprojection (FBP), which involves modifying the information in each profile prior to backprojection, it is possible to remove the worst of these artefacts (figure 6.87(*b*)) and to provide transaxial, sagittal and coronal sectional images (figure 6.88). The important step of filtering each linear profile was considered carefully in Chapter 4, to which the reader is referred for details of the convolution kernels involved. However, several important problems inherent in ECT must also be considered if clinically useful multisectional images are required. In particular, the effects of the non-space-invariant point-source response function (PSRF), photon absorption, scatter and statistical noise require special attention.

As discussed previously, the PSRF or spatial resolution of the system varies with the distance of the source from the collimator, and it is necessary to combine opposite projections to minimise this effect. This also enables the *photon absorption problem* to be solved approximately. We have already seen (in §6.6.4) that, for a point source in a uniform attenuating medium, the combined response from two opposed detectors is independent of source position and has a simple dependence on the total attenuating path length between the two detectors. Hence, provided an outline of the body can be obtained, a simple correction can be applied to the combined opposed projections. The correction factor (CF) is given by

$$CF = \tfrac{1}{2}\mu T \exp(\tfrac{1}{2}\mu T)/\sinh(\tfrac{1}{2}\mu T) \tag{6.44}$$

where μ is the linear attenuation coefficient, which for this correction must be assumed to be constant (which, of course, it is not) and T is the total attenuation path length, determined from the patient outline.

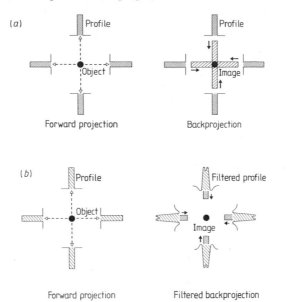

Figure 6.87 (*a*) Tomographic reconstruction using unfiltered back-projection and the production of star artefacts. (*b*) The removal of star artefacts using the filtered backprojection technique.

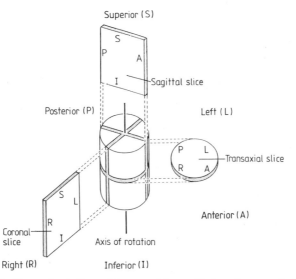

Figure 6.88 The orientation of transaxial, sagittal and coronal slices produced in tomographic imaging.

Equation (6.44) is derived by considering the attenuation within a uniformly distributed radioactive source. However, the correction factor is dependent on how the radioisotope is distributed in the object and, although the combination of opposed views helps to produce an approximately space-invariant PSRF, the application of equation (6.44) to provide a correction for photon attenuation provides only a simple approximation for a non-uniform radioisotope distribution and a non-uniform attenuation coefficient. A more accurate method of correcting for attenuation involves the use of iterative reconstruction techniques (§6.7.3). As will be seen later, the problem of attenuation correction in PET is easier, since two photons are detected in coincidence, and this leads to the technique being more accurately quantitative than SPECT.

The majority of SPECT attenuation correction methods require a determination of body outline. In some situations, a standard ellipse is used for all patient cross sections, while in others individual measurements may be made. A body outline can be determined either by placing a point source at the ends of the major and minor axes of the body, or by use of Compton-scatter data (§6.7.3) to provide a low-statistics body image (Jaszczak *et al* 1979b). A transmission scan offers the most accurate method, and this is used in studies with multicrystal PET systems, at the expense of increased imaging time and some increase in radiation dose to the patient. Because of the limitations of the simple attenuation correction procedures, the errors in patient outline determination are usually not the dominant errors.

Photon scatter in both the patient and the collimator gives rise to image blurring and can lead to inaccuracies in quantitation. With a standard 20% photopeak window, approximately 30% of counts in a ^{99}Tc^m image come from scattered photons (Webb *et al* 1986b). These events represent a serious problem in SPECT, particularly in making attenuation corrections more complicated. The simplest way of correcting for scatter is to use an empirical value of μ in the correction for photon attenuation (typically 0.12 cm^{-1} instead of 0.15 cm^{-1} for 140 keV photons of ^{99}Tc^m), thus preventing overcorrection of attenuation. This technique is, however, an oversimplified approach to the problem. Methods for removal of the scatter component from SPECT images usually involve either the subtraction of images taken with a 'scatter window' placed over the Compton part of the energy spectrum (Jaszczak *et al* 1984) or 2D/3D deconvolution of a PSRF either pre- or post-reconstruction (Yanch *et al* 1987). The former method is an oversimplistic approach and does not give an accurate correction for scattering. 3D deconvolution of a PSRF looks by far the most promising (figure 6.89), but this method would require a dedicated array processor for routine clinical use.

The *statistical noise* inherent in all radioisotope imaging poses a particular problem in ECT since the ramp filter used in FBP automatically amplifies the noise, which is predominantly at the high-frequency end of the spatial-frequency spectrum. Hence smoothing filters (§6.5.3) are used to reduce image noise at the expense of spatial resolution. This is shown in table 6.12, where the theoretical and measured spatial resolutions in an image of a point source are compared. The reasons why these do not always match are discussed by Webb *et al* (1983b).

(a) (b)

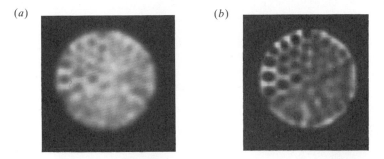

Figure 6.89 The use of 3D deconvolution of a point-source response function to remove the effects of scattered photons from tomographic images: (*a*) original tomogram; (*b*) after 3D deconvolution.

Table 6.12 Comparison of theoretical and measured spatial resolution for a rotating camera SPECT system. (Data taken from Webb *et al* (1983b).)

Reconstruction filter		Spatial resolution, FWHM of line-spread function (cm)	
Type	Parameters[a]	Measured	Theoretical
Ramp + Butterworth	$p = 10$, $Q = 0.5$	1.9	1.20
Ramp + Butterworth	$p = 10$, $Q = 0.4$	2.0	1.39
Ramp + Butterworth	$p = 10$, $Q = 0.33$	2.1	1.63
Ramp + Butterworth	$p = 10$, $Q = 0.25$	2.4	2.15
Ramp + Hanning	$F = 0.5$	2.4	2.36

[a] See equations (6.34) and (6.35) for definitions of p, Q and F.

A range of filters are usually provided within a tomographic software package and these allow pre- and/or post-reconstruction filtering as well as a choice of the smoothing process within FBP. Figure 6.90 shows the effect of four different reconstruction protocols and illustrates the inverse relationship between low noise (smoothness) and spatial resolution in the image.

The *sampling requirements* of SPECT need to be carefully determined in order to minimise statistical fluctuations. The number of projections required for ECT is more than $\pi D/2r$, where D is the object diameter and r is the image pixel size (note that this equation was also discussed (equation (4.27)) in the context of x-ray CT in Chapter 4). In addition, the linear sampling interval should be $(0.4-0.7)r$ (Huesman 1977). For a uniform disc of activity, this will lead to an image-count standard deviation σ given (Budinger *et al* 1977) by

$$\sigma(\%) = 120 \times (\text{number of pixels in object})^{3/4}$$
$$\times (\text{number of photons detected})^{-1/2}. \qquad (6.45)$$

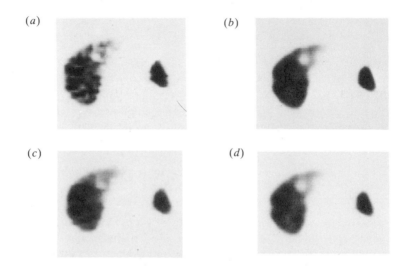

Figure 6.90 A single-slice liver tomogram produced using four different reconstruction protocols showing the effects of high-pass and low-pass filters on structures in the image: (*a*) ramp filter; (*b*) ramp + Butterworth filter ($Q = 0.4$ cm^{-1}, $p = 12.5$); (*c*) ramp + 3D median window filter; (*d*) ramp + linear 3D Hanning filter ($F = 0.83$ cm^{-1}). See §6.5.3 for definitions.

6.7.3 *Iterative reconstruction techniques in* SPECT

In the previous section, the production of SPECT images was discussed in terms of the reconstruction of multiview data using filtered backprojection (FBP). Alternative methods of image reconstruction, which may replace FBP in future with the rapid increase now available in computer power, are those based on iterative reconstruction techniques (Gilbert 1972, Herman *et al* 1973). These have already been introduced in the context of x-ray CT in Chapter 4. The basic method here is to make an initial image by, for example, setting each image pixel count to the mean pixel count (total counts/total number of pixels) and then to use an iterative procedure to alter this initial image gradually by comparing the resultant 'pseudo-projections' at each iteration with the actual raw (or experimental) data projections. In forming the pseudo-projections from each interim image, the effect of attenuation can be easily taken into account. By minimising the difference between pseudo- and true projections, it is possible to re-create an accurate image of the radioisotope distribution. A simple example of this process is the iterative least-squares technique (ILST) (Budinger and Gullberg 1974). If $A^n(i, j)$ is the radioactive concentration corresponding to image pixel (i, j), f_{ij}^θ is a factor accounting for geometry and attenuation, and $P_{k(\theta)}$ is the kth measured projection at angle θ, then the ray sum at angle θ ($R_{k(\theta)}^n$) after n iterations is given by

$$R_{k(\theta)}^n = \sum_{(i,j)\in k(\theta)} f_{ij}^\theta A^n(i, j). \tag{6.46}$$

(In equation (6.46), be careful to note that superscripts are labels, not powers. The symbol $\in$ means 'member of the set of' and $(i, j) \in k(\theta)$ means the summation is taken over all those pixels comprising ray $k(\theta)$.)

A_{ij} can be determined by minimising the value

$$\sum_\theta \sum_k [(P_{k(\theta)} - R_{k(\theta)}^n)^2/\sigma_{k(\theta)}^2] \tag{6.47}$$

where $\sigma_{k(\theta)}$ is the standard deviation of $P_{k(\theta)}$.

For high-count-rate data where $\sigma_{k(0)}^2$ is small, it may take many iterations to minimise this function and hence high-speed computing facilities are required. Note that the technique can automatically account for attenuation for each ray used in the image reconstruction, once again assuming uniform attenuation. For example, f_{ij}^θ can be given by

$$f_{ij}^\theta = f_g \exp(-\mu l_{ij}^\theta) \tag{6.48}$$

where f_g is a geometric factor and l_{ij}^θ is the attenuating path length between the pixels (i, j) and the object boundary at the angle θ.

Some workers have minimised the number of iterations required by using as the initial image the FBP reconstruction with no correction for attenuation. Obviously, the pseudo-projections (which allow for attenuation) formed from the initial image (which does not) will be different from the measured projections, but the difference will be smaller than if a uniform initial image had been used. This technique is known as iterative convolution (Walters *et al* 1976). The reconstruction time has been further reduced by a two-step procedure proposed by Chang (1978). Comparison of methods for attenuation correction in filtered backprojection and iterative methods have been reported (e.g. Webb *et al* 1983a).

6.7.4 Optimisation of SPECT acquisition and processing parameters

In order even to contemplate the use of SPECT in clinical nuclear medicine, there are several basic requirements without which it is impracticable to attempt to provide useful tomographic images. The gamma camera used must be extremely stable and have excellent uniformity and spatial linearity. Using the most modern camera fitted with an autotune and/or energy and linearity correction facilities, it is possible to minimise the production of image artefacts of the types shown in figure 6.91, which result from poor gamma-camera uniformity. Additionally, a choice of collimators that should include at least low-energy, high-resolution (LEHR) and low-energy, general-purpose (LEGP) collimators is necessary. There is presently some effort being applied to the manufacture of special-purpose SPECT collimators using asymmetric holes (Muehllehner and Colsher 1980), slant holes (Esser *et al* 1984), converging holes in the slice plane (Webb *et al* 1985a, 1986a) and biplanar converging holes (cone-beam tomography) (Jaszczak *et al* 1986a, b). As yet, special SPECT collimators have had little impact (Heller and Goodwin 1987).

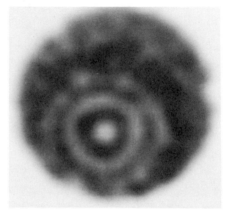

Figure 6.91 Circular artefacts in SPECT images caused by camera non-uniformity.

The camera gantry and patient couch are equally important—poor-quality mechanical/electrical gantry design has considerably hindered the development of SPECT and only recently have reliable systems become available. Finally, to complete the package, a high-speed computer system with a special-purpose SPECT software package, which includes quality control, reconstruction software, image processing and display facilities, is essential, preferably with the enhanced reconstruction features allowed by the addition of an array processor.

When there is available a state-of-the-art SPECT system, as described above, there are a whole range of acquisition and processing parameters, the choice of which will affect the quality of the final image, sometimes very stongly. The most important *acquisition parameters/ variables* are

(*a*) choice of collimator,
(*b*) radius of rotation,
(*c*) choice of energy window (position and width),

(*d*) number of acquired views,
(*e*) acquisition matrix size,
(*f*) angular range of acquisition,
(*g*) pixel size or image size,
(*h*) time per view, and
(*i*) administered activity levels.

Examples of the effect of these parameters are shown in figure 6.92.

The choice of collimator depends, obviously, on the radioisotope label being used and on the organ being studied. For a high-count-rate study (e.g. $^{99}Tc^m$-sulphur colloid in the liver), a low-energy, high-resolution collimator may be used; but for a low-count-rate study (e.g. $^{99}Tc^m$-pertechnetate in the brain), a higher-sensitivity collimator is preferable. The radius of rotation must be set to a minimum value, and the width and position of the energy window carefully chosen, as discussed in §6.3.6. Improved angular sampling can be achieved by increasing the number of views (which is typically 64) but at the expense of increased study time or reduced counts per view. Similarly, an increase in the matrix size above 64×64 is unreasonable given that the typical spatial resolution (approximately 18 mm) already spans three pixels for a 40 cm diameter camera. 180°-angle acquisitions have been shown to be valuable when imaging the heart (Coleman *et al* 1982, Tamaki *et al* 1982) and thyroid (Webb *et al* 1986a) but some image distortion results from this process. If the object to be imaged is much smaller than the field of view, zooming the projections can optimise the use of the computer matrix, but care in positioning the object is required to prevent movement out of the field of view during camera rotation. Finally, the longer the acquisition time per view, the more events are obtained but at the expense of possible artefacts from patient movement. A better solution would be to increase the administered activity, but this obviously has to be weighed against the consequent increase in radiation dose to the patient.

The most important *processing parameters/variables* are

(*a*) uniformity correction,
(*b*) centre-of-rotation correction,
(*c*) slice thickness/number of slices,
(*d*) choice of filter (pre-/post- and/or that used on FBP),
(*e*) reconstruction algorithm, and
(*f*) attenuation correction.

Examples of the effects of two of these parameters are shown in figure 6.93.

Non-uniformity of system response, whether from the camera or collimator, will show up as circular artefacts in the reconstructed image, as shown previously, unless removed by a uniformity correction using a high-count (approximately 3×10^7) flood source acquisition. These flood data must be acquired with the same parameters (such as photopeak window, zoom, collimator) as the clinical study.

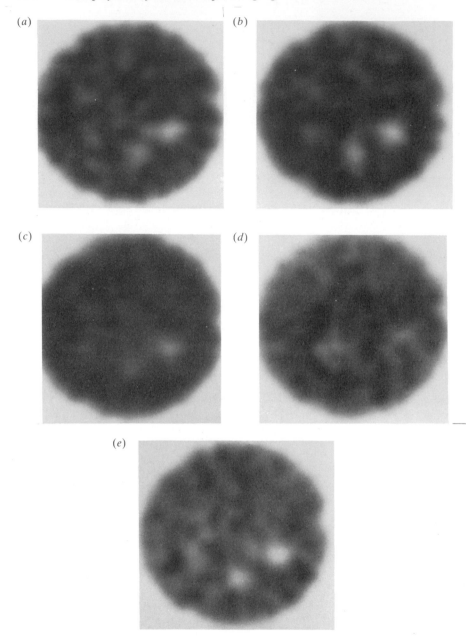

Figure 6.92 The effect of several acquisition parameters on a SPECT image of a phantom containing six 'cold' spheres. (*a*) The image produced using optimum acquisition conditions. (*b*)–(*e*) The effects of changing the acquisition parameters: (*b*) a general-purpose collimator compared with a high-resolution collimator in (*a*); (*c*) a radius of rotation of 34 cm compared with 18.5 cm in (*a*); (*d*) an off-centred energy window compared with a window centred on the photopeak in (*a*); (*e*) 32 acquired views compared with 64 in (*a*).

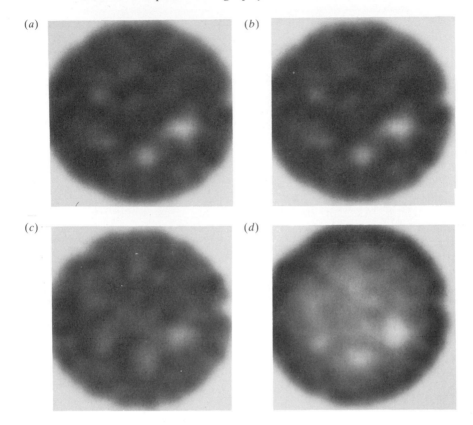

Figure 6.93 The effect of two processing parameters on SPECT images of the same phantom study illustrated in figure 6.92. (*a*) The image produced using standard reconstruction parameters. (*b*)–(*d*) The effects of changing the processing parameters: (*b*) a slice thickness of two pixels compared with one pixel in (*a*); (*c*) a slice thickness of four pixels; and (*d*) no attenuation correction compared with using the correction factor in (*a*) as defined in equation (6.44).

The centre of rotation used in the reconstruction process must be correct to avoid data misalignment and subsequent artefact production. A separate quality-control process to establish the correct centre-of-rotation coordinates is essential (see §6.8.8).

Large slice widths (in pixels) may be chosen to improve image statistics at the expense of partial-volume effects (see Chapter 4) and the subsequent loss of sensitivity to small abnormalities. The number of slices is usually chosen to fit the organ being imaged and varies from 10–15 for single-pixel slices of the brain to 20–30 for the liver.

Most modern SPECT software packages allow images to be filtered (smoothed) before, during and after FBP with a range of 2D/3D filters. The filter chosen is again study-dependent—a high(frequency)-pass filter being used for a high-statistics study but producing noisy images where acquired count rates are low.

The reconstruction algorithms of current commercial software packages vary very little from a standard FBP but some allow limited-angle reconstructions (e.g. 180°, 270°), which may be valuable in enhancing lesion contrast in some studies (Ott *et al* 1983a).

The use of an appropriate attenuation-correction algorithm is very important, particularly if quantitative information is needed. Most techniques allow user-defined variables, which may include a value for μ and a suitable choice of parameters to define a body outline for each slice.

Lastly, but also importantly, image-display options need the highest-quality hardware and software to maximise the visual impact of the SPECT data. As will be shown in §6.9, the ability to produce single/multiple-slice images and to interrelate transaxial, sagittal and coronal sections is fundamental to the optimal use of SPECT. In Chapter 14 we also discuss the role and importance of 3D shaded-surface display in SPECT.

6.7.5 *Recent developments and the future of ECT using a gamma camera*

Recent developments in SPECT with a rotating gamma camera have been aimed mainly at improved spatial resolution by using a smaller distance between the camera and the patient. There are several ways of achieving this for brain imaging:

(*a*) A special head support can be attached to the end of the patient couch.

(*b*) A slant-hole collimator and an angled camera head can be used (Esser *et al* 1984).

(*c*) A cut-away camera head can be used to prevent the shoulders from limiting the radius of rotation (Larsson *et al* 1984).

For body imaging, an elliptical or non-circular orbit (see §6.5.6) can result in improved resolution and uniformity (Todd-Pokropek 1983, Gottschalk *et al* 1983).

Developments in instrumentation and computational facilities have now reached the stage where dramatic improvements are unlikely. Counting rates and spatial resolution are intrinsically opposed and determined far more by the collimator used than by the intrinsic detector. Multiple-head systems (Lim *et al* 1984) provide improved sensitivity with no loss in resolution but are costly and unlikely to become generally available. Special-purpose collimators, e.g. fan-beam collimators (Jaszczak *et al* 1979a, Webb *et al* 1985a) and cone-beam collimators (Jaszczak *et al* 1986a, b), may increase sensitivity and/or resolution by small amounts (50% and a few millimetres) but the overall prospects are for improved images using more stable SPECT systems with better, more specific radiopharmaceuticals. The trend away from 'better detection' of abnormalities towards improved quantitative measurements of organ function is clearly the goal of ECT. This role, however, may be better satisfied by PET, as we shall see later (§§6.7.7 and 6.9).

6.7.6 *Special-purpose SPECT systems*

The rotating gamma camera has been designed essentially for planar imaging, but with suitable modifications and additions so that SPECT is available to the user as an option. In contrast, special-purpose SPECT devices have been designed usually for tomographic imaging alone, but occasionally with the option of performing planar imaging as well. These special-purpose SPECT systems usually produce single-section tomograms and have already been described (see §6.3.4 on MSPTS).

There are three basic advantages of the MSPTS over the rotating gamma camera. First, an increased sensitivity is achieved, since a larger crystal area is exposed for each slice imaged (see table 6.13). Secondly, an improved resolution is achieved by the use of focused collimators. Thirdly, the variation of resolution across the reconstructed field of view can be minimised by the careful design of the focused collimator (the longer the focal length the better). This last point is important since the basic BP reconstruction technique assumes that each projection value represents a line integral of activity elements along a line perpendicular to the detector. However, figure 6.94 shows how the field of view, as defined by a single parallel-hole and a focused collimator, varies with distance from the detector. By combining opposed views, the variation of resolution across the reconstructed image plane is reduced, but more successfully for the long-focal-length focused collimator than for the parallel-hole collimator (figure 6.95).

Table 6.13 Single-section sensitivity and exposed crystal area for three tomographic systems. (Data taken from Flower *et al* (1980).)

System	Single-section sensitivity[a] (cps kBq^{-1} ml)	Relative area of crystal exposed per section
Aberdeen Section Scanner	18	1.0
Cleon Brain Imager	260	25.0
Rotating gamma camera[b]	1.65	0.43

[a] Defined as the response to a single section of a 19 cm diameter uniform cylindrical source for a 34 cm diameter field of view.
[b] Three pixels of size 0.52 cm used to define a single section.

The Cleon Brain Imager (Stoddart and Stoddart 1979), with large-area detectors and unusually short-focal-length collimators, offered a unique approach to the reconstruction of tomograms from projections.

The assumption that each projection value represents the line integral of activity elements along the collimator axis is particularly invalid in this case. Instead, each projection value represents a bundle of line integrals, lying within the large solid angle subtended by the detector at the focal point. Hence the unique scanning motion, in which each of the 12 detectors performs a rectilinear scan in the plane of the slice being imaged, is necessary so that each point in the object is sampled by six detectors at 30° intervals. Although the total angular sampling was 360°, the inability to separate out the individual line integrals from their 30° bundles resulted in some inadequacies in the quantitative capabilities of this system (Flower and Parker 1980). This problem of undersampling is reduced in the new system (now marketed by Strichman), in which a second rectilinear scan is performed with the detectors rotated through 15°.

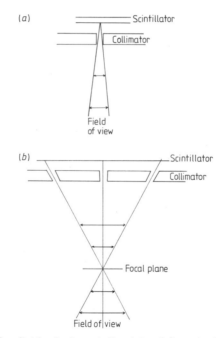

Figure 6.94 The field of view defined by (*a*) a single parallel-hole collimator and (*b*) a focused collimator.

Despite the improvements in resolution and single-slice sensitivity of the special-purpose SPECT systems when compared with the rotating gamma camera, the lack of versatility of the former, in particular their inability to perform dynamic acquisitions, has resulted in little success for those manufacturers who have produced purpose-built section scanners for nuclear medicine. Almost all SPECT systems sold today are rotating gamma cameras, but for special purposes new single-section imagers continue to be developed (Heller and Goodwin 1987).

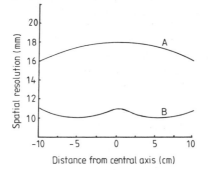

Figure 6.95 The variation of reconstructed spatial resolution with distance from the central axis of rotation for SPECT systems using a parallel-hole collimator (A) and focused collimators (B).

6.7.7 Positron emission tomography

Positron emission tomography (PET) systems enable absolute specific activities of positron-emitting radionuclides to be determined *in vivo* (Phelps 1986). Owing to their high cost, such systems are generally found only in large research institutions and have had few routine applications as yet in clinical nuclear medicine.

PET is based on the principle of *annihilation coincidence detection* (ACD). When a positron is emitted from a radionuclide within the body, it will travel only a short distance (about 1 mm) before annihilating with an electron to produce two 511 keV photons that are emitted in approximately opposite directions. Detection of these two photons in coincidence defines a line along which the annihilation event must have occurred (see figure 6.96). The range of the positron in tissue and also the slight deviation of the angle between the two photons from 180° sets a physical limit to the spatial resolution obtainable in PET. These effects are much smaller than the spatial resolution of systems presently in use; they may, however, become a factor limiting the performance of the next generation of scanners. Table 6.14 shows the spatial resolution that would be obtainable using a perfect detection system for some of the most commonly used positron-emitting isotopes.

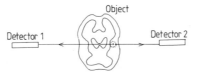

Figure 6.96 Schematic diagram showing how the coincidence detection of the two annihilation photons can be used to determine the emission line of the photons from the decay of the positron-emitting radionuclide.

The use of ACD to define a line along which a disintegration has occurred is inherently more suited to image formation than using a single-photon emitter and a collimator (Hoffman and Phelps 1976). This is illustrated in figure 6.97, which shows the line-spread functions (LSF) for both a typical single-photon collimated device and for an ACD device. In the single-photon case, the line-spread function varies both in shape and in magnitude as a function of distance from the detector. The shape of the line-spread function varies because of the geometric response of the collimator. The variation in magnitude is due to photon attenuation in the object. The fraction of photons reaching the detector is proportional to $\exp(-\mu x)$, where μ is the linear attenuation coefficient for the object and x is the depth in the object. In practice, these effects can be reduced to some extent by combining opposing views of the source distribution (see §6.6.4). In ACD the line-spread function is almost constant over the central third of the distance between the detectors. This is because (*a*) the geometric response is inherently more uniform (the line is defined at either end and is thus more 'rigid') and (*b*) the fraction F of photons detected in coincidence is given by

$$F \propto \exp(-\mu x)\exp[-\mu(T - x)] = \exp(-\mu T) \qquad (6.49)$$

where T is the thickness of the object. This is seen to be independent of the depth of the source within the object. The factor $\exp(-\mu T)$ can be easily measured and accurately corrected for by measuring the count rate obtained with and without the object present, with a suitable external source placed between the detectors. The procedure is much more complicated for the single-photon case, as we have seen earlier (in §6.7.2).

Table 6.14 The effect of positron range on spatial resolution in PET imaging.

Isotope	End-point energy (MeV)	Range in tissue (mm)	Intrinsic resolution, FWHM for ideal detectors 50 cm apart (mm)
^{11}C	0.959	0.28	1.3
^{18}F	0.633	0.22	1.2
^{68}Ga	1.898	1.35	2.0

A pair of detectors, as in figure 6.96, will measure coincidences from all the activity in a line joining them, i.e. for a general activity distribution they will measure (assuming that an attenuation correction has been performed) a line integral through the distribution. If all possible line integrals (or 'channels') through a given plane of the object are measured, then the distribution of activity in the plane may be reconstructed using a standard technique such as filtered backprojection (see Chapter 4). The data obtained for the ACD configuration are a much better approximation to the required line integrals than are those acquired in a single-photon/collimator combination. The space invariance of the line-spread function is a major reason why PET is more successful than single-photon techniques at providing quantitative measurements of radioisotope distributions.

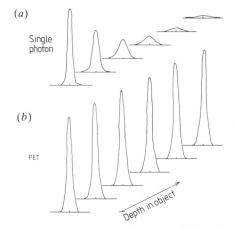

Figure 6.97 A comparison of the line-spread functions achieved for (*a*) a typical single-photon detector and (*b*) an annihilation-coincidence detector system.

Ideally an imaging device utilising the ACD technique should measure all possible channels through a plane of the object. There are many possible detector configurations that achieve this goal. By far the most popular design is to have a stationary circular ring of discrete detectors (usually small BGO scintillation crystals), each one operating in coincidence with many opposing detectors, as shown earlier in figure 6.20 (see §6.3.5). The attenuation correction for each channel is usually performed with an external ring source. A transmission scan is made with and without the object in place, and the ratio of the events in each channel gives the attenuation factor. A positron camera may also be configured as two large-area detectors, such as multiwire proportional chambers (see §6.3.7 and figure 6.98), that face each other and are rotated around the object to measure all channels required for a 3D reconstruction.

Reference to figure 6.20 shows the second major reason why PET is capable of providing quantitative measurements. The absence of a collimator means that the detection sensitivity (i.e. the fraction of all disintegrations that result in a recorded event) is in general between one and two orders of magnitude higher than that for a single-photon system (Budinger *et al* 1977). A higher sensitivity results in a higher statistical accuracy in the reconstructed image.

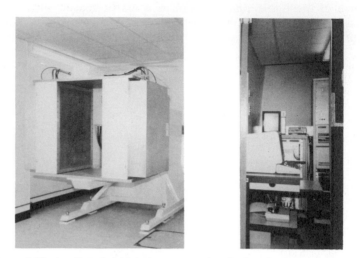

Figure 6.98 A clinical multiwire proportional chamber PET camera. (MUP-PET at the Royal Marsden Hospital, Sutton)

Only some of the photon pairs detected in coincidence by a PET system are true coincidence events and contribute to the formation of an image. Various sources of noise will always be present: in particular, accidental coincidences (or 'randoms') and scattered-photon events. A PET system is designed so as to maximise the true coincidence sensitivity whilst minimising the number of accidentals and scatters collected for a given imaging situation. Many play-offs are made between different performance parameters depending on the requirements of the system.

An expression for the *sensitivity* (*S*) of a given camera is in general very complex because it will depend on the source configuration. However, for the very simple case of a point source situated at the centre of the field of view, it is given by

$$S_p = \varepsilon^2 \phi / 2\pi \tag{6.50}$$

where ε is the quantum efficiency of one detector and ϕ is the solid angle subtended by the detector ring at the centre (neglecting the effects of attenuation and scatter). It is seen that the sensitivity can be increased by using scintillation crystals with a high quantum efficiency or by using a large opening angle. The latter is a less attractive option because it usually results in a large slice width and poor scatter rejection properties.

Accidental coincidences occur because of a finite uncertainty in the time of arrival of a photon at the detector. The decision that two photons have arrived 'simultaneously' (and are therefore connected with the same annihilation event) is made by opening an electronic gate of length τ when a photon is registered on one detector. If a photon is registered on an opposing detector within this time, then a coincidence is said to have occurred. The length of the gate is usually between about 4 and 10 ns. (Note that, since light travels at 30 cm ns^{-1}, there is not sufficient accuracy to localise the position of the annihilation along the line joining the detectors, but see the discussion of 'time-of-flight' systems in §6.3.5.) Figure 6.99 illustrates two classes of event for which only one 511 keV photon from a given annihilation is detected. The first is when annihilation occurs within the coincidence field of view but, due to the finite detector quantum efficiency, one of the photons is not registered (e.g. because it passes straight through the detector). The second is when annihilation occurs in a position that can only be 'seen' by one detector. If two of these events are detected on different detectors within the coincidence window by chance, then there is no way of distinguishing them from a genuine coincidence event. Clearly the line joining the two detectors does not correspond to a genuine annihilation and adds noise to the data. For a given detector pair, the rate at which two events falling within the coincidence window by chance (R_R) is easily shown (Knoll 1979) to be

$$R_R = 2\tau C^2 \qquad (6.51)$$

where C is the rate at which individual photons are detected at detector 1 and detector 2 (both rates are assumed to be roughly the same). For the reasons given above, the true coincidence rate (R_T) is in general much smaller than C, i.e.

$$R_T = kC \qquad k \ll 1. \qquad (6.52)$$

A typical value is $k \sim 0.02$ (Hoffman *et al* 1983a) depending on the system and the object being imaged. The true-to-random ratio is then given by

$$R_T/R_R = k/(2\tau C). \qquad (6.53)$$

Note that this ratio decreases as C (which is proportional to the source activity) increases. The above argument can easily be extended to a real system with more than one detector pair and many possible coincidence combinations. In a practical system, the signal-to-noise ratio is maximised by (*a*) using detectors with a short resolving time τ, and (*b*) increasing the value of k by using 'shadow shield' collimators (as in figure 6.99) to reduce the volume of the object that is seen by one detector only. Since the random fraction can never be reduced to zero (it is typically below 25% (Phelps 1986)), the random contribution to each detector pair (channel) is usually measured during data taking (by using a second delayed-coincidence window) and subtracted before performing image reconstruction. Whilst accounting for the distribution of randoms in the image, this procedure does not remove the added statistical uncertainty created (Hoffman *et al* 1981).

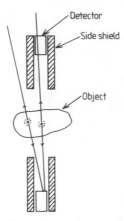

Figure 6.99 An illustration of two classes of photon-pair events seen by shielded coincidence detectors for which only one 511 keV photon from a given annihilation is detected.

Scattered coincidence events arise when one or both of the photons (still detected in coincidence) is scattered in the object, resulting again in a line that does not pass through the annihilation point. The distribution of scattered events is such that it does not degrade the spatial resolution in the reconstructed image but adds a low-frequency background that can upset quantitative measurements (Hoffman *et al* 1983a). Energy discrimination is not very successful at rejecting scattered 511 keV photons as a scattering angle of 30° corresponds to an energy loss of only about 60 keV, which is less than the energy resolution of most scintillation detectors at this energy. The 'shadow-shield' collimators and other geometrical factors (such as the ring diameter) are therefore the main ways in which scatter (from both inside and outside the coincidence field of view) is kept to a minimum (i.e. below 25% (Phelps 1986)). Since the scattered-event distribution cannot be measured directly, as randoms can, a software scatter correction is usually performed during image reconstruction (Bergström *et al* 1983b). An account of the complex process whereby quantitative information is derived from a series of images is given by Phelps (1986).

6.7.8 Summary

In conclusion, we can say that SPECT provides functional images with improved contrast when compared with planar imaging but usually at the expense of spatial resolution. As usual, there is a compromise between resolution and total image counts, but there is no doubt, as will be shown in §6.9, that SPECT provides the clarity of information usually missing from planar scintigraphy. For the detection of low-contrast variations in radiopharmaceutical distributions, it is presently the technique of choice. The advent of low-cost PET systems and the use of radioisotopes from generators or off-site cyclotrons should increase the importance of this technique in nuclear medicine in the next decade.

6.8 QUALITY CONTROL AND PERFORMANCE ASSESSMENT OF RADIOISOTOPE IMAGING EQUIPMENT

The basic requirements of a quality-control procedure are that the tests must be easy to perform, the measurements should be sensitive to important changes and the results easily recorded for reference. Various recommendations for quality-control and performance-assessment (QC/PA) procedures, particularly for gamma cameras, have come from a range of professional bodies (table 6.15). Some of these recommendations also cover electrical and mechanical safety requirements.

Table 6.15 Some recommendations for quality control and performance assessment of gamma cameras.

For the manufacturer	
British Standards Institution	(BSI 1977)
National Electrical Manufacturers' Association	(NEMA 1980, 1986)
International Electrotechnical Commission	(IEC 1980)
For routine hospital use	
Hospital Physicists' Association	(HPA 1978)
Department of Health and Social Security	(DHSS 1980, 1982)
Bureau of Radiological Health	(BRH 1976)
World Health Organisation	(WHO 1982)

There are two major classes of QC/PA tests performed. The first are those carried out immediately after installation of equipment, which provide both acceptance testing and baseline measurements for reference. The major questions here relate to whether the manufacturer's specifications have been met. Once the baseline measurements have proved to be satisfactory, the second category of tests are required. These include both routine daily and long-term tests to determine the overall performance and stability of the system for comparison with the baseline measurements. A series of routine daily, monthly and quarterly checks have been suggested by the British Hospital Physicists' Association (HPA).

The main QC/PA measurements for gamma-camera systems are

(*a*) non-uniformity,
(*b*) spatial resolution,
(*c*) energy resolution,
(*d*) spatial distortion,
(*e*) plane sensitivity,
(*f*) count-rate performance, and
(*g*) shield leakage.

Some of the tests can be made with the collimator removed and a lead mask positioned on the crystal surface to limit the area of the detector tested to that normally defined by the collimator—these are known as *intrinsic measurements.* If the collimator is in place when the tests are made, then they are referred to as *system* or *extrinsic measurements*. In most cases, an on-line computer system is required for quantitative measurements. Each of the measurements listed above will now be considered in turn. The methods described for quantitative measurements are those recommended either by the HPA or NEMA (see table 6.15).

6.8.1 Non-uniformity

The non-uniformity of response of the gamma camera can be determined by both global (integral) uniformity and local (differential) uniformity measurements.

Integral uniformity is an indication of the extreme range of variation in regional sensitivity over a defined field of view (FOV) and is given by NEMA as

$$U_I(\%) = 100 \times (\text{max} - \text{min})/(\text{max} + \text{min}) \qquad (6.54)$$

where 'max' is the maximum pixel count and 'min' is the minimum pixel count.

Differential uniformity indicates a maximum rate of change of sensitivity over a predetermined short distance within the FOV and is given by NEMA as

$$U_D(\%) = 100 \times (\text{hi} - \text{low})/(\text{hi} + \text{low}) \qquad (6.55)$$

where 'hi' and 'low' are the maximum and minimum pixel counts for all rows and columns within a localised line of five pixels for which hi − low is the largest deviation. (It should be noted that HPA definitions are similar but slightly different.)

Uniformity measurements are usually obtained for two FOV of the camera. NEMA definitions are as follows: the useful field of view (UFOV) is the area within 95% of the half-height radius of the count distribution from a uniform source, and the central field of view (CFOV) is that within 75% of this radius (figure 6.100). Both intrinsic and system non-uniformity measurements can be made. Intrinsic non-uniformity is typically (NEMA) measured by exposing the camera to a 20 MBq ^{99}Tc^m point source with the collimator removed. The minimum source–detector distance should be 1.5 m or five times the UFOV diameter and the count rate should be below 30 kcps with a photopeak window of 20%. For optimal measurements the maximum pixel count in a 64 × 64 matrix should be about 4000 and a simple nine-point smoothing function applied. System uniformity is typically measured (HPA) with a flood source containing approximately 80 MBq of ^{99}Tc^m. The source diameter should be greater than the geometric FOV. The mean pixel count should be more than 10^4, and if a 64 × 64 matrix is used, this means a study of 30 million counts. Figure 6.101 shows an example of integral and differential uniformities obtained from a modern digital gamma-camera system.

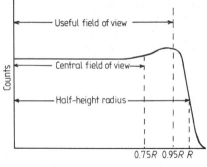

Figure 6.100 Schematic of count profile from flood image and the definition of useful field of view and central field of view for uniformity measurements with a gamma camera. (After NEMA (1980, 1986).)

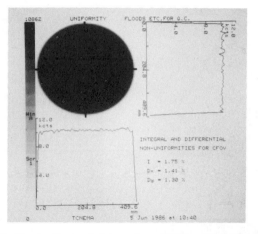

Figure 6.101 Illustrations of integral and differential non-uniformity obtained from a modern digital gamma-camera system (IGE, STAR-CAM).

6.8.2 *Spatial resolution*

Qualitative assessments of spatial resolution can be made using either transmission or emission phantoms. The former consist of lead foils in Perspex sheets and are uniformly irradiated by an appropriate source of γ-rays to provide images suitable for resolution measurements. Transmission phantoms (table 6.16) can be used for qualitative assessment of either intrinsic or system spatial resolution, with either a point source at a distance from, or a flood source immediately adjacent to, the phantom. Figure 6.102 illustrates some images obtained with a few of these phantoms. Emission phantoms (table 6.17) are made of Perspex containers filled with a radioactive liquid. Figure 6.103 illustrates some images obtained with a selection of these phantoms. As no collimation is provided within emission phantoms, they can only be used for assessment of system spatial resolution.

Table 6.16 Transmission phantoms for assessment of intrinsic or system spatial resolution and linearity.

Quadrant bar	(BRH 1976, HPA 1978, 1983)
Parallel-line equal-spacing bar	(BRH 1976, HPA 1978)
NEMA resolution and linearity	(NEMA 1980, 1986, WHO 1982)
Hine–Duley bar	(BRH 1976)
Orthogonal holes (OHTP)	(WHO 1982, HPA 1983)
Anger 'pie'	(BRH 1976, HPA 1978, 1983)

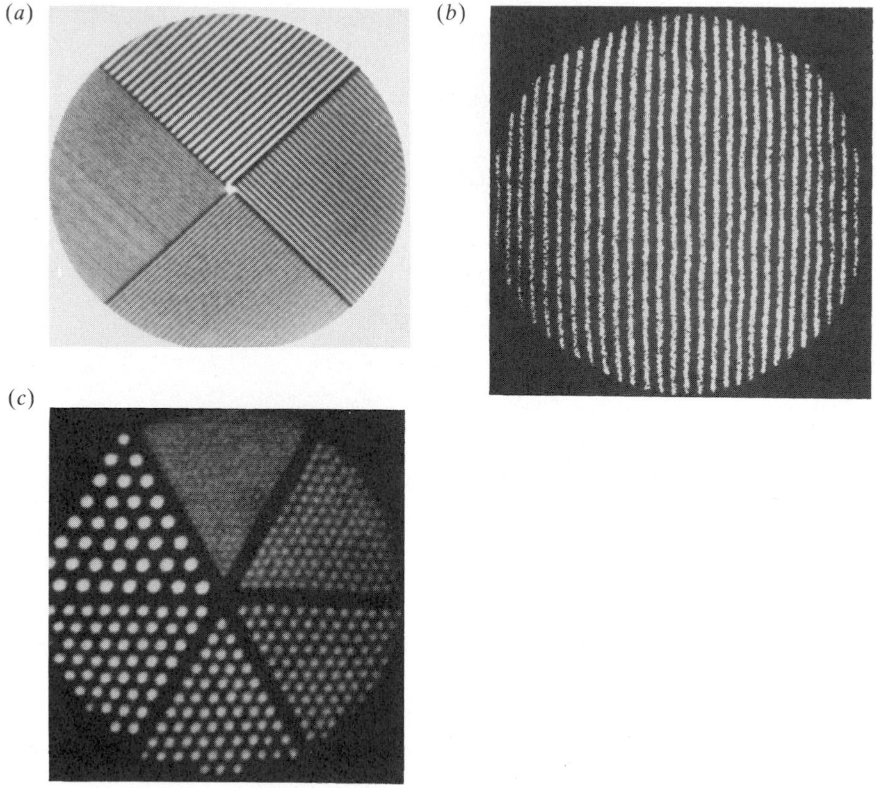

(a)

(b)

(c)

Figure 6.102 Spatial resolution images using: (*a*) a quadrant bar phantom; (*b*) a parallel-line equal-spacing bar phantom; and (*c*) an Anger pie phantom.

Table 6.17 Emission phantoms for assessment of system spatial resolution.

Picker thyroid	(BRH 1976)
Williams' liver slice	(BRH 1976)
Modified Williams' liver slice	(HPA 1978, 1983)
Shell	(DHSS 1980, HPA 1983)
London liver	(WHO 1982)
College of American Pathologists liver	(BRH 1976)
College of American Pathologists brain	(WHO 1982, BRH 1976)

Quantitative measurements of spatial resolution are made by determining the response of the detector to a radioactive line source. In real space this is referred to as the line-spread function (LSF), and in frequency space this is called the modulation transfer function (MTF) (see Chapters 2 and 12). In attempting to quote a single figure for comparative measurements of spatial resolution, the FWHM of the LSF or the 50% level of the MTF are often used. However, there are limitations (figure 6.104) in trying to represent the full LSF or MTF curves by a single parameter, and often the full width at tenth-maximum (FWTM) is quoted in addition to the full width at half-maximum (FWHM) as a compromise.

Quantitative measurements of *intrinsic* spatial resolution are made with a collimated radioactive line source placed 50 mm from the crystal surface (HPA). The source, containing approximately 40 MBq of $^{99}Tc^m$, is aligned with the x or y axis and the rest of the detector shielded with lead sheets (figure 6.105). Peak counts of approximately 10^4 are obtained, and LSF measurements repeated for different positions over the camera surface. The spatial resolution is given by the FWHM of the LSF. Special computing facilities are needed for this measurement since it is recommended that the pixel size must be small enough so that the FWHM contains at least 10 pixels.

Quantitative measurements of *system* spatial resolution are usually made (HPA) with an uncollimated radioactive line source. The source, approximately 40 mm long, is embedded in the face of a Perspex sheet approximately 240×240 mm^2 and 10 mm thick. Tissue-equivalent material can be used between the source and collimator to simulate the effects of scatter. The FWHM of the LSF is measured under a variety of conditions, namely

(*a*) at different positions in each plane;
(*b*) with different absorber thicknesses between detector and source (0–200 mm);
(*c*) with different collimators; and
(*d*) with different radioisotopes.

6.8.3 Energy resolution

Quantitative measurement of the energy resolution of a gamma camera

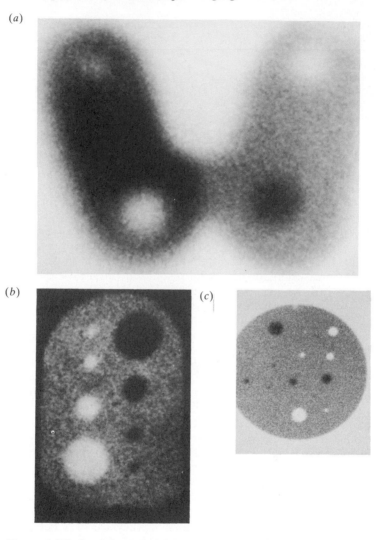

Figure 6.103 Spatial resolution images using: (*a*) the Picker thyroid phantom; (*b*) the Williams' liver-slice phantom; and (*c*) the Shell phantom.

requires the use of a multichannel analyser. The analyser is calibrated using standard γ-ray sources so that the energy spectrum can be measured in kiloelectronvolts. Measurements are usually made (HPA) of the intrinsic energy resolution using a point source. The analyser is usually set up so that there are at least 10 channels within the photopeak and 10^4 counts acquired in the peak channel. The energy resolution is given by

$$\Delta E(\%) = (\text{FWHM}_E/E) \times 100 \tag{6.56}$$

where FWHM_E is the width of the photopeak and E is the photopeak mean energy (see figure 6.30).

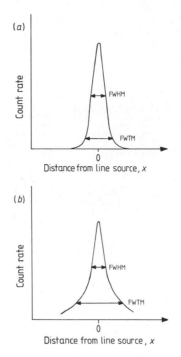

Figure 6.104 Line-spread function for (*a*) $^{99}\text{Tc}^m$ and (*b*) ^{131}I showing greater differences between full width at tenth-maximum (FWTM) than full width at half-maximum (FWHM).

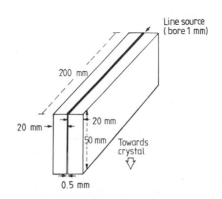

Figure 6.105 Schematic of a collimated line source for the measurement of intrinsic spatial resolution. (From HPA (1978).)

6.8.4 Spatial distortion

This term is used to describe the deviation of the image of a line source from linearity. *Qualitative assessment* can be made using a variety of linearity phantoms. For instance, a special linearity phantom, in the form of a set of line sources in an orthogonal pattern, can be used (figure 6.106). Alternatively, all transmission phantoms (table 6.16) can be used for simultaneous qualitative assessment of both spatial resolution and spatial distortion.

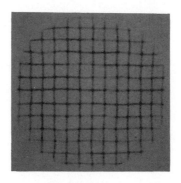

Figure 6.106 Image obtained using a tube-type emission phantom described in BRH (1976) to indicate spatial distortion in the *x* and *y* directions.

Quantitative measurements of spatial distortion require computer programs to determine the deviation of the image of a line source from a true straight line. The HPA describes the analysis of four images of a line source placed on the camera in positions that form a rectangle (figure 6.107) with sides parallel to the *x* and *y* axes, and corners within 20 mm of the edge of the UFOV. A least-squares fit of a straight line to each imaged line can determine the minimum and maximum deviations $(\Delta x, y)_{\text{min, max}}$ from the fitted line. The spatial distortion D is expressed as

$$(D_{x \text{ or } y})_{\text{min, max}}(\%) = \frac{(\Delta x \text{ or } y)_{\text{min, max}}}{L} \times 100 \qquad (6.57)$$

where L is the length of the line source. This measurement requires a large number of pixels to be included in the direction perpendicular to the line, to give an accurate determination of Δx or Δy.

6.8.5 Plane sensitivity

This is a measurement of the count rate obtained when imaging a given radioactive plane source and is expressed (HPA) as counts per second per megabecquerel (cps MBq^{-1}) for an area source of dimensions

100×100 mm². The source contains approximately 30 MBq of ⁹⁹Tcᵐ and is placed 100 mm from the collimator surface. The activity should be known to better than ±5%. Values for plane sensitivity can be determined for a range of radionuclide–collimator combinations.

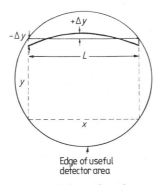

Figure 6.107 Line source positions for the measurement of spatial distortion. (From HPA (1978).)

6.8.6 Count-rate performance

Several measurements can be made to determine the maximum count-rate capacity and count-rate losses (see §6.3.6) of a gamma-camera system. These measurements are usually (HPA) made with a standard phantom (figure 6.108), which provides suitable scattering material.

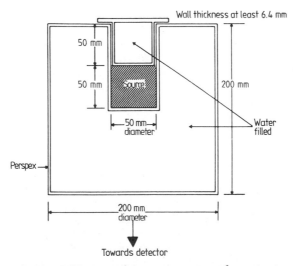

Figure 6.108 The BSI source for measurement of count-rate capability. (From HPA (1983).)

Examples of count-rate performance parameters are:

(*a*) the observed count rate at which the value measured by the gamma camera is 90% of that expected from calculations (HPA);

(*b*) maximum observed count rate (HPA); and

(*c*) observed count rates at which losses are 10%, 20% and 30% of expected rates (DHSS).

6.8.7 *Shield leakage*

The objective is to determine the gamma-camera count rates when exposed to a radioactive source outside the collimator field of view. A small cubic source of approximately 10 mm sides is used (HPA) to check count rates at various positions around the shield. These count rates are compared with the count rate recorded when the source is placed 100 mm from the collimator face on the axis of the collimator. Leakage count rates (L) are expressed as

$$L(\%) = \frac{\text{observed count rate}}{\text{reference count rate}} \times 100. \tag{6.58}$$

Common regions containing shield weaknesses are cable entry points, bolt holes and collimator–detector junctions. Many modern gamma cameras have very limited shielding to photons above 300 keV to minimise overall camera weight for tomography.

6.8.8 *Quality control and performance assessment for* ECT

In addition to the QC/PA procedures discussed above, extra tests and measurements are required if ECT is to be carried out (HPA 1983). As an example, those tests and measurements required for a rotating gamma-camera SPECT system are described in this section. Factors affecting the performance of SPECT have been reviewed by Heller and Goodwin (1987).

Mechanical tests include checking that the gamma camera rotates about a single well defined axis that is parallel to the front face of the collimator. In addition, measurements of angular precision and rotation speed are required. Errors in these parameters will produce gross ECT image distortion. Further simple checks of the imaging couch alignment are necessary to minimise patient positioning errors.

Electronic tests include measurements of the alignment of the x and y axes of the gamma camera with the axis of rotation, determination of any energy spectrum changes as a function of camera angle, and estimation of image pixel size. The latter is particularly important, as the pixel size can be used to monitor the stability of the analogue-to-digital converters used to form the digital images.

Quality-control procedures include a tomographic acquisition of a point source, which is used to determine the centre of rotation (COR) in the x direction. The COR acquisition data (figure 6.109) are used to align the subsequent projection data with the reconstructed image. Also included is the acquisition of a high-count (3×10^7) flood-source image used for subsequent uniformity correction. Both tests must be carried out prior to data reconstruction and under the same conditions as are subsequently used for the ECT acquisition, e.g. using the same collimator and pixel size.

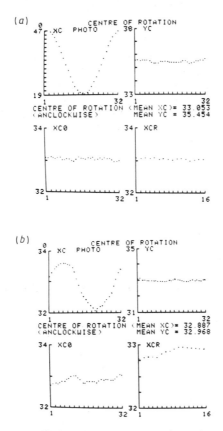

Figure 6.109 Processed data from a centre-of-rotation study, using a rotating gamma camera. (*a*) Output from a typical well set up gantry. (*b*) Output from a gantry showing slippage of the detector assembly. *Top left*: position of centre of source in *x* direction (XC) versus angular position of camera. *Top right*: position of centre of source in *y* direction (YC) versus angular position of camera. *Bottom left*: XCO = XC − sine wave fitted to curve for XC. *Bottom right*: XCR = average of xc(θ) and XC (180 + θ). XC should be sinusoidal. YC should be constant if source is imaged in single slice. XCO and XCR should be within 0.25 pixel. (From HPA (1983).)

Additional *performance assessment* for ECT may include the following:

(*a*) Measurement of the uniformity of a section through a volume source.

(*b*) Measurement of transaxial (in-slice) and longitudinal (slice thickness) spatial resolution.

(*c*) Determination of volume sensitivity in terms of total counts per second per megabecquerel per millilitre (cps MBq^{-1} ml^{-1}) using a Perspex cylinder uniformly filled with a radioactive liquid (Flower *et al* 1980).

(*d*) Estimation of contrast in the reconstructed image, which determines what count-rate difference is required between a small region of interest (ROI) and surrounding areas to allow the ROI to be resolved from the background radioactivity.

(*e*) Determination of the recovery coefficient (defined by Hoffman *et al* (1979) as the ratio of the apparent isotope concentration in the image divided by the true isotope concentration) as a function of object size and spatial resolution.

Additional performance assessment parameters for PET include:

(*a*) Timing resolution (τ), which is expressed as the FWHM of the coincidence pulses and affects the timing window selected for the coincidence count rates.

(*b*) Signal-to-noise ratio (SNR), which is the ratio of the true coincidence rates to the random rate.

(*c*) Scatter fraction, which is assessed by comparing coincident events with and without scattering material in the FOV.

(*d*) Figure of merit, which is defined as ε^2/τ, where ε is the detector quantum efficiency—however, there are limitations in the use of this parameter alone for comparing different PET systems, since it does not take into consideration other factors such as the spatial resolution, the scatter fraction and the detector field of view.

6.8.9 Summary

Quality control and performance assessment of radioisotope imaging systems are important in limiting the effects of detector-related artefacts in the resultant images. Much effort has been put into these areas in application to the hardware used but little quality control has been attempted in the use of computer software. This has led to a proliferation of algorithms for determination of functional parameters with no assessment of their accuracy—this is clearly an area of great need if the quantitative functional information produced is to be used clinically with any confidence.

6.9 CLINICAL APPLICATIONS OF RADIOISOTOPE IMAGING

6.9.1 Introduction

The previous sections in this chapter have discussed the physical principles of radioisotope imaging. We have seen how the development of photon detectors and associated data acquisition/processing computers allows functional images of the body to be made. Section 6.4

on radioisotopes has shown how they may be produced and selected for imaging. In this section, we will show how the science of radioisotope imaging can be applied to clinical studies to provide valuable information relating to the function of the body. Here we will discuss the production of specific radiopharmaceuticals and their uses over a selected range of organ studies. The examples chosen are typical and perhaps the most common studies performed in nuclear medicine, but they are by no means complete. It should be noted that the majority of radiopharmaceuticals available are not 'disease-specific' but show differential function between normal and abnormal tissues. The development of 'disease-specific' agents such as antibodies is discussed separately because of the potential differential nature of the diagnostic information together with their application as targeted-therapy agents.

No attempt is made to separate the information gained by conventional single-photon imaging from that of PET. Although the latter is not yet readily available to the majority of clinical nuclear medicine departments, the value of such studies is beyond question and the contribution of PET to the basic understanding of the physiology of disease is often far greater than from single-photon imaging. PET will play an increasingly important role in clinical nuclear medicine.

6.9.2 Radiopharmaceuticals

The great strength of nuclear medicine lies in its ability to image, qualitatively and quantitatively, dynamic physiological processes of the body. This ability is important because anatomical or morphological changes due to pathological conditions are often preceded by physiological or biochemical alterations, and to obtain such physiological or biochemical information one must be able to assess function. By using labelled compounds, it is possible to study various biological functions such as blood flow, glucose metabolism, the functional state of an organ (e.g. the function of the Kupfer cells of the liver), the fate of a drug (i.e. its distribution and metabolism), neurotransmitter receptor binding and more. In order to accomplish these tasks, a specific radiotracer or radiopharmaceutical must be employed, along with a suitable imaging system. This section will focus on some aspects of radiopharmaceutical chemistry, including design rationale and clinical uses.

A radiopharmaceutical is a radioactive compound (biomolecule or drug) that meets the criteria of sterility and apyrogenicity so that it can be safely administered to humans for the purpose of diagnosis, therapy or research. A radiopharmaceutical is usually made up of two components, the radionuclide and the unlabelled compound to which it is bound. However, elemental radiopharmaceuticals such as ^{133}Xe are also used.

The development of radiopharmaceuticals has followed various strategies (Burns 1978) and a large area of present-day radiopharmaceutical development is based on the design of agents that will successfully incorporate readily available, inexpensive radioisotopes with good imaging characteristics.

Radiopharmaceuticals are usually employed for one of two major reasons, either to study a particular physiological function of an organ or tissue (e.g. blood flow) or to localise specific disease (e.g. tumour). In either case, the radiopharmaceutical should meet certain requirements.

Since radiopharmaceuticals are used to study function, it is important that they impart no effect on the system under study, through pharmacological or toxicological effects. In most cases, the concentration of radiopharmaceutical (or radiotracer) is subpharmacological, usually of the order of nanomolar (10^{-9} M), and at this concentration imparts no measurable effect (see also §6.4). The radiotracer should also demonstrate a high target to non-target ratio (specificity) and an effective half-life suitable to the length of the study.

Practically, the radiopharmaceutical should be easily synthesized or labelled and should be readily available. The labelled material should be stable for a long enough period to permit optimal usage. Finally, the radiopharmaceutical should be pharmaceutically acceptable once labelled—that is, it should be sterile and pyrogen-free, as well as radiochemically and radionuclidically pure.

A number of factors are responsible for the ultimate distribution of a radiopharmaceutical or drug (Thompson 1983). These include the basic arrangement or shape of the molecule, the size (molecular weight) of the compound and the nature of the functional groups incorporated into the structure. Lipid solubility (hydrophobicity) is also important. Cell membranes are composed of proteins and lipids. The major lipids are phospholipids. The arrangement of the membrane is such that the water-soluble phosphate groups are at the outer and inner surfaces of the membrane. The lipid portion of the membrane is sandwiched between these two water-soluble surfaces. In order for a drug to cross a membrane, sufficient lipid solubility must be demonstrated.

The extent to which a compound is ionised (pK_a) at physiological pH (pH = 7.4) also affects lipid solubility and consequently the drug's ability to cross membranes. Membrane penetration can be passive or active. The former process relies on diffusion for the drug to enter into the cell; the latter requires a carrier or energy-dependent mechanism in order for the drug to be transported across the membrane. In some instances, there are highly selective 'receptors' that recognise a particular molecule and actively accumulate or transport this molecule onto or into the cell.

The distribution of the radiopharmaceutical is also governed by such factors as blood flow (per cent cardiac output/organ) and the drug's availability to tissue, and also depends on the proportion of the tracer that is bound to proteins in the blood. Ultimately, the fate of the radiotracer will be dictated by the way the body deals with the molecule (metabolism). If the molecule gains entry to cells of non-secreting organs, then the fate of the tracer may depend on whether it is utilised or broken down intracellularly and whether it is retained or exits from the cells (altered or intact) to be removed from the body by the kidneys or liver. The removal of substances from the body via the kidneys or liver depends on a number of factors, which will be discussed later.

In order to design a radiopharmaceutical with specific localising properties, one must take the factors mentioned above into consideration, as well as the choice of radionuclide. The choice of radionuclide is important because of the chemistry associated with its incorporation into the molecule as well as the overall effect it will have on the final compound.

While conventional nuclear medicine is currently limited to single-photon-emitting isotopes with photon energies of 50–300 keV, the radionuclides of the 'organic' elements ^{11}C ($T_{1/2} = 20.4$ min), ^{13}N ($T_{1/2} = 9.96$ min), ^{15}O ($T_{1/2} = 2.07$ min) and ^{18}F ($T_{1/2} = 109.7$ min) (though fluorine is a halogen, it is often included among the 'organic' radionuclides owing to the lack of steric perturbation associated with its substitution for –H or –OH groups) are restricted to use with PET. As these are the only potentially useful radioisotopes of the 'organic' elements that make up biological and pharmacological compounds such as sugars, proteins, vitamins, hormones, steroids, drugs, etc, one must look at ways of synthesising tracers with non-'organic' radionuclides attached.

Reviewing the list of medically useful radionuclides from §6.4, it can be seen that the predominant radioelements are either metals or halogens. As the object of this section is to draw attention to the possible constraints imposed by the choice of radionuclide, this discussion will compare and contrast the radiometal technetium and the radiohalogen iodine. This is not meant to be an exhaustive discussion on radiochemistry, but rather an illustration of the various considerations that one must be aware of when using various radionuclides.

Technetium is a group VIIB transition metal with an atomic number of 43 and a variety of oxidation states. Technetium has the electronic configuration $(Kr)4d^65s^1$. ^{99}Tc^m is produced via decay of ^{99}Mo and eluted from the Mo/Tc generator in the form of sodium pertechnetate, $Na^+TcO_4^-(Tc(VII))$, and must be reduced (i.e. the oxidation number lowered) in order to form reactive species. Reduction is usually accomplished by the use of stannous ions in the presence of the ligand and TcO_4^-. As with other metals, Tc forms coordinate covalent bonds (both shared electrons are donated by the other atom) with donor atoms such as N, O, P and S. For a more extensive review of Tc chemistry, the reader is referred to Srivastava and Richards (1983).

While simple complexes (the metal atom is bound to one donor atom of one molecule) of Tc are known to exist, the chemical properties of Tc(V), Tc(IV) and Tc(III) and many other transition metals (i.e. gallium and indium) are such that maximum complex stability is achieved via chelation (Cotton and Wilkinson 1980). A chelate is formed when a metal atom is bound to more than one donor atom of a complexing molecule or ligand, thereby forming a closed-ring structure. The added stability of the metal chelate is often required in order to resist oxidation, hydrolysis or the strong affinity that some metals (especially indium and gallium) display for the plasma protein transferrin. Furthermore, the chemical composition of the chelate will influence the *in vivo* behaviour of the radiometal–chelate complex. Diethylene-triaminepentaacetic acid (DTPA) was one of the first chelating agents to be used to complex $^{99}Tc^m$ (Richards and Atkins 1968). Tc-DTPA is primarily excreted from the body via the kidneys and is still used today to assess renal function. The reason why Tc-DTPA is effective as a renal imaging agent is probably due to its low molecular weight, low protein binding and its water solubility. Since the development of Tc-DTPA, at least a dozen Tc complexes have been described that demonstrate renal accumulation and/or clearance (Eckelman and Volkert 1982). Numerous other Tc complexes have also been developed that demonstrate accumulation in various other organs including the liver (Tc-iminodiacetic acid derivatives), brain (Tc-propylene amine oxime derivatives, Tc-diaminodithiol derivatives), heart (Tc-isonitriles, Tc-boronic acid adducts) and skeleton (Tc-phosphonates and phosphates). These and other agents will be discussed further in §6.9.3.

In general, the metal–chelate complexes have characteristic physicochemical properties (i.e. molecular weight, lipid solubility, charge, pK_a, proportion of protein binding, etc), which will govern their *in vivo* distribution. An important characteristic of these Tc chelates is the 'essential' nature of the Tc atom in determining their biodistribution. Burns *et al* (1978) have classified Tc radiopharmaceuticals into two categories: 'Tc-tagged' radiopharmaceuticals and 'Tc-essential' radiopharmaceuticals. The latter class includes the Tc chelates whose biodistribution is dependent upon some chemical property of the final complex, not necessarily the properties demonstrated by the uncomplexed ligand. In this case, the biological behaviour of the metal chelate will probably be governed by (1) the strength of the bonding donor groups, (2) the final charge of the un-ionised chelate complex and (3) the type of groups (e.g. $-CH_3$, $-NH_2$, $-SH$, $-OH$, $-Cl$, $-COOH$) that are able to interact freely with the surrounding environment (i.e. blood, intra- and intercellular fluids, cerebrospinal fluid, cell membranes, proteins, etc).

An example of a 'Tc-essential' radiopharmaceutical is $^{99}Tc^m$-HMPAO (hexamethylpropylene amine oxime). Tc-HMPAO has been demonstrated to cross the blood–brain barrier and localise in brain tissue in proportion to blood flow (Neirinckx *et al* 1987). However, the 'untagged' ligand (HMPAO) does not cross the blood–brain barrier, thereby precluding cerebral accumulation (McKenzie *et al* 1985).

This example demonstrates the intimate role that Tc plays in determining the biological fate of some chelate-based radiopharmaceuticals. Similarly, one should recognise the importance of the chelate structure in this metal–chelate complex.

Unlike 'Tc-essential' radiopharmaceuticals, 'Tc-tagged' radiopharmaceuticals often show no more than a minor alteration of biological behaviour after the addition of the Tc atom. One reason for this is that most of the materials to be labelled in this category are high-molecular-weight substances such as colloids, particles, proteins and cells (Burns *et al* 1978). The addition of the relatively small $^{99}Tc^m$ atom to these larger substances results in minimal alteration. However, the procedures one chooses for labelling may be harsh (i.e. extremes of pH, the presence of oxidising or reducing agents) and can adversely affect the substance to be labelled. Similarly, some biomolecules have 'critical' areas as part of their structure. These 'critical' areas may be responsible for the specific nature of their biochemistry. One example is the antibody molecule or immunoglobulin. The antibody is comprised of heavy and light amino-acid chains held together by disulphide bonds (see figure 6.110). There are areas in the heavy and light chains known as variable regions (N-terminus) in which the structural sequences of the chain vary from one antibody to the next. It is this N-terminus region that gives the antibody its unique ability to identify and bond with a particular antigen or foreign substance (Sikora and Smedley 1984). Radiolabelling procedures and/or the addition of the chelate-group label to the N-terminus can affect the antibody's ability to recognise or bind to the antigen for which it is specific (Paik *et al* 1983). It is therefore important that strict quality-control assays be performed when initiating a new labelling method, to ensure that the carrier substance is biochemically unaltered, i.e. the molecule retains its immunoreactivity.

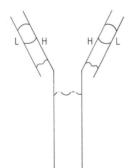

Figure 6.110 Schematic diagram of antibody molecule. H and L represent heavy and light amino-acid chains, respectively.

Recently, chelating agents such as ethylenediaminetetraacetic acid (EDTA), DTPA and desferrioxamine (DFO) have been chemically linked to proteins (antibodies, in particular) so that the radiometal will be strongly bound to the carrier molecule (Sundberg *et al* 1974, Krejcarek and Tucker 1977, Hnatowich *et al* 1983, Janoki *et al* 1983). While most of the 'bifunctional chelate' work has involved labelling using indium and gallium radioisotopes, attempts to label proteins in this way with Tc have also been reported (Arano *et al* 1987).

Similarly, bifunctional chelates have been used in an attempt to bind

Tc to other biomolecules of interest (i.e. fatty acids) that on their own would show little or no affinity to Tc (Eckelman *et al* 1975). However, the addition of a chelating agent may alter the ability of the parent structure to localise in the tissue of interest due to either increased molecular weight or steric bulk, altered lipid solubility or polarity, a change of ionisation constant, or variation in protein binding. The overall effect is likely to be due to a combination of factors.

While many Tc chelates have been prepared and are widely used in nuclear medicine, these agents, as yet, do not lend themselves to the investigation of more intricate biochemical processes, such as

(*a*) energy metabolism (e.g. glucose metabolism, fatty-acid metabolism),

(*b*) biochemical pathways (e.g. steroid synthesis, protein synthesis), and

(*c*) neuronal transmitter receptor studies (e.g. dopamine, histamine or serotonin receptors).

The reason for this void is probably due to the problems associated with incorporating $^{99}Tc^m$ and/or a chelating group into the required tracer without significant structural, electronic or hydrophobic alterations (Eckelman 1982).

An attractive alternative for labelling such biochemical tracers is to use the radiohalogens. The coordination number of the common halogens (fluorine, chlorine, bromine, iodine) is 1, in the -1 oxidation state. As the halogens (group VIIA) need only one electron to reach the noble-gas configuration, they easily form the -1 anion (X^-) or a single covalent bond (Cotton and Wilkinson 1980). This latter property is a distinct advantage over Tc when labelling low-molecular-weight biologically active molecules.

While extensive research effort has gone into the radiochemistry of the halogens (see Palmer and Taylor 1986), this section will focus on radioiodine in particular.

The radioisotopes of *iodine* (see tables 6.5–6.7 in §6.4.2) are of considerable interest in nuclear medicine due to their varied physical properties. The iodine atom (atomic number 53) occupies a similar volume to that of a methyl ($-CH_2$) or ethyl ($-C_2H_5$) group and the electronegativity of iodine (2.5) is similar to that of carbon (2.5). Therefore, iodine can be substituted for an alkyl group in an organic molecule without excessive perturbation of the molecule's polar or steric configuration (Coenen *et al* 1983). This potential iodine-for-alkyl-group substitution allows for the formulation of chemical congeners of a variety of parent structures. While the addition of iodine, or any halogen for that matter, is not without effect, the introduction of halogens into pharmacologically or biochemically active molecules is an accepted method of 'fine-tuning' *in vivo* behaviour or potency. Furthermore, the iodine-containing molecule can be synthesised on a macroscopic scale (not a simple or straightforward task for Tc radiopharmaceuticals), allowing for chemical, pharmacological and toxicological evaluation.

In addition to the development of iodine-containing analogues of small biochemically 'active' molecules, radioiodine can be incorporated into large substances such as proteins or peptides through the use of prelabelled reagents or the direct radioiodination of activated aromatic amino acids (i.e. tyrosine, histidine or tryptophane). The iodination of such activated aromatic rings is accomplished through electrophilic aromatic substitution: a positively charged iodine species (I^+) is formed (usually via oxidation) and replaces a hydrogen or other less electrophilic group on the aromatic ring (see de la Mare (1976) for further discussion).

The 'oxidising agents' are most widely used in the radioiodination of proteins or peptides, where the position of the iodine on the aromatic ring of the amino acid is not crucial. However, iodination can alter protein behaviour, and the molar ratio of iodine to protein should be considered. Furthermore, some antibodies may be particularly sensitive to iodination if the immunoreactive portion of the molecule contains a large proportion of iodination sites.

Iodination reagents or oxidising agents can also be used for incorporating iodine into smaller molecules, where the position of the iodine on the aromatic ring is of little consequence or where other 'activating' groups on the ring will 'direct' the iodine atom to a known position (Seevers and Counsell 1982).

However, when developing low-molecular-weight tracers for biochemical study, the position of the iodine atom often has an effect on the chemical and biological behaviour of the radiotracer (Coates 1981). Two iodinated compounds currently under clinical evaluation, *p*-iodoamphetamine and *m*-iodobenzylguanidine (*m*IBG), are such examples. Winchel *et al* (1980) demonstrated that the iodoamphetamine compounds iodinated in the *para* position had the highest brain uptake and brain-to-blood activity ratios of the isomers studied. Similarly, Wieland *et al* (1980) have demonstrated that the *para* and *meta* isomers of iodobenzylguanidine exhibit a greater ability to concentrate in the target tissue (adrenal medullae) than the *ortho* isomer. Their data also show that the *meta* isomer is more resistant to *in vivo* dehalogenation than the *para* or *ortho* isomers. Figure 6.111 shows the three forms of iodoisopropylamphetamine and iodobenzylguanidine.

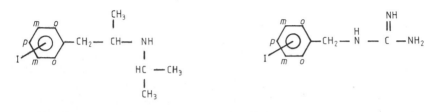

Iodoisopropylamphetamine   Iodobenzylguanidine

Figure 6.111 Molecular structures of iodoisopropylamphetamine and iodobenzylguanidine, showing various positional isomers (*o = ortho*, *m = meta*, *p = para*).

Because the position of the iodine atom is critical in the case of such biochemical tracers, and because some aromatic compounds do not possess 'activated' rings that favour easy iodination, exchange labelling (isotopic or non-isotopic) is a convenient method for radiolabelling (Baldwin 1986).

The simplest reaction is to exchange radioactive iodine for stable iodine. In this case, an iodo derivative is prepared using stable iodine incorporated at a known position in the molecule. The stable iodine-containing compound is mixed with radioactive iodine (usually as NaI) and heated to facilitate exchange. This type of exchange may be carried out in solvent (reflux) or in the molten state (melt), assuming the compound to be labelled can dissolve the radioiodide.

Iodine-for-iodine exchange may not always be rapid or efficient. For this reason, the use of catalysts such as ammonium sulphate, copper(I) salts and polymer-supported phosphonates has become increasingly popular. Iodine-for-iodine exchange also leads to low specific activity.

Brominated compounds can also be used for exchange labelling. The advantage of iodine-for-bromine substitution is that, providing the two compounds can be efficiently separated, high specific activities are theoretically possible. This technique has been used to prepare radioiodinated fatty acids (aliphatic compounds) as well as a variety of aromatic compounds (Stocklin 1977).

Iodine may also be exchanged for non-halogen groups such as diazonium salts and organometallic groups (e.g. organoboron, -tin, -silicon, -thallium and -mercury) (de la Mare 1976). These methods have the advantage of high specific activity and positional control of the iodine incorporation while labelling under relatively mild conditions (Coenen *et al* 1983).

The above discussion has demonstrated how the formulation of radiopharmaceuticals containing radiometals (e.g. ^{99}Tc^{m}) and radiohalogens (e.g. ^{123}I) offers a wide variety of compounds for the study of physiological functions.

Classification of radiotracers based on 'mechanism of localisation' has been proposed by Eckelman and Reba (1978) in an effort to promote further investigation into the biochemical and physiological processes that determine radiotracer localisation. Their classification is composed of 'substrate-non-specific' and 'substrate-specific' radiotracers. The latter classification includes those tracers which depend on specific biochemical or pharmacological interaction at the cellular level. Examples of 'substrate-specific' agents would include ^{18}F-fluorodeoxyglucose, ^{11}C-palmitic acid, ^{123}I-*m*-iodobenzylguanidine and, theoretically, radiolabelled monoclonal antibodies. All the above localise in their respective target tissue through a specific chemical reaction or take part in a ligand–substrate interaction. 'Substrate-non-specific' tracers, on the other hand, do not participate in specific biochemical reactions but rather localise via diffusion, capillary blockade, phagocytosis or compartmental-space dilution. Examples of tracers that localise via non-specific interaction include ^{99}Tc^m-labelled macroaggregates for lung perfusion imaging and ^{99}Tc^m-sulphur colloid for imaging of the reticuloendothelial system (RES).

Since their classification was published many tracers have been developed based on 'substrate-specific' mechanisms.

6.9.3 Quality control of radiopharmaceuticals

The previous section has highlighted some of the constraints placed on the development of radiopharmaceuticals labelled with ^{99}Tc^m and radioiodine. While the development of clinically useful radiotracers is not accomplished without great effort, a variety of tracers exist that make nuclear medicine imaging a viable diagnostic modality. The application of nuclear imaging to various disease states/organs will be discussed in more depth later. In order to guarantee that the radiopharmaceutical formulation is safe and will yield diagnostically useful information, a number of quality-control measures are required. These processes will help to determine the biological and radiopharmaceutical purity of the product.

Biological purity testing is required to assure that the radiopharmaceutical formulation and/or the components that make up the final preparation are safe for human administration. The tests that need to be carried out are those which determine sterility (absence of live microorganisms), apyrogenicity (absence of cellular debris of microorganisms) and toxicity.

For most radiopharmaceuticals in use today (excluding positron-labelled tracers), the components that go into the final product are produced at a manufacturer's facility. The manufacturer must assure that the components of any kit produced explicitly for human administration (especially non-oral formulations) are sterile and pyrogen-free.

Sterility may be tested by incubating appropriate samples of the component or radiopharmaceutical in approved culture medium (i.e. thioglycollate or soybean-casein digest) for 7–14 days at a specific temperature (see *United States Pharmacopeia* 1967 or *British Pharmacopoeia* 1980 for further details). If any growth occurs, the product fails sterility testing and cannot be used. Radiopharmaceuticals, especially those incorporating short-lived isotopes, cannot be tested in this manner due to the long incubation period required for confirmation of sterility.

Apyrogenicity may be tested by injecting three rabbits with the test material and then monitoring the rectal temperature of the animals over a 3 h period. If the animals collectively do not demonstrate a rise in temperature greater than 1.4° C or individually 0.6° C, then the material is considered non-pyrogenic. If one or all of the animals demonstrate temperature increases greater than that stated, the sample must be tested with five more rabbits. A temperature rise of 0.6° C in more than three of the eight animals or a rise of more than 3.7° C collectively would indicate the presence of pyrogens.

A more sensitive and rapid method for detecting pyrogens (endotoxins, in particular) has been developed, which is called the limulus amoebocyte lysate (LAL) test. The material that is responsible for the detection of pyrogens using this method is the lysate of amoebocytes from the blood of the horseshoe crab (*Limulus polyphemus*). The test material is mixed with the lysate and incubated for 1 h at 37° C. The formation of a gel (coagulation) indicates the presence of pyrogens. It should be noted that, while this method is more sensitive in the detection of pyrogens, a number of compounds can affect the formulation of the gel and therefore careful preparation is required to validate the technique. For further information, see Cooper and Avis (1975).

Radiopharmaceutical purity can be divided into three areas: radionuclidic purity, radiochemical purity and chemical purity.

Radionuclidic purity refers to the percentage of the total radioactivity of a sample that is in the desired radionuclidic form. One is interested in knowing to what extent the radiopharmaceutical is contaminated with other radionuclides. To assess purity of this type, γ-ray spectroscopy using a multichannel analyser is most often used. As each radionuclide has an emission spectrum that is unique, spectral data can allow for the determination of impurities. Other means of impurity detection include half-life determination and γ-ray filtration using attenuators known to block certain energies with high efficiency. Potential consequences of radionuclidic impurities include increased radiation dose and image degradation from collimator penetration and scattered photons.

Radiochemical purity refers to the percentage of total activity in a sample that is present in the desired chemical form. Conventional analytical techniques have been modified to detect the presence of radiochemical compounds. Chromatographic techniques (thin-layer, paper, liquid, high-performance liquid, etc) are by far the most widely used. These systems separate various components of a sample due to the varying affinity/solubility that these components display for the solid and liquid phases used. Other methods include electrophoresis and precipitation (for further discussion on radiochromatography, see Wieland *et al* (1986)).

Chemical impurities are those compounds that may be inadvertently added to a radiopharmaceutical preparation that (1) are not specified by definition or (2) pose potential formulation or toxicity problems. These impurities are rare with manufacturers' kits but may be potentially noteworthy when radiopharmaceuticals are prepared from generator eluants. Contamination of the pharmaceutical with breakdown products of the inorganic absorbent of the generator column may result in formulation problems. For example, spot tests for contaminants of aluminium (from the Mo/Tc generator column) can be performed using a colorimetric assay. Aluminium is a potential chemical impurity in the Mo/Tc generator eluant and can result in altered biodistribution of $^{99}Tc^m$-radiopharmaceuticals.

6.9.4 Studies of the brain and central nervous system

As the brain is the major control centre for many of the body's functions, the ability to assess brain and central nervous system function is of considerable interest to the medical practitioner. Radioisotope imaging allows for the study of various physiological parameters of the brain.

Techniques available to the nuclear medicine clinician have been applied to the brain, to determine at the simplest level the anatomical distribution of the radiopharmaceutical and, in the case of PET, to measure brain function quantitatively.

One of the most common nuclear medicine procedures prior to the development of x-ray CT (Chapter 4) was the visual assessment of the *integrity of the blood–brain barrier* (BBB). The BBB is a mechanism, both anatomical and physiological in nature, that maintains a selective filtration of the blood supply to the brain. It allows only a limited number of naturally occurring essential substances to enter brain tissue, such as water and glucose. It excludes a great many other compounds, such as those found in the peripheral blood, as well as a variety of drugs and toxic compounds. However, it is possible for drugs to enter the brain if they fulfil certain physicochemical criteria. These criteria will be discussed under the heading of cerebral blood-flow (CBF) agents.

When brain tissue is damaged, either through infarct or tumour, the BBB is often altered. This alteration affects the blood–brain barrier's selective filtration mechanism, thereby allowing the entry of agents that are usually excluded from normal brain tissue. It is therefore possible to image these defects in the blood–brain barrier by using radiolabelled compounds that do not usually cross the BBB.

Historically, planar scintigraphy using a labelled chelate such as $^{99}Tc^m$-glucoheptonate or $^{99}Tc^m$-DTPA has been the most common form of BBB study. Figure 6.112 shows the standard four views (anterior, posterior, left and right lateral) of the brain taken about 1 h after the administration of $^{99}Tc^m$-glucoheptonate. This agent does not traverse the intact blood–brain barrier but will enter the brain if a lesion penetrates this barrier. These images show uptake of the tracer in the scalp and the major blood vessels in the skull plus a localised area of uptake in a space-occupying lesion. The diffuse activity apparently in the rest of the brain is actually from superimposed activity in the periphery of the skull. This shows how the over/underlying activity 'blurs' the image and reduces the contrast between the lesion and the surrounding normal tissue. Figure 6.113 shows SPECT images of a patient with a lesion in the right parietal region of the brain. We can see how the tomographic study strips away unwanted information, improves the contrast between the lesion and surrounding tissue, and localises the lesion accurately in 3D. The mottled effect superimposed on these images is caused by the amplification of image noise during reconstruction with a sharp filter. Figure 6.114 shows how the image may be

(*a*)　(*b*)

(*c*)　(*d*)

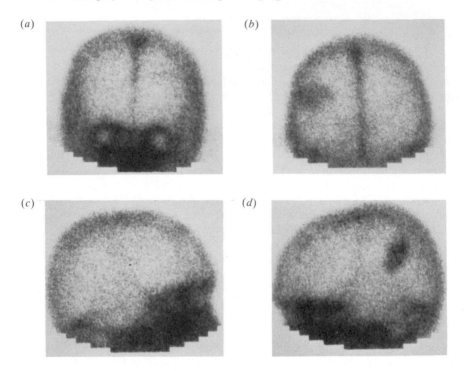

Figure 6.112 ^{99}Tc^m-glucoheptonate study of the brain showing a space-occupying lesion: (*a*) anterior; (*b*) posterior; (*c*) right lateral; (*d*) left lateral.

Figure 6.113 SPECT study of a patient with a space-occupying lesion in the right-parietal region, reconstructed with a high-pass filter.

smoothed, using a low-noise filter during reconstruction, at the expense of the spatial resolution. Although these images show, anatomically, the existence of a brain lesion, they give little functional or differential diagnostic information.

Figure 6.114 As for figure 6.113, but reconstructed with a low-pass filter.

It is believed that these compounds gain entry to diseased brain tissue through alteration in the blood–brain barrier. However, the mechanism by which these agents are retained in diseased tissue is uncertain. It is believed that pathological processes within the brain contain more extracellular water than most normal brain tissue. Therefore, the water-soluble chelate that gains entry into the lesion will equilibrate with this extracellular water, thereby showing increased activity in the area of the lesion.

Physiological function of the brain can be studied using tracers that measure blood flow, blood volume, glucose and amino-acid metabolism and receptor-binding properties. The majority of these measurements require PET, but blood flow and volume can be obtained using single-photon techniques.

One of the most common causes of death and disability in Western society is cerebral vascular disease (CVD). Most cerebral vascular illnesses are a result of atherosclerosis, hypertension or a combination of both. CVD can be categorised into three major types:

(*a*) cerebral insufficiency, due to transient blood-flow disturbances (TIA or transient ischaemic attacks);

(*b*) cerebral infarction, due to either embolism or thrombosis of the cerebral arteries; and

(*c*) cerebral haemorrhage.

A means of studying or assessing *cerebral blood flow* (CBF) would help in the accurate diagnosis of CVD, make it possible to screen for high-risk individuals, and aid in the assessment of therapies for CVD. Likewise, CBF measurements would allow for the determination of the role of CBF in both neurological and psychiatric disease.

It has been possible for some time to perform cerebral blood-flow imaging with positron emission tomography via the equilibrium measurement of ^{15}O-labelled water. However, ^{15}O is a short-lived (2.07 min half-life) radionuclide produced only in a cyclotron and therefore is not available to the great majority of nuclear medicine clinics as yet. Two other methods that would adapt themselves more readily to clinical practice would be the measurement of cerebral blood flow based on the rate of wash-out of non-metabolised substrate (Lassen *et al* 1963) and the uptake of a compound or compounds with high first-pass extraction (Sharp *et al* 1986).

The best-known tracers used for cerebral blood-flow imaging are the radioactive noble gases. ^{133}Xe with a half-life of 5.27 d is an inexpensive and chemically stable non-metabolisable radionuclide that is rapidly excreted in the lungs with minimal recirculation to the brain. Cerebral blood-flow measurements using xenon require either an intercarotid injection or an inhalation of xenon gas until equilibrium is reached between the blood and the brain tissue. The administration of the gas is stopped and the *rate of wash-out*, proportional to blood flow, is calculated. There are technical problems with this technique, one being the need for specially designed equipment to produce tomographic images (see figure 6.18 earlier) of the xenon distribution. Further, as the γ-ray energy of ^{133}Xe is very low (81 keV), problems from absorption and scatter arise. ^{127}Xe has a higher γ-ray energy but is not routinely available. Regional measurements of blood flow require careful calibration and, as a consequence, the technique has not been used widely in the clinic.

In view of the problems associated with the use of ^{133}Xe in the measurement of cerebral blood flow, other techniques have been attempted. These techniques involve the use of radiopharmaceuticals that demonstrate high *first-pass extraction* in the brain in proportion to blood flow. The advantage of using such a radiopharmaceutical is that a high percentage of administered activity enters the brain and is retained for a relatively long time, therefore allowing SPECT imaging with a gamma camera. The disadvantage of such a radiopharmaceutical is the prolonged retention time of the agent, which eliminates the possibility of sequential studies. Further, the problem of multicompartmentalisation arises, and modelling blood flow to the brain becomes complicated.

Recently, there has been considerable development effort to produce a suitable blood-flow agent labelled with a readily available radioisotope. These agents gain access to the brain tissue, that is to say they pass through the BBB due to their low molecular weight (less than 500 Daltons), their relative lipophilicity ($\log P \geqslant 1.0$) and their electrical neutrality. The mechanism of retention of these agents is as yet unknown. ^{123}I-iodoamphetamine (IMP) apparently localises in the brain in proportion to cerebral blood flow, as does ^{201}Tl-diethyldithio-carbonate (DDC). ^{99}Tc^{m}-HMPAO is also under investigation as a regional CBF agent, and particularly matches well with the gamma camera. It is a readily available agent, albeit at an enhanced cost when compared with ^{133}Xe. Figure 6.115 shows a SPECT image of ^{99}Tc^{m}-HMPAO in the brain identifying a region of low blood flow caused by a cerebral infarct (stroke). This agent has shown itself to be useful in the diagnosis of a wide range of cerebral diseases, including dementia and the measurement of neoplastic blood flow. Figure 6.116 shows a SPECT image of the distribution of HMPAO in the brain of a patient suffering from Alzheimer's disease, showing an abnormal distribution pattern. Figure 6.117

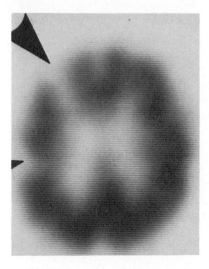

Figure 6.115 Single transaxial $^{99}Tc^m$-ʜᴍᴘᴀᴏ sᴘᴇᴄᴛ image of the brain of a patient suffering from stroke. Arrows indicate sites of reduced ᴄʙꜰ. (Courtesy of P J Ell.)

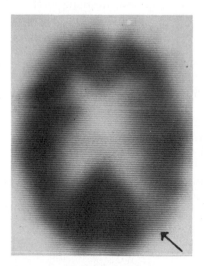

Figure 6.116 Single transaxial $^{99}Tc^m$-ʜᴍᴘᴀᴏ sᴘᴇᴄᴛ image of the brain of a patient suffering from Alzheimer's disease. Arrow indicates site of reduced ᴄʙꜰ. (Courtesy of P J Ell.)

shows ʜᴍᴘᴀᴏ sᴘᴇᴄᴛ images of the brain for patients with brain tumours. These images show a variable uptake pattern in the tumour, indicating differences in tumour perfusion.

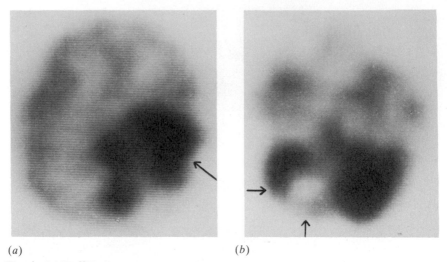

(*a*)   (*b*)

Figure 6.117 ^{99}Tc^m-HMPAO SPECT images of two patients suffering from brain neoplasms: (*a*) increased uptake at the site of a grade III astrocytoma; (*b*) decreased uptake at the site of a metastatic lesion in the cerebellum.

Cerebral blood volume, a measurement of the total volume of blood in the brain, can be determined by using a tracer such as ^{99}Tc^m-RBC (red blood cells). Again SPECT is required to provide regional measurements of blood volume.

Although blood–brain barrier and cerebral blood-flow imaging are now possible with conventional nuclear medicine techniques, a much wider range of functional imaging procedures are possible using positron tomography and appropriate radiopharmaceuticals labelled with positron-emitting radionuclides.

PET imaging of the brain has provided a vast reservoir of quantitative functional information, which has helped to quantify normal brain physiology and to highlight the changes in metabolism and function caused by cerebral disease. Much of this work is presently unavailable to clinical nuclear medicine because of the short half-life of the radioisotopes used (^{15}O, ^{13}N, ^{11}C, ^{18}F) (see table 6.7 in §6.4.2), although it may soon be feasible to transport ^{11}C- and ^{18}F-labelled tracers to hospitals in the vicinity of a cyclotron. There is an established battery of PET-based functional measurements to study the brain. Blood-flow measurements (made with C^{15}O$_2$) can be allied with oxygen (^{15}O$_2$) metabolism and blood volume (C^{15}O) to determine the oxygen-extraction efficiency of normal and abnormal brain tissue (figure 6.118). In addition, ^{18}F-fluorodeoxyglucose (FDG) can be used to mimic the glucose metabolism rates of the brain, and a comparison with oxygen extraction gives information on aerobic and anaerobic metabolic function of the brain. The use of FDG has provided considerable information relating to a wide range of normal and abnormal cerebral pathologies. The agent is metabolised similarly to glucose, which provides most of the cerebral energy. Hence, it is possible to identify which regions in the brain are functioning during a range of normal visual processes (figure 6.119), such as reading and viewing complex scenes. Clearly, abnormalities affecting any of these processes will be detectable by using a quantitative analysis of glucose metabolism. More recently, FDG has been shown to provide important information on cerebral defects including seizure, dementia, motor defects, neoplasms and other cerebral/neurological disorders. Figure 6.120 shows images of FDG uptake in the human brain for a patient suffering from Alzheimer's disease. One can surmise that the successful treatment of these disorders should lead to local functional changes in glucose metabolism, with an appropriate return to normal metabolism. There are claims that the pattern of FDG metabolism in the brain can be used to characterise disease. However, these claims must surely be treated guardedly, bearing in mind the complexity of cerebral function. It is unlikely that any *single* functional (or anatomical) test will be sufficient to identify unequivocally cerebral abnormalities. It is possible, however, that a battery of tests, as shown earlier and below, may elucidate this complex area of physiology.

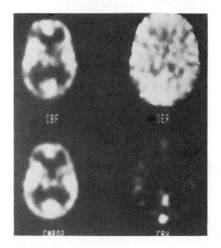

Figure 6.118 PET images of regional cerebral blood flow (CBF), regional cerebral oxygen metabolism (CMRO$_2$), regional cerebral oxygen extraction (OER) and regional cerebral blood volume (CBV) using C^{15}O$_2$, ^{15}O and C^{15}O. (From Phelps (1986).)

An example of this is the use of three separate tracers in the study of a primary brain neoplasm (glioma) (Bergström *et al* 1983a). Figure 6.121 shows how ^{68}Ga-EDTA can indicate a breakdown in the blood–brain barrier in a region showing an abnormality on an enhanced x-ray CT scan. Glucose metabolism, measured using ^{11}C-glucose, shows an increased uptake in the region of the lesion. However, ^{11}C-methionine, a radiolabelled amino acid used for the study of protein synthesis, shows increased uptake in a region somewhat larger than that indicated by previous studies. At *post mortem*, it was shown that the amino-acid image showed more closely the true extent of the tumour. The reasons for the variation in tests were thought to be due to variations in histopathology of different regions of the tumour, highlighting the state of differentiation of different parts of the tumour from normal brain tissue.

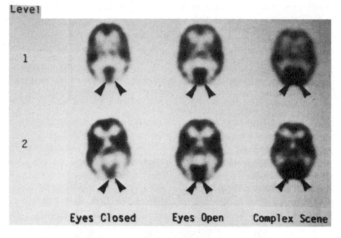

Figure 6.119 PET images at two different levels (1 and 2) of the regional distribution of ^{18}F-FDG in the brain and the effects of visual processes. (From Phelps (1986).)

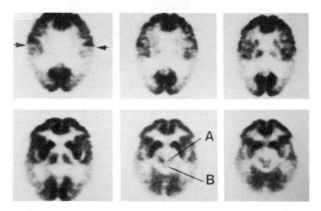

Figure 6.120 PET images of the regional distribution of ^{18}F-FDG in the brain of a patient suffering from Alzheimer's disease. Arrows indicate regions of hypometabolism. A and B indicate fine structure of brain. (From Phelps (1986).)

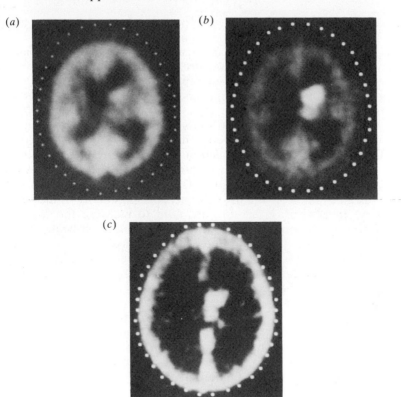

Figure 6.121 The distribution of (*a*) ^{11}C-glucose, (*b*) ^{11}C-methionine and (*c*) ^{68}Ga-EDTA in the brain of a patient with an astrocytoma. (From Litton *et al* (1984).)

Another growing area of study in the brain is the use of receptor-binding tracers to localise neurological function in the brain. Using tracers labelled with ^{11}C and ^{18}F, it may be possible to study problems associated with neuroreceptor abnormalities and gain insight into mental disease.

It can be seen from the discussion above that the majority of quantitative physiological information related to cerebral function has come from studies with PET. This contrasts with the more 'anatomical' nature of the conventional nuclear medicine studies and highlights the need for better physiological tracers for single-photon studies or improved access to the superior PET techniques.

6.9.5 Cardiac imaging

The function of the heart and the resultant effect on the other body mechanisms is, of course, one of the most important areas of study in medicine. It is self-evident that the enormous and continous stress applied to the heart to provide a blood supply sufficient for all the body's needs means that any abnormality, either in the heart muscle itself or in the supplying vessels, can lead to catastrophic failure and, as we know too readily in modern society, to death. There are two major areas of study of the heart which both provide ample scope for the application of radioisotope imaging and the production of quantitative functional information.

The most common radioisotope study of the heart is the measurement of the output of blood from the left ventricle to the body (*ventriculography*). Clearly, any significant reduction in the volume of blood supplied to the body can lead to serious malfunctions of other organs/systems. The left-ventricular ejection fraction (LVEF) has been discussed earlier (§6.6.7) as an example of the use of dynamic scintigraphy in the study of physiology. In general, ^{99}Tc^m-RBC are used to measure the fraction of blood in the circulation ejected by the heart during each beat. Measurements can be made through either the first-pass technique or using multigated acquisitions (MUGA). The importance of obtaining a good injection bolus and the correct positioning of the gamma camera during these studies should be emphasised. The effects of recirculation of the blood on the injected activity can produce spurious results. More importantly, because of the anatomical orientation of the heart with respect to the thorax, the central axis is oblique to the normally accepted imaging planes, making conventional planar images unacceptable. The camera must therefore be positioned to maximise information from the ventricles with the minimum of interference from either the atria or major vessels. Figures 6.122 and 6.123 show images of normal LV function during a MUGA study and the normal RV/LV system in a

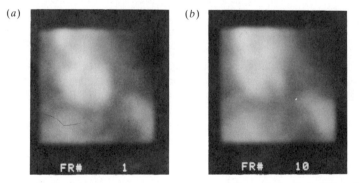

Figure 6.122 Blood-pool images of the left ventricle of the heart in a MUGA study using ^{99}Tc^m-RBC: (*a*) at end diastole; (*b*) at end systole.

first-pass study. The poor spatial resolution and sensitivity of the gamma camera means that it is difficult to delineate accurately regions of interest. Repeat studies of the same patient during rest–stress procedures and comparison of the results with a 'normal range' can give invaluable information relating to overall efficacy of cardiac function (figure 6.124). In addition, the use of ciné display of MUGA images can highlight wall-motion abnormalities, which also show up well in parametric images of amplitude and phase in the cardiac cycle (figure 6.125).

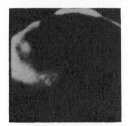

Figure 6.123 Blood-pool image of the left and right ventricles of the heart in a first-pass study using ^{99}Tc^m-RBC.

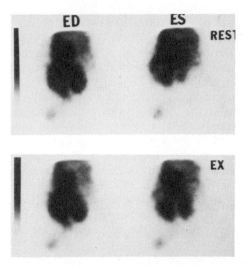

Figure 6.124 Left-ventricular images at end diastole (ED) and end systole (ES) comparing studies taken under rest and stress (EX) conditions. (From Thrall and Swanson (1983).)

(a) (b)

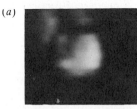 

Figure 6.125 (a) Amplitude and (b) phase images of the heart produced using a single harmonic fit to the left-ventricular ejection fraction.

The body's sodium/potassium pump maintains cardiac muscle tone and contractility, requiring an active expenditure of energy that is dependent on the supply of oxygen to the muscle. Hence blood flow is critical for *myocardial function*, since blood carries the necessary oxygen to the myocardium. One way to assess myocardial function, therefore, is to determine blood flow or perfusion (flow per unit volume) of the myocardium. Potassium is found intercellularly in the myocardium and the net extraction efficiency from blood is approximately 70%. As no useful potassium isotopes for medical imaging exist, we must look to the use of similar monovalent cations. The group III metal thallium (^{201}Tl) has an ionic radius of 0.144, similar to potassium, which has an ionic radius of 0.133. Therefore, the monovalent cation ^{201}Tl has found a place in imaging myocardial perfusion. Likewise ^{81}Rb and ^{82}Rb are potassium analogues useful for monitoring myocardial blood flow. However, ^{82}Rb is a positron emitter with a half-life less than 2 min and requires an infusion generator system (^{82}Sr/^{82}Rb).

Alternative myocardial blood-flow tracers incorporating ^{99}Tc^{m} are under development by a number of research groups (Jones *et al* 1982, Deutsch *et al* 1981, Narra *et al* 1987). The physical characteristics of ^{99}Tc^{m} are far superior to those of ^{201}Tl, and new compounds such as ^{99}Tc^{m}-isonitriles and ^{99}Tc^{m}-boronic acid adducts look promising. These compounds are currently under clinical evaluation.

Malfunctions in cardiac performance are usually related to lesions in the cardiac muscle, the majority of which are caused by impaired coronary blood supply. ^{201}Tl in conjunction with planar scintigraphy or SPECT has been used to show abnormalities in the heart-muscle function. SPECT studies can provide multislice oblique images of the cardiac muscle (figure 6.126), showing the familiar doughnut- or horseshoe-shaped distribution of the agent. In spite of the limited spatial resolution, it is usually not difficult to identify lesions by discontinuities in these distributions. It may even be possible to determine which of the coronary arteries are likely to be responsible for the cardiac malfunction.

Using ^{99}Tc^{m}-pyrophosphate, it is possible to image myocardial infarcts. ^{99}Tc^{m}-pyrophosphate was originally developed as a bone-seeking agent. It is hypothesised that, in infarcted heart tissue, calcification occurs, and the pyrophosphate probably interacts with and binds to the calcified tissue of the myocardium.

The contribution of PET to the study of cardiac function is steadily increasing. ^{82}Rb is available from a generator and its short half-life makes multiple studies of myocardial function possible over a short timespan. ^{13}NH$_3$ has also been used to monitor myocardial perfusion. Obviously, the increased spatial resolution and sensitivity of PET is a great advantage when imaging an organ containing important tissues of small dimensions. Figure 6.127 shows how the increased resolution of PET can achieve sharper and hence more useful images of the myocardium. Blood-pool imaging and ejection-fraction measurements are possible using ^{11}CO, which labels the red-blood cells. Figure 6.128 shows again the high quality of image available for analysis. Further contributions of PET to the study of muscle-energy metabolism measurements include the use of labelled palmitic or oleic acids and FDG. These show the muscle metabolising fatty acids in normal aerobic state and switching to glucose during the onset of anaerobic conditions.

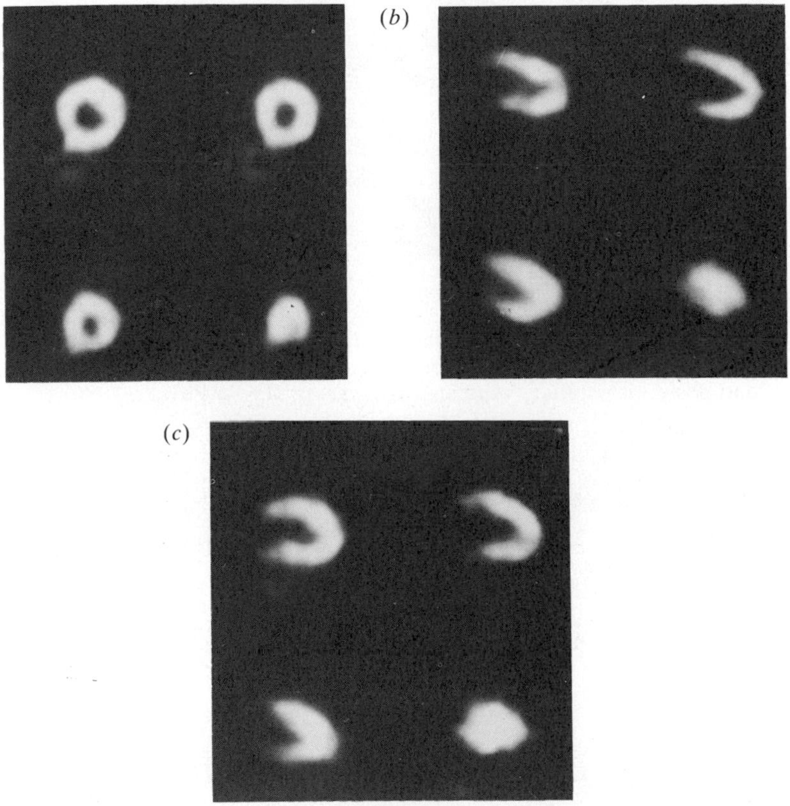

Figure 6.126 SPECT images of the myocardium using ^{201}Tl chloride in a patient with normal coronary arteries: (*a*) short axis; (*b*) superior long axis; (*c*) right anterior oblique long axis. (From Cerqueira *et al* (1987).)

Hence, although routine radioisotope imaging can be used as an indicator of cardiac function, the metabolic process involved in the myocardium can presently only be understood using PET.

6.9.6 *Renal imaging*

The kidneys regulate the volume and composition of extracellular fluid

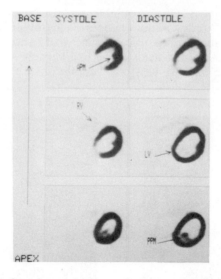

Figure 6.127 Gated PET images of the myocardium using $^{13}NH_3$, showing the left ventricle (LV), right ventricle (RV) and the anterior and posterior papillary muscles (APM, PPM). (From Phelps (1986).)

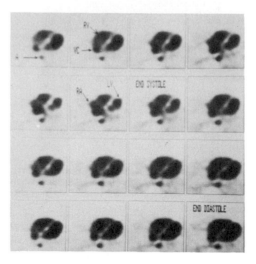

Figure 6.128 Gated PET images of the cardiac blood pool using ^{11}CO, showing the right atrium (RA), right ventricle (RV), left ventricle (LV), the vena cava (VC) and the aorta (A). (From Phelps (1986).)

in the body. They also have an important role in the body's blood circulation, as they receive 25% of the cardiac output. Blood supplied by the renal artery (figure 6.129) is filtered by the nephrons, the functional units of the kidney, and cleared via the renal vein. Both the glomerulus and tubules take part in this process. Removal of filtrate from the kidney is via the renal pelvis into the ureter to the bladder. Renal clearance C (ml min^{-1}), represented by blood volume cleared of any substance per unit time is given by

$$C = UV/P \tag{6.59}$$

where U and P are the concentrations of matter in the urine and in the plasma (mg ml^{-1}), and V is the flow rate of urine (ml min^{-1}).

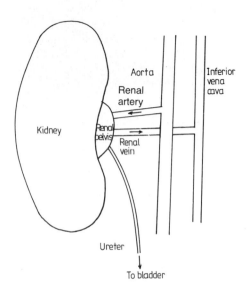

Figure 6.129 Schematic of the renal 'plumbing system'.

The glomerular filtration rate (GFR) of the kidney is best measured with chelating agents such as EDTA and DTPA. Renal imaging agents usually display low molecular weight, high hydrophilicity (water solubility) and low protein binding. However, the characteristics that determine whether the tracer is excreted, secreted or filtered are unclear.

Effective renal plasma flow (ERPF) is determined by using sodium *o*-iodohippurate (OIH), which is cleared 20% by the glomeruli and 80% by tubular secretion. In either case, renal function measurements are performed via a dynamic renogram, described in §6.6.8.

Kidney-function evaluation can be separated into several major areas. In the case of *renal transplant*, studies to determine the viability of the transplanted kidney may measure the perfusion rate of the kidney in comparison to the blood supply. Measurements are made over the first

30–40 s of the renogram and may be used to detect rejection, obstruction or ischaemic damage. The measurements of *divided kidney function* and detection of *renal scarring* are best performed using $^{99}\text{Tc}^{\text{m}}$-DMSA. This agent concentrates in the renal cortex over a period of a few hours and clears only slowly (40% in 24 h). Planar images of the kidneys 3 h after administration can be used to measure the relative kidney uptake (figure 6.130). Peripheral lesions can be highlighted by viewing the ciné display of reconstructed SPECT images.

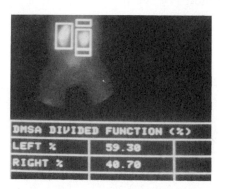

Figure 6.130 Image of dimercaptosuccinic acid (DMSA) uptake in kidneys, showing regions of interest used to determine relative divided function.

Dynamic renography using $^{99}\text{Tc}^{\text{m}}$-DTPA or ^{123}I-OIH can produce time–activity curves representing renal function and measurements of GFR or ERPF, and allow detailed examinations to be made of the function of the renal pelvises, ureters and bladder. Figure 6.131 shows examples of time–activity curves that would arise from a range of urinary problems. Clearance of urine from the bladder during micturition can also be studied to detect any 'reflux' of urine back into the kidneys—a common cause of kidney infection in young children, which may lead to kidney failure in adulthood.

Parametric imaging can play an important role in renography, as it allows a large quantity of information to be compacted into simple functional images. An example of its use is shown in figures 6.57 and 6.132. In this case, each image pixel within a kidney ROI has a 'time–activity curve' (a dixel) showing the characteristic shape of the renal-function curve. Pixels within the parenchymal region have a short time to peak (TTP) in their dixels, whereas those in the renal pelvis have a relatively long TTP. This allows a separation of the two regions by 'windowing' the display using TTP and, hence, separating out the role of those two regions in the kidney function. This may, in some cases, help to differentiate between different abnormal conditions, such as hydronephrosis and renal obstruction (Hyde *et al* 1988).

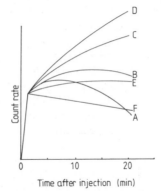

Time after injection (min)

Figure 6.131 Time–activity curves of renal function showing the effects of various disorders: A, earliest signs of obstruction; B and C, signs of increasing obstruction; D, complete obstruction; E, persisting obstruction; F, non-functioning kidney.

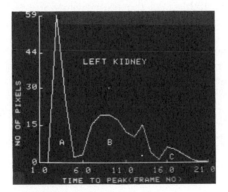

Figure 6.132 Histogram of time-to-peak values for dixels used to derive the parametric image shown in figure 6.57. The three regions in the histogram relate to the perfusion (A), parenchyma (B) and collecting system (C).

6.9.7 Skeletal imaging

Radionuclide imaging of the skeleton is a non-specific technique for evaluating acquired bone disease such as traumatic, neoplastic and degenerative problems. The localisation of skeletal disease with conventional bone-seeking radiopharmaceuticals is based on bone blood flow and the metabolic turnover of skeletal tissue. In general, bone production and increased blood flow associated with skeletal abnormalities lead to increased uptake of the radiopharmaceutical. In a few cases, where blood flow is impaired or bone has been replaced by tumour, decreased accumulation is seen.

Historically, isotopes of strontium (^{85}Sr, ^{87}Sr^m) and fluorine (^{18}F) have been used for skeletal imaging, but the emitted photon energies are only suitable for either a scanner or positron camera system, respectively. ^{99}Tc^m-labelled agents, suitable for gamma-camera imaging, were produced in the 1970s based on the use of ^{99}Tc^m-polyphosphates. However, phosphonates such as methylenediphosphonate (MDP) are now the agents of choice. Like ^{18}F, they clear from the body tissues in a few hours for patients with normal kidney function.

The ^{99}Tc^m-phosphonate and phosphate compounds are thought to localise in bone as a function of blood flow and affinity of the Tc complex towards the metabolically active bone mineral, hydroxyapatite.

The modern gamma camera, 40 cm in diameter, is still barely large enough to span the width of the human skeleton (figure 6.133). It is possible to scan the whole skeleton from head to foot using a standard collimator (figure 6.134). A diverging/parallel-hole collimator can be used to reduce imaging times for skeletal scanning, at the expense of some spatial distortion at the image edges. Generally, however, scanning with a gamma camera or linear scanner does not optimise the spatial resolution available. As the majority of the skeleton is peripheral, individual images of selected parts of the body are the method of choice. In general, up to 10 separate images, including anterior, posterior and some lateral views, are taken (figure 6.135), with the patient as close to the collimator surface as possible. Skeletal images are very 'anatomical' in nature due to the easy recognition of the structures and, with about half a million counts per image, excellent delineation of normal and abnormal skeleton is provided. For viewing localised tumours, SPECT bone scans can be useful, especially when the camera views are displayed in ciné mode to give a pseudo-3D effect.

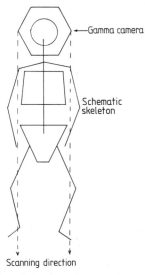

Figure 6.133 Schematic of the imaging of the human skeleton using a scanning gamma camera with a large field of view.

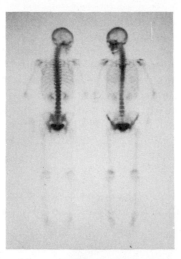

Figure 6.134 Whole-body skeletal images (posterior, anterior) using a scanning gamma camera. (Courtesy of Toshiba Medical Systems.)

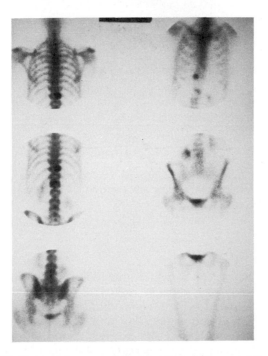

Figure 6.135 Six spot images of the skeleton using a gamma camera.

The major role for skeletal radioisotope imaging is in the staging of malignant disease, evaluation of primary skeletal tumours, diagnosis of inflammatory disease and the study of joints. Metastases in the skeleton are often difficult to detect radiographically, as a large percentage of mineral bone content must be lost before x-ray detection is possible. The majority of these lesions will be seen by skeletal scintigraphy, which may help in the staging of patients with breast, lung or other carcinomas. Imaging of joints is particularly valuable in the detection of inflammatory disease and for the detection of increased blood flow in diseases such as synovitis. Small-part imaging of knees, elbows and ankles can be improved by the use of a pinhole collimator, which, as well as improving spatial resolution, provides magnification (figure 6.136).

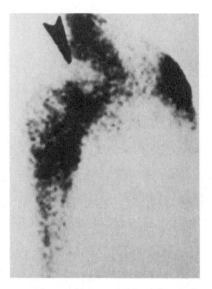

Figure 6.136 Image of an abnormal hip joint using a pinhole collimator, showing femoral necrosis (arrow). (From Harbert and Da Rocha (1984b).)

6.9.8 The respiratory system

The lungs are the organs in which the blood circulating in the body is reoxygenated and carbon dioxide removed. The blood/air interface in the lungs is populated with alveoli that provide the oxygen–carbon dioxide exchange. Blood supply to and from the lungs is via the pulmonary arteries and veins. The arteries divide into arterioles (blood vessels with a diameter of approximately 7–10 μm). Blood also reaches the lungs via the bronchial arteries (from the aorta) to nourish the tissues of the lung. Air enters the lungs via the pharynx, larynx and into the trachea or windpipe. This branches into main-stem bronchi, and then into smaller airways such as the bronchioles.

The major malfunctions of the lungs can cause either blood-flow or air-flow problems or, in some cases, both. The blood flow to each lung is divided into segments, and any abnormal flow will show up as abnormal perfusion in one or more lung segments. Similarly regional abnormal ventilation can show up as reduced air supply to a particular lung segment. However, the most common radioisotope studies of the lung are ventilation/perfusion scans.

Ventilation studies are carried out using freely diffusible gases such as xenon and krypton (figure 6.137). Generally, there are three lung ventilation examinations available for clinical diagnosis: single-breath studies, lung-volume measurements; and lung ventilation distribution, which is measured using clearance techniques. ^{133}Xe has been the radioisotope of choice for all these methods, but, with its long half-life (5.27 d) and low-energy photon emission (81 keV), it is not ideal for diagnostic gamma-camera imaging. ^{127}Xe is now readily available and has an acceptable photon emission energy of 203 keV, but a 36.4 d half-life. ^{81}Kr^m is available from a ^{81}Rb/^{81}Kr^m generator and, with a half-life of 13 s and a 190 keV γ-ray emission, is proving very useful when used in equilibrium, especially as it can be imaged on the same projections as the perfusion scans and so provide good registration.

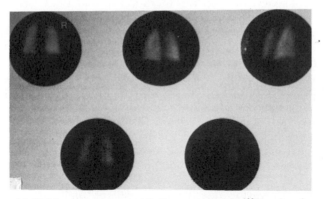

Figure 6.137 Normal lung ventilation scan with ^{133}Xe showing posterior, left and right posterior oblique views at equilibrium, and posterior wash-out images. (From Harbert and Da Rocha (1984b).)

Single-breath studies show the distribution of a radioactive bolus, whereas an equilibrium study shows the gas distribution that corresponds to lung volumes.

Perfusion studies, which show how well the blood supply to the lungs is distributed, are most easily made using radiolabelled microspheres or macroaggregated albumin (MAA) (figure 6.138). These particles are typically 30–40 μm across and pass out of the larger blood vessels and are impacted in the arterioles and capillaries. Their distribution is then closely related to blood flow in the pulmonary artery. Alternatively, xenon dissolved in saline and injected intravenously can also be used to measure lung perfusion.

Ventilation/perfusion (V/Q) scans in combination can show whether defects in blood flow are directly correlated to defects in ventilation. The intercomparison of the two scans can be a powerful diagnostic tool and can help to differentiate between different respiratory diseases. Figure 6.139 shows V/Q images of the same patient. There are many mismatches in the images, which indicate that ventilation is more abnormal than perfusion, a finding consistent with chronic obstructive airway disease. Figure 6.140 shows V/Q images where poor regional perfusion is matched to almost normal ventilation, an indication of pulmonary embolism. Other diseases where V/Q scans can be differentially diagnostic are emphysema, bronchitis, obstructive airway disease and cancer of the bronchus.

6.9.9 Liver/spleen imaging

The *liver* is the largest organ of the body and consists of several lobes, positioned under the right side of the rib cage, just inferior to the diaphragm. The major functions of the liver include detoxification of blood supply, formation of bile into the intestines, removal of foreign particles by phagocytosis and the metabolism and synthesis of a variety of proteins. Two types of cells make up the bulk of the liver mass: hepatocytes, which are responsible for numerous metabolic processes within the liver including the excretion of bile into the intestines; and Kupffer or phagocytic cells. The function of the phagocytic cells is to remove foreign particulate matter from the circulation. Using appropriate radiopharmaceuticals, it is possible to image both types of cell function within the liver. It should be noted that the phagocytic cells in the liver are part of a larger system known as the reticuloendothelial system (RES), which also includes the spleen and bone marrow.

In order to image the reticuloendothelial system or the phagocytic cells of the liver, spleen and bone marrow, a radiolabelled colloid is used. A variety of radiocolloids have been used for the assessment of RES function. However, due to its excellent imaging characteristics, $^{99}Tc^m$ is the radionuclide of choice for liver/spleen imaging. Distribution of the radiolabelled colloid is related to the blood supply to the various organs containing phagocytic cells. In a normal adult, approximately 85% of the injected colloid will be taken up in the liver, 10% in the spleen and the remainder will be taken up into the RES within the bone marrow. It is possible to influence the distribution of the radiocolloid by altering its size. However, it is not possible to make a size-specific colloid for imaging either the liver, spleen or bone marrow in total isolation. In order to be removed by the phagocytic cells of the RES, the colloid must be smaller in size than the capillary junctions in the lungs. This size allows the agent to pass through the lung capillary bed after intravenous injection and into the left of the heart, where it will then be distributed to the organs containing the phagocytic cells in proportion to blood flow.

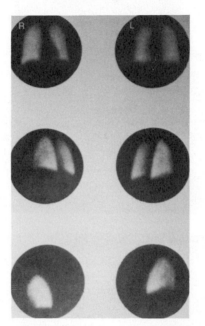

Figure 6.138 Normal lung perfusion scan of the same patient as in figure 6.137 using ^{99}Tc^{m}-MAA showing anterior, posterior, left and right posterior oblique, and left and right lateral views. (From Harbert and Da Rocha (1984b).)

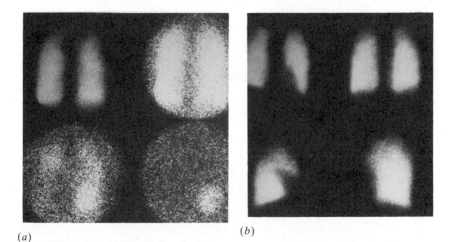

(*a*) (*b*)

Figure 6.139 (*a*) Ventilation and (*b*) perfusion scans of a patient showing the effects of bronchial obstruction. (From Harbert and Da Rocha (1984b).)

(a)

(b)

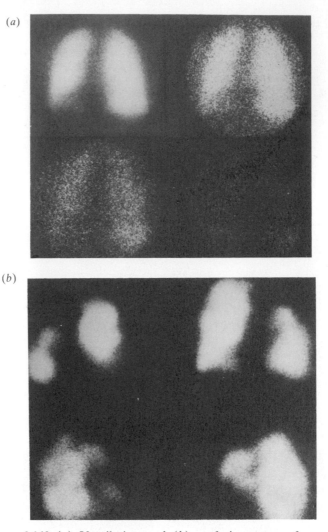

Figure 6.140 (*a*) Ventilation and (*b*) perfusion scans of a patient with pulmonary embolism. (From Harbert and Da Rocha (1984b).)

Images of the liver are usually obtained 20 min after intravenous injection. A variety of views are required, including anterior, posterior and right lateral views (figure 6.141). Because the left lobe of the liver is very anterior, it is often not visible in a posterior view due to photon absorption in the spine. Other views may be required, depending on the results of the initial images, or if there is a clinical indication. In most cases, only 80–150 MBq of ^{99}Tc^m-colloid is used for liver/spleen imaging because the high localisation in the liver and spleen allows for good count rates and short imaging times.

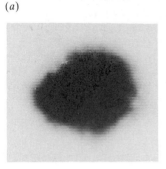

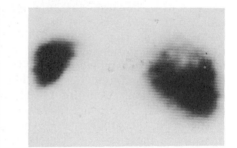

(a)

(b)

(c)

Figure 6.141 Planar images of the liver and spleen (using $^{99}Tc^m$-sulphur colloid) of a patient with a space-occupying lesion: (a) anterior; (b) posterior; (c) right lateral.

Liver/spleen imaging using $^{99}Tc^m$-colloids aid in the diagnosis of a variety of liver ailments, including hepatitis, cirrhosis, tumours, metastatic lesions and conditions affecting the anatomy of the liver. Figure 6.142 shows a SPECT image of a liver with space-occupying lesions from breast cancer and indicates clearly how SPECT improves the visualisation of these lesions when compared with figure 6.141. Additionally, diseases of the spleen such as splenomegaly can also be diagnosed using this technique.

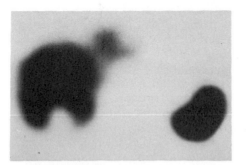

Figure 6.142 A transaxial SPECT image of the study shown in figure 6.141. The lesion is clearly shown in the posterior liver.

Using another range of radiopharmaceuticals, it is possible to assess the hepatocyte function of the liver. One of the first agents used for the assessment of hepatobiliary function was ^{131}I-labelled rose bengal. This radiolabelled dye was first used as a hepatobiliary agent because of its high extraction from the blood into the hepatocyte cells, followed by rapid clearance into the bile system. However, this ^{131}I-labelled tracer has been all but replaced by ^{99}Tc^m-labelled iminodiacetic acid (IDA) derivatives. IDA compounds have been shown to bind to ^{99}Tc^m and form stable compounds that are useful for assessing hepatobiliary function (figure 6.143). There are a variety of IDA compounds that exhibit a spectrum of hepatobiliary specificity. While in a normal patient a majority of these IDA compounds perform sufficiently to image the gall bladder, only a few allow for visualisation of the hepatobiliary system in patients with jaundice or high bilirubin levels in the blood.

^{67}Ga-citrate has been used to identify hepatoma (primary liver cancer) and amoebic abscess in the liver. Both conditions cause increased gallium accumulation at locations showing decreased uptake of ^{99}Tc^m-colloid. Liver metastases and regenerating liver tissues do not show this increased gallium accumulation and hence the agent can be powerfully diagnostic in these cases.

6.9.10 The endocrine system

The organs of the endocrine system are involved in the production of hormones, which are used to control a wide range of body metabolic functions. Much of the study of the function of this system is based on *in vitro* assays, which are used to determine the production levels of hormones and to relate these levels to normal/abnormal function. However, the thyroid gland and, more recently, the parathyroid glands and adrenal glands have been the subject of growing radioisotope imaging procedures to determine the distribution of functions within these organs and to perform radiotherapy using radiopharmaceuticals.

The *thyroid gland* secretes two hormones, thyroxin (T$_4$) and triiodothyronine (T$_3$), which are used to regulate cell metabolism. The thyroid also produces thyrocalcitonin, which is involved in inhibition of calcium resorption in bone. Anatomically, the gland is conveniently placed in the neck (figure 6.144), so that imaging is not difficult, there being little surrounding tissue exhibiting strong metabolic function. However, the organ is small and the lobes of the thyroid gland are comparable in size to the spatial resolution of SPECT. Hence, although simple planar scintigraphy can provide gross qualitative functional information, thyroid SPECT is more difficult and of little use in the detection of intrathyroid malfunction.

The most important feature of thyroid function from a diagnostic and therapeutic viewpoint is the high rate of metabolism of iodine. A normal thyroid will contain 25% of the body's iodine and an overactive or hyperfunctioning gland may concentrate almost all of the available iodine.

Historically, ^{131}I has been used, in small quantities, to study thyroid function. In large quantities, ^{131}I can be used therapeutically to provide

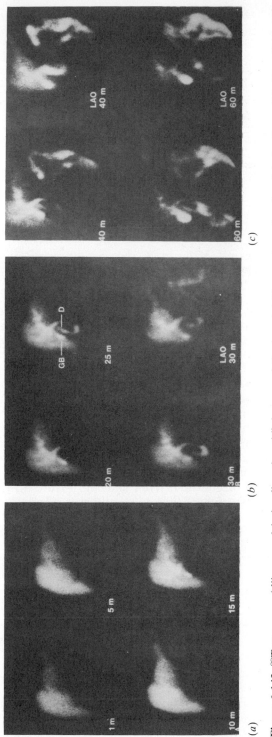

Figure 6.143 ⁹⁹Tcᵐ-ᴅꜱɪᴅᴀ (diisopropyl iminodiacetic acid) images of the hepatobiliary system showing (*a*) the liver, (*b*) the gall bladder and (*c*) duodenal function over a period of 60 min. ɢʙ is gall bladder, ᴅ is duodenal bulb. (From Harbert and Da Rocha (1984b).)

partial or complete ablation of the gland, and this presents the most outstandingly successful use of radiation therapy using internal radionuclides in the treatment of disease.

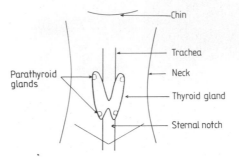

Figure 6.144 Schematic of the neck showing the position of the thyroid and parathyroid glands.

From a diagnostic viewpoint, ^{131}I is totally unsuitable, of course, being a β^- emitter and producing high-energy γ-rays (see table 6.5 in §6.4.2). ^{123}I has been used successfully as a diagnostic replacement, as the emitted photon energy (159 keV) is more appropriate for gamma-camera imaging, and the absence of $\beta^\pm$ emission produces a lower radiation dose to the organ. However, this isotope is relatively expensive and not readily available. ^{99}Tc^m in the form of pertechnetate is used to measure the trapping mechanism of the thyroid, although it cannot be used to monitor the full metabolic pathway. This radiopharmaceutical is now used in more than 90% of all diagnostic thyroid studies and only occasionally is the use of an iodine isotope necessary. Table 6.18 shows a comparison of administered activity and radiation doses for the use of these radioisotopes in diagnostic thyroid imaging. Figure 6.145 shows an image of a normally functioning thyroid gland using ^{99}Tc^m.

Table 6.18 Radioisotopes used for diagnostic thyroid imaging and their related radiation doses.

Radioisotope and form	Decay mode	Principal γ-ray energy (keV)	$T_{1/2}$	Usual dose (MBq)	Radiation dose (mGy) Whole body	Thyroid
Na^{131}I	$\beta\gamma$	364	8.04d	1–2	0.2–0.4	250–500
Na^{123}I	EC	159	13h	2–20	0.3–3.0	5–40
Na^{99}Tc^{m}O$_4$	IT	140	6h	80–160	0.8–1.6	2–4

The most common diseases of the thyroid are hyperthyroidism, hypothyroidism, thyroiditis and cancer. Hyperthyroidism can take several forms, which include Graves' disease and Plummer's disease. Figure

6.146 shows images of thyroid glands illustrating these two pathologies—the former shows as a fairly uniform general increase in thyroid function, the latter as a toxic goitre. Hashimoto's thyroiditis is a virally produced inflammatory disease of the thyroid, which affects organification of iodine and can often be differentiated most simply from hyperthyroidism by radioisotope-uptake measurements. Cancer of the thyroid always shows up clearly as a hypofunctioning area in a thyroid image (figure 6.147), although differentiation between cancer and cystic disease requires an ultrasound examination.

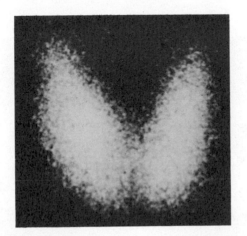

Figure 6.145 ^{99}Tc^m-pertechnetate image of normal thyroid function.

(a)   (b)

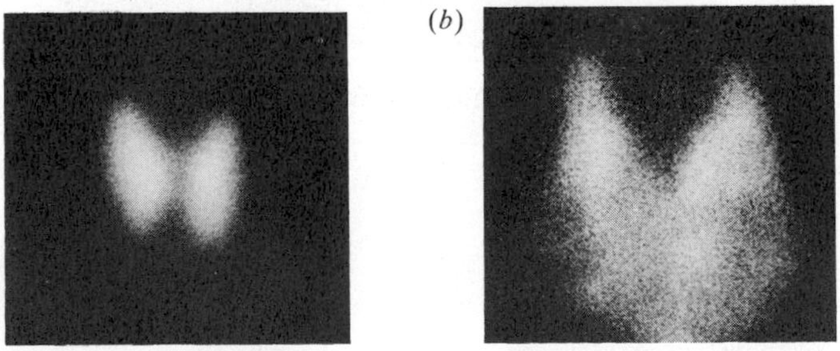

Figure 6.146 Na^{131}I images of thyroid glands showing (a) Graves' disease and (b) Plummer's disease.

In addition to the diagnostic use of radioisotopes, it is possible, as mentioned before, to use ^{131}I as a therapeutic agent. A modest dose of 75–400 MBq of ^{131}I can be used to treat hyperactive glands. The exact dose required has been difficult to determine because of the wide variation in gland size and uptake. Overdosing of the thyroid can lead to hypothyroidism and a lifelong dependency on thyroid hormone tablets. Recent measurements (Ott *et al* 1987) using [124]I and PET have indicated a potential variation by a factor of 5 in the radiation dose achieved by this therapy. These studies show that high-resolution PET imaging may have a very important role to play in the measurement of functioning organ volume (figure 6.148) and, in particular, in determining the radiation dose to an organ/tumour from radioisotope therapy (table 6.19). These studies have also shown that Graves' disease may not be as uniform in nature as conventional scintigraphy demonstrates (figure 6.149). Thyroid volume may also be assessed using ^{99}Tc^m-pertechnetate and SPECT (Webb *et al* 1986a).

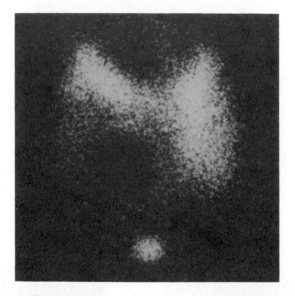

Figure 6.147 ^{99}Tc^m-pertechnetate image of a patient with follicular thyroid cancer.

(*a*)

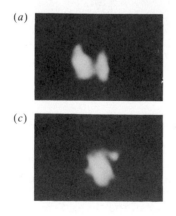

(*c*)

(*b*)

Figure 6.148 Na^{124}I PET images of a hyperactive thyroid (3 mm thick sections) showing pyramidal lobe: (*a*) coronal; (*b*) transaxial; (*c*) sagittal.

Table 6.19 Estimated radiation doses to a range of hyperactive thyroids from a single dose of 75 MBq of [131]I. (Data from Ott *et al* (1987).)

Thyroid mass (kg)	Thyroid uptake (%)	Radiation dose[a] (Gy)
0.029	86	48
0.034	87	41
0.046	86	30
0.033	44	22
0.069	98	23
0.036	40	18
0.053	35	11

[a] Assumes an effective half-life of the radioisotope in the thyroid of 6±1 d.

 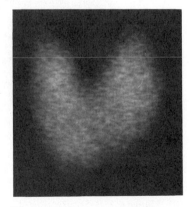

(*a*)   (*b*)

Figure 6.149 (*a*) PET image of a hyperfunctioning thyroid showing non-uniform isotope distribution compared with (*b*) uniform uptake shown in the conventional gamma camera image.

Similarly, the treatment of thyroid cancer with ^{131}I, although very successful, can lead to radiation-induced leukaemia (Pochin 1967). Treatment of primary thyroid cancer is by surgery and then ^{131}I therapy to ablate any remnant normal and abnormal tissue. Metastases, mostly in the skeleton or the lungs, can only be treated after complete ablation of normal thyroid tissue. Little attempt is made to determine the radiation dose to these tumour sites. Recent use of a rectilinear scanner and PET has shown that, again, dosimetry is possible and may be valuable in determining the efficacy of successive treatments.

The *parathyroid glands* consist of four separate glands close to the posterior surface of the thyroid gland (see figure 6.144). The glands produce parathyroid hormone, which is used to regulate the concentration of calcium in the body fluids. The most common pathological abnormalities in this case are hypo- and hyperfunction, the latter often associated with a benign adenoma or, infrequently, carcinoma. In the past ^{75}Se-selenomethionine has been used to localise hyperfunctioning parathyroid glands. More recently ^{201}Tl chloride has been used in conjunction with ^{99}Tc^{m}-pertechnetate to localise parathyroid adenoma. The ^{201}Tl chloride localises in both the thyroid and hyperfunctioning parathyroid gland, whereas ^{99}Tc^{m}-pertechnetate is trapped only in the thyroid. Careful use of the dual-isotope subtraction technique (Ferlin *et al* 1983) can remove the thyroid activity from the ^{201}Tl image and highlight the parathyroid adenoma (figure 6.150).

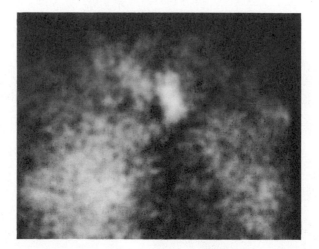

Figure 6.150 Subtraction (^{201}Tl−^{99}Tc^{m}) image of the parathyroid showing enhanced localisation of hyperfunctioning parathyroid gland.

The *adrenal glands*, located close to the upper pole of each kidney (figure 6.151), have two main component tissues, the cortex and the medulla. The cortex is principally involved in the production of hydrocortisone, a steroid that affects organ metabolism and electrolyte/water balance. The hormones of the adrenal cortex are derivatives of cholesterol, which is believed to be the precursor substrate for *in vivo* synthesis. Iodinated cholesterol (^{131}I-NP59) has been used for imaging the adrenal cortex, and shown to be effective in the assessment of Cushing's syndrome and tumours of the adrenal cortex. The medulla produces adrenaline (epinephrine), which affects the heart rate, cardiac metabolism and fat/carbohydrate metabolism. Also produced is noradrenaline (norepinephrine), which acts as a vasoconstrictor/oppressor. The recently developed adrenergic neuron-blocking agent, *m*-iodobenzylguanidine (*m*IBG), has demonstrated localisation in adrenergic tissues and, in particular, in the neuroectodermally derived tumours, pheochromocytoma and neuroblastoma. Labelled with ^{131}I, *m*IBG has been successfully used to localise both primary and secondary sites related to these neoplastic conditions, and, more recently, targeted radiotherapy has been successful in some cases.

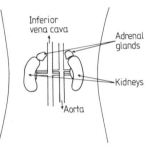

Figure 6.151 Schematic of the anatomical location of the adrenal glands.

Figure 6.152 shows the uptake of a tracer dose (20 MBq) of ^{123}I-*m*IBG in a 10-year-old boy with neuroblastoma. Figure 6.153 shows sequential images of the uptake of therapeutic levels of ^{131}I-*m*IBG (3.7–7.4 MBq per dose). These images were used to determine the radiation dose to the tumour, the liver and the whole body (table 6.20), and illustrate the long-term goal of quantitative imaging.

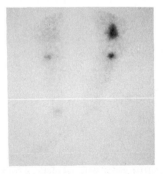

Figure 6.152 ^{123}I-*m*IBG diagnostic image of a child with neuroblastoma showing distant skeletal metastasis, which was not seen on conventional imaging with ^{99}Tc^{m}-MDP (methylene diphosphonate).

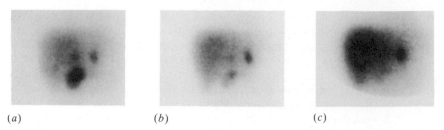

(*a*)   (*b*)   (*c*)

Figure 6.153 Sequential ^{131}I-*m*IBG therapy images of a child with neuroblastoma showing decreased tumour uptake with time: (*a*) first therapy; (*b*) second therapy; (*c*) third therapy.

Table 6.20 Estimated radiation doses to a child with neuroblastoma from sequential doses of ^{131}I-MIBG.

Dose number[a]	Activity given (GBq)	Target	Uptake (%)	$T_{1/2}^{\text{eff}}$ (h)	Dose (Gy)
1	3.7	Whole body	30	47	0.7
		Liver	15	30	5.6
		Tumour	3.5	30	100
2	5.1	Whole body	54	29	1.1
		Liver	16	30	7.5
		Tumour	0.6	20	20
3	5.1	Whole body	54	29	1.0
		Liver	7.5	33	4.5
		Tumour	0.07	35	4.0

[a] Doses given approximately five weeks apart.

6.9.11 Tumour and inflammatory lesion imaging

A major goal of nuclear medicine research has been the development of highly sensitive tumour-specific radiopharmaceuticals (Kim and Haynie 1984). While many radiotracers in use today delineate tumours (hot- and cold-spot imaging), none can be called 'tumour-specific'. The goal of tumour imaging is basically two-fold:

(*a*) detection of primary disease and assessment of its spread (secondary disease), and
(*b*) monitoring the response of tumour therapy.

The most commonly used agent for general-purpose tumour imaging is ^{67}Ga-citrate. Upon administration of the weak gallium chelate, ^{67}Ga becomes bound to the plasma protein, transferrin. Because of the high level of gallium binding with transferrin, blood clearance is slow. After administration, high levels of ^{67}Ga are found in the skeleton (including marrow), liver, kidney and spleen (figure 6.154). The mechanism of uptake into neoplastic tissue is still unresolved; however, several factors are thought to play a role in tumour localisation, including

(a) high tumour vascularity,
(b) high vascular permeability,
(c) local pH effects,
(d) lysosomal accumulation, and
(e) transferrin receptor binding.

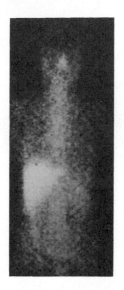

Figure 6.154 ^{67}Ga-citrate distribution in a normal patient (anterior view).

For imaging studies, approximately 70–370 MBq of ^{67}Ga-citrate is injected intravenously. Images are taken 48–72 h post-injection. In order to eliminate excessive bowel activity, the use of cathartics is recommended. Normal ^{67}Ga distribution is seen in figure 6.154, while figure 6.155 shows abnormal uptake in the region of the lower right lung. Gallium imaging has been used for the detection of Hodgkin's disease, lymphoma, lung cancer, hepatoma, metastatic melanoma and sarcoma, as well as for a variety of benign disorders, especially sarcoidosis and acute pyrogenic abscesses (Larson and Carrasquillo 1984).

Radiolabelled bleomycin has also been used for general tumour imaging. Bleomycin, an antineoplastic agent, is known to form metal complexes with cobalt, indium and gallium. The ^{111}In-bleomycin is administered intravenously, with imaging performed 48–72 h post-injection. The ability of radiolabelled bleomycin to detect tumours has been demonstrated in tumours of the head and neck, brain, breast, lungs and gastrointestinal tract, as well as melanomas. However, In-bleomycin also suffers from non-specificity, with difficulty arising from increased activity seen in inflammatory processes.

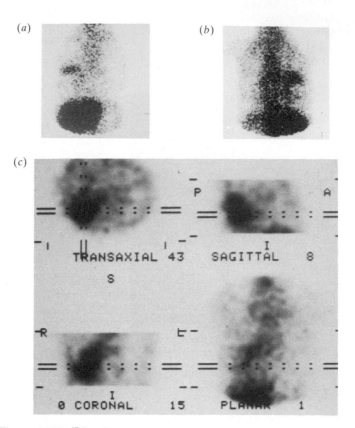

Figure 6.155 ^{67}Ga-citrate uptake in a carcinoma of the bronchus shown in both conventional planar images and SPECT: (*a*) anterior view; (*b*) posterior view; (*c*) SPECT images.

A variety of non-specific imaging agents also exist. This group includes the organ-specific radiopharmaceuticals such as the $^{99}Tc^m$-phosphonates (skeleton) and $^{99}Tc^m$-colloids (liver/spleen). These agents give non-specific information regarding the function of the organ under investigation. Often, tumour involvement within the organ system will produce abnormal scans; however, other processes besides tumour involvement can be the cause of such findings. More recently, $^{99}Tc^m(v)$-dimercaptosuccinic acid (DMSA) has been shown to concentrate in a variety of head and neck tumours, as well as in medullary carcinoma of the thyroid (Ohta *et al* 1985). The mechanism of action is yet unknown, and the clinical investigation of this compound is still under way. The move towards disease-specific agents will be discussed in §6.9.12.

Imaging of *inflammatory processes* has been performed using ^{67}Ga-citrate as described for tumour imaging. However, owing to the non-specificity of ^{67}Ga-citrate, alternative radiopharmaceuticals have been investigated. The most successful method for imaging acute inflammatory processes utilises ^{111}In-labelled leukocytes. Since the introduction of ^{111}In-leukocytes by McAfee and Thakur (1977), a number of investigators (Doherty *et al* 1978, McDougall *et al* 1979) have proven the clinical utility of this technique.

For imaging inflammatory processes, approximately 18 MBq of ^{111}In-leukocytes are injected, with imaging 4–24 h post-injection. The uptake of the ^{111}In-leukocytes is thought to be due to migration of the labelled leukocytes to the area of inflammation. Figure 6.156 shows abnormal accumulation of ^{111}In-leukocytes indicative of an inflammatory process.

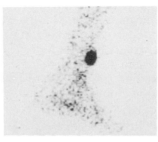

Figure 6.156 ^{111}In-leukocyte images showing abnormal uptake in the lower leg indicative of inflammatory disease.

6.9.12 *'Disease-specific' imaging*

A pharmaceutical may become 'disease-specific' if, for a variety of reasons, the uptake in diseased tissue greatly exceeds that in *any* normal tissue in the body. This is an unlikely situation, but certainly there are a few agents that appear to fall loosely into this category. If the diseased tissue contains cells that display, as part of their abnormality, receptor sites that are less frequently present in normal tissue, then it may be possible to produce agents that localise specifically to these sites and so provide 'disease-specific' imaging.

One class of agents in this category is antibodies, proteins produced naturally in the body, which recognise and destroy invading foreign proteins, known as antigens. Antigens may be introduced as isolated proteins or may be associated with specific cells.

Polyclonal antibodies are produced by inoculation of an animal with an antigenic substance. The animal's immune system recognises the antigen as foreign and the B-lymphocytes of the immune system produce antibodies directed against the antigen. The antibodies are then harvested by periodic bleeding of the animal with subsequent purification of the plasma proteins. This method produces small quantities of a range of antibodies with varying degrees of specificity and avidity.

A more promising approach for the production of highly specific antibodies in large quantities is the use of monoclonal antibody (Mab) technology. Monoclonal antibodies are produced by the same initial inoculation and immunisation technique, but, in order to increase production and specificity, the B-lymphocytes of the spleen, which are responsible for the production of antibodies, are fused with myeloma cells. The myeloma cells are transformed lymphocytes (i.e. malignant lymphocytes), which have the ability of indefinite reproduction. The fused cells are separated from unfused cells and screened for the type and specificity of antibody produced. This hybrid cell, known as a hybridoma, allows for the potential production of a highly specific antibody.

Both radiolabelled polyclonal and monoclonal antibodies have been used for the detection of tumours, but more recently research efforts have focused primarily on the use of Mab (see Srivastava and Mausner 1987). This is probably due to the theoretical increased specificity possible with Mab. However, it should be noted that most tumour antigens are not totally unique to tumour tissue, and the Mab will recognise the antigen-binding site anywhere in the body. Furthermore, the antibody, being a large complex molecule, typically of mass 50 000–200 000 Daltons, has other non-specific receptor sites, which allows non-specific binding to non-tumour cells.

Nevertheless, the possibility remains that, by appropriate molecular engineering, a suitable tumour-specific antibody may become available for both diagnostic and therapeutic uses. An example of such 'molecular engineering' is the use of appropriate enzymes to 'chop up' the antibody into fragments (Fab', Fab$_2$) (figure 6.157), which are smaller in size. Whilst still containing the antigen-recognition properties of the parent antibody, these fragments have lost some of the undesirable non-specific facilities. The importance of the size of the molecule can be seen from the rate of circulation of naturally occurring proteins in the body. Radioisotope-labelled human serum albumin (HSA) has a clearance time of more than 24 h from the blood, and several organs, in particular the liver, take up this protein non-specifically as part of the natural filtering process. Hence, if a labelled antibody is used for imaging, as might be expected, it clears the blood pool slowly and inhibits the visualisation of abnormalities until 24–48 h or more after injection (figure 6.158). Frustratingly, the *in vivo* distribution of radioisotope-labelled Mab seems to depend on the radioisotope used.

The most common labels are ^{131}I, ^{111}In and ^{67}Ga, although shorter-lived radionuclides such as ^{123}I and [99]Tc^m may be useful with Fab (Eckelman *et al* 1980). There are several well defined methods (Mather 1986) for the iodination of protein, and hence the use of ^{131}I, ^{123}I or even the positron emitter ^{124}I is relatively straightforward. Specific activities of about $400\,kBq\,\mu g^{-1}$ of antibody are possible using standard labelling techniques. This allows for a diagnostic scan with about 200 μg of antibody and 80 MBq of radioactivity. Care is needed to prevent the iodination process affecting the antigen-binding properties, because the position of the label may affect the antibody binding site. In addition, there is considerable evidence of dehalogenation *in vivo*, giving rise to free iodide in the circulating blood pool. This produces high background count rates and, in addition to the loss of labelled antibody available for localisation, leads to poor imaging (and a potential thyroid hazard).

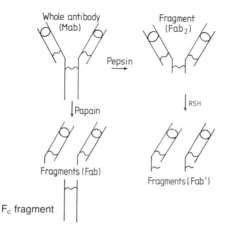

Figure 6.157 Schematic diagram of Mab, Fab, Fab′ and Fab$_2$ antibodies (RSH indicates a disulphide reducing agent).

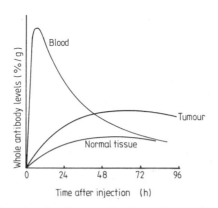

Figure 6.158 Levels of whole antibody in blood, tumour and the whole body as a function of time after injection.

The high levels of radioactivity in the blood pool have led to many groups (Ott *et al* 1983b) resorting to dual-isotope techniques to subtract out the blood-pool contribution from the antibody image. The most common agent used has been ^{99}Tc^m-HSA, which is given 15–20 min prior to imaging to provide a protein blood-pool image. In some cases, ^{99}Tc^m-pertechnetate has also been used to simulate the distribution of free iodide in the blood pool. However, as discussed in §6.6.3, dual-isotope subtraction can produce significant artefacts (figure 6.159) in images and, in general, has been of little use in generating *new* diagnostic information.

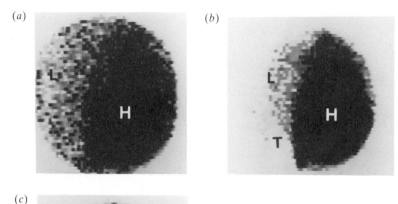

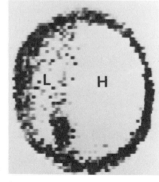

Figure 6.159 Planar images of the distribution of (*a*) ^{131}I-CEA (carcinoembryonic antigen) and (*b*) ^{99}Tc^m-HSA, and (*c*) a difference image (*a*)–(*b*) showing the production of artefacts by dual-isotope subtraction. H = heart, L = limb, T = tumour.

As a sensible alternative, SPECT has been carried out at a few centres but, again, as shown in §6.7, ^{131}I images suffer all the problems associated with high-energy photon imaging in addition to the special problems occurring with Mab.

More recently, metal radioisotopes, such as ^{111}In, ^{67}Ga and ^{99}Tc^m, have been used as antibody labels in attempts to overcome some of the problems of iodinated antibodies. The most common technique used here is the bonding of the isotope to the antibody via a suitable bifunctional chelating agent, such as EDTA and DTPA. This can be done at any time prior to use, and the protein–chelate complex can be frozen to maintain its viability. Subsequently, the labelling of the complex with the metal radioisotope is relatively straightforward, and the formation of the initial complex has proven to be stable *in vivo* and little of the complex is apparently damaged. Again, specific activities of about 400 kBq μg^{-1} are obtained with ^{111}In-DTPA and similar levels with ^{67}Ga-DFO (desferrioxamine). ^{99}Tc^m, with its much shorter half-life, is only useful for use with Fab, where the clearance from the body is rapid and imaging can take place at 6 h post-injection. The complex of choice is presently ^{111}In-DTPA, which is stable and provides good planar and SPECT images (figure 6.160). One major property of the metal chelate–antibody complex is the high non-specific uptake in the liver (10–30% of injected dose). The mechanism for this uptake is not yet clear but this results in the blood-pool activity clearing rather more quickly than with the halogen-labelled antibody. This leads to improved image quality in the body, except in the liver, where the high uptake precludes its use as a detector of liver disease.

Ultimately, the image quality obtained with radioimmunoscintigraphy is rather poor. For example, even if the tumour/normal tissue uptake ratio is 10:1, this is degraded to below about 3:1 in a planar image (Ott *et al* 1983b). This must be set against tumour/blood-pool ratios of 1:3 to 1:1 with a whole antibody.

Furthermore, estimates of the count rates from a small tumour (10 g) at depth in the torso are discouraging. For a typical tumour uptake of 0.005% dose per gram, the count rate will be a few counts per second, only a fraction (less than 1%) of the total image count rate.

The tumour-uptake level quoted is typical for antibody studies and highlights the low number of cell-binding sites localised in these studies. This, of course, is a direct consequence of the use of monoclonal antibodies, where only very specific antigens are localised. There have been suggestions for the use of cocktails of antibodies, where each localises a different antigen, but there has been little action on this front. Presumably, the increased sensitivity will only scale in proportion to the number of antibodies used. A further large factor (roughly 10^{-4}) in sensitivity is lost by the use of a gamma camera. Obviously PET scanning can win back perhaps a factor of 50, giving rise to much higher count rates from both the tumour and the normal tissue. Statistically, the image quality will improve, and so efforts are now being turned to the labelling of antibodies with ^{124}I, ^{55}Co and ^{66}Ga, all suitable positron-emitting radioisotopes. A further advantage here is that the same radioisotope can be used for diagnostic and therapy purposes (therapy dose about 100 times diagnostic dose).

Much effort is going into developing techniques of removing antibody from non-target organs more rapidly using either Fab or second-antibody techniques. However, until the specific binding properties are *improved* by a factor of 10 (more than 0.05% dose per gram) and the non-specific binding levels *reduced* by a similar factor, diagnostic radioimmunoscintigraphy and therapy applications will remain anecdotal.

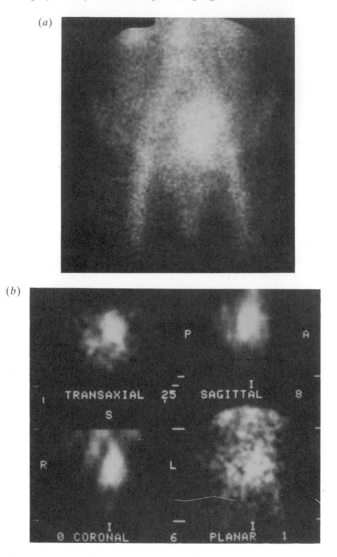

Figure 6.160 (*a*) Planar image and (*b*) SPECT images of the uptake of ¹¹¹In-DTPA–M8 antibody in an ovarian carcinoma.

6.9.13 Summary

In this section we have discussed a large number of clinical applications of radionuclide imaging. The areas discussed are illustrative and are not complete by any means; further examples may be found in the clinical reference works. Much of what has been discussed here refers to the diagnostic information available from nuclear medicine. However, with the increasing emphasis on quantitative imaging, the true value of radioisotope imaging will be in the determination of absolute function, monitoring the effects of therapeutic regimes and, as shown above, in radiation (and perhaps, in future, chemotherapy) dosimetry.

REFERENCES

ANDERSON D F, BOUCLIER R, CHARPAK G and MAJEWSKI S 1983 Coupling of a BaF_2 scintillator to a TMAE photocathode and a low-pressure wire chamber *Nucl. Instrum. Meth.* **217** 217–23

ANGER H O 1958 Scintillation camera *Rev. Sci. Instrum.* **29** 27–33

—— 1964 Scintillation camera with multichannel collimators *J. Nucl. Med.* **5** 515–31

—— 1969 Multiplane tomographic gamma ray scanner *Medical Radioisotope Scintigraphy* (Vienna: IAEA) pp203–16

ARANO Y, YOKOYAMA A, FURUKAWA T, HORIUCHI K, YAHATA T, SAJI H, SAKAHARA H, NAKASHIMA T, KOIZUMI M, ENDO K and TORIZUKA K 1987 Technetium-99m-labeled monoclonal antibody with preserved immunoreactivity and high *in vivo* stability *J. Nucl. Med.* **28** 1027–33

ATTIX F H and ROESCH W C 1966 *Radiation Dosimetry* (New York: Academic Press)

BALDWIN R M 1986 Chemistry of radioiodine *Appl. Radiat. Isot.* **37** 817–21

BATEMAN J E, CONNOLLY J F and STEPHENSON R 1985 High speed quantitative digital beta autoradiography using a multistep avalanche detector and an Apple II microcomputer *Nucl. Instrum. Meth. Phys. Res.* A **241** 275–89

BATEMAN J E, CONNOLLY J F, STEPHENSON R, TAPPERN G J and FLESHER A C 1984 The Rutherford Appleton Laboratory's Mark I multiwire proportional chamber positron camera *Nucl. Instrum. Meth.* **225** 209–31

BECK R N 1964 A theory of radioisotope scanning systems *Medical Radioisotope Scanning* vol I (Vienna: IAEA)

BENDER M A and BLAU M 1963 The autofluoroscope *Nucleonics* **21** 52–6

BERGER H J, MALTHAY R A, PYTLIK L M, GOTTSCHALK A and ZARET B 1979 First-pass radionuclide assessment of right and left ventricular performance in patients with cardiac and pulmonary disease *Semin. Nucl. Med.* **9** 275–95

BERGSTRÖM M, COLLINS V P, EHRIN E, ERICSON K, ERIKSSON L, GREITZ T, HALLDIN C, VON HOLST H, LANGSTROM B, LILJA A, LUNDQUIST H and NAGREN K 1983a Discrepancies in brain tumour extent as shown by computed tomography and positron emission tomography using [^{68}Ga]EDTA, [^{11}C]glucose and [^{11}C]methionine *J. Comput. Assist. Tomogr.* **7** 1062–6

BERGSTRÖM M, ERIKSSON L, BOHM C, BLOMQVIST G and LITTON J 1983b Correction for scattered radiation in a ring detector positron camera by integral transformation of the projections *J. Comput. Assist. Tomogr.* **7** 42–50

BIRKS J B 1964 *The Theory and Practice of Scintillation Counting* (Oxford: Pergamon)

BIZAIS Y, ZUBAL I G, ROWE R W, BENNETT G W and BRILL A B 1983 Potentiality of D7PHT for dynamic tomography *J. Nucl. Med.* **24** P75

BOWLEY A R, TAYLOR C G, CAUSER D A, BARBER D C, KEYES W I, UNDRILL P E, CORFIELD J R and MALLARD J R 1973 A radioisotope scanner for rectilinear, arc, transverse section and longitudinal section scanning (ASS—the Aberdeen Section Scanner) *Br. J. Radiol.* **46** 262–71

BRH (BUREAU OF RADIOLOGICAL HEALTH) 1970 *Radiological Health Handbook* DHEW Publ. no. 2016 (Rockville, MD: BRH)

—— 1976 *Workshop Manual for Quality Control of Scintillation Cameras in Nuclear Medicine* DHEW Publ. FDA-76-8039 (Rockville, MD: BRH)

BRIGHAM E O 1974 *The Fast Fourier Transform* (Englewood Cliffs, NJ: Prentice-Hall)

BRITISH PHARMACOPAEIA 1980 vol II (London: HMSO)

BRITTON K E and BROWN N J G 1971 *Clinical Renography* (London: Lloyd-Luke)

BSI (BRITISH STANDARDS INSTITUTION) 1977 *Draft Standard Method of Measuring and Defining the Characteristics of Radionuclide Imaging Devices* Document 77/26291DC (London: BSI)

BUDINGER T F 1983 Time-of-flight positron emission tomography: status relative to conventional PET *J. Nucl. Med.* **24** 73–8

BUDINGER T F, DERENZO S E, GULLBERG G T, GREENBERG W L and HUESMAN R H 1977 Emission computer assisted tomography with single-photon and positron annihilation photon emitters *J. Comput. Assist. Tomogr.* **1** 131–45

BUDINGER T F and GULLBERG G T 1974 Three-dimensional reconstruction in nuclear medicine emission imaging *IEEE Trans. Nucl. Sci.* **NS-21** 2–20

BURNS H D 1978 Design of radiopharmaceuticals *The Chemistry of Radiopharmaceuticals* ed N O Heindel, H D Burns, T Honda and L W Brady (New York: Masson) ch 3

BURNS H D, WORLEY P, WAGNER H N JR, MARZILLI L and RISCH V 1978 Design of technetium radiopharmaceuticals *The Chemistry of Radiopharmaceuticals* ed N O Heindel, H D Burns, T Honda and L W Brady (New York: Masson) ch 17

CASEY M E and NUTT R 1986 A multicrystal two-dimensional BGO detector system for positron emission tomography *IEEE Trans. Nucl. Sci.* **NS-33** 460–3

CASSEN B, CURTIS L, REED C and LIBBY R 1951 Instrumentation for ^{131}I used in medical studies *Nucleonics* **9** 46–50

CERQUEIRA M D, HARP G D and RITCHIE J L 1987 Evaluation of myocardial perfusion and function by single photon emission computed tomography *Semin. Nucl. Med.* **17** 200–13

CHANG L T 1978 A method for attenuation correction in radionuclide computed tomography *IEEE Trans. Nucl. Sci.* **NS-26** 638–43

CHO Z H and FARUKHI M R 1977 Bismuth germanate as a potential scintillation detector in positron cameras *J. Nucl. Med.* **18** 840–4

COATES E A 1981 Quantitative structure–activity relationships *Radiopharmaceuticals: Structure–Activity Relationships* ed R P Spencer (New York: Grune and Stratton) pp8–10

COENEN H H, MOERLIN S M and STOCKLIN G 1983 No-carrier added radiohalogenation methods with heavy halogens *Radiochim. Acta* **34** 47–68

COLEMAN R E, JASZCZAK R J and COBB F R 1982 Comparison of 180° and 360° data collection in thallium-201 imaging using single photon emission computerised tomography (SPECT): concise communication *J. Nucl. Med.* **23** 655–60

COOPER J F and AVIS K E 1975 Control and detection of microbial contamination in short-lived radiopharmaceuticals *Radiopharmaceuticals* ed G Subramanian, B A Rhodes, J F Cooper and V S Sodo (New York: Society of Nuclear Medicine) pp254–7

COTTON F A and WILKINSON G 1980 *Advanced Inorganic Chemistry* (New York: Wiley) pp71–81

CROFT B Y 1986 *Single-Photon Emission Computed Tomography* (Chicago: Year Book Medical)

DE LA MARE P B D 1976 *Electrophilic Halogenation: Reaction Pathways Involving Attack by Electrophilic Halogens on Unsaturated Compounds* (Cambridge: Cambridge University Press)

DERENZO S E 1984 Initial characterization of a BGO-photodiode detector for high resolution positron emission tomography *IEEE Trans. Nucl. Sci.* **NS-31** 620–6

DEUTSCH E, BUSHONG W, GLAVAN K A, ELDER R C, SODD V J, SCHOLZ K L, FORTMAN D L and LUKES S J 1981 Heart imaging with cationic complexes of technetium *Science* **214** 85–6

DHSS (DEPARTMENT OF HEALTH AND SOCIAL SECURITY) 1980 *Performance Assessment of Gamma Cameras* part I, Report no. STB 11 (London: DHSS)

—— 1982 *Performance Assessment of Gamma Cameras* part II, Report no. STB 13 (London: DHSS)

DIFFEY B L, HALL F M and CORFIELD J R 1976 The Tc-99m DTPA dynamic renal scan with deconvolution analysis *J. Nucl. Med.* **17** 352–5

DOHERTY P W, BUSHBERG J T, LIPTON M S, MEARES C F and GOODWIN D A 1978 The use of ^{111}In-labelled leukocytes for abscess detection *Clin. Nucl. Med.* **3** 108–10

ECKELMAN W C 1982 Radiolabelled adrenergic and muscarinic blockers for *in vivo* studies *Receptor Binding Radiotracers* vol I, ed W C Eckelman (Boca Raton, FL: CRC Press) pp77–8

ECKELMAN W C, KARESH S M and REBA R C 1975 New compounds: fatty acids and long chain hydrocarbon derivatives containing a strong chelating agent *J. Pharm. Sci.* **64** 704–6

ECKELMAN W C, PAIK C H and REBA R C 1980 Radiolabelling of antibodies *Cancer Res.* **40** 3036–42

ECKELMAN W C and REBA R C 1978 The classification of radiotracers *J. Nucl. Med.* **19** 1179–81

ECKELMAN W C and VOLKERT W A 1982 *In vivo* chemistry of ^{99}Tc^m-chelates *Int. J. Appl. Radiat. Isot.* **33** 945–51

ESSER P D, ALDERSON P O, MITNICK R J and ARLISS J J 1984 Angled-collimator SPECT (A-SPECT): an improved approach to cranial single photon emission tomography *J. Nucl. Med.* **25** 805–9

EVANS N T S, KEYES W I, SMITH D, COLEMAN J, CUMPSTEY D, UNDRILL P E, ETTINGER K V, ROSS K, NORTON M Y, BOLTON M P, SMITH F W and MALLARD J R 1986 The Aberdeen Mark II single-photon-emission tomographic scanner: specification and some clinical applications *Phys. Med. Biol.* **31** 65–78

EVERETT D B, FLEMING J S, TODDS R W and NIGHTINGALE J M 1977 Gamma radiation imaging system based on the Compton effect *Proc. IEE* **124** 995–1000

FERLIN G, BORSATO N, CAMERANI M, CONTE N and ZOTTI D 1983 New perspectives in localising enlarged parathyroids by technetium–thallium subtraction scan *J. Nucl. Med.* **24** 438–41

FLEMING J S 1979 A technique for the absolute measurement of activity using a gamma camera and computer *Phys. Med. Biol.* **24** 176–80

FLOWER M A, ADAM I, MASOOMI A M and SCHLESINGER T 1986 Special collimators for quantitative imaging of high activity levels of iodine-131 *Br. J. Radiol.* **59** 836–7

FLOWER M A and PARKER R P 1980 Quantitative imaging using the Cleon emission tomography system: recent developments *Radiology* **137** 535–9

FLOWER M A, PARKER R P, COLES I P, FOX R A and TROTT N G 1979

Feasibility of absolute activity measurements using the Cleon emission tomography system *Radiology* **133** 497–500

FLOWER M A, ROWE R W and KEYES W I 1980 Sensitivity measurements on single-photon emission tomography systems *Radioakt. Isot. Klin. Forsch.* **14** 451–62

FREEDMAN G S 1970 Tomography with a gamma camera—theory *J. Nucl. Med.* **11** 602–4

GATTI E, REHAK P, LONGONI A, KEMMER J, HOLL P, KLANNER R, LUTZ G, WYLIE A, GOULDING F, LUKE P N, MADDEN N W and WALTON J 1985 Semiconductor drift chambers *IEEE Trans. Nucl. Sci.* **NS-32** 1204–8

GERBER M S, MILLER D W, SCHLÖSSER P A, STEIDLEY J W and DEUTCHMAN A H 1977 Position sensitive gamma ray detectors using resistive charge division readout *IEEE Trans. Nucl. Sci.* **NS-24** 182–7

GILBERT P 1972 Iterative methods for the three dimensional reconstruction of an object from projections *J. Theor. Biol.* **36** 105–17

GOTTSCHALK S C, SALEM D, LIM C B and WAKE R H 1983 SPECT resolution and uniformity improvements by noncircular orbit *J. Nucl. Med.* **24** 822–8

GRAHAM L S and NEIL R 1974 *In vivo* quantitation of radioactivity using the Anger camera *Radiology* **112** 441–2

HARBERT J and DA ROCHA A F G 1984a *Textbook of Nuclear Medicine* vol I *Basic Science* (Philadelphia: Lea and Febiger)

—— 1984b *Textbook of Nuclear Medicine* vol II *Clinical Applications* (Philadelphia: Lea and Febiger)

HELLER S L and GOODWIN P N 1987 SPECT instrumentation: performance, lesion detection and recent innovations *Semin. Nucl. Med.* **17** 184–99

HERMAN G T, LENT A and ROWLAND S W 1973 ART: mathematics and applications—a report on the mathematical foundations and on the applicability to real data of the algebraic reconstruction techniques *J. Theor. Biol.* **42** 1–32

HILSON A J W, MAISEY M N, BROWN C B, OGG C S and BEWICK M S 1978 Dynamic renal transplant imaging with Tc-99m DTPA (Sn) supplemented by a transplant perfusion index in the management of renal transplants *J. Nucl. Med.* **19** 994–1000

HISADA K I, OHBA S and MATSUDAIRA M 1967 Isosensitive radioisotope scanning *Radiology* **88** 124–8

HNATOWICH D J, LAYNE W W and CHILDS R L 1983 Radioactive labelling of antibodies: a simple and efficient method *Science* **220** 613–15

HOFFMAN E J, HUANG S C and PHELPS M E 1979 Quantitation in positron emission computed tomography: 1. Effect of object size *J. Comput. Assist. Tomogr.* **3** 299–308

HOFFMAN E J, HUANG S C, PHELPS M E and KUHL D E 1981 Quantitation in positron emission tomography: 4. Effect of accidental coincidences *J. Comput. Assist. Tomogr.* **5** 391–400

HOFFMAN E J and PHELPS M E 1976 An analysis of some of the physical aspects of positron transaxial tomography *Comput. Biol. Med.* **6** 345–60

HOFFMAN E J, PHELPS M E and HUANG S C 1983a Performance evaluation of a positron tomograph designed for brain imaging *J. Nucl. Med.* **24** 245–57

HOFFMAN E J, PHELPS M E, HUANG S C, MAZZIOTTA J, DIGBY W and DAHLBOHM M 1987 A new PET system for high-resolution 3-dimensional brain imaging *J. Nucl. Med.* **28** 758

HOFFMAN E J, RICCI A R, VAN DER STEE L M A M and PHELPS M E 1983b ECAT III—Basic design considerations *IEEE Trans. Nucl. Sci.* **NS-30** 729–38

HPA (Hospital Physicists' Association) 1978 *The Theory, Specification and Testing of Anger-type Gamma Cameras* Topic Group Report no. 27 (London: HPA)

—— 1983 *Quality Control of Nuclear Medicine Instrumentation* ed R F Mould (London: HPA)

Huesman R H 1977 The effects of a finite number of projection angles and finite lateral sampling of projections on the propagation of statistical errors in transverse section reconstruction *Phys. Med. Biol.* **22** 511–21

Hyde R J, Ott R J, Flower M A and Meller S T 1988 A simple method of producing parenchymal renograms using parametric imaging *Clin. Phys. Physiol. Meas.* in press

IEC (International Electrotechnical Commission) 1980 *Characteristics and Test Conditions of Radionuclide Imaging Devices* Draft Document 80/22888DC (London: British Standards Institution)

Janoki G Y A, Harwig J F, Chanachai W and Wolf W 1983 [^{67}Ga]desferrioxamine-hsa: synthesis of chelon protein conjugates using carbodiimide as a coupling agent *Int. J. Appl. Radiat. Isot.* **34** 871–7

Jaszczak R J, Chang L T and Murphy P H 1979a Single photon emission computed tomography using multi-slice fan beam collimators *IEEE Trans. Nucl. Sci.* **NS-26** 610–11

Jaszczak R J, Chang L T, Stein N A and Moore F E 1979b Whole-body single-photon emission computed tomography using dual large-field-of-view scintillation cameras *Phys. Med. Biol.* **24** 1123–43

Jaszczak R J, Floyd C E Jr, Greer K L, Coleman R E and Manglos S H 1986a Cone beam collimation for spect: analysis, simulation and image reconstructions using filtered backprojection *Med. Phys* **13** 484–9

Jaszczak R J, Floyd C E Jr, Manglos S H, Greer K L and Coleman R E 1986b Cone-beam spect: experimental validation using a conventionally designed converging collimator *J. Nucl. Med.* **27** 930

Jaszczak R J, Greer K L, Floyd C E Jr, Harris C C and Coleman R E 1984 Improved spect quantification using compensation for scattered photons *J. Nucl. Med.* **25** 893–900

Jeavons A, Hood K, Herlin G, Parkman C, Townsend D, Magnanini R Frey P and Donath A 1983 The high-density avalanche chamber for positron emission tomography *IEEE Trans. Nucl. Sci.* **NS-30** 640–5

Jones A G, Davison A, Abrams M J, Brodack J W, Kassis A I, Goldhaber S Z, Holman B L, Stemp L, Manning T and Hechtman H B 1982 Investigations on a new class of technetium cations *J. Nucl. Med.* **23** P16

Kim E E and Haynie T P 1984 *Nuclear Imaging in Oncology* (Norwalk, CT: Appleton-Century-Crofts)

Knoll G F 1979 *Radiation Detection and Measurement* (New York: Wiley)

Krejcarek G E and Tucker K L 1977 Covalent attachment of chelating groups to macromolecules *Biochem. Biophys. Res. Commun.* **77** 581–5

Kuhl D E and Edwards R Q 1963 Image separation radioisotope scanning *Radiology* **80** 653–61

Kuhl D E, Edwards R Q, Ricci A R, Yacob R J, Mich T J and Alavi A 1976 The Mark IV system for radionuclide computed tomography of the brain *Radiology* **121** 405–13

Lacy J L, LeBlanc A D, Babich J W, Bungo M W, Latson L A, Lewis R M, Poliner L R, Jones R H and Johnson P C 1984 A gamma camera for medical applications, using a multiwire proportional counter *J. Nucl. Med.* **25** 1003–12

LARSON S M and CARRASQUILLO J A 1984 Nuclear oncology 1984 *Semin. Nucl. Med.* **14** 268–76

LARSSON S A 1980 Gamma camera emission tomography *Acta Radiol.* Suppl. **363**

LARSSON S A, BERGSTRAND G, BERGSTEDT H, BERG J, FLYGARE O, SCHNELL P O, ANDERSSON N and LAGERGREN C 1984 A special cut-off gamma camera for high-resolution SPECT of the head *J. Nucl. Med.* **25** 1023–30

LASSEN N A 1985 Cerebral blood flow tomography with xenon-133 *Semin. Nucl. Med.* **15** 347–56

LASSEN N A, HOEDT-RASMUSSEN K, SORENSEN S C, SKINHOJ E, CRONQUIST S, BODFORSS B and INCUAR D H 1963 Regional cerebral blood flow in man determined by [85]Krypton *Neurology* **13** 719–27

LAVAL M, MOSZYNSKI M, ALLEMAND R, CORMORECHE E, GUINET P, ODRU R and VACHER J 1983 Barium fluoride—inorganic scintillator for subnanosecond timing *Nucl. Instrum. Meth.* **206** 169–76

LEAR J L 1986 Principles of single and multiple radionuclide autoradiography *Positron Emission Tomography and Autoradiography* ed M E Phelps, J C Mazziotta and H R Schelbert (New York: Raven)

LIEBERMAN D E (ed) 1977 *Computer Methods: The Fundamentals of Digital Nuclear Medicine* (St Louis: C V Mosby)

LIM C, GOTTSCHALK S, SCHREINER R, WALKER R, VALENTINO F, COVIC J, PERUSEK A, PINKSTAFF C and JANZSO J 1984 Triangular SPECT system for brain and body organ 3-D imaging; design concept and preliminary imaging result *J. Nucl. Med.* **25** P6

LITTON J, BERGSTROM M, ERICKSSON L, BOHM C, BLOMQVIST G and KESSELBERG M 1984 Performance study of the PC-384 positron camera system for emission tomography of the brain *J. Comput. Assist. Tomogr.* **8** 74–87

McAFEE J G and THAKUR M L 1977 Survey of radioactive agents for *in vitro* labelling of phagocytic leukocytes *J. Nucl. Med.* **17** 980–7

McCREADY V R, PARKER R P, GUNNERSON E M, ELLIS R, MOSS E, GORE W G and BELL J 1971 Clinical tests on a prototype semiconductor gamma-camera *Br. J. Radiol.* **44** 58–62

McDOUGALL I R, BAUMERT J E and LANTIERI R L 1979 Evaluation of ^{111}In-leukocyte whole body scanning *Am. J. Roentgenol.* **133** 849–54

McKENZIE E H, VOLKERT W A and HOLMES R A 1985 Biodistribution of ^{14}C-PnAO in rats *Int. J. Nucl. Med. Biol.* **12** 133–4

MAISEY M 1980 *Nuclear Medicine: A Clinical Introduction* (London: Update)

MALLARD J R and PEACHEY C J 1959 A quantitative automatic body scanner for the localisation of radioisotopes *in vivo Br. J. Radiol.* **32** 652–7

MARSDEN P K, BATEMAN J E, OTT R J and LEACH M O 1986 The development of a high efficiency cathode converter for a multiwire proportional chamber positron camera *Med. Phys.* **13** 703–6

MATHER S J 1986 Radioiodinated monoclonal antibodies: a critical review *Appl. Radiat. Isot.* **37** 727–33

MAYNEORD W V and NEWBERY S P 1952 An automatic method of studying the distribution of activity in a source of ionizing radiation *Br. J. Radiol.* **25** 589–96

MOORE S C, DOHERTY M D, ZIMMERMAN R E and HOLMAN B L 1984 Improved performance from modifications to the multidetector SPECT brain scanner *J. Nucl. Med.* **25** 688–91

MUEHLLEHNER G 1969 A diverging collimator for gamma ray imaging cameras *J. Nucl. Med.* **10** 197–201

—— 1971 A tomographic scintillation camera *Phys. Med. Biol.* **16** 87–96

MUEHLLEHNER G and COLSHER J G 1980 Single photon imaging: new instrumentation and techniques *Extended Synopses of Int. Symp. on Medical Radionuclide Imaging* SM-247/202 (Heidelberg: IAEA)

MULLANI N A, GAETA J, YERIAN K, WONG W H, HARTZ R K, PHILIPPE E A, BRISTOW D and GOULD K L 1984 Dynamic imaging with high resolution time-of-flight PET camera—TOFPET I *IEEE Trans. Nucl. Sci.* **NS-31** 609–13

MURPHY P H, BURDINE J A and MAYER R A 1975 Converging collimation and a large field-of-view scintillation camera *J. Nucl. Med.* **16** 1152–7

NAHMIAS C, FIRNAU G and GARNETT E S 1984 Performance characteristics of the McMaster positron emission tomograph *IEEE Trans. Nucl. Sci.* **NS-31** 637–9

NARRA R K, KUCZYNSKI B L, FELD T, NUNN A D and ECKELMAN W C 1987 A comparison of the pharmacokinetics of a new Tc-99m-labelled myocardial imaging agent *J. Nucl. Med.* **28** 674

NEIRINCKX R D, CANNING L R, PIPER I M, NOWOTNIK D P, PICKETT R D, HOLMES R A, VOLKERT W A, FORSTER A M, WEISNER P S, MARRIOTT J A and CHAPLIN S B 1987 Tc-99m *d,l*-HMPAO: a new radiopharmaceutical for SPECT imaging of regional cerebral blood perfusion *J. Nucl. Med.* **28** 191–202

NEMA (NATIONAL ELECTRICAL MANUFACTURERS' ASSOCIATION) 1980, 1986 *Performance Measurements of Scintillation Cameras* Standards Publ. no. NU1 (Washington, DC: NEMA)

OHTA H, ENDO K, FUJITA T, NAKASHIMA T, SAKEHARA H, TORIZUKA K, SHIMIZU Y, ISHII Y, MAKIMOTO K, HATA N, HORIUCHI K, YOKAYAMA A and ISHII M 1985 Imaging of head and neck tumours with technetium(V)-99m DMSA. A new tumour seeking agent *Clin. Nucl. Med.* **10** 855–60

OTT R J 1986 Emission computed tomography *J. Med. Eng. Technol.* **10** 105–14

OTT R J, BATEMAN J E, BATTY V, CLACK R, FLOWER M A, LEACH M O, MARSDEN P K, MCCREADY V R, WEBB S, SHARMA H and SMITH A 1986 3D positron emission tomography: preliminary results *Br. J. Radiol.* **59** 419–22

OTT R J, BATTY V, WEBB S, FLOWER M A, LEACH M O, CLACK R, MARSDEN P K, MCCREADY V R, BATEMAN J E, SHARMA H and SMITH A 1987 Measurement of radiation dose to the thyroid using positron emission tomography *Br. J. Radiol.* **60** 245–51

OTT R J, FLOWER M A, KHAN O, KALIRAI T, WEBB S, LEACH M O and MCCREADY V R 1983a A comparison between 180° and 360° data reconstruction in single photon emission computed tomography of the liver and spleen *Br. J. Radiol.* **56** 931–7

OTT R J, GREY L J, ZIVANOVIC M A, FLOWER M A, TROTT N G, MOSHAKIS V, COOMBES R C, NEVILLE A M, ORMEROD M G, WESTWOOD J H and MCCREADY V R 1983b The limitations of the dual radionuclide subtraction technique for the external detection of tumours by radioiodine labelled antibodies *Br. J. Radiol.* **56** 101–9

PAIK C H, MURPHY P R, ECKELMAN W C, VOLKERT W A and REBA R C 1983 Optimization of the DTPA mixed-anhydride reaction with antibodies at low concentration *J. Nucl. Med.* **24** 932–6

PALMER A J and TAYLOR D M (ed) 1986 Radiopharmaceuticals labelled with halogen isotopes *Appl. Radiat. Isot.* **37** (8) Special issue

PHELPS M E 1986 Positron emission tomography: principles and quantitation *Positron Emission Tomography and Autoradiography* ed M E Phelps, J C Mazziotta and H R Schelbert (New York: Raven)

POCHIN E E 1967 Prospects for the treatment of thyroid carcinoma with radioiodine *Clin. Radiol.* **18** 113–25

RICHARDS P and ATKINS H L 1968 [99m]Technetium labelled compounds *Proc.*

7th Ann. Mtg. Jap. Soc. Nucl. Med., Tokyo 1967 (Jap. Nucl. Med. **7** 165)

ROGERS W L, HAN K S, JONES L W and BEIERWALTES W H 1972 Application of a Fresnel zoneplate to gamma-ray imaging J. Nucl. Med. **13** 612–15

SCHELBERT H R, HENZE E and PHELPS M E 1980 Emission tomography of the heart Semin. Nucl. Med. **10** 335–73

SEEVERS R H and COUNSELL R E 1982 Radioiodination techniques for small organic molecules Chem. Rev. **82** 575–90

SHARP P F, DENDY P P and KEYES W I 1985 Radionuclide Imaging Techniques (London: Academic Press)

SHARP P F, SMITH F W, GEMMELL H G, LYALL D, EVANS N T S, GVOZDANO-VIC D, DAVIDSON J, TYRELL D A, PICKETT R D and NEIRINCKX R D 1986 Technetium-99m HMPAO stereoisomers as potential agents for imaging regional cerebral blood flow: human volunteer studies J. Nucl. Med. **27** 171–7

SHORT M D 1984 Gamma-camera systems Nucl. Instrum. Meth. Phys. Res. **221** 142–9

SIKORA K and SMEDLEY H M 1984 Monoclonal Antibodies (Oxford: Blackwell Scientific) pp1–12

SINGH M and DORIA D 1983 An electronically collimated gamma camera for single photon emission computed tomography. Part 1, Theoretical considerations in design criteria; Part 2, Image reconstruction and preliminary experimental measurements Med. Phys. **10** 421–35

SORENSON J A and PHELPS M E 1980 Physics in Nuclear Medicine (New York: Grune and Stratton)

SRIVASTAVA S C and MAUSNER L F (ed) 1987 Radiolabelled monoclonal antibodies: chemical, diagnostic and therapeutic investigations Nucl. Med. Biol. **13** no. 4 (Special issue)

SRIVASTAVA S C and RICHARDS P 1983 Technetium labelled compounds Radiotracers for Medical Applications vol I, ed G V S Rayudn (Boca Raton: CRC Press)

STOCKLIN G 1977 Bromine-77 and iodine-123 radiopharmaceuticals Int. J. Appl. Radiat. Isot. **28** 131–47

STODDART H F and STODDART H A 1979 A new development in single gamma transaxial tomography: Union Carbide focussed collimator scanner IEEE Trans. Nucl. Sci. **NS-26** 2710–12

STOKELY E M, SVEINSDOTTIR E, LASSEN N A and ROMMER P 1980 A single photon dynamic computer assisted tomograph (DCAT) for imaging brain function in multiple cross sections J. Comput. Assist. Tomogr. **4** 230–40

SUNDBERG M W, MEARES C F and GOODWIN D A 1974 Chelating agents for the binding of metal ions to macromolecules Nature **250** 587–8

TAMAKI N, MUKAI T, ISHII Y, FUJITA T, YAMAMOTO K, MINATO K, YONE-KURA Y, TAMAKI S, KAMBARA H, KAWAI C and TORIZUKA K 1982 Comparative study of thallium emission myocardial tomography with 180° and 360° data collection J. Nucl. Med. **23** 661–6

THOMPSON J H 1983 Drug absorption, distribution and excretion Essentials of Pharmacology ed J A Bevan and J H Thompson (Philadelphia: Harper and Row)

THRALL J H and SWANSON D P 1983 Interventional aspects of nuclear medicine Nuclear Medicine Annual 1983 ed L M Freeman and H S Weissmann (New York: Raven)

TODD-POKROPEK A (ed) 1982 The use of computers in nuclear medicine IEEE Trans. Nucl. Sci. **NS-29** 1272–367

—— 1983 Non-circular orbits for the reduction of uniformity artefacts in SPECT

Phys. Med. Biol. **28** 309–13

TOWNSEND D, FREY P, DONATH A, CLACK R, SCHORR B and JEAVONS A 1984 Volume measurements *in vivo* using positron tomography *Nucl. Instrum. Meth.* **221** 105–12

TOWNSEND D, SCHORR B, JEAVONS A, CLACK R, MAGNANINI R, FREY P, DONATH A and FROIDEVAUX A 1983 Image reconstruction for a rotating positron tomograph *IEEE Trans. Nucl. Sci.* **NS-30** 594–600

UNITED STATES DISPENSATORY AND PHYSICIANS' PHARMACOLOGY 1967 Osol A. *et al* (eds) (Philadelphia: Lippincott)

VOGEL R A, KIRCH D, LEFREE M and STEELE P 1978 A new method of multiplanar emission tomography using a seven pinhole collimator and an Anger scintillation camera *J. Nucl. Med.* **19** 648–54

WAGNER H N JR (ed) 1975 *Nuclear Medicine* (New York: H P Publishing)

WALTERS T E, SIMON W, CHESLER D A, CORREIA J A and RIEDERER S J 1976 Radionuclide axial tomography with correction for internal absorption *Information Processing in Scintigraphy—Proc. 4th Int. Conf., Orsay, 1975* ed C Raynaud and A Todd-Pokropek (French Atomic Energy Authority) pp333–42

WEBB S 1987 Significance and complexity in medical images: space variant texture dependent filtering *Proc. 10th Information Processing in Medical Imaging Conf., Utrecht* ed M A Viergever and C N de Graaf (New York: Plenum)

WEBB S, BRODERICK M and FLOWER M A 1985a High resolution SPECT using divergent geometry *Br. J. Radiol.* **58** 331–4

WEBB S, FLOWER M A, OTT R J, BRODERICK M D, LONG A P, SUTTON B and McCREADY V R 1986a Single photon emission computed tomographic imaging and volume estimation of the thyroid using fan-beam geometry *Br. J. Radiol.* **59** 951–5

WEBB S, FLOWER M A, OTT R J and LEACH M O 1983a A comparison of attenuation correction methods for quantitative single photon emission computed tomography *Phys. Med. Biol.* **28** 1045–56

WEBB S, FLOWER M A, OTT R J, LEACH M O, FIELDING S, INAMDAR C, LOWRY C and BRODERICK M D 1986b A review of studies in the physics of imaging by SPECT *Recent Developments in Medical and Physiological Imaging* ed R P Clark and M R Goff (London: Taylor and Francis) (*J. Med. Eng. & Tech. Suppl.* 132–46)

WEBB S, FLOWER M A, OTT R J, LEACH M O and INAMDAR R 1983b The spatial resolution of a rotating gamma camera tomographic facility *Br. J. Radiol.* **56** 939–44

WEBB S, LONG A, OTT R J, FLOWER M A and LEACH M O 1985b Constrained deconvolution of SPECT liver tomograms by direct digital image restoration *Med. Phys.* **12** 53–8

WEBB S, PARKER R P, DANCE D R and NICHOLAS A 1978 A computer simulation study for the digital processing of longitudinal tomograms obtained with a zoneplate camera *IEEE Trans. Biomed. Eng.* **BME-25** (2) 146–54

WHITE W 1979 Resolution, sensitivity and contrast in gamma camera design: a critical review *Radiology* **132** 179–87

WHO (WORLD HEALTH ORGANISATION) 1982 *Quality Assurance in Nuclear Medicine* (Geneva: WHO)

WIELAND D M, TOBES M C and MANGNER T J (ed) 1986 *Analytical and Chromatographic Techniques in Radiopharmaceutical Chemistry* (New York: Springer)

WIELAND D M, WU J-L, BROWN L E, MANGNER T J, SWANSON D P and

Beierwaltes W H 1980 Radiolabeled adrenergic neuron-blocking agents: adrenomedullary imaging with [^{131}I]iodobenzylguanidine *J. Nucl. Med.* **21** 349–53

Williams E D (ed) 1985 *An Introduction to Emission Computed Tomography* Institute of Physical Sciences in Medicine Report no. 44 (London: IPSM)

Winchell H S, Baldwin R M and Lin T H 1980 Development of ^{123}I-labeled amines for brain studies: localization of ^{123}I-iodophenyl alkyl amines in rat brain *J. Nucl. Med.* **21** 940–6

Yanch J, Webb S, Flower M A and Irvine A T 1987 Constrained deconvolution to remove resolution degradation caused by scatter in SPECT *Proc. 10th Information Processing in Medical Imaging Conf., Utrecht* (New York: Plenum) ed M A Viergever and C N de Graaf

Yano Y 1985 Radionuclide generators: current and future applications in nuclear medicine *Radiopharmaceuticals* ed G Subramanian, B A Rhodes, J F Cooper and V J Sodd (New York: Society of Nuclear Medicine) pp236–45

CHAPTER 7

DIAGNOSTIC ULTRASOUND

J C BAMBER AND M TRISTAM

7.1 INTRODUCTION

Our attention is now turned away from those imaging techniques that make use of ionising radiation and to the subject of diagnostic ultrasound. Ultrasound is a form of radiation that, like x-rays, is useful for medical imaging because it both penetrates and interacts with the body. Information about the body structures is encoded on the transmitted and scattered radiation, and it is the job of the imaging system to decode it. Unlike x-rays, but by analogy with light, ultrasound suffers refraction and reflection at interfaces between media of different acoustic refractive indices, to an extent that makes it possible to build focusing systems. However, as we shall see later, the relative dimensions of wavelength and typical focusing apertures are such that wave (rather than geometric) optics is applicable, and phenomena such as diffraction and interference become limiting factors.

From the point of view of the way in which one makes use of a radiation for imaging, there are also marked differences between ultrasound and both light and x-rays. Ultrasonic waves propagate sufficiently slowly that, for the distances travelled in the body, transit times are easily measurable and radar-like pulse–echo methods may be used to create images. Conversely, the propagation speed is fast enough that all of the data needed for a complete image may be gathered and reconstructed within about 80 ms, making it possible (for example) to view moving images of structures in the living heart. Coupled with the apparent very low risk of hazard of the examination (see also Chapter 15) and the low cost of the equipment (being nearly all electronic), these features have eventually, over the 25–30 years of development of the technique, led to ultrasound being one of the most frequently used imaging methods in diagnostic medicine.

Another difference, relative to other forms of radiation, is that ultrasound is a coherent radiation, and, as with laser light, pronounced interference effects tend to dominate the images. Unlike even laser light, where the sensor is usually sensitive to optical intensity, ultrasound receivers are usually amplitude-sensitive and are able to generate coherent interference-like fluctuations in apparent received magnitude, due to integration over a non-planar received wave.

Ultrasound is usually defined as sound of frequency above the bandwidth of human hearing. In the context of 'imaging' techniques, this frequency range covers a considerable variety of applications; from underwater sonar and animal echo location (up to about 300 kHz), through medical diagnosis, therapy and industrial non-destructive testing (0.8–15 MHz), to acoustic microscopy (12 MHz to above 1 GHz).

7.2 BASIC PHYSICS

7.2.1 Wave propagation and interactions in biological tissues

7.2.1.1 Speed of sound. To a good approximation, with the exception of bone, biological tissues act, for ultrasound propagation, like fluids. They are not able to support transverse waves to any great extent and the longitudinal wave speed is given approximately by

$$c = (K/\rho_0)^{1/2} \qquad (7.1)$$

where ρ_0 is mean density and K is adiabatic bulk elastic modulus. Typical ranges of values for speed of sound are shown in figure 7.1. Note that the average speed, and the value often assumed by ultrasonic instrument designers, for soft tissues is about 1540 m s^{-1}, with a total range of ±6%. This variation, which is so small that it is neglected by conventional image reconstruction methods, is thought to be due mainly to fluctuations in elasticity rather than density. Practically, c is not a strong function of frequency f (the dispersion is less than 1% over the medical range of frequencies), except in bone. Since $c = f\lambda$, typical wavelengths encountered range from 1.5 mm at 1 MHz to 0.1 mm at 15 MHz. The speed of sound is temperature-dependent, the temperature coefficient being positive for non-fatty tissues/organs and negative for fatty tissue. Fat is also the only soft-tissue component that has a speed of sound lower than that for water, except for lung, which has a value close to that for air but dependent on the degree of inflation. For soft tissues other than fat and lung, there is an inverse relationship between speed of sound and water content, but a direct relationship with structural protein (collagen) content—hence the very high values for tendon and cartilage.

7.2.1.2 Acoustic field parameters. Both energy and momentum are transferred by the compressional wave, but there is no net transfer of matter unless this is induced as a result of momentum transfer. As the wave passes any particular point in the medium, oscillatory changes in a number of related parameters occur at that locality. For a sound pulse emerging from a typical diagnostic ultrasound scanner, the peak longitudinal displacement of the medium induced by the wave is only about

8×10^{-8} m. The peak velocity is of the order of $0.5\,\mathrm{m\,s^{-1}}$ and the maximum acceleration is a massive 3×10^5 times the acceleration due to gravity! The peak local pressure increase would be about 8 atm. All of these figures are calculated using approximate expressions, assuming that at the moment of peak amplitude the pulse is equivalent to a plane sine wave of intensity $2 \times 10^5\,\mathrm{W\,m^{-2}}$.

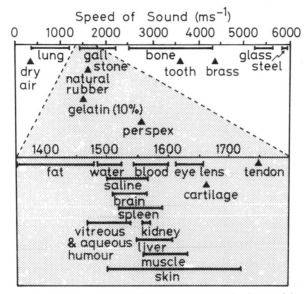

Figure 7.1 Ranges of measured values for speed of sound in various biological and non-biological media. The data for soft tissues and biological liquids, which fall within a narrow range, are shown using an expanded scale in the shaded portion of the figure. (After Bamber (1986b), which also contains the references to the original sources of data and methods of measurement.)

The wave intensity I ($\mathrm{W\,m^{-2}}$), which is the energy flowing per unit time per unit area, is related to the oscillating incremental pressure p and velocity u via the sum of the kinetic energy density and the potential energy density ($\mathrm{J\,m^{-3}}$), i.e.

$$I = \frac{c}{2}\left(\rho_0 u^2 + \frac{p^2}{\rho_0 c^2}\right). \tag{7.2}$$

Note that (under ideal conditions of a perfectly plane sine wave in an infinite uniform medium), at the moment of peak velocity U, the pressure is at ambient level and the potential energy term disappears. Similarly, at the time of peak pressure P, the displacement is also maximal but the velocity is zero, and so too is the kinetic energy term. In this situation the pressure and velocity fluctuations are $90°$ out of phase with each other.

We shall see later that diagnostic systems use very short sound pulses, which are repeated relatively infrequently. Because of this, typical time-averaged diagnostic intensities are very much lower than the corresponding instantaneous peak intensities (see §7.4.2).

7.2.1.3 Acoustic impedance. A derived field parameter, analogous to electrical impedance, is the specific acoustic impedance $Z_{sp} = p/u$, which in general is a complex quantity dependent on the relative phase of p and u, which in turn may be a function of spatial position, and is dependent on the type of wave field and propagation conditions. This is distinct from the characteristic impedance $Z = \rho_0 c = (\rho_0 K)^{1/2}$, which is a property of the medium only, and is equal to Z_{sp} only for the perfect plane-wave conditions mentioned above.

7.2.1.4 Attenuation. All media attenuate ultrasound, so that the intensity of a plane wave propagating in the x direction decreases exponentially with distance as

$$I_x = I_0 \exp(-\mu x) \qquad \text{or} \qquad \mu = - (1/x) \ln(I_x/I_0). \quad (7.3)$$

Similarly, for any of the amplitude parameters, P, U, etc (represented by Q),

$$Q_x = Q_0 \exp(-\alpha x) \qquad \text{or} \qquad \alpha = - (1/x) \ln(Q_x/Q_0) \quad (7.4)$$

where μ is the intensity attenuation coefficient and α is the amplitude attenuation coefficient. Since $(I_x/I_0) = (Q_x/Q_0)^2$, we see that $\mu = 2\alpha$. The units of μ and α are cm^{-1}, but are usually called nepers cm^{-1} (the word emanating from Naperian logarithm). Practically, the intensity and amplitude ratios are often expressed in decibels (dB). Hence, equation (7.3) becomes

$$\mu\,(\text{dB cm}^{-1}) = - (1/x)\, 10 \log_{10}(I_x/I_0)$$
$$= - (1/x) \ln(I_x/I_0)\, 10 \log_{10} e = 4.343\mu \;(\text{cm}^{-1}). \quad (7.5)$$

Similarly,

$$\alpha\,(\text{dB cm}^{-1}) = - (1/x)\, 20 \log_{10}(Q_x/Q_0)$$
$$= 8.686\alpha \;(\text{cm}^{-1}) = \mu \;(\text{dB cm}^{-1}).$$

The intensity attenuation coefficient μ (and likewise the amplitude attenuation coefficient α) has contributions from absorption and scattering,

$$\mu = \mu_a + \mu_s \qquad\qquad\qquad (7.6)$$

where μ_a is the intensity absorption coefficient and μ_s is the intensity scattering coefficient. In practice, if one attempts to measure μ or α in a simple transmission loss experiment, additional losses termed 'diffraction losses' (and sometimes corresponding gains) will be observed due to the diffraction field of the sound source (see §7.2.4). The relative contributions of μ_a and μ_s to μ are not known for many tissues, although it is believed that for normal liver μ_s might be between 10% and 30% of μ.

Attenuation of ultrasound increases with frequency. Many soft tissues of the body attenuate ultrasound to a similar degree, and display a nearly linear frequency dependence (see figure 7.2). This gives rise to the ultrasonic instrument designer's rule of thumb for soft tissues, which is

$$\alpha = A \text{ dB cm}^{-1} \text{ MHz}^{-1} \tag{7.7}$$

where, for a wide range of soft tissues, $A \simeq 1$.

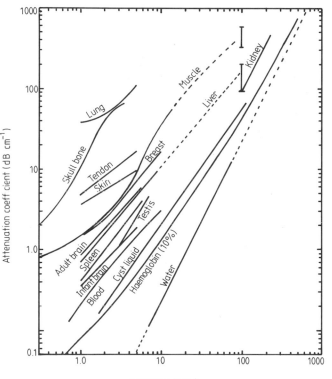

Figure 7.2 Illustration of the general trends observed for the variation of the ultrasonic attenuation coefficient (and its frequency dependence) over various biological tissues and solutions. (After Bamber (1986a), which also contains the references to the original sources of data.)

7.2.1.5 Absorption. Absorption results in the conversion of the wave energy to heat, and is responsible for the temperature rise made use of in ultrasound-induced hyperthermia. There are many mechanisms by which this may happen, although they are often discussed in terms of three classes. Classical mechanisms, which for tissues are small and involve mainly viscous losses, give rise to an f^2 frequency dependence. Molecular relaxation, in which the temperature or pressure fluctuations associated with the wave cause reversible alterations in molecular configuration, are thought to be predominantly responsible for absorption in tissue (except bone and lung), and, because there are likely to be many such mechanisms simultaneously in action, produce a variable frequency dependence close to, or slightly greater than, f^1. Finally, relative motion losses, in which the wave induces a viscous or thermally damped movement of small-scale structural elements of tissue, are also thought to be potentially important. A number of such loss mechanisms might also produce a frequency dependence of absorption somewhere between f^1 and f^2. Generally, however, one can say that, for simple solutions of molecules, increasing molecular complexity results in increasing absorption. For tissues, a higher protein content (especially structural proteins such as collagen), or a lower water content, is associated with greater absorption of sound.

The temperature dependence of absorption is complicated, tending to be increasingly negative at higher frequencies (above 1 or 2 MHz) and positive at low frequencies. Fatty and non-fatty tissues do, however, appear to behave similarly.

7.2.1.6 Scattering. Structures within tissues, which may potentially scatter ultrasound, range over at least four orders of magnitude in size, from cells (at about 10 μm, or 0.03λ at 5 MHz) to organ boundaries (up to 10 cm, or 300λ at 5 MHz). Different kinds of scattering phenomena occur at different levels of structure. These are classified in table 7.1 and will now be discussed in more detail.

Table 7.1 Types of scattering interaction classified according to the scale of the characteristic dimension a of the scattering structure relative to the wavelength λ of sound for frequencies typical of those used in medical imaging.

Scale of interaction	Frequency dependence	Scattering strength	Examples
$a \gg \lambda$ Geometrical region, ray theory for reflection and refraction	f^0	Strong	Diaphragm, large vessels, soft tissue/bone, cysts, eye orbit
$a \sim \lambda$ Stochastic region (diffractive)	Variable	Moderate	Predominates for most structures (even for examples in the other two categories)
$a \ll \lambda$ Rayleigh region	f^4	Weak	Blood

Geometrical region   At a large-scale boundary, representing the interface between two homogeneous media, the usual law of reflection, and Snell's law for refraction, apply to predict the direction of the reflected and refracted sound, i.e. (referring to figure 7.3)

$$\frac{\sin \theta_i}{\sin \theta_t} = \frac{c_1}{c_2} \qquad \text{and} \qquad \theta_i = \theta_r. \qquad (7.8)$$

The intensity of the reflected sound beam, relative to the incident intensity, is given by the intensity reflection coefficient

$$R = \left(\frac{Z_2 \cos \theta_i - Z_1 \cos \theta_t}{Z_2 \cos \theta_i + Z_1 \cos \theta_t} \right)^2. \qquad (7.9)$$

At normal incidence, and neglecting density variations, this becomes

$$R = [(\sqrt{K_2} - \sqrt{K_1})/(\sqrt{K_2} + \sqrt{K_1})]^2. \qquad (7.9a)$$

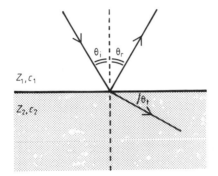

Figure 7.3 Geometrical scattering (reflection and refraction) at a plane boundary between two media, of characteristic acoustic impedances Z_1 and Z_2 and speed of sound c_1 and c_2.

There are no absolutely plane, smooth interfaces in the body, but the diaphragm has been noted to behave somewhat like a mirror, causing a second image of structures that are actually in the liver to appear where the lungs are situated. Other structures, such as large round cysts and the eye, produce refractive effects that can be successfully modelled using ray acoustics. Values for R, computed from differences in values for speed c, are useful in providing an upper limit for, and an intuitively helpful measure of, relative magnitude of echoes from various interfaces (see figure 7.4). As one would expect, interfaces between media that are separated by the greatest difference in speed of sound in figure 7.1 provide the largest reflection coefficient, and it is easy to see why lung attenuates sound so much, and why it is difficult for ultrasound to penetrate (and visualise beyond) bone or gas (as in lung or sometimes in the gastrointestinal tract). It might not, however, have been obvious that the range of echo levels from boundaries between different soft tissues and liquids could span at least 30 dB, given the relatively narrow range of sound speeds for these tissues. Reflection and refraction are determined by inhomogeneities in the speed of sound (c) and the characteristic impedance (ρc). Changes in c and ρc are not always simultaneous. Echo imaging, therefore, is a technique with an inherently excellent contrast for depicting the boundaries between media with different sound speeds.

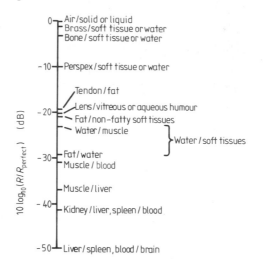

Figure 7.4 Calculated reflection coefficients (in decibels relative to a perfect reflector) for sound at normal incidence to a variety of hypothetical boundaries between biological and non-biological media.

Rayleigh region Examples of this kind of scattering structure are blood (predominantly from red cells) and cells in solid tissues (which contribute to the frequency dependence of scattering in tissues). The scattering strength is very weak, is proportional to the volume of the scatterer and follows an f^4 frequency dependence, and the angular distribution of scattering is fairly uniform but predominantly backwards.

Stochastic region This kind of interaction tends to predominate in internal regions of organs, and modifies reflections from boundaries (i.e. rough surfaces). Scattering in this regime is characterised by a variable frequency dependence and average angular distribution—although for some tissues (like liver) the average scattering has been measured to be somewhat forwards. The scattering is highly anisotropic, and interference of scattered waves gives rise to 20–30 dB fluctuations in measured scattered energy with angle, position, orientation and frequency.

7.2.2 Movement effects

Figure 7.5 depicts the situation of a scatterer, such as a red blood cell, moving with velocity of magnitude v and direction θ with respect to the direction of the transmitted sound wave whose amplitude variations are

$$q_t = Q_0 \cos(\omega_0 t).$$

If the round-trip distance to the scatterer is $2x$, then the received wave will be

$$q_r = Q\cos[\omega_0(t + 2x/c)].$$

However, x is changing at a rate $v\cos\theta$. Therefore, substituting for $x = tv\cos\theta$, we get

$$q_r = Q\cos[\omega_0(t \pm 2tv\cos\theta/c)]$$

which (if $v \ll c$) is approximately equal to

$$Q\cos[\omega_0(1 \pm 2v\cos\theta/c)t].$$

Putting this in the form $q_r = Q\cos(\omega t)$ produces

$$2\pi f = 2\pi f_0(1 \pm 2v\cos\theta/c).$$

Therefore, the Doppler shift (or difference frequency) for backscattered radiation is given by

$$\delta f = f - f_0 = (\pm 2f_0 v\cos\theta)/c. \tag{7.10}$$

Note that the Doppler shift is double that which arises for a stationary observer and a moving source emitting rather than reflecting sound waves. For f_0 in the range 1–10 MHz, and $v\cos\theta$ in the range 0–1 m s^{-1}, δf is 0–13 kHz, which is within the audiofrequency range. Thus a common method for subjective interpretation of Doppler-shift signals from moving blood is to amplify and listen to the signals.

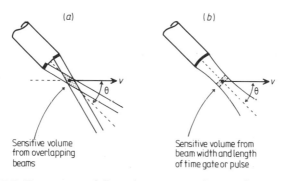

Figure 7.5 Two types of Doppler system, the continuous wave (*a*) and the pulsed wave (*b*) illustrating the angle θ, which is of significance in the backscatter Doppler equation (7.10).

It will be mentioned in §7.2.5 that the echo signal (or A-scan as we shall refer to it in §7.3.2), which forms one line in a pulse–echo image, is the result of a convolution between the point-spread function (PSF) of the transducer and the backscattering impulse response of the object. It will also be mentioned that coherent interference of waves scattered by a uniform medium that does not contain resolvable structure results in a randomly fluctuating echo amplitude, called speckle. If a speckle-producing object is in motion, then the correlation coefficient computed between two A-scans obtained apparently from the same position but at different times is, for small movements of the tissue, inversely related to the distance moved (the constants of the equation being determined by the PSF). Correlation processing provides an alternative ultrasonic method of monitoring tissue movement, the applications of which are only just beginning to be explored.

7.2.3 Parameters for imaging

Of the wave propagation characteristics mentioned above, those which fundamentally govern the fate of the sound wave are ρ_0, K and μ_a. The ideal acoustic imaging system might well aim to produce maps of the spatial distribution of these quantities, plus other information associated with movement. Practically, however, this is difficult, if not impossible, to achieve. Nevertheless, to a degree dependent upon the circumstances and access to the part of the body being imaged, one may be able to create images of parameters associated with the speed of sound, attenuation coefficient, scattering coefficient and movement information. Although these are derived quantities, and practical imaging methods are often only able to provide relatively crude (and sometimes somewhat qualitative) representations of them, they often represent measures of relatively independent aspects of tissue structure and function.

7.2.4 Acoustic radiation fields

7.2.4.1 Continuous-wave and pulsed excitation. The distribution of acoustic field parameters (intensity, pressure, etc), as a function of both space (three dimensions) and time (one more dimension), in front of a radiating source of ultrasound, and the corresponding spatial and temporal sensitivity pattern of a receiver, are factors on which the performance of any ultrasonic imaging system is critically dependent. Prediction of acoustic radiation fields is often possible by calculation, particularly for many of the simple shapes of sources and detectors used in practice. Two alternative theoretical approaches have been used.

First, direct application of Huygens' principle to a distribution of elemental point sources of appropriate strength and phase covering a source vibrating with a continuous sinusoidal displacement leads to the Fresnel theory. Analytical solutions exist for the on-axis field variations, and for the field distribution far from the source, for a plane circular transducer (see figure 7.6), and for other simple shapes such as a line source. For other field positions, and for more complicated transducer shapes, numerical integration of the contributions from each element is necessary. Although the model of a continuous-wave, plane, circular source is not often applicable to realistic imaging situations, since both focusing and pulsed excitation are often employed, it does help to understand the general features of a radiation field pattern. One of the main features is the (somewhat arbitrary) division of the radiation field pattern into two regions. One is close to the transducer (often referred to as the Fresnel zone or 'near field'), where pronounced interference maxima and minima occur but where most of the energy is confined within a transducer radius of the central axis. The other is further away

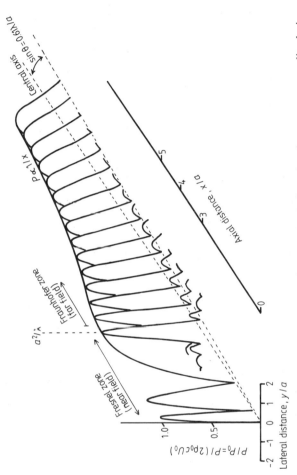

Figure 7.6 Fraunhofer and (on-axis) Fresnel solutions for the pressure amplitude in front of a plane circular source of radius a radiating a continuous sine wave.

from the transducer (often called the Fraunhofer zone or 'far field'), where the wave field is more uniform but tends towards a spherically divergent wave whose amplitude is modulated as shown in the figure. The on-axis variation is such that, starting with a value of zero at infinity, as one moves closer to the source, alternate maxima and minima occur at distances x_m given by

$$x_m = (a^2/m\lambda) - (m\lambda/4) \tag{7.11}$$

where a is the radius of the transducer and m is an odd integer for maxima and an even integer for minima. The position of x_1, whose approximate value is a^2/λ for $a \gg \lambda$ (the so-called 'last axial maximum'), is usually regarded as the boundary between the two zones. In the far field most of the energy is contained within the main lobe, defined by the first off-axis minimum, which occurs at a divergence angle θ given by

$$\sin \theta \approx 0.61\lambda/a. \tag{7.12}$$

Note that the quantity x_1 has no physical meaning if $a < \lambda/2$, in which case $x_1 \leqslant 0$ and θ is $90°$; i.e. there is no near field and the transducer behaves much like a point source.

Secondly, an alternative interpretation of the diffraction field (due to Young) is to consider that it is the result of superimposing just two waves, rather than the infinite number of spherical Huygens' wavelets, ideally considered in the classical Fresnel summation. In this case, the two waves considered are (i) a wave emerging with identical spatial extent and phase as the radiating aperture, and (ii) a wave (which has a very special phase variation) spreading out in all directions from the edge of the transducer. The diffraction field is then the result of interference of these two waves. For a plane circular transducer, the two waves are a plane wave and a hemi-toroidal edge wave, as illustrated in figure 7.7. This approach helps to understand the nature of pulsed acoustic fields, calculated using the impulse response method described below.

It is possible to compute the field distribution associated with a sound pulse emerging from a given aperture by adding all of the continuous-wave solutions, with weights given by the components of the frequency spectrum of the original pulse (i.e. an approach analogous to modulation transfer function (MTF) characterisation of imaging systems). The impulse response method, however, is more suitable for producing results that permit direct inspection of the temporal shape of the pulse waveform at each point in the field (directly analogous to the point-spread function (PSF) method of characterising imaging systems). Also, it is a two-stage calculation such that, once the four-dimensional pressure impulse response h' of the source has been calculated, one can obtain the acoustic field distribution for any source velocity–time waveform by a convolution,

$$p(x, y, z, t) = \rho_0 u(t) * h'(x, y, z, t). \tag{7.13}$$

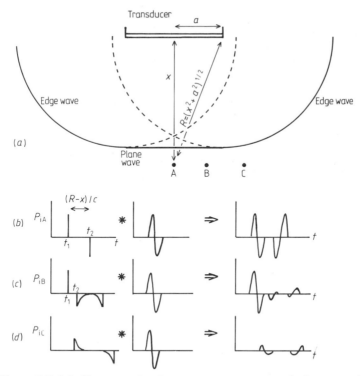

Figure 7.7 (*a*) Plane- and edge-wave components of the acoustic field generated by a plane circular source. The plane wave, and portions of the edge wave that are in phase with the plane wave, are shown as full curves. The broken portions of the edge wave indicate that the wave in this region is 180° out of phase with the plane wave. (*b*)–(*d*) The impulse response functions at field points A, B and C, respectively, and the corresponding waveforms resulting from convolution (represented by *) with a single-cycle sine-wave source driving function.

The pressure impulse response may be derived from a graphical solution, in which the instantaneous response at a given point (at position *x, y, z*) is given by the time differential of the value obtained from a surface integration of those source elements which lie on an arc of equal phase, for the situation of a source that undergoes a theoretical unit velocity impulse, i.e. a step in position. At any on-axis point (A in figure 7.7(*a*)) the plane-wave impulse will arrive first, followed by the simultaneous arrival of all parts of the antiphase portion of the edge wave. This is depicted in figure 7.7(*b*). The time separation ($t_2 - t_1$) is given by the difference in acoustic paths from the centre and edges of the transducer. The rest of figure 7.7(*b*) shows the final pulse waveform at point A, resulting from the convolution of the impulse response with a single-cycle sine wave. Figures 7.7(*c*) and (*d*) illustrate the same sequence of operations for field points B and C, which are off-axis but inside the geometrical beam and outside the geometrical beam, respectively.

From figure 7.7, it can be seen that very complex pulse shapes can occur, particularly close to the source and off-axis, where (if the driving pulse is short) relative path differences may be sufficient for several pulses to be received when only one was sent out. In practice, however, such pulse splitting is difficult to observe unless a receiver with extremely good time resolution is used. Large changes in pulse structure occur moving down the axis and off-axis, corresponding to the summation of continuous-wave solutions for all the frequency components present in the original pulse; i.e. high frequencies are concentrated near the central axis and extend further down the axis before diverging, whereas low-frequency components spread off-axis much sooner. When the same transducer is used to detect the echoes from a point scatterer, similar (reciprocal) properties apply to the time sensitivity of the transducer as a receiver. The result is to introduce still more complexity into the received pulse shape.

It is also possible to explain features of the continuous-wave field of figure 7.6, using the impulse response plus edge-wave descriptions. Take, for example, the response at the on-axis point A (figure 7.7(b)). At $x = \infty$ the plane wave and antiphase edge wave overlap and exactly cancel each other. As we move closer to the source, these components gradually separate until, at a distance x_1 (corresponding to $R - x = \lambda/2$), they are a half-wavelength apart and constructively interfere. At closer distances, the other on-axis minima and maxima discussed earlier can be predicted to occur.

It is somewhat interesting that, although the diffraction field of the aperture has a complicated effect on the structure of the pulse as it propagates, the use of shorter pulses has a simplifying effect on the overall structure of the sound beam, if it is assessed in terms of some single field parameter such as the peak pressure amplitude within the pulse. This effect can be thought of as being the result of averaging the continuous-wave field distributions for each frequency component, which produces considerable smoothing of the near-field interference structure and far-field sidelobes. The position of the last axial maximum also becomes blurred, so that it is no longer easy to define a boundary between the near field and far field. Note, however, that the precise distribution observed depends on the pulse parameter used as a measure of the acoustic field at each point (i.e. peak positive pressure, peak negative pressure, integrated energy in the pulse, etc).

Many factors can influence the ultrasonic field, and not all can easily be included in a calculation (e.g. very complex transducer shapes, non-uniform mounting and clamping of the piezoelectric element, electrode connection). It is therefore preferable, for a good understanding of the behaviour of a particular device, to be able to make direct measurements of the acoustic field distribution in space and time. There are many methods by which this may be done. The subject is beyond the intended scope of this chapter, but some of them are mentioned in §7.3.5.

7.2.4.2 Effect of focusing. If the source is focused (as is usually the case in medical imaging), so that a spherically curved wavefront is launched rather than the plane one discussed above, the acoustic field distribution is modified in a number of ways, some of which will now be described. The continuous-wave on-axis solution is modified such that the position of the last axial maximum of the equivalent plane source (sometimes referred to as the 'natural focus' of the aperture) is always moved to a position (called the true focus) where it is closer to the transducer than either the geometrical focus (given by the radius of curvature of the initial wavefront, R) or the distance x_1. There is also an increase in the on-axis maximum pressure amplitude at the focus, accompanied by a decrease in the width of the sound beam at the same position. The shape of the field distribution off-axis, around the position of the focus and beyond, is similar to that in the far field of the plane transducer depicted in figure 7.6. However, since the wavefront has been made to converge towards the focus, it diverges beyond the focus more rapidly than does the wavefront in the far field of the equivalent plane-wave radiator. This brings about the depth-of-field compromise mentioned in §7.3.1, i.e. the stronger the focusing, the narrower the beam width at the focus but the shorter the depth range over which a narrow beam is maintained. It is often convenient to talk about transducers in terms of their 'strength of focusing'. Strong, medium and weak focusing are sometimes defined, respectively, as $R < 0.25x_1$, $0.25x_1 \leqslant R < 0.5x_1$ and $R \geqslant 0.5x_1$. Note that, even if R is made equal to x_1, the position of the true focus becomes $0.6x_1$, and the on-axis intensity at this point is four times as large as that at the natural focus of the equivalent plane source. This is sometimes quite a useful focusing design, since it results in a moderately narrow but fairly long, parallel beam around the focal region.

Another, often used measure of focusing strength is the 'f-number' of the aperture, n_f, given by the ratio of the radius of curvature to the diameter of the source, $R/2a$. In combination with the wavelength, the f-number provides a convenient method of assessing lateral resolution, as defined below. The first off-axis minimum, at the true focal distance, occurs at radial distances equal to $0.61\lambda n_f$. About 84% of the total power radiated by the source is contained within the main peak defined by this radius.

7.2.4.3 Resolution. There can be no absolute simple definition of the resolution of ultrasonic imaging systems, since in practice too many variables affect the displayed resolution. Even for a particular transducer, there are focusing schemes other than the spherically curved wavefronts discussed above. Thus a resolution measure such as the full width at half-maximum (FWHM), which is commonly used in other medical imaging systems (see Chapter 12), will produce a result that depends on the shape of the response function. Nevertheless, such variations are often ignored. The point response function of ultrasonic imaging systems may be highly asymmetrical, the equivalent length of the sound pulse usually being smaller than the beam width. Hence it is common to talk separately about axial and lateral resolution of pulse–echo systems:

axial resolution   =   half the length of the pulse envelope at some defined level below the peak, multiplied by the sound speed

lateral resolution = full width of the beam at some defined level below the maximum.

It is common to use the FWHM, but many other levels are in use, e.g. 3 dB, 6 dB, 10 dB. Note that FWHM refers simply to the received signal, whatever it may be (usually the amplitude of the echo from some small point-like reflector), whereas the method of calculation of the decibel levels depends on whether the distribution is in intensity or in amplitude. This can be a source of confusion if one does not state precisely how the measure of resolution has been calculated.

For the spherical focusing system described above, the lateral resolution at the focus, defined by the FWHM of the intensity distribution (i.e. the 3 dB level), is given by

$$\text{FWHM}_{\text{sphere}} \approx 1.1\lambda n_f. \tag{7.14}$$

Since the echo amplitude distribution of a transducer is effectively proportional to the square of the one-way pressure amplitude distribution, equation (7.14) also provides an effective measure of the lateral resolution defined in terms of pulse–echo amplitude.

On-axis, the depth range d_x over which the intensity remains within 3 dB of the maximum is given approximately by

$$d_x \approx 15(1 - 0.01\phi)\text{FWHM} \tag{7.15}$$

where $\phi = \sin^{-1}(a/R)$, expressed in degrees. Although d_x does not provide a measure of the change in beam width with depth, it does provide a helpful relative measure of depth of focus.

A very important aspect of resolution in ultrasonic systems is that (at least for the simple beam-forming systems so far discussed) the axial and lateral resolutions are highly spatially variant. The component of this variation, due to the diffraction field of the source aperture, is present even in a homogeneous, loss-less medium. In tissues, extra beam divergence and deviation occur due to scattering phenomena. However, the major distorting effect is often due to frequency-dependent attenuation. Selective attenuation of the higher frequencies causes the beam to widen and the pulse to lengthen. The situation is a complicated one, being influenced by size of aperture, focal length, tissue properties, depth in tissue, initial pulse bandwidth and the mismatch of the speed of sound between the body and any coupling medium. As an example of an empirical result, a 5 MHz transducer with a 6 dB bandwidth extending from 2 to 6 MHz, a focus at 10 cm and an aperture of 2 cm produces a FWHM (at the focus) of 3 mm in water, but 7 mm after propagating 5 cm in breast tissue (see Foster and Hunt 1979). Typically, for such distances of propagation, one observes a degradation in lateral resolution, compared to that observed in water, of a factor of 2.3 for breast and 1.3 for liver.

7.2.5 *Physical principles and theory of image generation*

Chapter 12 covers the subject of formal mathematics of imaging in some detail. This section is primarily for comparison purposes and serves only to highlight those aspects of ultrasound imaging which distinguish it from other imaging methods.

7.2.5.1 Pulse–echo scanning. The reflected waveform is a convolution of the wave field incident on the tissue structure being imaged and an impulse response associated with the scattering properties of the tissue. For the sake of simplicity, we shall state this as a two-dimensional convolution, although in reality both the tissue structure and sound beam exist in three dimensions:

$$g(x, y) = h_1(y)*h_2(x)*f(x, y) \qquad (7.16)$$

where $g(x, y)$ is the image before envelope detection, $h_1(y)$ is the axial pulse–echo impulse response of the system (RF pulse shape), $h_2(x)$ is the lateral pulse–echo impulse response (beam profile) and $f(x, y)$ is the backscattering impulse response of the tissue.

A considerable amount of signal processing may be applied to $g(x, y)$ before it is displayed (as we shall see in §7.3.2), so that the point-spread function (PSF) is the (four-dimensional) impulse response of the imaging system (part of which is due to the transducer pressure impulse response). Major differences with other areas of imaging occur in that:

(a) the processing is non-linear;
(b) the PSF is spatially variant (radiation field pattern);
(c) the PSF depends on the object (speed, attenuation and scattering);
(d) the radiation is coherent and the detector is phase-sensitive (results in interference between different parts of the image);
(e) the PSF has, and therefore so will the image have, negative contributions;
(f) the PSF is not circularly symmetric; and
(g) the PSF is not separable (especially in the near field), although it is often assumed to be.

Equation (7.16) makes many assumptions, e.g. uniform speed of sound, no attenuation, no multiple scattering, no system noise, and PSF spatially invariant but separable into axial and lateral contributions.

If $f(x, y)$ is an impulse (i.e. a single point scatterer), then the image is the PSF of the system. It is instructive to use this situation to see how the different stages of convolution, followed by envelope detection, lead to the image of a point (see figure 7.8). More usually we have distributed or extended targets. As more scatterers are introduced, and the same process is repeated, it becomes easy to see how interference occurs and how the speckle patterns mentioned below might build up (figure 7.9).

The really difficult part in this model of ultrasonic image formation is that no one yet knows exactly what the tissue backscattering impulse response should look like. Various models exist, but none have proven to be generally applicable. For example, it is popular (though not realistic) to model tissue as a random distribution of point scatterers of a given number per unit volume. Another alternative (which is more instructive) has been to consider that the density and elastic modulus vary continuously in the object. Then one (of many) expression that has been derived for the backscattering impulse response (in the 2D approximation) is

$$f(x, y) = \frac{1}{4} \frac{\partial^2}{\partial x^2} \left(\frac{\delta\rho(x, y)}{\rho_0} + \frac{\delta K(x, y)}{K_0} \right) \tag{7.17}$$

where $\delta\rho$ and δK represent small fluctuations in density and elastic modulus about their mean values, and x is the direction of axial pulse propagation. The main point to note is that the scattering results from the second spatial derivative of density and elasticity fluctuations.

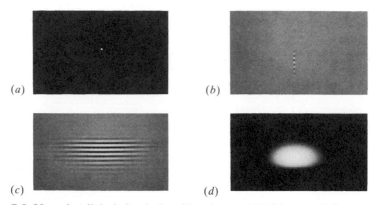

(a) (b)

(c) (d)

Figure 7.8 Use of a digital simulation (Bamber and Dickinson 1980) to illustrate the two stages of convolution, first with the RF pulse (b) and then with the beam profile (c), followed by envelope detection, leading to an image (d) of a single point (a).

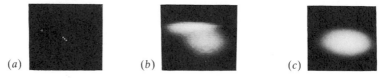

(a) (b) (c)

Figure 7.9 Use of the same imaging model as in figure 7.8 to illustrate the generation of artefactual detail, or speckle, (b) from coherent interference between waves from closely spaced scatterers (a). Image (c) simulates the blurring produced by an incoherent imaging system with the same PSF. It shows no artefactual fine detail.

7.2.5.2 Speckle. For any uniform region of an object in which many scattering sources exist within a resolution cell (defined by the PSF), the image will have a magnitude that varies apparently randomly with position, due to constructive and destructive interference. This leads to the characteristic speckled effect in ultrasound images and contributes a form of visual noise or clutter.

7.2.5.3 Compound scanning and image decoherence. Averaging of multiple images of the same object region, obtained under different imaging conditions, is known as compound imaging. The result is an averaging of the speckle pattern and consequent improvement of signal (structure) to noise (speckle) ratio or contrast resolution. The images to be averaged may be obtained by varying either frequency, position (time if the object is moving), or relative orientation, all of which lead to some loss of spatial resolution.

7.2.6 Non-linear effects

This is an important but extremely complicated subject. However, for the present purposes, it is sufficient to be aware that everything discussed so far in this chapter (and later) assumes that the medium in which the sound wave is travelling responds linearly to the mechanical stresses imposed by the wave, and that this is only an approximation (valid for waves of very small amplitude). All media are (in differing degrees) non-linear, and a consequence of this is that the wave travels at a speed dependent on the local wave amplitude (e.g. particle velocity). This causes a progressive change in the wave shape as it propagates—the distortion takes the form of a gradual transferral of energy in the fundamental to higher harmonics, which are then attenuated due to the frequency dependence of the attenuation coefficient. The distortion is greater for larger inherent non-linearity of the medium, higher frequencies, lower speed of sound, larger distances (in a non-attenuating medium), higher initial wave amplitude and lower attenuation coefficient. There are a multitude of practical consequences of non-linear propagation, including increased absorption of the wave energy, greater spatial variation of the PSF and increased dependence of the beam shape both on the field parameter observed and on the initial pressure amplitude at the source.

7.3 ENGINEERING PRINCIPLES OF ULTRASONIC IMAGING

7.3.1 Generation and reception of ultrasound—Transducers

The transducer is probably the single most important component in an ultrasonic imaging system. Its function is to convert applied electrical signals to pressure waves, which propagate through the medium, and to generate an electrical replica of any received acoustic waveform. A well designed transducer will do this with high fidelity, with good conversion efficiency and with little introduction of noise or other artefacts. Also, it is primarily through transducer design that one has control over the system resolution and its spatial variation.

7.3.1.1 Conventional construction (single element). A good-quality transducer will possess the following design features, as illustrated in figure 7.10.

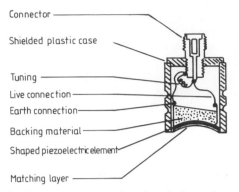

Connector
Shielded plastic case
Tuning
Live connection
Earth connection
Backing material
Shaped piezoelectric element
Matching layer

Figure 7.10 Typical components in the design of a conventional single-element ultrasonic transducer.

Piezoelectric element. This is cut and shaped from a piezoelectric ceramic (usually lead zirconate titanate, PZT) or plastic (polyvinylidine difluoride, PVDF). Silver electrodes are deposited on the front and back faces, and the element is permanently polarised across its thickness. It then has the property that any voltage applied across the electrodes produces a proportional change in thickness and, conversely, pressure applied across its two faces produces a potential difference between the electrodes. The speed of sound in PZT is approximately $4000\,\mathrm{m\,s^{-1}}$, which gives a fundamental resonance ($\lambda/2$) at frequency f and thickness T related by

$$T\,(\mathrm{mm}) \approx 2/f\,(\mathrm{MHz}).$$

For example, at 5 MHz, $T \approx 0.4$ mm.

The characteristic acoustic impedance Z for PZT is approximately 14 times that for water and soft tissue, i.e. the equivalent reflection coefficient (calculated from equation (7.9) as $10\log_{10}R$), $R_{\mathrm{PZT}} \approx -1$ dB, relative to R_{perfect}. On the other hand for PVDF, Z is only about 1.5 times that of water, so that $R_{\mathrm{PVDF}} \approx -14$ dB. Thus PVDF couples its energy to tissue much better than does PZT, and correspondingly has a lower mechanical Q (resonance factor). This makes it a material with a much wider bandwidth, with a flatter frequency response, than PZT but it also suffers from a somewhat poorer conversion efficiency and thus lower sensitivity.

Focusing may be applied by a number of methods: (i) a shaped concave (bowl) ceramic as shown in the diagram; (ii) a lens (concave for materials such as epoxy or Perspex, or convex for silicone rubber) plus a plane ceramic disc; (iii) a bowl plus a defocusing lens; and (iv) overlapping beams from two elements (as in continuous-wave Doppler systems).

Matching layer This attempts to overcome the above-mentioned acoustic mismatch between the element and the tissue, to increase efficiency of energy transfer. The theory is only strictly applicable to continuous waves, in which case the ideal impedance and thickness of the layer are

$$Z_{\text{matching}} = (Z_{\text{element}} \times Z_{\text{tissue}})^{1/2}$$
$$T_{\text{matching}} = \lambda/4. \tag{7.18}$$

A good example is aluminium powder in Araldite, which has $T \approx 0.14$ mm for the above 5 MHz element. A matching layer is not required for PVDF elements.

Backing medium This is usually required for mechanical support, but should be minimal (preferably air only) for maximum efficiency (high Q). For pulse–echo imaging, short pulses are necessary (Q generally about 2–4). Both electrical and mechanical Q contribute to the overall Q. Appropriate mechanical damping to reduce the Q factor is partly provided by the $\lambda/4$ layer but also by choosing Z_{backing} as close to Z_{element} as possible. This involves a compromise between wide bandwidth and sensitivity. Typically, tungsten powder in epoxy resin produces a $Z_{\text{backing}} \approx \frac{1}{2} Z_{\text{element}}$. The energy that enters the backing should be absorbed and not reflected back into the element. To accomplish this, plasticised epoxy, embedded fine scatterers and shaped backs have all been used. Note that PVDF has a naturally low mechanical Q and does not generally need special backing to improve the bandwidth.

Casing This should be electrically screened and acoustically decoupled from the element (otherwise the dynamic range is reduced due to either acoustic ringing or electrical interference). This either means a plastic case with a screening layer, or a metal case with an acoustic insulator.

Electrical tuning This is often used to filter out low-frequency radial-mode vibrations of the element, and to manipulate the electrical Q for the best compromise between sensitivity and resolution. The capacitance of the transducer element C_t is given by the 'parallel-plate' formula

$$C_t = \varepsilon A_t / T \tag{7.19}$$

where ε is the dielectric constant of the element and A_t is the element area. In short-pulse applications, a single shunt inductance (as shown in figure 7.10),

$$L \approx 1/(2\pi f)^2 C_t$$

is sometimes used to reduce the electrical Q, but again at the expense of sensitivity. Expressing equation (7.19) in terms of c and f, for thickness $T = \lambda/2$, and using C_t to calculate the reactance $\chi_t = \omega L$, we obtain

$$\chi_t = c/\varepsilon(2\pi f r_t)^2 \tag{7.20}$$

where r_t is the radius of a plane disc ($=(A_t/\pi)^{1/2}$). For PZT-5A, f in MHz and r_t in mm:

$$\chi_t \approx 6.4 \times 10^3/(r_t f)^2. \qquad (7.21)$$

For example, if $f = 5\,\text{MHz}$ and $r_t = 10\,\text{mm}$, then $\chi_t \approx 3\,\Omega$, which is a bad match to the transmitting and receiving electronics if it uses the usual transmission impedances of 50 or 75 Ω. Generally, at above 3–4 MHz, transformer matching will improve this electrical coupling, though with some loss of bandwidth.

7.3.1.2 Multiple-element transducers. Single-element transducers, as described above, are not often used in modern scanning equipment, although the other basic aspects of design still apply to multiple-element systems. More than one element may be required (in the simplest case) to permit the use of continuous waves (as in a Doppler system), where separate transmitting and receiving elements are required. In pulse–echo imaging, multiple elements may be used for electronic (and rapidly changing) beam forming and/or focusing, and for fast electronic beam translation and/or steering. In this context there exists a great variety of element shapes and corresponding field patterns.

The general principles of beam forming, focusing, scanning and steering are illustrated schematically, and in two dimensions only, in figure 7.11. By simultaneously exciting a group of tiny elements, one can, if each element behaves like a Huygens' source, synthesise a plane wavefront, which emerges from the aperture formed by the spatial extent of that group of elements (figure 7.11(a)). The sound beam may then be translated from position 1 to position 2 by exciting a different, but overlapping, group of elements. To focus (figure 7.11(c)) or steer (figure 7.11(b)) the sound beam one must be able to excite each element via a variable delay. Systems using this technology have come to be known as 'phased-array' systems, the term (and indeed the method) being adopted from the world of radar.

By applying the same group switching and delay programming, similar receiving sensitivity patterns can also be generated. In fact, within the time taken for a complete sequence of echoes to return (as a result of a single transmitted sound pulse), it is possible to adjust the focusing delays continuously so that the system has a receiving directivity pattern that is maximally sensitive, and therefore focused on each echo position as it arrives from each and every depth. This approach is known as 'swept (or dynamic) focusing' and, clearly, can only be applied to the received signal. A cheaper system is to switch between a number of fixed delay combinations, providing a sequence of 'focal zones'. Changes in the focal properties of the transmitted beam can only be made over successive sound pulses, with a consequent trade-off of rate of data capture. Generally, one finds combinations of these approaches: scanning plus focusing, and steering plus focusing.

Phased arrays are potentially a very powerful method of creating ultrasonic images, providing:

(a) beam steering without motion,

(*b*) dynamic focusing,

(*c*) arbitrary sequencing of directions,

(*d*) arbitrary frame-rate advantages,

(*e*) easy digital control, and

(*f*) possibility of parallel processing, which may be used to reduce speckle by compounding or to increase either resolution or frame rate.

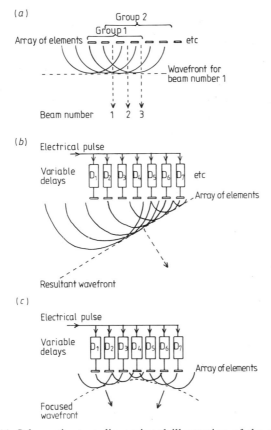

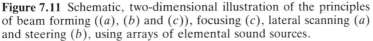

Figure 7.11 Schematic, two-dimensional illustration of the principles of beam forming ((*a*), (*b*) and (*c*)), focusing (*c*), lateral scanning (*a*) and steering (*b*), using arrays of elemental sound sources.

Their disadvantages, which might be overcome in the future, are:

(*a*) sampling problems (grating lobes, quantisation errors—dealt with in next subsection), and

(*b*) system cost and complexity.

Typical array configurations are shown in figure 7.12. The annular array (figure 7.12(*a*)) permits the use of dynamic phase focusing to improve depth of focus over that of a single-element circular transducer. However, to produce an image one must steer or translate the beam by mechanically moving the whole transducer assembly. Linear (switched or phased) arrays (figures 7.12(*b*) and (*c*)) allow rapid beam translation or steering with no moving parts. Phase focusing may also be used, but only in the plane of the scan—fixed mechanical (lens) focusing is usually applied in the orthogonal plane.

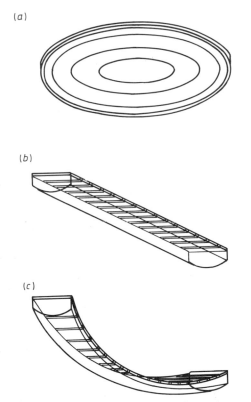

(*a*)

(*b*)

(*c*)

Figure 7.12 Two main transducer array configurations currently in use: (*a*) annular array (often prefocused by shaped ceramic or a lens, so that the maximum electronic delay between elements is minimised) and (*b*) linear array (shown with a fixed focus in the plane orthogonal to the scan plane). (*c*) The linear array is sometimes laid down on a curved surface to produce a sector-type image without phase-controlled beam steering.

Construction of arrays is usually accomplished by cutting (using a diamond saw or laser trimmer) one large piece of PZT, after plating, polarising and mounting. Annular arrays generally consist of a fairly small number of elements (five or six), providing a similar number of overlapping focal zones. Linear arrays are usually one of two kinds:

small-aperture and phase-steered, or large-aperture and switch-translated. Phase focusing may be applied in either case. For the phase-steered systems the number of array elements is usually in the range 20 to 30, although is reported to be 128 in one new product. The large-aperture systems may easily contain 300 elements. Ideally the pitch between the elements should be as small as possible and the width of an element should be less than $\lambda/2$, if the elements are to behave as omnidirectional sources (e.g. 0.3 mm at 5 MHz), so that steering and focusing are possible. These are very difficult to manufacture for high frequencies.

A problem that occurs with these systems is that the array acts as a diffraction grating; it is not possible to make the distance between element centres small enough to represent a continuous distribution of Huygens' sources. The result is that other sound beams (grating lobes) are generated at various angles to the intended main beam. If one considers that the far-field, or Fraunhofer, diffraction pattern is in fact obtained from the Fourier transform of the source aperture, then it will be seen that this phenomenon is equivalent to the sampling theorem (see Chapter 12), i.e. sampling of the transmitting or receiving aperture produces repetition of the angular spectrum at angular intervals given by the reciprocal of the sampling distance. In terms of conventional Bragg diffraction theory for the grating, one obtains sound beams in directions θ for

$$\sin \theta = n\lambda/a. \qquad (7.22)$$

The primary beam corresponds to $n = 0$. The secondary beam, at $n = 1$, is often referred to as the first 'grating lobe'. If the images are to be free from artefactual echoes, which may occur if a grating lobe interacts with a strongly reflecting structure, the aperture should be sampled finely enough such that no grating lobes exist. This condition is just met if $\theta = \pi/2$ for $n = 1$, which, from equation (7.22), gives $a = \lambda$.

For steered arrays the problem is actually much more serious, since the sensitivity varies with the steering angle. As the main beam moves off-axis, the grating lobes move closer to the axis. The maximum usable deflection of the main beam is

$$\sin \theta_{\max} \approx \lambda/2a \qquad (7.23)$$

which is the angle at which the main lobe and the first grating lobe become equal in magnitude, e.g. for $\theta_{\max} = \pi/2$, $\lambda/a = 2$.

Much of the improvement in image quality in recent years has been due to advances in transducer array construction, and electronic processing to avoid grating lobe artefacts.

7.3.2 *Pulse–echo techniques (echography)*

Figure 7.13 shows a generalised scheme for signal generation and processing in a pulse–echo imaging system, with a pictorial representation of the signals to be expected at each stage. We now discuss, in turn, each of the components of figure 7.13.

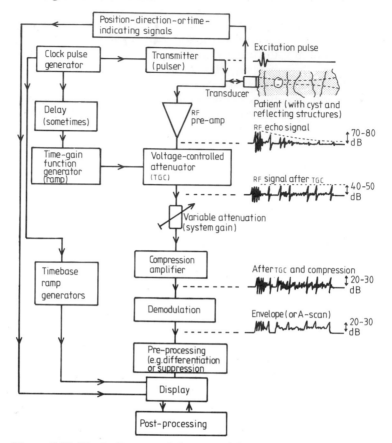

Figure 7.13 Block diagram of the essential components in the signal-processing chain of an ultrasound pulse–echo imaging system. On the right of the figure are labelled schematic versions of the kind of signal one might expect to observe at the points indicated, in response to a single transmitted sound pulse.

7.3.2.1 Transducer and frequency. The choice of frequency is a compromise between good resolution and deep penetration. One often finds frequencies in the range 3–5 MHz used to image the liver and other abdominal organs, the uterus and the heart, whereas more superficial structures such as thyroid, carotid artery, breast, testis and various organs in infants would warrant the use of somewhat higher frequencies (4–10 MHz). The eye, being extremely superficial and exhibiting low attenuation, has been imaged with frequencies in the range 7–15 MHz.

The short pulse of sound that emerges from a diagnostic transducer or array is generally no more than 3–4 cycles in length, and is generated (at the transmitter) by applying to the transducer either a momentary voltage step or a single-cycle gated sine wave of frequency equal to the resonant frequency of the transducer (see below).

7.3.2.2 Clock pulse. This triggers the excitation of the transducer and acts, in some circumstances, to synchronise the display. Following the emergence of one sound pulse, a stream of echoes returns, so that all of the echo information for one image line is contained in one clock cycle. High repetition rates are desirable, however, for fast scanning or for following moving structures. The maximum pulse repetition rate (PRR_{max}) is limited by the maximum depth (D_{max}) to which one wishes to image, according to

$$PRR_{max} \text{ (kHz)} \approx \frac{1.5}{2D_{max}} \times 10^3 \qquad (7.24)$$

where D_{max} is millimetres and the factor of 1.5 is the approximate speed of sound (often conveniently expressed in $mm \, \mu s^{-1}$ for such calculations). If this repetition rate is exceeded, echo ambiguity occurs, since echoes from deep structures associated with a previous pulse become coincident in time with those from more superficial structures for the current pulse. Typically, for abdominal imaging, the PRR is often in the region of 1 kHz.

7.3.2.3 Transmitter. The voltage pulse applied to the transducer is usually below 500 V and often in the range 100–200 V. Its shape is equipment-dependent, but it must have sufficient frequency components to excite the transducer properly; e.g. for frequency components up to 10 MHz, a pulse risetime of less than 25 ns is required.

7.3.2.4 Linear RF amplifier. This is an important component, since noise generated at this point may well limit the performance of the complete instrument. Clever design is required, since, whilst maintaining low noise and high gain, the input must be protected from the high-voltage pulse generated by the transmitter. Any circuits that do this must have a short recovery time, and the amplifier itself should possess a large dynamic range and good linearity. As indicated in the diagram, the radiofrequency echo signal returning from the object may have an overall dynamic range (from the noise level to the largest echoes) of some 70 to 80 dB.

7.3.2.5 Time gain control. This is provided by a voltage-controlled attenuator. Some form of time-varying function, synchronised with the main clock and triggered via a delay, is used as a control voltage so that the system gain roughly compensates for attenuation of sound in the tissue. The simplest function used is a logarithmic voltage ramp, usually set to compensate for some mean value of attenuation. The effect that this has on the time-varying gain of the system is illustrated schematically in figure 7.14. The delay period indicated is often adjustable, so that the attenuation compensation does not become active until echoes begin returning from attenuating tissue. This might be used, for example, to scan through a full bladder or through a water bath (see later). As a general result of this attenuation correction, the dynamic range of the signal at the output of the time gain control (TGC) section has often been reduced to some 40 to 50 dB.

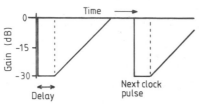

Figure 7.14 A simple, but common, form of time-varying gain for a pulse–echo imaging system.

An artefact of this type of attenuation correction is that echoes posterior to a structure that attenuates sound more than the average for the surrounding tissue tend not to be amplified enough. This causes the appearance of a shadow behind such structures. Conversely, the simple TGC function will overcompensate for structures that attenuate sound less than average, so that a kind of negative shadow, or 'enhancement', appears on images posterior to such structures (usually fluid-filled, such as cysts). Even though these features are rather crude artefacts of a simple attenuation correction system, they have, over the years, come to represent important diagnostic features, which often permit experienced diagnosticians to understand more of the nature and consistency of discrete lesions. Various attempts at providing some form of automatic time gain compensation have been tried, but fully automatic systems have failed to gain widespread acceptance.

7.3.2.6 Compression amplifier. There is a wide range of gain characteristics in use but a general feature is that the gain decreases with increasing input signal. One example is an amplifier with a logarithmic response. This allows the remaining 40–50 dB echo range to be displayed as a grey scale on a cathode ray tube (CRT), which might typically have only a 20–30 dB dynamic range (cf amplitude windowing, which is inappropriate in ultrasound imaging because of the speckle modulation, used to view x-ray CT images, which also possess a wide dynamic range (see Chapter 4)).

7.3.2.7 Demodulation. At this point the envelope of the RF echo signal is extracted, usually by simple full- or half-wave detection followed by smoothing with a time constant of about 1.5λ. The signal depicted at this point is known as a detected A-scan (representing the amplitude of the echoes).

7.3.2.8 Pre- and post-processing. It is common to pre-process the A-scan echo signal both before and after it is digitised and stored in the display memory (see below). Examples are various forms of edge detection (usually differentiation), and further adjustments to the gain characteristic, such as suppression, which is dynamic range restriction by rejecting from the display echoes that are below an operator-defined threshold (noise suppression). In modern machines, the term 'suppression' is generally not used, but the same result may be effected by various processing options, often provided for dynamic range manipulation. These facilities are of debatable utility, probably because of the presence of the large fluctuations in grey level caused by the coherent speckle phenomenon. These make it difficult to appreciate subtle changes in grey level and break up the continuity of regions that would otherwise be uniform and of limited dynamic range.

7.3.2.9 Display. As the sound beam is scanned across the object being imaged, a sequence of A-scans is generated. Scanning methods and the images they produce will be described below, but as the sequence of scan lines arrives they must be stored and geometrically reconstructed to form an image. The simplest method used to be to build up the image on photographic film in a camera with the shutter left open whilst the brightness-modulated spot of a CRT moved across the $x-y$ display. A much better method is to use a device known as a scan converter. In modern machines, this is essentially a digital memory with specially designed 'read' and 'write' systems. Data are read out continuously, in conventional television format, but are updated with new scan data by writing to memory pixels in the order, and in the coordinate system, in which the A-scan data arrive. Hence, at some stage in the signal processing, the A-scan is digitised. This is usually after envelope detection, but there is a trend in recent systems to digitise earlier. After digitisation, the data are written into the two-dimensional digital memory along a vector calculated from coordinates of the position-sensing system. This involves a coordinate transformation (e.g. from the polar coordinates of a sector scan to the Cartesian coordinates of the display memory), and some interpolation to fill display memory locations that do not have a vector passing through them. There is plenty of opportunity for errors to occur in this process, and much effort has gone into optimising the scan-conversion algorithms (e.g. to avoid Moiré patterns due to the limited display resolution undersampling the vector lines). Minimum memory requirements for a modern display are now considered to be 512×512 pixels by six bits (though many systems use eight bits to avoid contouring effects).

7.3.2.10 Additional sections. The design of echo signal-processing schemes varies widely from manufacturer to manufacturer, and is often proprietary information. Even within the simplest of (effectively idealised) processing schemes discussed here, other amplification stages would be present, providing an overall gain of between 70 and 100 dB. In particular, we have neglected to discuss the (often quite complicated) signal processing that takes place in sound beam, and echo, synthesis for multi-element transducer array systems.

7.3.2.11 Scanning and display format. There are three commonly used formats for displaying ultrasound echo data, and one rarely used format. The simplest is the *A-mode* of display, referred to above as the A-scan, which is a one-dimensional display of echo amplitude versus time. This is most used in eye examinations, when precise measurements of axial length of the eye are required, but used to be applied to the detection of space-occupying abnormalities in the head, by observing the position of the echo from the brain midline. This application has now been completely superseded by x-ray computed tomography and radionuclide imaging.

The *M-mode* (also called *TM-mode*) is used for observing changes in the A-scan as a function of time. In this the A-scan, for echoes arriving from along a constant line of interest through the body, is used to brightness-modulate a display in the vertical direction. A slow timebase in the horizontal direction then produces the TM (time–motion) recording. A typical example of an application of M-mode scanning is the examination and quantification of the pattern of movement of heart-valve leaflets.

The *B-mode* is by far the most commonly used mode, and is indeed a true image format. A B-scan is what one normally refers to implicitly when talking of ultrasonic pulse–echo imaging. There are many designs of scanning arrangement, but a general feature is that the A-scan is used to brightness-modulate a display, which may have two (x and y) fast-timebase signals operating simultaneously. The starting point, and the direction, of the resulting image line vector is determined from the position and direction signals at the transducer. The two-dimensional image is built up by moving the sound beam in the xy plane.

Finally, a lesser known and rarely used format is the *C-mode*, or constant-depth scan. In this, an electronically gated portion of the A-scan signal is used in a brightness-modulated display of the projected two-dimensional position of the beam in a plane at a fixed distance from the transducer plane. The method enables a wide-aperture, strongly focused, transducer to be used and the image to be produced in the plane of the focus. It does, in principle, also allow otherwise inaccessible scan planes to be visualised with optimised resolution. A major disadvantage is that out-of-plane attenuating structures may influence the received echo level, and the data take a considerably longer time to acquire. However, future designs of three-dimensional scanners might well find it useful to provide an option for selecting a display format equivalent to this one-time experimental scan mode.

7.3.2.12 Scanning methods. These are illustrated in figure 7.15. A major classification exists between simple scans and compound scans; compounding refers to the process of summing multiple simple scans onto the same image (see §7.2.5). Two types of simple scan are often referred to. The simple linear scan requires a large window of access to the body but gives the best images, with a wider field of view near the skin. The simple sector scan has the advantage that it is easier to move the beam rapidly if the scanning is mechanical, and a much narrower access window may be used. The narrower field of view near the skin may be a disadvantage, but this form of scanning is often the only way to get a wide enough field of view at depth. Typical examples of this are imaging of the heart or some parts of the liver through intercostal spaces (between the ribs), or imaging the infant brain through the fontanel (a space between component bones of the skull).

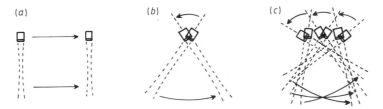

Figure 7.15 Some of the scanning methods used in echographic imaging: (*a*) simple linear scan, (*b*) simple sector scan and (*c*) compound scan formed by integrating a number of sector scans.

Compound and simple scanning methods have a number of relative advantages and disadvantages. These features are summarised in table 7.2. The two methods thus provide complementary information.

Table 7.2 Advantages and disadvantages of compound and simple scanning, demonstrating the complementary nature of the two approaches.

Method	Advantages	Disadvantages
Compound scan	More complete display of boundaries Improves effective lateral resolution for single targets (in homogeneous medium) by averaging Reduces speckle by averaging and therefore improves image resolution and grey-level discrimination	Requires large window of access to the object Complex—difficult to scan fast Registration of individual scans may not be good if inhomogeneities cause refraction or if body moves during scan. Tends to offset improved resolution and speckle averaging
Simple scan	Provides some information related to the diffractive scattering properties of tissues (seen as the 'texture' of the image) Portrays regions of differing attenuation coefficient (and sometimes velocity) by changes in posterior echo level (or position) Takes less time (real-time imaging) Image distortion due to refraction, or body movement, is less of a problem than with compound imaging	Only boundaries or structures that lie normal to the beam are clearly delineated Speckle makes it difficult to perceive low-contrast targets

Although simple scanning is now almost always used (because it can be done easily in real time), in those instruments which are capable of compound imaging both methods are often employed sequentially on the same object (as, for example, in water-bath scanning of the female breast).

A further division occurs between contact and water-bath scanning. Again, there are advantages and disadvantages to both approaches, which are summarised in table 7.3. In contact scanning, the transducer or array surface is placed effectively in contact with the patient's skin, the junction being lubricated with a water-based gel or viscous oil (the main function of which is to provide good acoustic coupling). This class may also cover a number of scanners for which the transducer is housed within a small self-contained water (or oil) bath, although some of the advantages of water-bath scanning are possessed by these devices, e.g. the ability to make use of larger-aperture more tightly focused transducers. Because of its inconvenience water-bath scanning is tending gradual-

Table 7.3 Comparison between contact scanning and automated water-bath scanning methods.

Method	Advantages	Disadvantages
Contact scanning	Very versatile (completely arbitrary scan plane) Relatively simple to use Relatively cheap For simple sector scan, only a small access window is required	Large-aperture (strongly focused) transducers not easily applied (especially 3D) Superficial tissue structure may not be imaged well due to recovery time and near field (poor focus)—better on recent instruments May distort tissues while scanning due to contact pressure Requires medically trained operator on many occasions
Automated water-bath scanning	Can move transducer without distorting tissue Can use wide-aperture transducers with strong (3D) focusing Easier and better imaging of superficial structures Easily adapted for compound scanning Suitable for a rapid, automatic and systematic study of an organ by serial tomography May be operated by non-medically trained personnel	Large—occupies a lot of floor space Relatively expensive Complex mechanical systems may be prone to failure Cumbersome for patient handling Requires more maintainance by the user Reduced access to some deeper parts of the body Multiple echoes occur between skin and transducer face—limits maximum penetration depth and pulse repetition rate

ly to go out of fashion, but is an approach where the part of the body to be imaged is immersed in a bath of water (saline or oil), which acts as a propagation delay.

7.3.2.13 Real-time (rapid) scanning. It is important that the image refresh (scanning) rate is fast enough such that one can (*a*) obtain information about moving structures, (*b*) eliminate movement artefacts from single frames, and (*c*) permit rapid coverage during a fine search through a large volume of any stationary organ such as the liver (although even organs like the liver are never entirely stationary). One of the reasons why M-mode scanning is still much used for the detailed examination and quantification of heart-valve motion is that the frame rate for whole images cannot be as high as that required (of the order of 100 Hz), whereas the M-mode sampling rate may be as fast as the PRR itself. In general, the frame rate is limited by the number of lines (A-scans) in the image, N_{lines}, and the maximum depth to be visualised. Using the same terminology as for equation (7.24) we have

$$\text{PRR} = N_{\text{lines}} \times \text{frame rate.}$$

Hence

$$\text{max. frame rate (Hz)} \approx \frac{1.5 \times 10^6}{2D_{\text{max}}N_{\text{lines}}} \tag{7.25}$$

where again D_{max} is measured in millimetres.

Many systems exist for rapidly translating or steering the sound beam. Purely electrical systems, using linear and curved-linear arrays, are indeed popular because of their lack of moving parts and consequent inherent reliability and convenience. Mechanical systems, however, use circular transducers to provide good lateral resolution obtained ortho-gonal to, as well as within, the scan plane.

7.3.3 Doppler methods

7.3.3.1 Choice of frequency (f_0). As with pulse–echo imaging, the choice of operating frequency is governed by a compromise between resolution (but this time we include both spatial and flow velocity resolution) and depth of penetration. However, the f^4 dependence of scattering from blood helps a little and provides the basis for a quantitative analysis, based on optimisation of the signal-to-noise ratio. This leads to an equation that can best be expressed graphically, as in figure 7.16. The analysis depends on many factors, including the frequency dependence of α. Many assumptions are made, and hence the resulting relationship between optimum frequency and depth is an approximation. Nevertheless, it is a useful general guide to the choice of frequency for a particular application.

7.3.3.2 Notes on the 'Doppler equation'. Equation (7.10) is often refer-red to as the Doppler equation for (backscattered) ultrasound in medical applications. Before we proceed to a discussion of practical ultrasound

Doppler measurement systems, it is useful to note some of the features, limitations and approximations which are associated with this simple equation.

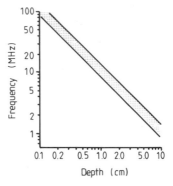

Figure 7.16 Estimated range of frequencies, as a function of depth in muscle tissue, that will maximise the signal-to-noise ratio for detection of ultrasound scattering from blood (plot of the equation $f_0 = 3/2d$). (After Baker *et al* (1978).)

(*a*) One needs to know $\cos \theta$ and c in order to measure v in absolute terms. It is often not easy to determine θ accurately, and this makes accurate determination of flow velocity difficult.

(*b*) It is indeed a measure of the velocity that is obtained, in spite of the commonly used terminology 'flow meter'.

(*c*) The information about velocity is encoded by frequency modulation, which helps reduce noise. However, low noise is vital since the echoes from blood are completely buried in the backscattered signal from stationary tissue. The latter part of the signal is often called 'clutter', and it is one of the functions of the signal processing to reject it.

(*d*) In general, there will be many moving scatterers, each producing a different δf. This results in a Doppler-shift audio *spectrum*, which varies with time (during the cardiac cycle), as depicted schematically in figure 7.17. This 3D distribution contains a great deal of potentially useful information. The ability to observe the distribution of flow velocities within a blood vessel, and time variations to within a relatively small fraction of the cardiac cycle, is a unique feature of ultrasound Doppler techniques. The specific shape of the Doppler-shift spectrum at a given instant is a complicated function of the beam shape, the shape and size of the vessel, and the distribution of flow velocities within the vessel. Much effort has been put into understanding these relationships and, given simple shapes for the sound beam (e.g. Gaussian) and blood vessel (cylindrical), it is possible to infer the radial distribution of flow velocities (e.g. parabolic or otherwise) from the measured Doppler spectrum. Much use is made clinically of qualitative observations of spectral-shape changes as a function of longitudinal distance along a blood vessel, for detection of flow disturbance induced by lesions and constrictions of the blood vessel. Generally, the disturbed flow results in an increased range of Doppler frequencies, known as spectral broadening. The time envelope of the maximum Doppler-shift frequency is also useful since its shape is influenced by arterial-wall elasticity and downstream vascular impedance.

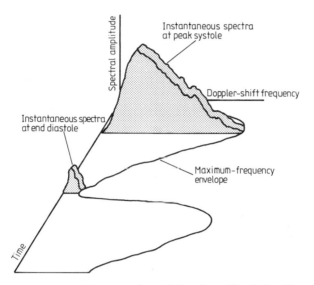

Figure 7.17 Schematic illustration of the three-dimensional nature of the Doppler-shift signal from pulsatile, flowing blood.

(*e*) The simple Doppler equation neglects to take into account the finite width of the sound beam, and predicts that a single scatterer travelling at a constant velocity should produce a line spectrum. In reality, scattering structures move across the sound beam and thus their echoes are amplitude-modulated. This amplitude modulation broadens the Doppler line spectrum. Another way of thinking of this phenomenon is to say that, to define v for a given scatterer by measuring δf, we would need access to the backscattered signal for an infinite amount of time, whereas we only possess data lengths of the time taken for the scatterer to traverse the beam, i.e. this is a Fourier-domain 'leakage' phenomenon (cf Chapter 4). There exists an uncertainty compromise here, in that if the beam width is reduced to improve lateral resolution, this is at the expense of the frequency (velocity)-resolving properties of the Doppler system, and vice versa.

(*f*) Continuous-wave transmission is assumed. This will not be the case for many systems—particularly those used also for imaging, as we shall see later.

7.3.3.3 Signal processing in Doppler systems

Continuous-wave flow detector   In order to extract δf (positive or negative) from the returned echo signal, which includes the clutter from stationary structures, some signal processing is required. The achievements in this respect are directly related to the sophistication of the system. A number of levels of sophistication may be defined. The simplest is the continuous-wave (CW) flow detector, illustrated in figure 7.18. A continuous acoustic wave is transmitted from one element of the transducer whilst a second element receives the echoes. After amplification, the difference frequencies are extracted by the demodulator, amplified and made audible directly or processed further, prior to some other form of presentation. Many different designs of demodulator exist. The simplest mixes the echo signal with the reference frequency (producing both the sum and the difference frequencies), and then low-pass filters the result to extract only the difference frequencies. Such a simple system does not distinguish between flow towards or away from the sound source. An improvement is the heterodyne system, in which the signal is mixed with an offset reference frequency (f_R), producing a signal where forward flow is represented as frequencies above f_R and reverse flow as frequencies below f_R. Such a signal is not so suitable for audible presentation (because of the constant frequency for zero flow) and places a severe limit on the maximum measurable reverse-flow velocity. The best method is (arguably) the so-called 'quadrature-phase demodulator' with 'phase-domain' processing—this provides two audio signals, one for forward flow and the other for reverse flow (sometimes used with stereo headphones). Excellent directional separation is possible with this system, although it suffers from the disadvantage that further processing (or recording) must proceed with two channels, with consequent duplication of electronics.

Methods of further processing (other than listening), analysis and display of the time-varying Doppler-shift signal (figure 7.17) have, in the past, also varied greatly. With the reduction in the price of semiconductor memory and other integrated electronic components, the best method, which is real-time spectrum analysis, has also become the preferred method. The usual form of display is known as the 'sonogram', a plot of frequency (ordinate) and amplitude (grey level) versus time (abscissa). Derived characteristics of the instantaneous spectrum (such as the mean frequency, peak frequency, or some form of flow disturbance index) may also be computed, and their time variation displayed, usually after obtaining several cardiac cycles' worth of sonogram data. With increased processing speed of modern devices, or by using analogue circuits, real-time display of some of these characteristics has been possible.

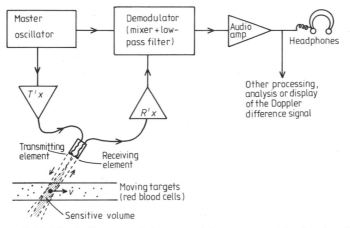

Figure 7.18 Block diagram of the essential components in the signal-processing chain of a cw ultrasound Doppler blood-flow detector.

A cw Doppler system may be turned into an imaging device by mounting the ultrasonic transducer on an arm with sensors that can resolve the transducer position in two dimensions. These position signals are used to control the $x–y$ position at which the intensity or colour of the image is modulated by a characteristic of the Doppler signal, such as the mean frequency. The images produced are projections onto a plane approximately parallel with the skin. Such images must be built up very slowly by scanning manually for periods of many cardiac cycles. These systems are little used now, and have given way to more advanced Doppler imaging systems described later.

Pulsed-wave flow detector   As indicated by figure 7.5, the cw Doppler system possesses no resolution in the depth direction (other than an increased sensitivity over the depth range covered by the focus) and would be unable to distinguish between the signals from two blood vessels lying within the sound beam but at different depths (unless, of course, the blood in each of the vessels was flowing in opposite directions). This problem is resolved by use of increased sophistication, in signal coding and processing to preserve range information as well as frequency information. The usual method of achieving this is embodied in the pulsed-wave Doppler system, as illustrated in figure 7.19, which possesses attributes similar to both the pulse–echo imaging and cw Doppler signal-processing schemes already described. So as to aid the description of the pulsed Doppler method, figure 7.19 also depicts a schematic representation of the signal that might be observed at the various numbered points in the circuit block diagram.

At point 1 the master oscillator generates a sine wave similar to the cw system above. This is gated under the control of a clock, so that at point 2 one would observe a train of short pulses (a few cycles long) repeated at intervals determined by the PRR, as for echography. These pulses are amplified and used to excite a single-element transducer, which also acts as a receiver. After preamplification, the echo signal, which (at point 3) looks just like the exponentially decaying A-scan prior to TGC correction and envelope detection, passes through another gate. This gate may be of adjustable length but, in all systems, it is triggered at a time which is adjustable relative to the clock that triggered the transmitted pulse. The gate output, at point 4, is therefore a depth-selected portion of the RF echo signal. Demodulation of this sampled echo signal, by one of the methods described for the CW system above (e.g. mixing and smoothing), results in a sampled Doppler-shift signal (point 5) where the sampling frequency is the PRR. A sample-and-hold amplifier, synchronised to the delayed clock, is used to convert this to a continuous, though quantised, signal, as shown for point 6. Further smoothing then produces the final audio signal at point 7.

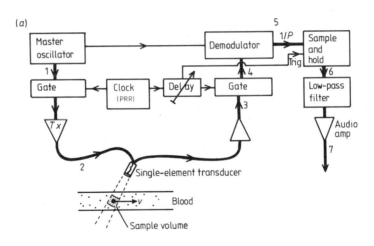

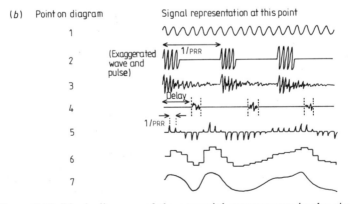

Figure 7.19 Block diagram of the essential components in the signal processing for a pulsed Doppler system (*a*) and schematic versions of the signal one might expect to observe at the points labelled (*b*).

An important problem arises with all pulsed Doppler systems. As may be seen from the representation of the sampled Doppler signal at point 5 in the above description, unless $\delta f < \text{PRR}/2$, aliasing (see Chapters 4 and 12) of the detected audiofrequencies will occur. Equation (7.24) from pulse–echo imaging, when combined with the Doppler equation (7.10), produces the 'range–velocity' compromise for the maximum velocity v_{max} that may be unambiguously observed for a given maximum depth of the gate, D_{max}: $(v \cos \theta)_{max} = c^2/8f_0 D_{max}$, i.e.

$$v_{max} \approx 2.8 \times 10^5/f_0 D_{max} \tag{7.26}$$

(f_0 in MHz, D_{max} in mm and v_{max} in mm s^{-1}). The rapid rate of decrease in v_{max} with either f_0 or D_{max}, due to the inverse relationship, imposes severe limits on the applicability of pulsed Doppler methods to studying deep-lying blood vessels. Figure 7.20 illustrates the limits of range and velocity, obtained from equation (7.26), which must be observed if range–velocity ambiguities are not to be present. A common method of maximising system performance at all depths is to link the PRR to the delay that controls the position of the gate in the receiving circuit. This causes other complications in the design, such as the necessity for a sharp cut-off smoothing filter at point 6, whose cut-off frequency is variable and linked to the PRR. Another approach to this problem is mentioned later when discussing linked pulsed Doppler and echographic imaging (§7.3.4).

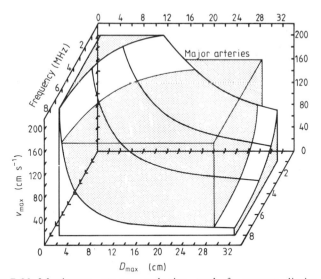

Figure 7.20 Maximum range, velocity and frequency limits for pulsed Doppler, compared with depths and flow velocities of major arteries. Points above the depicted surface produce range/velocity ambiguities. (Adapted from Baker *et al* (1978).) The solid box shows the range of depths and frequencies (assuming $\theta = 0$) for the major arteries in the body. Aliasing can often be avoided by control of θ.

The range-gating process of the pulsed Doppler method introduces another Fourier-domain leakage problem, in addition to that described earlier due to the finite width of the sound beam. As a result of the limited range over which the sample volume extends, good frequency resolution may not be maintained. If the pulse or gate length is reduced to improve range resolution, it is at the expense of frequency resolution, and vice versa. As with spectral broadening due to the width of the sound beam, this effect is not usually a serious limitation.

It is possible slowly to build up a 2D tomogram of the spatial distribution of flow velocity, in the plane of the skin if the scanning is like that for cw Doppler imaging system, but a constant gate delay is maintained. It is also possible to obtain tomographic images in any other plane if the gate delay is swept in time, during scanning. If one considers the demodulation and smoothing part of the signal processing of a swept-gate system (i.e. essentially a moving window from which the result at a given depth is obtained from a smoothed sum of products of the echo signal within the window and a reference waveform), one may see a clear resemblance of such a process to that of computing the cross-correlation function between the received echo waveform and the transmitted pulse. Real-time cross-correlation (and related) methods have, in very recent years, been applied to the component RF A-scans of real-time echographic images, to produce real-time images of the changing spatial distribution of blood-flow velocity. The frame rate, given by equation (7.25), imposes such severe limitations on range and velocity that aliasing is almost inevitable, even though such images are generally created over a limited field of view (using a small number of lines) and flexible pulse-sequencing systems are used, so that these image regions are refreshed more frequently than the rest of the image. The resulting images are more qualitative than even conventional ultrasound Doppler methods. It is likely, however, that depiction in real time of the changing spatial relationships of flow velocities, and turbulence if present, will tend to compensate for lack of quantitation. Such images are certainly very impressive to observe, especially when combined as a colour overlay with real-time echography.

7.3.4 Current trends in commercial instruments

Having described the basic methods of forming images and extracting Doppler-shift information from ultrasonic echo signals, it is now worth mentioning very briefly some of the forms in which commercial ultrasound instrumentation appear. In particular, commercial instrumentation tends to follow trends, as many manufacturers move simultaneously in a similar direction. Such semi-global market changes may be driven by a variety of events, including the availability of cheaper or new electronic components (permitting smaller/cheaper products or a more advanced specification), the transference of new ideas and principles from research, or the identification of new specific clinical requirements. In this regard, many of the trends are divergent, tending to split the market.

Some divergent trends, easily identified, are increased specialisation compared with increased versatility, or reduction of cost compared with enhanced performance. In addition, for the higher-priced machines it is becoming common to include multiple facilities and to have displays that use colour coding to superimpose information of different kinds. Each of these trends is discussed below.

7.3.4.1 Specialisation. So-called 'small-parts' imagers (somewhat mis-named) have been built to supply a need for very-high-resolution imaging of superficial structures, mainly to look for localised pathology in the neck (thyroid) and scrotum, and for atherosclerotic plaque in the carotid artery. The eye, the breast and the new-born infant are also sometimes classified in this category. These applications permit a high ultrasonic frequency to be used (up to 13 MHz) with a low-*f*-number (less than 2.5) transducer. Often the scanning is performed by mechanically moving a spherically focused transducer, permitting good resolution to be obtained transverse to, as well as within, the scan plane. Imaging is likely to be performed through a water or oil delay, integral to the scanning head, thus permitting good visualisation of all structures including the skin. Very recently, even higher frequencies have been employed (up to 40 MHz) to provide echo imaging over depths of only half a centimetre or so, for dermatological applications.

Another kind of specialised scanner is the automated water-bath scanner for breast imaging. Although not new in a clinical research environment, potential interest in the use of echography as a breast-cancer screening method prompted a number of manufacturers to produce dedicated whole-breast imaging systems that were able systematically to explore the breast by gathering a large number of uniformly spaced slice images. In spite of the provision of sophisticated image-review systems, the problems of reading up to a hundred images per breast were never entirely solved. Combined with the high cost of the systems, and insufficiently high diagnostic accuracy rate, these problems have led to the disappearance from the market of most (though not all) of these systems. The complementary nature of pulse–echo imaging and x-ray mammography (see Chapter 2) has, however, ensured a reasonably permanent place for ultrasound in the examination of specific breast problems, although the use of hand-held real-time scanners, operating at the low-frequency (i.e. 5–7.5 MHz) end of the range for 'small-parts' scanners, is now the most popular type of instrument for breast imaging.

Specialised intracavity (or endoscopic) scanners are also available. The use of body cavities in this way enables one to place the ultrasound transducer physically closer to the organ of interest (thus enabling better resolution to be obtained by using higher frequencies) and may permit, in some cases, the examination to be performed with less discomfort, or an organ to be imaged that is otherwise inaccessible because it is obscured (from surface imaging) by gas or bone. Examples of applications are the imaging of the prostate and rectal wall from within the rectum, the uterus and ovaries from the vagina, the bladder wall from the bladder (entered via the urethra) and the stomach wall from within the stomach (entered via the oesophagus). Doppler instruments have also been built to operate with catheter-tip transducers, although this is more of a research activity. Some combination instruments are beginning to appear, incorporating both ultrasound and optical endoscopic imaging devices. Various scanning systems are in use, including linear arrays mounted on one edge of a long cylinder. A scanning system unique to endoscopic application, however, is the 360° mechanically rotating transducer, producing a radar-like 'plan-position' display. Figure 7.21 illustrates one of these devices and provides an example of this form of image display.

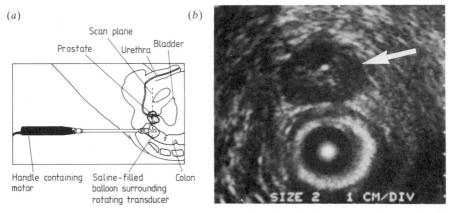

(a)

Scan plane
Prostate   Urethra   Bladder

Handle containing   Saline-filled   Colon
motor   balloon surrounding
                     rotating transducer

(b)

Figure 7.21 Diagram of assembly and mechanics of one type of transrectal ultrasound probe (a) and a typical image, showing cancer of the prostate (indicated by white arrow), obtained with this system (b). (Diagram adapted from sales brochure of Bruel and Kerr Ltd; grey scale image courtesy of D O Cosgrove.)

7.3.4.2 Scan versatility. Many general-purpose real-time scanners are being designed to accept a very wide range of linear or sector probes, with a wide range of frequencies, including, in some cases, those normally associated with the small-parts systems described above. Some systems will also permit one form of probe (usually a linear array) designed for endoscopic work to be used.

Furthermore, probes consisting of a line of rectangular transducer elements on a curved base are now popular since they possess many of the advantages of an electronic sector scanner without the disadvantages of the phase-steering method (see §7.3.1 and figure 7.12).

7.3.4.3 Cost reduction. The last few years have seen the introduction of relatively cheap real-time systems, which have limited facilities but are nevertheless effective (and are often highly portable). At the present rate of cost and size reduction, it will soon be possible for district hospitals in developing countries, or a small group of general practitioners to be able to afford one, although whether the latter would be desirable is a matter currently being debated.

7.3.4.4 Enhanced performance. In 1983 the first of a new generation of improved scanners appeared. The improvements were due largely to a dramatic increase in the number (128 as compared to the usual 13 or so) of transmitting and receiving channels (elements) used to form and focus the sound beam, such that, under some circumstances, the whole array may be used as a very wide acoustical aperture. The result is greatly improved lateral resolution and spatial uniformity of point-spread function within the scan plane. Many of the images that illustrate §7.4 were obtained with such a system. The penalty that one pays for such electronic sophistication is in the cost of the system. It seems, however, from the growth in this section of the market, that the benefits of improved image quality are well worth the increased financial cost. There is now available a range of these enhanced imaging systems, offering differing compromises in the cost versus performance trade-off, with the number of channels varying from 128 down to 32.

7.3.4.5 Multiple facilities. Another kind of increased sophistication has been to put together multiple facilities within one instrument. A particular example is the combination of echo imaging with pulsed Doppler in what are commonly called 'Duplex Scanners'. This is in the form of a real-time sectional echo image on which is positioned a visibly defined user-selectable line of site and range gate for the pulsed Doppler. This approach offers advantages in two ways: the Doppler system can enhance ability to identify correctly vascular anatomy and, more importantly, the imaging system can make the Doppler examination less ambiguous and the information more quantitative. The increased quantitation results from the fact that, when imaging and studying the blood flow in a well defined vessel (e.g. the carotid artery), the imaging system permits specification of the angle θ in equation (7.10) so that the Doppler-shift spectrum may be expressed in terms of actual flow velocities $(m\,s^{-1})$. The diameter of the vessel is also available, in principle, for volume flow-rate calculation, although in practice the errors in estimating the active cross-sectional area by this method may be too large for it to be useful. The range–velocity ambiguity (equation (7.26)) can also now be overcome, and the PRR can be kept constant as the depth of the Doppler gate is varied. This is done by displaying on-screen markers for all of the possible locations from which the Doppler signal might have arisen and using the image, plus knowledge of anatomy, to sort out the ambiguity.

7.3.4.6 Superposition of information. An extension of the provision of multiple facilities is to superpose the different kinds of information onto one display, using colour coding for interpretation. The 'colour-flow mapping' systems described in the previous section do exactly this by combining a two-dimensional real-time monochrome echo image with a two-dimensional colour-coded real-time Doppler image. It can be expected that in the future other kinds of information, perhaps related to the tissue characterisation procedures described in §7.5.2, will be treated in a similar manner.

7.3.5 *Calibration and performance checks for clinical equipment*

This is an extensive subject in its own right and only a brief description can be given here. More extensive information is given in the works cited in the references and further reading section at the end of the chapter.

Performance comparison for equipment purchase, and acceptance testing, is difficult to achieve solely by objective physical methods. It is generally necessary for the equipment to be subjectively assessed during a trial period of genuine clinical use. Many physical methods for checking equipment performance are used as a means of early detection of changes that indicate the development of subtle faults. Only pulse–echo imaging systems are discussed in this section, since approaches to calibration and performance assessment of Doppler systems is still largely at an experimental stage. In general, there are three classes of methods, as follows.

7.3.5.1 Test targets and phantoms. It is possible to measure some instrument parameters by using the instrument to obtain an image of a specially constructed object. Such objects, as in other areas of medical imaging, are known as test phantoms. No single phantom is able to provide all the information of interest, but at some time or other all of the characteristics listed in table 7.4 have been described as being assessed using a phantom (although many are as relative measures). Phantoms also find limited application in areas that have little to do with performance assessment. Examples are in the training of personnel in the use of ultrasound equipment for imaging a particular part of the body (e.g. breast phantoms), and for research into a better understanding of the physics of ultrasound propagation and interaction in tissues. Related to the latter is the use of phantoms for testing and optimising tissue-characterisation procedures and computer algorithms.

The following are a few representative examples of commercially available and/or 'standard' designs for test phantoms. The AIUM (American Institute of Ultrasound in Medicine) standard test object (referred to by many of the references), which is essentially a high-contrast resolution and geometrical accuracy test, consists of a grid of wires in a water–alcohol mixture (used to obtain the correct speed of sound). Nuclear Associates Ltd produce a 'multipurpose' tissue-equivalent phantom, which provides similar facilities but has a more realistic background scattering material that attempts to simulate the scattering from parenchymal regions of the liver. This phantom also incorporates negative-contrast inclusions in the form of non-scattering cylinders, which appear like cysts on an echo image. The 'Cardiff test system' marketed by Diagnostic Sonar Ltd (several phantoms consisting of wires and various objects in a background scattering gel) provides facilities for a wide range of measurements dealing with a large number of the characteristics listed in table 7.4, and incorporates some (limited) ability for three-dimensional assessment of resolution. Nuclear Associates Ltd also produce an ultrasound contrast/detail phantom (Smith *et al* 1985), an interesting device consisting of a series of cones, each of which is made from a scattering gel but of different scattering strength from cone to cone, embedded in a uniform background scattering gel. A comparative assessment of low-contrast detectability is obtained by scanning each of the cones in cross section and determining the cross-sectional diameter at which the presence of the cone is just detectable. Repetition of the process for cones of varying scattering contrast, relative to the background gel, enables the plotting of a contrast versus diameter curve for the threshold of detectability (called the 'contrast–detail' curve—see Chapter 13). The better the imaging system, the closer the curve lies to the origin on either axis.

Table 7.4 Performance characteristics of ultrasound pulse–echo imaging equipment that have been assessed using imaging phantoms. (After McCarty (1986).)

Scan uniformity	Dynamic range
Caliper accuracy	Shape of compression curve
Image aspect ratio	Signal-processing options
Linearity/distortion	System noise
Ringdown (excitation pulse length)	Paralysis (dead time)
Axial resolution	Penetration depth (low and high contrast)
Lateral resolution within scan plane	Registration error
Lateral resolution orthogonal to scan plane (slice width)	System artefacts
	Low-contrast detectability
	Movement-related parameters

7.3.5.2 Acoustic signal injection. Possibly unique to ultrasound imaging is the availability of signal-injection systems as a means of equipment assessment. In this approach the transducer of the system under test is coupled directly to a separate signal-injection transducer. One might wish to think of this as an acoustically active (as opposed to passive) phantom. The transmitted sound pulse from the scanner under test is detected by the test system and used to initiate a sequence of pulses, which are fed back, through both transducers, to be imaged (as what appear like echoes) on the scanner display. The test pattern is a persistent image because it repeats after each cycle of the PRR, and it is displayed over an apparent depth that is controlled by timing circuits in the test system. Signal-injection test systems are commercially available, and have been used to check the factors listed in table 7.5.

Table 7.5 Performance characteristics of ultrasound pulse–echo imaging systems that have been assessed using acoustic signal-injection methods. (After Duggan and Sik (1983).)

Dynamic range	Grey-scale transfer characteristic
TGC curves	and linearity of pre- or post-
Velocity calibration and depth	processing options
marker/caliper accuracy	System gain
Acoustic output	Frequency response
'Tissue-characterisation'	
algorithms	

7.3.5.3 Acoustic output measurement and relative beam plotting. Although they are of use as a method for performance monitoring, absolute acoustic output measurements are important more as a means of ensuring compliance with safety standards. The availability of figures for the acoustic output levels is becoming a prerequisite for sale of diagnostic ultrasound equipment in some countries. As may be appreciated by reviewing §§7.2.1, 7.2.4 and 7.2.6, complete and accurate specification of acoustic output is far from being a simple matter. In continuous-wave systems, one may measure the spatial distribution of either intensity or pressure. The intensity may be integrated over the beam area to provide the total power output of the system. For pulsed systems, one must consider both the instantaneous and the time-averaged versions of these quantities. Finally, the non-linear characteristics of the test medium (water) will mean that the pulse shape and beam shape are amplitude-dependent, and that measurement of different acoustic field parameters (e.g. intensity, peak positive pressure, peak negative pressure) may result in different observed spatial distributions.

Complete agreement (a necessarily imperfect result of many compromises) is only just beginning to be reached by the various international standards organisations, regarding exactly which acoustic field parameters should be used to characterise diagnostic equipment. There are generally four parameters, all based on intensity: the spatial-peak temporal-peak (SPTP), spatial-peak temporal-average (SPTA), spatial-peak pulse-average (SPPA) and spatial-average temporal-average (SATA) intensities. The last of these is obtained from the total power output divided by the transducer area or, if the intensity distribution is measured directly in the focal plane, the total power may be obtained by multiplying the SATA intensity by the beam area (specified at the 20 dB level) in the plane of measurement. It has also been suggested that the SPTP intensity be measured twice, being derived from (see equation (7.2)) the peak-positive and peak-negative acoustic pressures, because the negative pressure may be more relevant to the likelihood of initiating cavitation (see §7.4.4) but the positive pressure is usually the larger of the two. The pressure-distribution measurements themselves, and therefore the derived intensity distributions, are made by scanning over the transmitted acoustic field with a small receiving transducer, called a hydrophone, or by using an array of hydrophones. Various designs of hydrophone exist, but their important properties are small size for adequate spatial resolution, non-interference with the acoustic field being measured, adequate time resolution (i.e. wide bandwidth) and a completely flat response over their useful bandwidth, so that a faithful reproduction of the transmitted acoustic signal (or pulse) is obtained. The bandwidth must, in fact, be much greater than that of the likely range of transmitter frequencies to be measured because the hydrophone must be able to reproduce the high-frequency harmonic components that are introduced as a result of non-linear propagation. These devices need to be calibrated against an absolute standard in order for them to be used to measure acoustic pressure, so a further important property is that their sensitivity is stable over a substantial period of time. If only the total power output of the diagnostic ultrasound system is required, then a device known as a radiation force balance may be used. This is essentially a very sensitive balance that measures the force exerted by the sound beam on a vane (target) specially designed as an acoustic absorber or (more usually) reflector. The theory underlying the phenomenon is complex (although it results in a relatively simple practical equation), but the method is also one of those used as a primary standard for the calibration of hydrophones.

Common methods of obtaining a picture of the relative acoustic field distribution (often called a 'beam plot') are the use of a scanned hydrophone or hydrophone array, a scanned small target with the system operating in pulse–echo mode, or a very sensitive optical method (such as a schlieren system), which enables the visualisation of the relative change in refractive index, induced as the density fluctuates with the passage of the acoustic wave. Beam plots are generally applicable to evaluating the performance of the transducer only, are time-consuming and use facilities that are expensive to set up, relative to the phantoms and signal-injection systems described above.

7.4 CLINICAL APPLICATIONS AND BIOLOGICAL ASPECTS OF DIAGNOSTIC ULTRASOUND

7.4.1 *Ultrasound in clinical diagnosis*

Ultrasound imaging and Doppler methods have found vast application throughout the breadth of medical diagnostic investigations. It is impossible, in this brief survey, to do justice to the extent to which they are used, or to their value. Most of the major uses have been mentioned, but inevitably, there are omissions.

7.4.1.1 Obstetrics. Ultrasound plays a major role in management of both normal and complicated pregnancies. It is used to monitor development of the foetus and to diagnose maternal and foetal disorders.

Early pregnancy The main applications are:

(*a*) Confirmation of pregnancy: this can be demonstrated even a few days after the missed menstruation.

(*b*) Assessment of gestational age, based on the size and structure of the gestational sac.

(*c*) Multiple pregnancy: can be assessed with confidence by the end of the first trimester.

(*d*) Differential diagnosis of failed or abnormal pregnancy, in particular ectopic pregnancy.

Second and third trimesters   Ultrasound is used in the following examinations:

(*a*) Evaluation of foetal maturity and growth in such standardised measurements as biparietal diameter (BPD), crown–rump length (CRL), head and abdomen circumference, length of limbs, etc.

(*b*) Amniocentesis, i.e. aspiration of a sample of amniotic fluid for cytological and biochemical analysis, performed under ultrasonic control (disorders such as Down's syndrome can be diagnosed).

(*c*) Imaging of foetal abnormalities such as microcephaly, hydrocephaly, spina bifida, renal agenesis, urinary and bowel obstruction, skeletal and cardiac anomalies and malignant tumours.

(*d*) Assessment of foetal movement: foetal activity such as pseudo-breathing, trunk movement and response to stimuli (e.g. vibration) is characteristic of a healthy foetus and is decreased in a hypoxic foetus.

In figure 7.22, two examples of B-scans of the normal foetus at different ages are shown.

Depending on gestational age and severity of foetal disorders, termination of pregnancy might be considered. In some cases, surgical procedures are performed on foetus *in utero*, under control of ultrasound.

In most of the above applications, examinations are performed transabdominally using real-time B-scanners, equipped with sector or linear array probes operating at frequencies of 3.5–5 MHz. Recently, it has been found that, for early or suspected pregnancy, transvaginal scanning has some advantages, including the ability to use higher frequencies (5–7 MHz) for better resolution and the removal of the need for an (uncomfortable) full bladder, which acts as an acoustic window for transabdominal imaging. M-mode scans are sometimes used in the assessment of the foetal heart, while Doppler methods find use in measurements of placental and foetal blood flow. Continuous-wave Doppler is used for foetal-heart monitoring during labour.

7.4.1.2 Gynaecology. In comparison with obstetrics, the use of ultrasound in gynaecology is more limited. It is used mainly as an adjunct diagnostic procedure, and the following are the most common applications:

Uterus and cervix
(*a*) Size and shape assessment.
(*b*) Intra-uterine devices: ultrasonic examination is useful in demon-

strating the presence and position of the contraceptive coil.

(c) Diagnosis and follow-up of malignant tumours of the uterus.

Ovaries.

(a) Management of infertility; follicle puncture for ovum collection.

(b) Detection of ovarian cysts.

(c) Ovarian tumours: radiotherapy planning and monitoring of response to treatment.

In gynaecological applications, real-time B-scanning is the method most commonly used. The examination techniques are either transabdominal scanning, using the full bladder as an acoustic window, or transvaginal scanning.

(a)

(b)

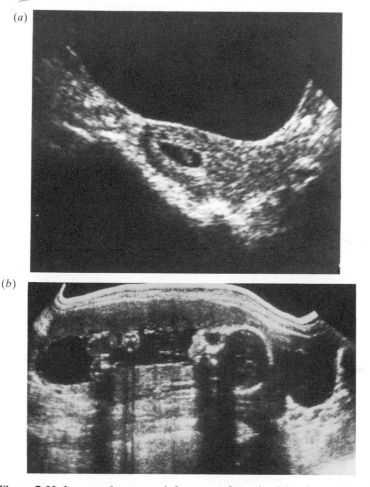

Figure 7.22 Image of a normal foetus at 5 weeks (a) taken with a modern electronic real-time scanner and another at 22 weeks (b) taken with a contact manual B-scan system. (Courtesy S G Shoenberger and D O Cosgrove).

7.4.1.3 Abdomen. In the upper abdomen, the scanning method most useful diagnostically is again the real-time B-scan, which has superseded the older static scanning. The probes used are sector or linear arrays, operating at frequencies of 3.5–5 MHz. In the pelvis, and for parts of the gastrointestinal tract, intracavity probes (transvaginal, transrectal, transurethral and oesophageal) are also used.

Gastrointestinal tract Stomach and bowel are normally filled with varying amounts of gas, fluid and solid faecal material, which render transabdominal ultrasonic scanning difficult. Ultrasonography is used selectively in assessment and managing of intestinal obstruction, inflammatory changes and tumours.

Liver and biliary tree
(*a*) Assessment of liver size and shape: hepatomegaly often accompanies liver disease, while change of shape (e.g. an increased roundness in ascites) might also be indicative of pathology.
(*b*) Focal disorders: diagnosis of primary and secondary malignant tumours, abscesses and haematomas.
(*c*) Diffuse diseases: diagnosis of fatty change, cirrhosis and rarely malignant infiltration.
(*d*) Detection of gall stones and inflammation of the gall bladder.
(*e*) Diagnosis of obstructive jaundice.

Pancreas
(*a*) Diagnosis of pancreatic carcinoma (guidance of needle biopsy).
(*b*) Management of pancreatitis (acute and chronic) and pancreatic trauma.

Spleen and lymphatics
(*a*) Assessment of spleen size: splenomegaly often accompanies inflammatory disease and it is also common in many malignancies, as a manifestation of the immune response to a tumour.
(*b*) Management of spleen trauma.
(*c*) Detection and follow-up of lymphadenopathy.

Kidneys
(*a*) Detection of congenital anomalies (agenesis, hypoplasia, etc).
(*b*) Diagnosis of renal cysts (e.g. polycystic disease) and tumours.
(*c*) Management of renal trauma.
(*d*) Assessment of complications (rejection) of renal transplants.
(*e*) Detection of kidney stones and of hydronephrosis.

Bladder
(*a*) Assessment of volume (residual urine volume), size and shape: these factors are indicative of pathology in several disorders, e.g. neurological disturbances.
(*b*) Bladder tumour diagnosis.

The full bladder is readily examined by contact scanning through the abdominal wall, using a real-time sector B-scanner. Ultrasound endoscopic visualisation of tumours via the urethra, however, is superior to transabdominal.

Prostate
(*a*) Diagnosis of prostatic benign and malignant tumours.
(*b*) Diagnosis of prostatitis.
(*c*) Cancer management: insertion of radioactive implants under ultrasound guidance.

Abdominal visualisation of the prostate gland is difficult in obese patients; generally, transrectal B-scanning is a more reliable method of examination.

7.4.1.4 Cardiovascular system

Heart In contrast with the majority of abdominal applications, a very common scanning technique in echocardiography is the M-scan. In fact, the two techniques, the B-mode and M-mode, are complementary: although the M-mode is ideally suited to investigation of valve and wall motion, information can be obtained only from along lines of sight defined by the ultrasonic beam. A better method for examining the spatial relationships within the heart is a two-dimensional scan, which, on the other hand, provides less quantitative information about motion. Modern echocardiography permits examination of anatomy in motion by utilising both techniques simultaneously. The need for this scan flexibility, and to scan through small acoustic windows (either intercostally or subcostally), has resulted in phased-array sector scanning (2.5–5.0 MHz) being the most commonly employed technology.

An example of a cardiac echogram, together with a corresponding M-mode, is shown in figure 7.23.

Additionally, haemodynamic effects of structural and functional abnormalities are assessed in Doppler measurements, of which colour-coded blood-flow imaging is one of the more recent developments. The main applications are:

(*a*) Diagnosis of valve disorders: mitral stenosis and regurgitation, rheumatic aortic valves, etc, management of malfunctioning prosthetic valves.
(*b*) Assessment of left-ventricular function (wall thickness and movement, diastolic and systolic volume, etc).
(*c*) Diagnosis of pericardial effusions.
(*d*) Evaluation of congenital heart disease (largely a problem of infancy and childhood).
(*e*) Diagnosis of cardiac tumours: these are rare, but often benign, and early diagnosis can lead to effective surgery.

Arteries and veins In clinical investigation of the vascular system, Doppler techniques, often in conjunction with ultrasonic imaging methods, are used. The main fields of application are:

(*a*) Cerebrovascular disease: real-time imaging, combined with Doppler spectral analysis, provides a non-invasive and accurate method of screening for carotid artery disease; detection of signs premonitory to impending stroke might justify surgical intervention.

(*b*) Investigation of lower-limb arterial disease (detection and measurement of stenotic lesions).

(*c*) Imaging of large abdominal vessels (e.g. to assist identification of vascular anatomy or to identify aortic aneurysms).

(*a*)

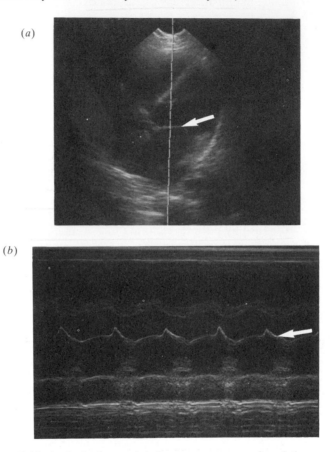

(*b*)

Figure 7.23 A single frame (*a*) from a sequence of real-time sector scans of a normal heart, indicating the line of site used for an M-mode recording of mitral valve motion (*b*). Arrows indicate mitral valve. (Courtesy M Blaszczyk.)

In figure 7.24, an example is shown of a B-scan of a vessel, together with corresponding Doppler signal.

7.4.1.5 Superficial structures. In ultrasonography of superficial structures (and peripheral vessels), i.e. breast, thyroid and testes, high-resolution real-time probes are used, of frequencies 5–13 MHz. The method of examination is either contact scanning or water-delay technique, where a polythene bag filled with water is placed between the skin and the probe, or the organ (breast, scrotum) hangs in a water bath in which the probe is also placed.

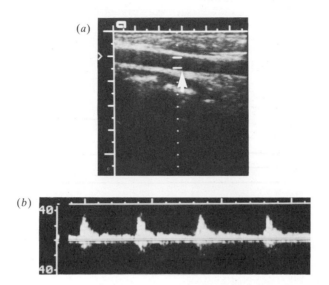

Figure 7.24 Linear B-scan image of the common carotid artery (*a*), with the maximum velocity–time waveform (*b*) of a pulsed Doppler signal obtained from the sample volume indicated in (*a*).

Breast In the female breast, ultrasound plays an important role (complementary to palpation and mammography) in diagnosis of cysts, fibrocystic disease, and malignant and benign tumours. In particular, in differential diagnosis of breast carcinoma in the symptomatic breast, it helps to establish whether a palpable mass is solid and, if so, exhibits features which suggest malignancy (illustrated in figure 7.25). B-scans are used for imaging of breast structure (and, using real time, also assessment of mobility and elasticity) and Doppler examinations for blood-flow measurements. Ultrasound is of particular value in the young or dense breast, during pregnancy, and in evaluating the radiographically evident but non-palpable mass. Ultrasound guidance of fine-needle aspiration, for draining cysts and for cytological diagnosis, is also of great value, particularly for non-palpable lesions.

Thyroid and parathyroid. In thyroid echography, ultrasound provides structural information, which complements radionuclide functional studies (see §6.9.10). The chief contribution of ultrasonic examinations is differentiation of cystic from solid nodules. The following are some of the applications:

(*a*) Demonstration of cysts and tumours (adenoma, carcinoma), especially those lesions which appear cold on radioisotope scans.

(*b*) Diffuse diseases: diagnosis of multinodular goitre.
(*c*) Parathyroid tumours.

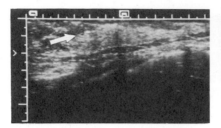

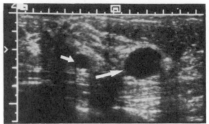

(*a*)

(*b*)

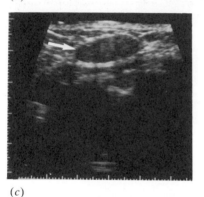

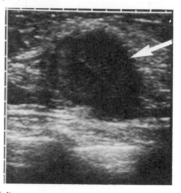

(*c*)

(*d*)

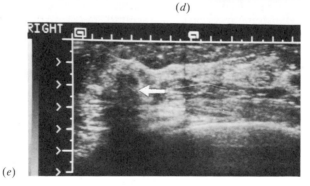

(*e*)

Figure 7.25 Single frames from real-time studies of the breast, showing (*a*) normal breast gland, (arrow indicates boundary between glandular tissue and subcutaneous fat), (*b*) two cysts (note the sharp borders, lack of internal echoes and posterior enhancement), (*c*) a fibroadenoma (well defined regular border, uniform echoes), (*d*) a large carcinoma (irregular fuzzy border, heterogeneous internal echoes and slight enhancement), and (*e*) a smaller carcinoma (irregular fuzzy border and shadowing). (Courtesy of M Blaszczyk.)

Testes Ultrasonography is useful in portraying the scrotal anatomy and abnormalities, such as cysts, tumours, abscesses, haemorrhage, etc. It allows differentiation of cystic and solid masses.

7.4.1.6 Eye and orbit. Ophthalmic ultrasonography has become an essential diagnostic tool. Because of the small size of ocular and orbital structures, high resolution without deep penetration is required. Equipment is designed specifically for ophthalmic use and commonly used frequencies are 8–13 MHz, although frequencies up to 20 MHz are used for some applications (biometry). The eye is scanned either in direct contact or using a water bath. A-scans are used to provide precise information about the localisation of reflecting interfaces and to make measurements of the axial length of the eye. Relatively cheap, semi-automatic A-scan instruments are available for this purpose. Real-time B-scans are used to provide topographical and motional information, and represent the primary diagnostic method.

Some of the applications are:

(*a*) Detections of foreign bodies, trauma and haemorrhage.
(*b*) Diagnosis of intra-ocular and orbital tumours.
(*c*) Diagnosis of retinal detachment (illustrated in figure 7.26).
(*d*) Biometry: the required power of intra-ocular or contact lenses can be calculated.

7.4.1.7 Brain. Echographic visualisation of adult-brain anatomy is made difficult by the attenuating and refracting properties of the skull. In infants, however, an acoustic window (the anterior fontanel) allows scanning during the early months of life and has become an important method of diagnosing such disorders as cerebral haemorrhage in premature babies, hydrocephalus and congenital malformations. Specially designed small transducers are used, in the frequency range 3.5–7 MHz.

7.4.1.8 Lungs. Ultrasound does not penetrate normal lung, which is filled with air. When a fluid or solid mass is present, ultrasonic examination might help in differential diagnosis of opacities seen on a chest radiograph, e.g. confirmation of pleural effusion.

7.4.1.9 Ultrasound in invasive procedures. Although one of the main advantages of ultrasonography is the non-invasive character of examinations, ultrasound also finds diagnostic applications in some invasive procedures, such as needle puncture and intra-operative scanning.

Needle puncture techniques The ability of ultrasonic imaging to visualise the depth and position of internal tissue structures makes it a useful tool in guiding the insertion of biopsy needles. Tips of needles penetrating the tissue will be seen as bright reflections and can be guided precisely to the position desired for obtaining material for cytological and histopathological evaluation and for drainage of abscesses. The procedure, carried out under local anaesthesia, involves minimal hazard and discomfort to the patient.

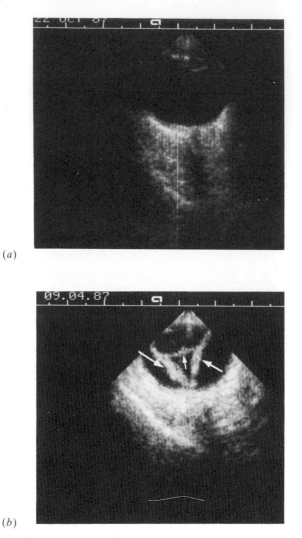

(a)

(b)

Figure 7.26 Sector B-scan of the normal eye (*a*) and another depicting multiple retinal detachments (*b*). (Courtesy of M Restori.)

Puncture techniques are used in diagnosis of lesions of many organs, including breast, liver, pancreas, kidney, retroperitoneum and prostate. Collection of fluid—as in amniocentesis—and aspiration of cysts (e.g. in liver, breast, etc) is easily performed under control of ultrasound. For non-palpable breast malignancies, echography may be useful in guiding the placement of wires with hooking tips, which aid the location of a mass during surgery.

Specially designed transducer heads, equipped with needle holders, are sometimes used in these techniques.

Intra-operative applications Internal lesions can be visualised by ultrasonic imaging pre-operatively, but the intra-operative use of ultrasound in direct contact with an organ gives better resolution and provides the surgeon with additional information, aiding the process of decision-making. Some of the applications are:

(*a*) Neurosurgery: localisation of tumours, controlled biopsy and abscess aspiration.

(*b*) Abdominal surgery: pre-operative examination of liver, pancreas and kidney (tumours), and search for stones in kidneys and the common bile duct.

Intra-operative transducers are of special design, satisfying the requirements of smaller size and easy maintainance of sterility.

7.4.2 *Biological effects and safety*

The fact that ultrasound techniques have gained such an important place in diagnostic medicine is due, in large part, to their presumed lack of hazard to the patient. Indeed, ultrasound has been in extensive clinical use in obstetrics, the application where there is greatest potential cause for concern over some biological effect, for about 25 years. Over that time, numerous investigations have been undertaken in an endeavour to detect adverse effects. None of these studies have shown that ultrasound at diagnostic intensities as used today has led to any deleterious effect to the foetus or the mother. It is extremely unlikely that, over this time, any deterministic (systematic) effects to the foetus would not by now have been detected and reported. Nevertheless, there is continuing awareness that lack of evidence that a hazard exists is not proof of its non-existence. Ultrasound should be used prudently, with minimum exposure time and acoustic intensity, and only when medical benefit is expected.

Indeed, at levels of acoustic power higher than those currently used diagnostically, biological effects are well known, understood to some extent, and made use of in medical therapy. Ultrasonic diagnostic equipment manufacturers are constantly striving, in a competitive market, to improve the quality of the information provided by their systems. This may sometimes be achieved by increasing acoustic power output of the device and is definitely associated with increased spatial peak intensities due to improved focusing methods. Thus, before comprehensive standards can be defined for safe equipment output levels and measurement methods, it is necessary to gain an understanding of the biological effects that occur, the mechanisms by which they are made to occur, the degree to which they may constitute a hazard, and the appropriate dose–effect relationships. This subject is still a matter of much debate and research. The matter is further considered in the context of potential hazards from other imaging modalities in Chapter 15.

From the point of view of the physics and biophysics of the interaction of ultrasound with biological media, many potential mechanisms of ultrasound effects on tissues may be identified. This complicates the issue of dose–effect relationships, since different mechanisms may be presumed to predominate under different exposure conditions.

Thermal mechanisms arise because the molecular relaxational, viscous and relative motional loss mechanisms of acoustic absorption result in a local temperature rise. Intensity and time are the exposure parameters relevant for thermal mechanisms. It is, however, generally believed that diagnostic levels of the SPTA intensity are much too low to be able to generate a temperature rise in tissue that could constitute a hazard.

Cavitation, which is the process of acoustically induced growth and vibration of gas bubbles, is a phenomenon commonly associated with high-power ultrasound and continuous-wave systems, although there is mounting evidence that, in some limited form (which need not be hazardous), it may be observed even for the small number of cycles of the wave present in a diagnostic pulse. Peak negative pressure and pulse length are believed to be the exposure parameters of interest, since it is in the rarefaction phase of the wave that gas is transferred from solution into the bubble, causing it to grow. Certainly, the SPTP pressures in diagnostic pulses would be easily large enough for cavitation to occur, if they were maintained over a large enough number of cycles. In this regard, the increased use of pulsed Doppler systems, with their slightly longer acoustic pulses (and higher PRR) than those normally used for echography, is regarded as something to monitor carefully. Most observations of cavitation have required a gassy, liquid environment. It is believed that cavitation *in vivo* is less likely to occur, but it is not impossible. Damage from cavitation at very high intensity levels in continuous-wave fields is due to mechanical effects on structures such as the cell membrane (e.g. rupturing) from shock waves and streaming forces, and chemical reactions initiated by the release of hydrogen and hydroxyl ions. These effects are not present at low intensities and there is no evidence to suggest that there is danger from the mild forms of cavitation which might, or might not, exist under diagnostic conditions *in vivo*.

Other, non-thermal potential mechanisms include the steady radiation force exerted on all tissue structures (some of which may be set in motion), streaming agitation of acoustically absorbing liquids, shear stress of objects present in a streaming liquid, and the direct oscillatory force of the sound field on all structures.

There is a serious lack of long-term, prospective epidemiological studies—of the kind that would be required to detect some subtle stochastic effect to the foetus. Indeed, the chance of being able to carry out a thorough study of this kind becomes slimmer as time passes, because the increasing use of ultrasound in obstetrics will make it very difficult to set up a control group of pregnant women who will not be scanned. The search for experimental evidence of hazard, and the acquisition data that will enable more precise specification of allowable levels of ultrasound exposure, is taking place, therefore, mainly in the laboratory, by observing diverse effects, *in vivo* and *in vitro*, of potential clinical significance. From this work, most of which is carried out at exposure levels somewhat higher than those used diagnostically, a great many new biological effects of ultrasound have been identified, some of them actually beneficial. Of greatest potential importance, however, must be those experiments which aim to discover whether ultrasound irradiation can have genetic effects. Unfortunately, some of the *in vitro* experiments involving DNA are prone to artefacts if not very carefully executed, and the conditions of exposure may permit mechanisms of interaction that need not be the same *in vivo*. To date, only two studies (from a very large number) have shown that ultrasound may produce chromosomal damage, and these could not be substantiated when the experiments were repeated by other workers. The vast majority of experiments show no effect. This is backed up by a large series of *in vivo* experiments of ultrasonic exposure of mammalian embryos *in utero*—demonstrating no adverse effects (in terms of teratogenic or developmental changes) at intensities used diagnostically. However, the search continues.

Based on a systematic review of published reliable data, the American Institute of Ultrasound in Medicine (AIUM) has made (in 1977) a statement to the effect that: (*a*) no independently confirmed significant biological effects have, as yet, been demonstrated in mammalian tissues exposed to SPTA intensities below $100 \, \text{mW cm}^{-2}$, and (*b*) even at higher intensities no substantial effects have yet been demonstrated where the product of SPTA intensity and time is less than $50 \, \text{J cm}^{-2}$, where, for pulsed operation, time is the total time including that during which the system is receiving rather than transmitting. These figures are currently regarded as a guide to a safe upper limit of acoustic output, although in practice acoustic exposure is reduced by attenuation in intervening tissue and by the fact that the process of scanning reduces the amount of time that any particular tissue volume is in the sound field. In any event, the AIUM statement is inadequate, and awaits extension and/or possible modification based on more relevant (and sensitive) experiments, and improved knowledge regarding the effect of factors such as pulsing conditions and frequency.

Actual measurements on diagnostic equipment demonstrate that a very wide range of outputs exists from one machine to another, the values sometimes exceeding the recommended limit. Continuous-wave Doppler instruments tend to have greater SPTA intensities (20–$800 \, \text{mW cm}^{-2}$) than do pulse–echo scanners (0.07–$680 \, \text{mW cm}^{-2}$), whilst the latter may possess extremely high SPPA (0.4–$1100 \, \text{W cm}^{-2}$) and SPTP (0.7–$2800 \, \text{W cm}^{-2}$) intensities.

7.5 RESEARCH TOPICS

To conclude this introduction to the physical, engineering and clinical aspects of ultrasonic imaging, we now look briefly at three selected research topics. A topic is defined as being in the category of research in the present context if it has not been put into general clinical use, via commercial supply to a broad range of hospitals outside the centres in which the original work was done.

7.5.1 Synthetic-aperture and other tomographic methods

The process of forming an image by focusing and pulse–echo scanning described in §§7.2.4, 7.3.1 and 7.3.2 appears, at first glance, to be fundamentally different from the manner in which x-ray or NMR tomograms, for example, are reconstructed. It is, however, a special (analogue) case of image reconstruction that is fast and convenient, but does not make maximum use of all the potentially available information. More general approaches exist, which, by spatially sampling the amplitude *and phase* of many wavefronts that have travelled through the same object via many different paths, can treat the ultrasonic signals as projection data for digital image reconstruction. If a sufficient number of such measurements (i.e. in three dimensions) could be made, and if a sufficiently good model were available for the physics of the ultrasound-scattering interactions, it would, in principle, be possible to reconstruct images of the distribution of density and compressibility. Much theoretical work has been published on this subject, but it is not a method that one is likely to see used for practical purposes in the very near future. Here we consider a range of methods that lie between these two extremes, and provide a multitude of alternative techniques for imaging a variety of the derived acoustic characteristics of tissues mentioned in §7.2.3. The principal disadvantage of all of these methods is the time needed to collect the projection information and/or to reconstruct the image. Fortunately, the seemingly inevitable advancement of technology is continuously eroding these problems, and it is likely that many of these methods will one day find their way into commercial medical instrumentation.

Figure 7.27 attempts to illustrate some of the great variety of approaches to digital ultrasound image reconstruction. The simplest is probably the so-called time-of-flight (TOF) method for reconstructing the distribution of speed of sound in the object. Consider the case of speed of sound and attenuation coefficient distributions, $c(x, y)$ and $\alpha(x, y)$, in the imaging plane within an object surrounded by water of speed of sound c_0 and negligible attenuation coefficient (figure 7.27(a)). A sound pulse is transmitted in the y' direction and travels a distance D to be received after a measured time t. If, in the region of the water only, $t = t_0$, then the shift parameter τ is defined as $\tau = t_0 - t$. From the figure

$$t_0 = \int_0^D (1/c_0) \, \mathrm{d}y' = D/c_0. \tag{7.27}$$

Scanning both the transmitter and the receiver in the x' direction (at angle θ to the x axis) produces a 'shift projection' at orientation θ:

$$\tau(x', \theta) = (D/c_0) - \int_0^D [1/c(x, y)] \, \mathrm{d}y'. \tag{7.28}$$

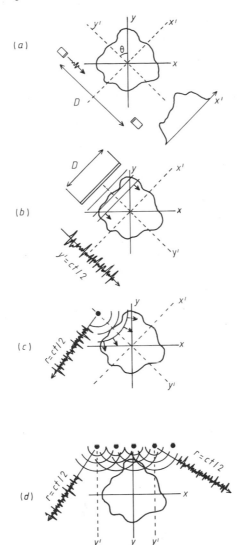

Figure 7.27 Schematic illustration of coordinate systems, transducer arrangements and lines of projection data, for ultrasound CT reconstruction: (*a*) speed of sound or attenuation-coefficient reconstruction from transmission projections; (*b*) coherent plane-wave backscatter reconstruction; (*c*) coherent cylindrical-wave backscatter reconstruction; (*d*) limited(linear)-aperture cylindrical-wave coherent backscatter reconstruction (also called synthetic-aperture imaging).

The problem of reconstruction is then to estimate $c(x, y)$ from a number of shift projections at various angles θ. The various methods available for doing this are as described in Chapter 4.

By measuring signal magnitude, instead of TOF, the same approach can be used to reconstruct images of the attenuation coefficient $\alpha(x, y)$. The projections would be described by

$$A(x', \theta) = A_0 \exp\left(-\int_0^D \alpha(x, y) \, dy'\right) \qquad (7.29)$$

where A_0 is a constant (the magnitude of the transmitted wave). The need to be able to pass sound completely through the object is a major problem for transmission methods. Applications are limited to accessible organs containing no bone, such as the female breast. Indeed, the distribution of fat in the breast is imaged extremely well by speed of sound images, which tend to complement the attenuation and conventional pulse–echo images, as far as the nature of the information displayed is concerned. Other difficulties with transmission reconstruction methods are: (*a*) the resolution tends to be poor, being limited by the width of the transmitted sound beam and the receiver sensitivity pattern; (*b*) sound refraction causes image degradation by invalidating backprojection along straight lines; and (*c*) phase cancellation at the receiver tends to produce measurement errors, adding object-dependent noise to the attenuation images (mainly at edges) and making them qualitative rather than quantitative.

Referring to figure 7.27(*b*), the backscattered signal from a large, plane-wave source/receiver can be viewed (neglecting attenuation, amongst others things) as a projection of line integrals across the width of the wavefront, i.e. the projection itself now occurs as a function of y', scaled in terms of time and speed of sound:

$$A(y', \theta) = A_0 \int_0^D \alpha_{bs}(x, y) \, dx' \qquad (7.30)$$

where $\alpha_{bs}(x, y)$ is the image to be reconstructed, which can be referred to as the amplitude backscattering coefficient of the tissue. Note that A is now the signal amplitude and may be positive or negative.

Linear backprojection of such coherent projection data is essentially a digital method of creating a focusing system having a completely circular aperture. To avoid corruption of the projection data by phase cancellation, it is necessary for the object to lie in the far field of the source, meaning that large distances, small objects or low frequencies (or all) must be used. This approach is seldom used. More commonly, a two-dimensional point source (i.e. a line source) is used to generate cylindrical waves, as in figure 7.27(*c*). Thus, far-field conditions are automatically satisfied and the source is easier to manufacture. Reconstruction, however, is no longer possible by backprojection along straight lines. Curved lines, corresponding to the shape of the wavefront, must be used (as shown in the figure). The advantage of this system arises with limited-angle reconstruction, when new projection data are obtained by linearly shifting the reference frame, instead of by the usual method of rotating it. As illustrated in figure 7.27(*d*), with ultrasound this can be achieved simply by making use of linear-array technology, using each element of the array as a source for gathering a backscatter projection. Digital reconstruction by backprojection along the paths of the wavefronts will then form an image in which the whole array has been used to focus for each and every point in the image. This method of focusing, and extensions of it, has been used in many fields of remote imaging where the phase of the wave may be measured as well as its magnitude. Examples of applications occur in underwater acoustics, radar satellite imaging, and geophysics (seismology). It passes by various names, including holography and *synthetic-aperture* imaging.

7.5.2 *Ultrasound quantitation and tissue characterisation*

Ultrasonic tissue characterisation is simply the process of describing tissue structure or function in terms of ultrasonically derived information. This description, if strongly related to specific normal and/or diseased tissue classes, may then be used for purposes of improving diagnosis or for other aspects of patient management, such as monitoring response to treatment. In fact, subjective feature analysis, as, for example, performed (either subconsciously or consciously) by a radiologist 'reading' ultrasound images, is very much a tissue-characterisation process, albeit a non-quantitative one.

Although non-quantitative, this example does highlight two basic aspects of the tissue-characterisation process. First, for the technique to be of some practical value, the tissue/disease classes of interest must be sufficiently well defined and separated, such that there is no great variation in measured characteristics by any one class and there is not a great overlap in the characteristics of any two or more such classes that one wishes to differentiate. Secondly, the process of identifying the presence of features with a statistically acceptable level of certainty must involve examining a spatially extended region of an image—the larger the sample, the greater the statistical significance of the judgment. There is thus an uncertainty relationship, or trade-off, between the statistical accuracy with which elemental regions of tissue may be assigned to particular histological categories and the size of such elemental regions (i.e. the spatial resolution). In the trade-off between resolution and classification accuracy, two factors are influential: (*a*) the statistical uncertainty introduced by system artefacts (blurring by the psf, speckle modulation, poor reconstruction due to inhomogeneous speed of sound, etc); and (*b*) an uncertainty arising from the spatial uniformity of the tissue properties themselves.

Research in ultrasonic tissue characterisation generally aims to make the above process quantitative, by extracting quantitative image features and making use of multivariate statistical methods, such as discriminant analysis, to automate the classification process. The potential advantages of a quantitative and automated approach include better control over studying the performance of the diagnostic technique (and therefore for improving it), improved consistency of results by reducing dependence on the (variable) diagnostician, and the ability to include a very large number of features, some of which may not be available to the human observer. In addition, human observers of information are generally excellent at making comparative judgments, over short distances of space or time, but they find it more difficult to interpret absolute values of a number of quantities that may vary slowly. Thus diffuse disease of the liver and monitoring of tumour response to therapy have often been chosen subjects for quantitative ultrasound research. The aim is not to replace the decision-making role of the diagnostician, rather to provide machine-based assistance in the manner of what in recent years have come to be known as expert systems. As such, another of the potential functions of such systems is to assist with the training of new diagnostic personnel.

At the simplest level of quantitation, one might, for instance, visually examine an image and make a judgment on the presence or absence of a number of features (such as shadowing or an irregular boundary), assigning ones or zeros accordingly. Other meaningful (though more quantitative) features, such as speed of sound or attenuation coefficient, might be extracted either directly from an image (such as the transmission CT images described in the previous section) or by computer analysis of the echo signals from relatively large regions of homogeneous organs. In fact, all of the 'parameters for imaging' mentioned in §7.2.3 are quite suitable features for tissue characterisation. Finally, many features that may have little, or no, intuitively obvious meaning may be extracted digitally. Examples lie in the area of so-called B-scan texture analysis, in which from the digitised B-scan are extracted features that are related to the coarseness or the relative contrast of the pattern of echoes from uniform regions of an image.

Figure 7.28 illustrates the principle of multivariate discrimination of two groups of subjects, based on a number of features, F_1, F_2, F_3, etc. Each subject to be classified is represented by a point (or feature vector) in this so-called feature space, which is two-dimensional in this example because higher-dimensional spaces become confusing when represented on the printed page. With a single feature, the process of classification simply involves deciding whether the value of the feature measured for a given individual is greater than or less than some specified mid-value, known as the decision value. With two features, as illustrated, the task is to determine on which side of a decision line the point in question lies. In three dimensions the line becomes a plane, or surface, which is generalised in higher-dimensional spaces to a hyperplane, or multidimensional surface. The decision surface must be drawn using a training set of cases, which are known to belong to a particular class. Many algorithms have been devised for this purpose, and the subject is now highly evolved and complex. Nevertheless, the simple situation illustrated in figure 7.28 does demonstrate that, under some circumstances, multiple features can achieve a more accurate classification of subjects than any one feature alone.

The shear diversity of approaches to ultrasonic tissue characterisation is, at present, as much a disadvantage as it is an advantage, since it has made it difficult for workers in this area to adopt a systematic approach to the subject. Another major problem, yet to be overcome, is that many of the methods yield features that, although quantitative in a relative sense, are highly instrument-dependent. This also impedes collectively systematic work and slows the process of teaching the classifying algorithms, since data collected from different establishments are, in general, incompatible.

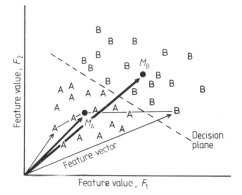

Figure 7.28 Illustration of the potential utility of multivariate discrimination of two classes of tissue/disease (hypothetical data). Neither feature alone achieves the classification accuracy that would be achieved by using both features and the discriminant line indicated. The probability that a particular point belongs to a specific class can be obtained using a measure of distance of its feature vector from the centroid of the cluster of vectors for the class of interest.

A recent development in this subject, which may help with the instrument-dependence problem, is to study the spatial variation of features rather than (or in addition to) their absolute values. This can be done either by displaying the features themselves as images—often called parametric images—or, since there may be many features, by using a multivariate statistical process of some kind to combine the features into an image of a more meaningful quantity. One way in which this can be achieved may be illustrated by referring again to figure 7.28, but now treating each point in feature space as representing the measured feature values at a particular spatial location in one subject. A new image, which one might term a class image, may then be created by assigning colour, or grey level, to the display according to the region of feature space occupied by the measured features for that position. It is immediately apparent that the uncertainty principle referred to earlier will severely restrict the spatial resolution of the class image, if the assigned colours or levels are to be statistically meaningful.

Another method of image display from multivariate data is to create a probability, or similarity, image. Again referring to figure 7.28, an image of probability of belonging to class A may be created by making the grey level proportional to some scaled version of the distance of the measured feature vector to the centre of the cluster of points known to belong to class A.

Application of this last idea has led recently to a novel form of processing to remove speckle from pulse–echo images, without the need for compounding over a wide range of angles (see §7.2.5). This works because the texture features measured for fully developed speckle are characteristic of the scanner and, for constant settings of a given instrument, occupy a specific region of feature space. The location of this region of feature space may be defined by analysing the texture of a scan from a reference object, which contains fine structure designed to generate a speckle image. Figure 7.29(a) shows a simulated B-scan of two strongly scattering point targets and a uniform circular region of low backscattering strength in a background speckle pattern associated with a moderate level of backscattering. Figure 7.29(b) is an image where the grey level is proportional to $P = 1 -$ the probability of being speckle, and figure 7.29(c) is an image produced by using this probability image to control the amount of smoothing applied to the original B-scan, so that regions of full speckle are replaced by a more precise (average) estimate of the local backscattering coefficient, whereas regions of structural significance are preserved. Finally, figure 7.29(d) shows how the structural information is lost if a simple spatially invariant low-pass filter is applied to the original image using the maximum amount of smoothing applied in figure 7.29(c).

7.5.3 Acoustic microscopy

Figure 7.30 indicates two (of many) possible arrangements of the scanning acoustic microscope (SAM), in which both transmitted amplitude and phase may be used to create an image. Forms of acoustic microscope other than SAM also exist. Systems are commercially available but have yet to find clearly defined medical applications outside the research laboratory—non-destructive testing (particularly of integrated electronic circuits) and metallurgy are two established applications outside medicine.

Acoustic microscopy can compete with optical microscopy in a number of ways, often complementing and extending the optical instrument by providing different information derived from the acoustic absorption and elastic properties of the material, as opposed to an image of the degree to which different histological structures take up an optically visible stain. Indeed, one of the potential advantages of acoustic microscopy is that there is sufficient intrinsic acoustic imaging contrast for no staining process to be required. This also means that living (even moving) specimens can be viewed microscopically and in real time.

Microscopes have been constructed to operate at frequencies within a very wide range, from 12 MHz to many gigahertz. The resolution

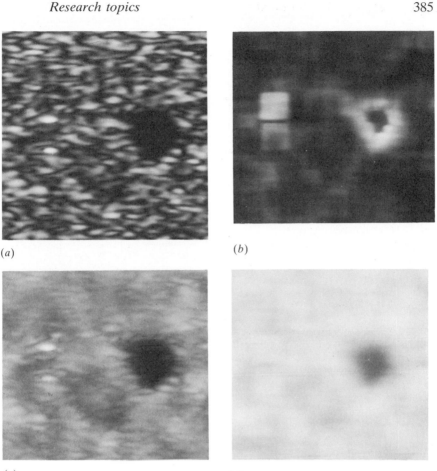

(*a*)

(*b*)

(*c*)   (*d*)

Figure 7.29 Example of the use of multivariate methods to process an ultrasound image and (in this case) suppress the speckle artefact: (*a*) original (computer-synthesised) image of a test phantom which contains two point targets of differing contrast on the left and a circular region of low scattering level on the right, all within a uniform speckle background; (*b*) an image where the grey scale is a measure of the probability that the local texture does not belong to the class speckle (assessed using two texture features); (*c*) the final (reduced-speckle) image obtained when the probability image is used to control the amount of smoothing applied to the original image; and (*d*) the result obtained when the maximum amount of smoothing used in (*c*) is applied uniformly to the whole image.

obtained depends on the wavelength in the coupling medium, and on absorption losses, which limit the highest frequency of operation. One advantage is that the difference in velocity of sound between lens and coupling medium is often so large that spherical aberration is negligible, even for apertures of very low *f*-number. With water as the coupling medium, a resolution of about 1 μm is relatively easily attained.

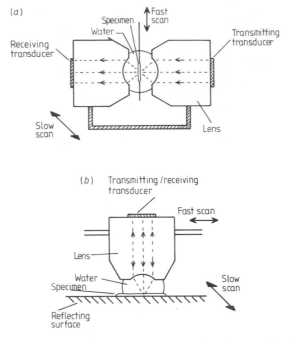

Figure 7.30 Schematic illustration of the modes of operation for two common forms of scanning acoustic microscope: (*a*) transmission microscope and (*b*) reflection microscope.

REFERENCES AND FURTHER READING

General physics and engineering (plus some clinical/research)
BAMBER J C 1986a Attenuation and absorption *Physical Principles of Medical Ultrasonics* ed C R Hill (Chichester: Ellis Horwood)
—— 1986b Speed of sound *Physical Principles of Medical Ultrasonics* ed C R Hill (Chichester: Ellis Horwood) pp220–4
BAMBER J C and DICKINSON R J 1980 Ultrasonic B-scanning: a computer simulation *Phys. Med. Biol.* **25** 463–79
FOSTER F S and HUNT J W 1979 Transmission of ultrasound beams through human tissue—focusing and attenuation studies *Ultrasound Med. Biol.* **5** 257–68
FRY F J (ed) 1978 *Ultrasound, Its Applications in Medicine and Biology* (New York: Elsevier)
GOOBERMAN G L 1968 *Ultrasonics: Theory and Application* (London: English Universities Press)
HILL C R (ed) 1986 *Physical Principles of Medical Ultrasonics* (Chichester: Ellis Horwood)
HILL C R and KRATOCHWIL A (eds) 1981 *Medical Ultrasonic Images: Formation, Display, Recording and Perception* (Amsterdam: Excerpta Medica)
HUSSEY M 1985 *Basic Physics and Technology of Medical Diagnostic Ultrasound* (London: Macmillan)

KINSLER L E, AUSTIN R F, COPENS A B and SANDERS J V 1982 *Fundamentals of Acoustics* (New York: Wiley)

LONGHURST R S 1968 *Geometrical and Physical Optics* (London: Longmans, Green)

McDICKEN W N 1981 *Diagnostic Ultrasonics: Principles and Use of Instruments* (New York: Wiley)

MOORES B M, *et al* (eds) 1981 *Physical Aspects of Medical Imaging* (Chichester: Wiley)

OTSON B W 1979 *Medical Imaging Techniques: IEE Monographs* (Exeter: Peter Peregrinus)

PAIN H J 1968 *The Physics of Vibrations and Waves* (Chichester: Wiley) ch 5

WELLS P N T 1977 *Biomedical Ultrasonics* (London: Academic Press)

WOODCOCK J P 1979 *Ultrasonics: Medical Physics Handbook* (Bristol: Adam Hilger)

Doppler methods (see also above general texts)

ATKINSON P A and WOODCOCK J P 1982 *Doppler Ultrasound and Its Use in Clinical Measurement* (London: Academic Press)

BAKER D, FORSTER F K and DAIGLE R E 1978 Doppler principles and techniques *Ultrasound, Its Applications in Medicine and Biology* ed F J Fry (New York: Elsevier) part I, ch III

JAFFE C C (ed) 1984 *Vascular and Doppler Ultrasound* (New York: Churchill Livingstone)

TAYLOR K J W, BURNS P N and WELLS P N T (eds) 1988 *Clinical Applications of Doppler Ultrasound* (New York: Raven)

Calibration and performance checks

CHIVERS R C (ed) 1978 *Methods of Monitoring Ultrasonic Scanning Equipment* HPA Topic Group Report no. 23 (London: Hospital Physicists' Association)

DUGGAN T C and SIK M J 1983 Assessment of ultrasound scanners by acoustic signal injection *Ultrasound '82* ed R A Lerski and P Morley (Oxford: Pergamon) pp179–84

EVANS J A (ed) 1986 *Physics in Medical Ultrasound* IPSM Report 47 (London: Institute of Physical Sciences in Medicine)

LOCKE D J 1986 *Quality Assurance in Medical Imaging* IOP Short Meetings Series no. 2 (Bristol: Institute of Physics) pp67–76

McCARTY K 1986 *Quality Assurance in Medical Imaging* IOP Short Meetings Series no. 2 (Bristol: Institute of Physics) pp77–98

MADSEN E L, ZAGZEBSKI J A, FRANK G R, GREENLEAF J F and CARSON P L 1982 Anthropomorphic breast phantoms for assessing ultrasonic imaging system performance and for training ultrasonographers: I and II *J. Clin. Ultrasound* **10** 67–86

SMITH S W, LOPEZ H and BODINE W J Jr 1985 Frequency independent ultrasound contrast–detail analysis *Ultrasound Med. Biol.* **11** 467–77

ZAGZEBSKI J 1983 Ultrasound quality assurance *Textbook of Diagnostic Ultrasound* ed S Hagen-Ansert (St Louis: C V Mosby)

Clinical applications

BARNETT E and MORLEY P 1985 *Clinical Diagnostic Ultrasound* (Oxford: Blackwell Scientific)

BASSETT L W, GOLD R H and KIMME-SMITH C 1984 *Hand Held and Automated Breast Ultrasound* (Thorofare, NJ: Slack)

CALAN P W 1983 *Ultrasonography in Obstetrics and Gynaecology* (Philadelphia: W B Saunders)

CosGrove D O and McCready V R 1982 *Ultrasound Imaging: Liver, Spleen, Pancreas* (Chichester: Wiley)

de Vlieger M, *et al* (eds) 1978 *Handbook of Clinical Ultrasound* (New York: Wiley)

Goldberg B B (ed) 1981 *Ultrasound in Cancer* (New York: Churchill Livingstone)

Goldberg B B and Wells P N T 1983 *Ultrasonics in Clinical Diagnosis* (Edinburgh: Churchill Livingstone)

Haller J O and Shkolnik A (eds) 1981 *Ultrasound in Pediatrics* (New York: Churchill Livingstone)

Hansmann M, Hackeloer B J and Staudach A 1985 *Ultrasound Diagnosis in Obstetrics and Gynaecology* (Berlin: Springer)

Hill C R, McCready V R and Cosgrove D O 1978 *Ultrasound in Tumour Diagnosis* (Tunbridge Wells: Pitman Medical)

Holm H H and Cristenson J K (eds) 1985 *Interventional Ultrasound* (Copenhagen: Munksgaard)

Kossoff G D and Fukuda M (eds) *Ultrasonic Differential Diagnosis of Tumours* (New York: Igaku Shoin)

Morley P, Donald G and Sanders R 1983 *Ultrasonic Sectional Anatomy* (Edinburgh: Churchill Livingstone)

Oto R Ch and Wellauer J (eds) 1985 *Ultrasound Guided Biopsy and Drainage* (Berlin: Springer)

Simeone J F (ed) 1984 *Coordinated Diagnostic Imaging* (New York: Churchill Livingstone)

Taylor K J W (ed) 1985 *Atlas of Ultrasonography* (New York: Churchill Livingstone)

Biological effects and safety

Dunn F and O'Brien W D Jr (eds) 1977 *Ultrasonic Biophysics* Benchmark Papers in Acoustics, vol 7 (Stroudsburg, PA: Dowden, Hutchinson and Ross)

Hill C R and ter Haar G 1988 Ultrasound *Non-Ionizing Radiation Protection* 2nd edn (Copenhagen: World Health Organisation)

NCRP (National Council on Radiation Protection and Measurement) 1983 *Biological Effects of Ultrasound: Mechanisms and Clinical Implications* NCRP Document no. 74 (Washington, DC: NCRP)

Wells P N T (ed) 1987 The safety of diagnostic ultrasound *Br. J. Radiol.* Suppl. 20

Williams A R 1983 *Ultrasound: Biological Effects and Potential Hazards* (London: Academic Press)

Research topics

Greenleaf J F (ed) 1986 *Tissue Characterisation with Ultrasound* (Boca Raton, FL: CRC Press)

Lee H and Wade G (eds) 1986 *Modern Acoustical Imaging* (New York: Institute of Electrical and Electronics Engineers)

CHAPTER 8

SPATIALLY LOCALISED NUCLEAR MAGNETIC RESONANCE

M O LEACH

8.1 INTRODUCTION

Having discussed transmission and emission imaging with ionising radiation, and imaging with ultrasound, the use of nuclear magnetic resonance (NMR) to measure and image tissue properties is considered.

This chapter describes the physics of nuclear magnetic resonance and the associated phenomena of relaxation together, and also explains how these methods are used to obtain images that are a function of proton spin density and relaxation times. Methods of measuring the spatial distribution of other nuclei and metabolites using NMR imaging and spectroscopy are discussed.

Although NMR imaging has become one of the diagnostic tools available at many major medical centres, it is still at a relatively early stage of evolution, and new techniques are developing rapidly. Spatially localised NMR spectroscopy, using whole-body spectrometers, has only become practicable within the last few years, and techniques for obtaining localised signals are still in their infancy.

In common with cross-sectional x-ray tomography (Chapter 4), NMR can provide three-dimensional image data sets providing precise anatomical displays. However, rather than providing a map of electron density, related principally to physical density, NMR proton images provide information relating not only to proton density but also to the freedom of hydrogen-containing molecules to rotate and to the proportion of water contained in different body-fluid compartments. By appropriate manipulation of the pulse sequences, the relative contrast of different anatomical structures can be varied and quantitative measurements of tissue relaxation times and water diffusion coefficients can be obtained. Whilst providing anatomical information, the NMR image can also provide functional information more akin to nuclear medicine

(Chapter 6) and ultrasound (Chapter 7), by using paramagnetically labelled tracers to provide selectively increased contrast, and by providing direct measurements of blood flow and water diffusion. ^{31}P NMR spectroscopy also allows tissue energy metabolism to be monitored non-invasively.

The chapter is written at a level appropriate to postgraduate courses in imaging and medical physics, but also contains information relevant to the operation of NMR units. References to more advanced texts and other general reference works are given below. As part of a book based on work at one institution, examples and illustrations have been primarily drawn from work in progress at the Royal Marsden Hospital, Sutton, using a 1.5 T Siemens Magnetom imaging and spectroscopy system.

The development of NMR is described in §8.2. NMR theory is discussed in §8.3. More advanced references include Andrew (1958), Slichter (1978) and Abragam (1983). A number of textbooks, such as Burcham (1973) on nuclear physics, Bleaney and Bleaney (1976) on electricity and magnetism, and Kittel (1966) on solid-state physics, have useful sections. Basic pulse sequences are described in §8.4 and the phenomena and measurement of relaxation in §8.5. The principles of NMR imaging, together with a description of the different methods and their applications, are given in §8.6. Imaging principles are covered in detail by Mansfield and Morris (1982) and Morris (1986).

Spatially localised NMR spectroscopy is described in §8.7, including an introduction and a short discussion of *in vivo* applications. Useful advanced reading on NMR spectroscopy includes Farrer and Becker (1971), Dwek (1973), James (1975), Shaw (1976), Gunther (1980), Fukushima and Roeder (1981) and Gadian (1982). Section 8.8 is concerned with instrumentation and §8.9 discusses the potential hazards and the safety of NMR. Several volumes provide an overview of the applications of NMR, including Partain *et al* (1983), Lerski (1985), McCready *et al* (1987) and an issue of the *British Medical Bulletin* (1984, vol 40, no. 2). A wide range of introductory articles and reviews have also been published, including Pykett (1982), Gordon (1985) and Leach (1987). The latter includes a more comprehensive bibliography of introductory articles.

8.2 THE DEVELOPMENT OF NUCLEAR MAGNETIC RESONANCE

Nuclear magnetic resonance (NMR) was discovered independently by Bloch *et al* (1946) and by Purcell *et al* (1946). This was soon followed by the discovery of chemical shift, allowing nuclei in different chemical environments to be identified as a result of the small change in resonant frequency caused by the electron cloud of the molecule.

Although high-resolution NMR has developed as a versatile tool for studying the chemistry and structure of solids and liquids, the major biochemical and medical interest has arisen from the possibilities of making non-invasive measurements in living tissue. Initial measurements of ^{31}P in intact blood cells were carried out by Moon and Richards (1973). Measurements of frog sartorius muscle followed, but earlier experiments were limited by the small bore of the available magnets. Developments in magnet technology permitted phoshorus studies to be extended, initially to small animals and, recently, with the advent of wide-bore high-field magnets, to studies of humans. Proton and fluorine spectroscopy can also be performed on these instruments. As *in vivo* NMR spectroscopy of animals and humans has become possible, methods of obtaining spatially localised signals from a well defined region of tissue have been developed.

In parallel with the development of spectroscopic techniques, methods of imaging the distribution of protons in tissue evolved. These techniques again depended on the spatial localisation of the NMR signal, although in this case with a much higher spatial resolution. In 1973 the principle of utilising the shift in resonant frequency resulting from the imposition of a magnetic-field gradient was proposed by Lauterbur (1973) and by Mansfield and Grannell (1973). The early images that were formed were limited to small objects, but the first whole-body image was published in 1977 by Damadian *et al* (1977). These early results have been followed rapidly by technical and commercial developments, producing a variety of techniques that allow proton images and sodium images to be acquired, providing information on spin density and T_1 and T_2 relaxation times. Fast and 'real-time' imaging methods have been developed, as well as methods for separating water and fat in proton images, and techniques of measuring blood flow *in vivo*. Whole-body systems are now available that permit both imaging and spectroscopy. Despite the rapid rate of development in recent years, NMR is still at a relatively early stage in its development, and many further advances are likely.

8.3 PRINCIPLES OF NUCLEAR MAGNETIC RESONANCE

8.3.1 Classical description

The behaviour of the net nuclear magnetic moment of materials can usually be sufficiently described by using the classical theory of magnetism. For the moment, consider an isolated proton. This possesses a charge $+e$ and has angular momentum I. An electric charge circulating in a conducting loop produces a magnetic field through the loop normal to the plane of current rotation. The charge on a proton can be considered as being distributed and rotating about a central axis as a result of the angular momentum. This gives rise to a field and magnetic dipole moment m_p antiparallel (for a proton) to the angular-momentum vector and therefore normal to the plane of charge circulation, as shown in figure 8.1, i.e.

$$m_p = \gamma I \tag{8.1}$$

where γ is the gyromagnetic ratio. For a simple model of the proton

$$\gamma = e/2m \qquad (8.2)$$

where m is the proton mass.

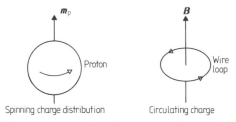

Figure 8.1 The spinning electric charge distribution of a proton generates a magnetic moment in a similar manner to the magnetic flux density produced by electric charge circulating in a wire coil.

An external magnetic flux density B_0 will exert a couple C on a magnetic dipole moment, causing the angular momentum to change at a rate equal to the torque or couple, i.e.

$$C = m_p \times B_0 \qquad (8.3)$$

$$= dI/dt. \qquad (8.4)$$

Substituting for I from equation (8.1) gives

$$dm_p/dt = \gamma m_p \times B_0. \qquad (8.5)$$

Equation (8.5) is the Larmor equation, which describes the precession of m_p about B_0 with angular velocity

$$\omega_0 = -\gamma B_0. \qquad (8.6)$$

The magnetic dipole moment m_p can be resolved into a component m_{pz} parallel to B_0 (B_0 defines the z direction) and a component m_{pxy} in the plane perpendicular to B_0. If $m_{pxy} = 0$, equation (8.5) will be equal to zero and no couple will act on the nucleus. The angle between m_p and B_0 is not changed by this precession, and depends only on the orientation of m_p when B_0 was established. Figure 8.2 shows the effect of couple acting on the magnetic moment.

In a sample comprising many nuclei, the net magnetic moment M will be derived from the vector sum of all the nuclear magnetic moments. Classical theory is inadequate to describe fully the behaviour of single spins, because of the quantum nature of nuclear properties, but for a large ensemble of nuclei, the classical description is normally adequate. In a magnetic flux density B_0, the net magnetic moment M at equilibrium will be aligned with B_0. As only the xy component of M gives a measurable signal at equilibrium, M, which is then equal to M_z, cannot be measured directly. In order to measure M, it must be tilted away from the B_0 or z direction, to produce a measurable component in the

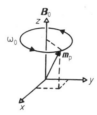

Figure 8.2 The interaction of a magnetic moment m_p with a magnetic flux density B_0 results in the magnetic moment experiencing a couple C which produces motion (precession) with angular velocity ω_0 about B_0.

xy plane. This is the basis of NMR measurements. If a magnetic field of flux density B_1 oriented in the *xy* plane, and rotating at the Larmor angular frequency ω_0, is applied, M will experience a second torque, which will rotate it into the *xy* plane as shown in figure 8.3. The angle α that M moves through from the *z* axis will depend upon the magnitude and duration of B_1 (see next section). The flux density B_1 has a frequency in the radiofrequency (RF) range for NMR measurements, and a 90° RF pulse is one that tilts M through 90° into the *xy* plane (figure 8.4), such that $M_z = 0$. The *xy* component of M is a rotating RF field that can be detected with a suitable coil. A typical signal, known as a free induction decay (FID), that is obtained following a 90° pulse is shown in figure 8.4. The signal decays due to a loss of phase coherence, which will be discussed further below.

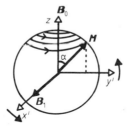

Figure 8.3 If a rotating magnetic flux density B_1 is applied in the *xy* plane in the presence of a static field of flux density B_0 orientated along the *z* axis, M will experience a torque D (in addition to the couple C which gives rise to precession about B_0) moving M through an angle α from the *z* axis. (*x'* and *y'* are axes rotating at ω_0, the Larmor frequency (see §8.3.1.1).)

The motion of M under the influence of an arbitrary applied field with flux density B can be described, as in the case of equation (8.5) for a single proton, by the expression

$$\mathrm{d}M/\mathrm{d}t = \gamma M \times B. \tag{8.7}$$

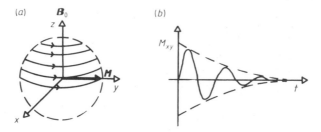

Figure 8.4 (*a*) The motion of *M* in the laboratory frame following a 90° RF pulse and (*b*) the resultant FID that can be measured from the *xy* component of the sample magnetisation.

8.3.1.1 Rotating reference frame. The behaviour of *M* can often be more simply described by considering a reference frame rotating with angular velocity ω in the *xy* plane. The velocity of *M* in the rotating frame DM/Dt can be related to its velocity in the laboratory frame dM/dt as follows. A point *r* fixed in the rotating frame will move a distance

$$\delta r = \omega \delta t\, r \sin \theta$$

$$= \boldsymbol{\omega} \times \boldsymbol{r} \delta t \tag{8.8}$$

in time δt, with respect to the laboratory frame. Hence the velocity of *r* is

$$\lim_{\delta t \to 0} (\delta r / \delta t) = dr/dt = \boldsymbol{\omega} \times \boldsymbol{r}. \tag{8.9}$$

If *r* is not fixed but has a velocity Dr/Dt in the rotating frame, then by addition of the velocities

$$dr/dt = Dr/Dt + \boldsymbol{\omega} \times \boldsymbol{r}. \tag{8.10}$$

This result is discussed in more detail by Kibble (1966). It simply says that the motion in the laboratory frame is the motion of the rotating frame relative to the laboratory frame plus the motion within the rotating frame. Hence the motion of *M* in the laboratory frame can be related to the rotating frame by

$$dM/dt = DM/Dt + \boldsymbol{\omega} \times \boldsymbol{M}. \tag{8.11}$$

Substituting from equation (8.7) we obtain

$$DM/Dt = \gamma \boldsymbol{M} \times \boldsymbol{B}_0 - \boldsymbol{\omega} \times \boldsymbol{M} \tag{8.12}$$

$$= \gamma \boldsymbol{M} \times \boldsymbol{B}_0 + \boldsymbol{M} \times \boldsymbol{\omega} \tag{8.13}$$

$$= \gamma \boldsymbol{M} \times (\boldsymbol{B}_0 + \boldsymbol{\omega}/\gamma). \tag{8.14}$$

This is equivalent to equation (8.5) with $\boldsymbol{B}_0$ replaced by the effective field $(\boldsymbol{B}_0 + \boldsymbol{\omega}/\gamma)$, the sum of the laboratory flux density and a fictitious flux density $\boldsymbol{B}_f$.

In the rotating frame, M can be expressed as components $M_{x'}i' + M_{y'}j' + M_{z'}k'$ with respect to the axes x', y', z' rotating at $\boldsymbol{\omega}$. The laboratory-frame axes are x, y and z. If a circularly polarised RF field of frequency ω (in the laboratory frame) and with magnetic flux density B_1 is applied in the plane perpendicular to B_0 the magnetic flux density B_1 and the electric field E will rotate, as shown in figure 8.5. In the rotating plane it will have a fixed orientation as shown in figure 8.6. As this will have an arbitrary direction, it will be considered as defining the x' axis. The effective flux density in the rotating frame will then be

$$B_e = B_0 + \omega/\gamma + B_1 \qquad (8.15)$$

which can be resolved into components

$$B_e = (B_0 + \omega/\gamma)k' + B_1 i'. \qquad (8.16)$$

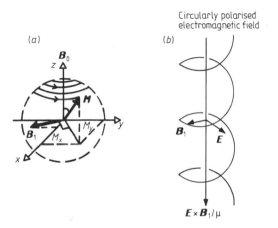

Figure 8.5 (*a*) A circularly polarised RF field applied in the plane perpendicular to B_0 such that the magnetic component B_1 lies in the xy plane. (*b*) The direction of propagation of the RF field is shown by the Poynting vector $E \times H_1 = E \times B_1/\mu$, where μ is the relative permeability. The effect of B_1 on M is shown in the laboratory frame.

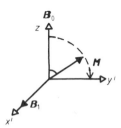

Figure 8.6 The effect of an $\alpha°$ RF pulse on M is shown in the rotating frame.

The effect of B_1 will be to act as an additional couple on M and then, for the frame rotating at the Larmor frequency, $B_0 = -\omega_0/\gamma$,

$$\mathrm{D}M/\mathrm{D}t = \gamma M \times B_1 \tag{8.17}$$

which has the effect of rotating M about the direction B_1 (the x' direction) with an angular frequency $\omega_1 = \gamma B_1$. If the RF pulse is applied for time t, M will rotate through an angle $\alpha = \omega_1 t = \gamma B_1 t$. When $\alpha = \pi/2$, M will be directed along the y' axis and will have been subjected to a 90° pulse. A 180° pulse will rotate M along the $-z$ axis, antiparallel to B_0. In the laboratory frame, the motion will be seen to be more complex, as M will be subject to precession about B_0 as well as about x'. Thus M will spiral down the surface of a sphere whilst the RF signal is applied. Figure 8.7 shows the rotation of M during an RF pulse.

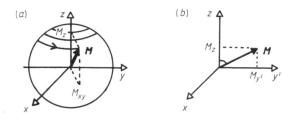

Figure 8.7 (*a*) The motion of M in the laboratory frame, following an $\alpha°$ RF pulse, and (*b*) the same motion in the rotating frame, in both cases showing the resultant components of M along the z axis and in the xy plane.

Precession about x' only occurs during the RF pulse, and α depends upon both B_1 and t, and so, in principle, either can be varied to give the desired angle. As α also depends upon the gyromagnetic ratio γ, the RF power or duration of pulse required to obtain a given tilt angle α will depend upon the nucleus under observation. For instance, phosphorus will require some 2.5 times the RF power that hydrogen requires to tip M through a given angle, and this can be important when considering RF power deposition and NMR safety.

In most NMR equipment, a plane-polarised RF field is applied, as this is technically easier and sometimes the only possible method. A plane-polarised field with flux density $2B_1 \cos(\omega t)$ can be considered as the superposition of two circularly polarised fields with flux density B_1 rotating in opposite directions. The field rotating in the same sense as the precession will interact with M, whilst the counter-rotating field will have negligible effect. The linear or plane-polarised field is, however, less efficient, as the power deposited is twice that required. For this reason, some NMR imaging systems have been designed to transmit and receive circularly polarised signals.

If B_1 has angular velocity $\boldsymbol{\omega}$, the effective field will be (from equation (8.16))

$$B_e = (B_0 + \omega/\gamma)k' + B_1 i' = -\omega_e/\gamma \tag{8.18}$$

where ω_e is the angular velocity of the Larmor precession of M about B_e,

$$\omega_e = - [(\omega_0 - \omega)^2 + \omega_1^2]^{1/2} \gamma/|\gamma| \qquad (8.19)$$

and if the angle between B_e and B_0 is θ, then

$$\tan \theta = \omega_1/(\omega_0 - \omega). \qquad (8.20)$$

From this, it can be seen that M will only be significantly reoriented if θ becomes large, which occurs when $|\omega - \omega_0|$ becomes close to $|\omega_1|$, the latter usually being very much smaller than ω_0. Thus B_1 must have a frequency close to the resonant frequency to have a significant effect.

An adiabatic rapid-passage experiment is one in which B_1 is sufficiently large that effectively no relaxation (see §8.5) takes place whilst the B_1 field is applied, meaning that the sweep is rapid, but at the same time the nuclear precession about B_e is always fast compared with the rotation of B_e, and this is the adiabatic term. This means that the value of M is unchanged during the experiment whilst the magnetisation always has the same relationship with B_e.

8.3.2 *Quantum-mechanical description*

The above is an approximate description for isolated nuclei, and is of more value when considering the net effect on an assembly. In the individual nucleus, a quantum-mechanical description is required. Consider a nucleus with spin angular momentum I and a magnetic moment μ. The magnitude of the nuclear angular momentum

$$|I| = \hbar[I(I + 1)]^{1/2} \qquad (8.21)$$

where I is the nuclear spin quantum number and $\hbar = h/2\pi$, where h is Planck's constant. From the uncertainty principle, the direction of I cannot be established. If an axis of quantisation z is defined by applying an external magnetic field with flux density B_0, the z component of I will be $m_I \hbar$, where m_I is the magnetic quantum number, taking values $\pm I, \pm (I - 1), \ldots$, giving $(2I + 1)$ possible orientations (degeneracies) of angular momentum.

For the proton, $I = \frac{1}{2}$ and therefore m_I can take values $\pm \frac{1}{2}$, which correspond to spin up and spin down. For a more complex nucleus than hydrogen, m_I may have more than two values. For ^{31}P and ^{19}F, $I = \frac{1}{2}$, but for ^{23}Na, $I = \frac{3}{2}$, giving $m_I = -\frac{3}{2}, -\frac{1}{2}, \frac{1}{2}, \frac{3}{2}$. Nuclei with $I > \frac{1}{2}$ have nuclear quadrupole moments that can produce splitting of the resonance lines or line-broadening effects. Figure 8.8 shows the possible orientation of I for a nucleus, and it can be seen that I is oriented at a semi-angle β to the z axis such that

$$\cos \beta = m_I \hbar/|I|. \qquad (8.22)$$

The energy of these states is

$$E = -\mu \cdot B_0 \qquad (8.23)$$

$$= -\gamma \hbar m_I B_0. \qquad (8.24)$$

For $m_I = \pm\frac{1}{2}$, the energy separation between the two states is ΔE, where

$$\Delta E = \gamma\hbar B_0. \qquad (8.25)$$

The quantity $\gamma\hbar$ is known as the nuclear magneton. Transitions can take place from the lower state (spin down) to the upper state if a quantum of energy $\gamma\hbar|B_0|$ is absorbed (figure 8.9), where

$$\hbar\omega_0 = \gamma B_0\hbar. \qquad (8.26)$$

Thus energy at the Larmor frequency must be supplied.

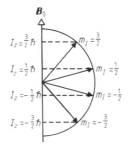

Figure 8.8 The available orientations of I for a spin-$\frac{3}{2}$ nucleus, showing the corresponding values of $m_I\hbar$.

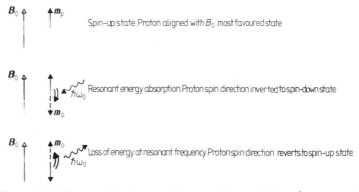

Figure 8.9 Change of spin state for a nucleus with $I = \frac{1}{2}$, as a result of absorption or emission of a quantum of energy.

As the magnetic moment of the proton is allowed to occupy only two states, spin up and spin down, it is not always evident how the experimentally observed (and important) transverse component of nuclear magnetisation can arise. The behaviour of the proton magnetic moment in quantum mechanics is described in terms of the probability of the proton occupying a given state at any one time. It can occupy either of the two states, and this occupancy can vary from observation to observation. It can be shown (Abragam 1983) that the expectation value taken over the wavefunction of a free spin is consistent with the classical equation. The relationship is linear, so large assemblies of independent nuclei will also behave classically, and the net magnetic moment can be described classically, provided there is negligible interaction between spins.

8.3.3 Statistical distribution of spin states

In the absence of a B_0 flux density, there is no degeneracy in spin states, so spin up and spin down cannot be defined. If a field with flux density B_0 is applied, spins will be distributed between the two states, and, after a period dependent on the T_1 relaxation time (see §8.5), the population of the two levels will reach equilibrium, with the spin-up state being more favoured (figure 8.10).

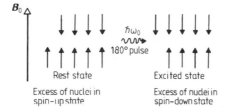

Figure 8.10 The distribution of spin-$\frac{1}{2}$ nuclei between the two states at equilibrium, and following a 180° RF pulse.

The population difference between the two states is described by the Boltzmann distribution, where

$$\frac{n(\text{spin up}, + \frac{1}{2})}{n(\text{spin down}, - \frac{1}{2})} = \exp\left(\frac{\Delta E}{kT_s}\right) \tag{8.27}$$

where k is the Boltzmann constant, T_s is the absolute temperature of the spin system (which is equal to the lattice temperature when the sample is at equilibrium), n is the number of spins in a given state (spin up is spin aligned with B_0, which is the low-energy state) and

$$\Delta E = \gamma \hbar B_0 \tag{8.28}$$

is the energy difference between the two states. At room temperature and equilibrium in a field of flux density 1 T, there is an excess of protons in the spin-up state of about 3×10^{-6}. This explains the relatively low sensitivity of NMR, where only some three in a million protons can be measured, compared with, for instance, isotope measurements, where under some circumstances a single decay can be measured.

The concept of spin temperature is sometimes useful. If the population of spins is disturbed, for instance by supplying energy, $n(-\frac{1}{2})$ will increase, changing the population difference between the states. As ΔE is fixed, for equation (8.27) to hold, T_s (the temperature of the spin system) must increase. Energy loss is characterised by T_s returning to its equilibrium value.

Unlike γ-radiation, where energy loss via γ-ray emission is a spontaneous process, energy loss in NMR from the high- to the low-energy state is via stimulated emission. This is a result of the probability of spontaneous emission being proportional to frequency cubed. The formula for the lifetime of a dipole radiating in free space is

$$\tau(\text{s}) = \frac{3}{2} \frac{c^3}{\omega_0^3 \gamma^2 \hbar} \tag{8.29}$$

where c is the speed of light *in vacuo*. For a proton in a magnetic field of flux density 1.5 T, this corresponds to approximately 10^{24} s. Stimulated emission occurs as a result of coupling with a magnetic field fluctuating at the Larmor frequency. Non-coherent coupling with a radiation field in thermal equilibrium is again a very weak energy-loss mechanism, and the major energy-loss mechanism is coupling with the lattice. This is further discussed in the section on relaxation.

8.3.4 Bloch equations

Bloch *et al* (1946) proposed a set of equations that, for most purposes, accurately describe the behaviour of the nuclear magnetic moment of a sample of non-interacting or minimally interacting spins, such as a liquid sample. In a homogeneous field with flux density B, the equation of motion is given by equation (8.7), which is

$$\mathrm{d}\boldsymbol{M}/\mathrm{d}t = \gamma \boldsymbol{M} \times \boldsymbol{B}.$$

In a static field of flux density $B_z = B_0$, the return of the z component of magnetisation, M_z, following a stimulus, such as an RF pulse, to its equilibrium value M_0 can be described by

$$\mathrm{d}M_z/\mathrm{d}t = -(M_z - M_0)/T_1 \tag{8.30}$$

where T_1 is the longitudinal relaxation time. If the nuclear magnetisation has a component perpendicular to the z direction, this transverse magnetisation will decay due to interactions with local spins, with the rate of change being described by

$$\mathrm{d}M_x/\mathrm{d}t = -M_x/T_2 \tag{8.31}$$

$$\mathrm{d}M_y/\mathrm{d}t = -M_y/T_2 \tag{8.32}$$

where T_2 is the transverse relaxation time. Relaxation times are discussed in §8.5.

If it is assumed that the motion due to relaxation can be superimposed on the motion of the free spins under the influence of a static field and a much smaller RF field, the behaviour of the magnetisation can be described by

$$\frac{\mathrm{d}\boldsymbol{M}}{\mathrm{d}t} = \gamma \boldsymbol{M} \times \boldsymbol{B} - \frac{(M_x \boldsymbol{i} + M_y \boldsymbol{j})}{T_2} - \frac{(M_z - M_0)}{T_1} \boldsymbol{k} \tag{8.33}$$

where $\boldsymbol{i}$, $\boldsymbol{j}$ and $\boldsymbol{k}$ are unit vectors in the laboratory frame. As in §8.3.1, the behaviour of M in the rotating frame can be described by

$$\frac{dM}{dt} = \gamma(M \times B_e) - \frac{(M_{x'}i' + M_y j')}{T_2} - \frac{(M_{z'} - M_0)}{T_1} k' \quad (8.34)$$

with i', j' and $k' = k$ being unit vectors in the rotating frame. Equation (8.34) can be rewritten to provide expressions for the three orthogonal components of M: $M_{x'}$, $M_{y'}$ and $M_{z'}$ in the rotating frame. To evaluate the components, we make use of equation (8.18) for B_e:

$$dM_{x'}/dt = -M_{x'}/T_2 + \Delta\omega M_{y'} \quad (8.35)$$

$$dM_{y'}/dt = -\Delta\omega M_{x'} - M_{y'}/T_2 - \omega_1 M_{z'} \quad (8.36)$$

$$dM_{z'}/dt = \omega_1 M_{y'} - (M_{z'} - M_0)/T_1. \quad (8.37)$$

Here, $\Delta\omega = \omega - \omega_0$, where ω is the rotation frequency of the rotating reference frame, ω_0 is the Larmor frequency, and $\omega_1 = -\gamma B_1$. These equations can then be solved under a variety of boundary conditions to describe the time domain or k-space behaviour of the NMR signal following a stimulus.

Although the signal initially acquired from the transverse component of magnetisation, as a free induction decay (FID) or spin echo (see §8.4.3), is a time signal, the frequency signal, obtained following Fourier transformation, is of most interest in both imaging (being proportional to displacement; see §8.6) and spectroscopy (giving chemical shift; see §8.7.1). Terms frequently met in NMR texts are the absorption and dispersion signals. These are the two signals acquired following quadrature detection (§8.8.5) and Fourier transformation of a free induction decay, the two components also being known as the real and imaginary parts, the indentity being arbitrary. The absorption signal is

$$v = -(\gamma/|\gamma|) M_{y'} = \pm M_y' \quad (8.38)$$

$$= -\pi|\gamma| B_1 M_0 F(\Delta\omega) \quad (8.39)$$

for negligible saturation, where $F(\Delta\omega)$ is the normalised Lorentz shape function, with halfwidth at half-intensity $1/T_2$. The absorption mode signal has the same form as the absorption of RF signal by the spins, and the Lorentzian lineshape is given by

$$S_{abs}(\Delta\omega) = \frac{M_0 T_2}{1 + T_2^2(\Delta\omega)^2}. \quad (8.40)$$

The dispersion mode signal is given by

$$S_{disp}(\Delta\omega) = \frac{M_0 T_2^2 \Delta\omega}{1 + T_2^2(\Delta\omega)^2}. \quad (8.41)$$

The two signals are shown in figure 8.11. The dispersion mode signal B is $90°$ out of phase with the absorption signal, and is an odd, as opposed to an even, function of $\Delta\omega$, going to 0 at ω_0. Often the modulus of the two signals is taken to give the absolute signal, with a consequent loss of phase information.

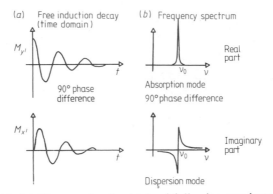

Figure 8.11 (*a*) The FID signals obtained following quadrature detection, and (*b*) the absorption and dispersion mode signals obtained after Fourier transformation.

8.4 NUCLEAR MAGNETIC RESONANCE PULSE SEQUENCES

The preceding sections have shown how an NMR signal can be generated by manipulating M with suitable RF pulses. This section describes some of these pulse sequences, which are important components of NMR imaging and spectroscopy.

8.4.1 Saturation recovery

In a saturation-recovery sequence, a 90° RF pulse is applied along the x' axis to rotate M into the xy plane (see §8.3.1.1). This process is shown in figure 8.12. Once magnetisation is tipped into the xy plane, a free induction decay (FID) signal is generated, which can be observed using a suitable receive coil. The initial amplitude of the FID will be proportional to the amplitude of M_z, the z component of magnetisation, immediately preceding the pulse sequence. Provided a perfect 90° pulse was used throughout the sample, M_z following the RF pulse should be 0. It is not practicable to measure the FID until after the end of the RF pulse, due to the large signal induced in the receiving coil by the RF pulse. The FID can then be digitised and Fourier-transformed to provide a frequency spectrum. This is the normal mode of measurement in spectroscopy. Alternatively, as described in §8.4.3 below, the xy magnetism can be rephased with a 180° pulse to produce an echo, and this spin echo can be digitised as a measure of M_z.

When the spin system is saturated, the population of the two energy levels is equal, as is indicated by the absence of the z component of magnetisation. Following the 90° pulse, the M_z component will recover with relaxation time T_1 (see equation (8.46)). The exponential envelope of the FID results from a decay in M_{xy} due to a loss of phase coherence in the xy plane. This results from spins changing relative phase with one another due to processes such as T_2 relaxation effects (including spin exchange and changes in the local molecular magnetic field, and hence resonant frequency), inhomogeneities in the B_0 field, and effects such as local diffusion and blood perfusion in capillaries, which move the spins into a region with a different local magnetic field. It is therefore characterised by T_2^*, which includes the effect of T_2, and these additional processes are discussed further in §8.5. Figure 8.13 shows how the recovery of M_z can be monitored by further 90° pulses, following the initial pulse, at increasing delay times T_D. In a practical NMR sequence, the pulse sequence will often be repeated many times, with a characteristic repetition time T_R. If T_R is less than about $5T_1$, this will result in a reduction in M_z compared with M_0, as a result of partial saturation.

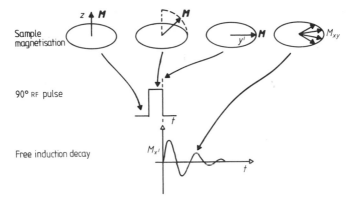

Figure 8.12 The effect of a 90° RF pulse on the sample magnetisation M, and the resultant FID showing loss of net magnetisation in the xy plane due to dephasing.

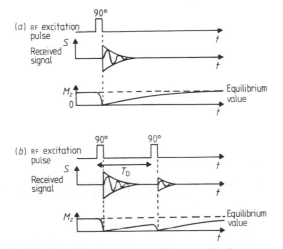

Figure 8.13 (*a*) The M_z component of magnetisation recovers to its equilibrium value following an initial 90° RF pulse of a saturation-recovery sequence; (*b*) the magnetisation M_z at time T_D can be monitored with a second 90° RF pulse in a saturation-recovery sequence.

8.4.2 Inversion recovery

As is apparent from the name of this sequence, the z component of magnetisation, M_z, is inverted to lie along the $-z$ axis, by using a 180° RF pulse. This process is shown in figure 8.14. If an accurate 180° pulse is used, there should be no xy component of M. Thus this inverted magnetisation can only be observed by interrogating M_z by following the 180° inversion pulse by a 90° pulse at a time T_I known as the inversion time. In returning to equilibrium, the z component of magnetisation, M_z, must first pass through zero and then continue until $M_z = M_0$. It therefore takes longer for equilibrium to be restored, and a longer repetition time T_R is required between sequences (see equation (8.47)).

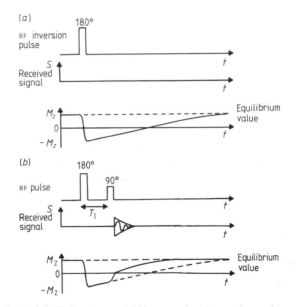

Figure 8.14 (*a*) Following a 180° RF pulse in an inversion-recovery sequence, the z component of magnetisation is inverted and subsequently returns to its equilibrium value; (*b*) by applying a second 90° RF pulse, the value of M_z at time T_I can be measured in a saturation-recovery sequence.

8.4.3 Spin-echo sequence

This sequence is shown in figure 8.15. As in the saturation-recovery sequence, a 90° pulse interrogates the z axis magnetisation. After an interval $T_E/2$ (T_E is the echo time), a second 180° pulse is applied that refocuses the xy magnetisation, producing an echo centred at time T_E after the 90° pulse. In the rotating frame (figure 8.16), spins at flux density B_0 and thus rotating at the resonant frequency will appear stationary. Some spins will rotate faster, and others more slowly, due to inhomogeneities in the magnetic field. Those at lower fields will lose phase compared to ω_0 and will appear to move anticlockwise. Those at higher frequencies will move clockwise, causing a dephasing of the xy signal. If the spins are then rotated through 180° about the x' axis, the spins will still be subject to the same fields. However, the direction of motion is reversed, and the spins will now be converging on the $-y'$ axis, to produce an echo. A useful analogy is to consider a group of runners, some fast, some slow and some of average speed. At time $t = 0$ they start running for a time T. At time T they are now out of phase and reverse direction and retrace their steps at the same speeds. At $t = 2T$, the runners are then all in phase again.

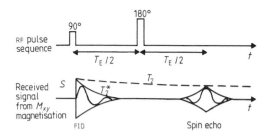

Figure 8.15 In a spin-echo sequence, a 90° RF pulse rotates M_z into the xy plane. Following the loss of coherence due to T_2^* relaxation, the M_{xy} magnetisation is refocused with a 180° RF pulse to produce an echo.

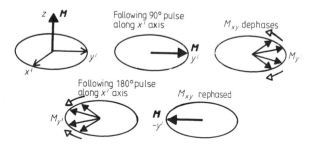

Figure 8.16 The M_{xy} component of magnetisation, produced following a 90° RF pulse oriented along the x' axis in the rotating frame, dephases due to B_0 inhomogeneity. A 180° RF pulse oriented along x' rotates the y' component of magnetisation through 180°, causing the spins to rephase.

This pulse sequence is the basis for many of the current imaging sequences. Further echoes can be obtained by repeating the 180° pulse. Again the whole sequence can be repeated with repetition time T_R. There are several different multi-echo sequences, and these are discussed in more detail in §8.5.4.

8.5 RELAXATION PROCESSES AND THEIR MEASUREMENT

8.5.1 *Longitudinal relaxation, T_1*

Longitudinal relaxation is often known as T_1 or spin–lattice relaxation. It describes the transfer of energy to or from the spin system, resulting in spin-state changes. As noted above, changes in spin state must be excited by coupling and exchanging energy with a lattice. For spin-$\frac{1}{2}$ nuclei, fluctuating magnetic fields that can interact with the nuclear dipole moment provide a stimulus. For spins having an electric quadrupole moment $(I > \frac{1}{2})$, electric-field gradients must also be considered. The major mechanism, in the absence of paramagnetic nuclei with unpaired electrons, is the magnetic field generated by the dipole moments of molecules such as water (see figure 8.17). The magnetic fields due to this dipole moment will have components in the xy plane, some of which will vibrate at ω_0 and can therefore stimulate emission or absorption of energy. This energy is transferred between the lattice and the spin system. The de-excitation of spin states results in a reduction in the spin temperature. The rate at which energy is lost indicates how closely the spins are coupled to the lattice or how tightly the spins are bound in the liquid. For randomly tumbling molecules, there will be a frequency distribution

$$J(\omega) \propto \frac{\tau_c}{1 + (\omega\tau_c)^2} \qquad (8.42)$$

where τ_c is the correlation time of the molecule. The correlation between the position of the molecule at time t and time $t = 0$ is proportional to $\exp(-t/\tau_c)$, so a molecule undergoing rapid motion will have a short correlation time, and vice versa. For small molecules such as glucose in solution, $\tau_c \sim 10^{-12}$ to 10^{-10} s. Correlation times are discussed in more detail in James (1975). The temperature dependence of τ_c can often be described by a form of Arrhenius equation

$$\tau_c \propto \exp(-E_a/kT_{\text{latt}}) \qquad (8.43)$$

where E_a is the activation energy, T_{latt} is the lattice temperature and k is Boltzmann's constant. Often T_1 has a temperature dependence of this form. Also T_1 depends on $J(\omega)$ at the Larmor frequency and on the flux density B_{xy} arising from molecular motion, so

$$1/T_1 \propto B_{xy}^2 J(\omega). \qquad (8.44)$$

By inducing vibrations at ω_0, for instance by applying an ultrasound field, spin-state changes can also be induced.

The presence of paramagnetic species, with unpaired electrons and therefore magnetic moments of a factor of about 1000 greater than the nuclear magnetic moment, will also lead to enhanced relaxation rates. This effect is proportional to r^{-6}, where r is the distance between the nucleus and the paramagnetic ion. Relaxation effects due to paramagnetic agents are considered in detail by Dwek (1973), the relaxation produced by paramagnetic sites being described by the Solomon–Bloemberger equations (Solomon 1955, Bloemberger 1957). Paramagnetic spin-enhancement agents are discussed in §8.6.6.6.

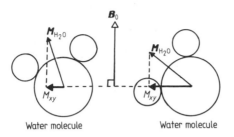

Figure 8.17 The xy component of the dipole moment of a water molecule varies with the orientation of the molecule.

T_1 relaxation is defined by the Bloch equation (see also §8.3.4)

$$dM_z/dt = (M_0 - M_z)/T_1. \tag{8.45}$$

The T_1 relaxation time provides valuable information about vibrational motion in the lattice, which in tissues is generally water, and therefore is a measure of the ability of water molecules in the neighbourhood of the spins being measured to tumble and rotate. As a result of this, it is often related to the water content of the tissue and appears also to be related to the degree to which water is bound or absorbed onto surfaces in the tissue.

8.5.2 Measurement of longitudinal relaxation times

T_1 can be measured with a saturation-recovery $90°$—T_D—$90°$ sequence (see figure 8.13) where the signal (FID) following a $90°$ pulse at time T_D is given by

$$A(\tau) = A(0)[1 - \exp(-T_D/T_1)]. \tag{8.46}$$

If the sequence is repeated for varying τ, the function $\ln[A(0) - A(T_D)]$ can be plotted against T_D. For a monoexponential, a straight line will be obtained. In a heterogeneous tissue, multiple exponents may be observed.

An alternative method is to use an inversion-recovery sequence (see figure 8.14). Here the signal intensity is described by

$$A(T_D) = A(0)[1 - 2\exp(-T_D/T_1)]. \tag{8.47}$$

Again a curve can be plotted to obtain T_1. A rapid, although less precise, method is to determine the time T_D where $A(T_D) = 0$. At this point, $T_1 = T_D/\ln 2$.

It is important to note that in a given sample, although the T_1 relaxation time is only defined for exponential processes, the relaxation may not be accurately described by an exponential function, and this will only be evident if the recovery of the longitudinal magnetisation is measured at a number of different times. The saturation-recovery method of measuring T_1 has a number of advantages compared with the inversion-recovery method. Although the inversion-recovery method has twice the dynamic range of the saturation-recovery method, it is necessary, as $A(0)$ cannot be sampled during each repetition, to ensure that M_z has reached equilibrium prior to each repetition of the pulse sequence. However, with saturation-recovery sequences the FID can be measured after both 90° pulses, and thus the second FID, $A(t)$, can be referred to the first FID, $A(0)$, for each repetition of the sequence, allowing shorter repetition times to be used if desired and also compensating for drifting electronics. If the pulse is not exactly 90°, provided that it is reproducible, the slope of the sample magnetisation as a function of M_0 will still permit the T_1 relaxation time to be calculated. A discussion of differing methods of measuring relaxation times is given in Fukushima and Roeder (1981).

8.5.3 Transverse relaxation, T_2

Transverse, T_2 or spin–spin relaxation is the loss of magnetisation from the xy plane. This results both from a loss of phase coherence in the xy plane and also as a result of the longitudinal relaxation causing a net loss in signal from the xy plane. In tissue samples, T_2 relaxation is usually considerably faster than T_1 relaxation, but when $\omega_0 \tau_c \ll 1$, the T_2 relaxation time approaches the T_1 relaxation time; in other words the dephasing processes are not significant when compared with the longitudinal relaxation processes. This occurs as a result of the rapid motion of small molecules in a dilute solution, and thus gadolinium-doped water solutions often have T_2 values approaching the T_1 values of the solution, and therefore do not completely serve to mimic the behaviour of tissues.

The molecular magnetic fields referred to above also usually give rise to a z component in addition to the xy component, which has the effect of slightly altering the effective field B_0 experienced by the nucleus, causing the frequency of precession to vary slightly depending on the relationship between the nucleus and the molecular magnetic field. This causes a loss of phase and therefore a reduction in the net xy magnetisation. Another effect that gives rise to loss of phase is an exchange of the spin state between two nuclei, with no net loss of energy from the spin system, but a loss of phase information.

For a Lorentzian lineshape, where T_2 relaxation is the only significant process producing a loss of transverse magnetisation, the T_2 relaxation time is inversely proportional to the full width at half-maximum (FWHM) of a spectral line, i.e.

$$\text{FWHM} f(\omega) = 1/\pi T_2. \tag{8.48}$$

In practice, factors other than T_2 relaxation also cause loss of transverse magnetisation. These additional features together with the T_2 relaxation time affect the decay of the envelope of the FID or spin echo, and the decay of the envelope is described by T_2^*, the effective T_2 relaxation time. This is primarily influenced by magnetic-field inhomogeneities, which give rise to a spread in frequencies for a given nucleus and therefore a loss in phase coherence, i.e.

$$1/T_2^* = 1/T_2 + \gamma \Delta B_0/2. \tag{8.49}$$

Here ΔB_0 is the variation in magnetic field due to inhomogeneities. These effects also give rise to further broadening in the spectral lines, which have a width that is inversely proportional to T_2^*.

8.5.4 Measurement of T_2 relaxation time

As the envelope of the FID decays with T_2^* rather than T_2, the T_2 relaxation time cannot be measured from the shape of the FID or spin echo. The method usually adopted is to use a spin-echo sequence with multiple Hahn echoes (Hahn 1950). During formation of the spin echo by rephasing the transverse magnetism, the effect of field inhomogeneities is removed, providing that protons do not move into regions of different field strength. Figure 8.15 showed a simple $90°$—τ—$180°$ spin-echo sequence. In practice, diffusion of the spins also occurs, causing the nuclei to experience different magnetic fields in an inhomogeneous field. The expression for the amplitude of the spin echo at time $2\tau = T_E$ therefore contains a term to describe the effects of diffusion,

$$M_{xy}(2\tau) = M_{xy}(0) \exp(-2\tau/T_2 - \tfrac{2}{3}\gamma^2 G^2 D\tau^3) \tag{8.50}$$

where G is the spatial magnetic-field gradient and D is the diffusion coefficient of the spins, which for proton imaging will be predominantly water molecules. It will be evident that the importance of the second term in the exponential function becomes more important as τ increases. As the value of $M_{xy}(0)$ is not measured directly from this sequence, the T_2 measurement has to be obtained by carrying out a number of measurements using different values of τ. In sequential measurements there is obviously scope for errors to arise due to electronic drift and incorrect values of pulse amplitude or length.

This problem was overcome by the development of the Carr–Purcell (CP) spin-echo train (Carr and Purcell 1954), which by repeating $180°$ pulses at intervals of 2τ generates a train of pulse echoes of alternating polarities (see figure 8.18). In this sequence, the $90°$ and $180°$ pulses are all applied along the x' axis and the sequence can be described as

$$90°—\tau—(180°—2\tau)_n$$

where for any echo the xy component of magnetisation is described by

$$M_{xy}(t) = M_{xy}(0) \exp(-t/T_2 - \tfrac{1}{3}\gamma^2 G^2 D\tau^2 t). \tag{8.51}$$

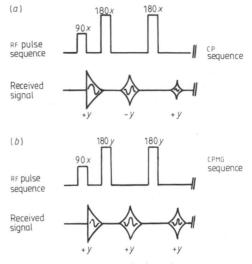

Figure 8.18 (*a*) The Carr–Purcell (CP) pulse sequence, and (*b*) the CP sequence as modified by Meiboom and Gill (CPMG).

For short values of τ the second term in the exponential function becomes negligible. Thus this sequence considerably reduces the effects of diffusion on the measurement. A curve can be plotted for the many echo values obtained, allowing a T_2 measurement to be carried out within a single repetition of the sequence. This method suffers from the requirement of an accurate 180° pulse length. As the 180° pulse is always applied in the same sense about the x' axis, if the pulse length is incorrect, it will cause a cumulative error in the projection of M onto the xy planes. This method is also therefore sensitive to inhomogeneities in B_1 and often results in a measured value of T_2 that is too short.

The sequence was modified by Meiboom and Gill (1958) to give the CPMG sequence in which all of the 180° pulses are shifted 90° in phase with respect to the initial 90° pulse. In other words, they are oriented along the $\pm y'$ axis. This results in all of the echoes being formed along y' regardless of the exact tip angle thus avoiding cumulative errors.

8.5.5 Calculation of relaxation times and other parameters in imaging sequences

In NMR spectrometers it is common practice to obtain a number of points for a T_1 or T_2 measurement and to derive the relaxation time by plotting a curve of the different values. In imaging measurements, to be described later, the relaxation times are often calculated from a pair of images differing either in repetition time or in echo time.

For T_1 measurement, two images measuring $A(\tau)$ with differing τ are often obtained (see equation (8.46)). If T_R was sufficiently long that M_z completely recovered $(T_R > 5T_1 \text{ max})$, this would allow T_1 to be calculated on a pixel-by-pixel basis. In practice, T_R is shorter, and considerable suppression of M_z occurs. Algorithms provided by manufacturers attempt to overcome this by, for example, calculating a table of intensity ratios for different T_1 values at the given T_R values used, and interpolating between the nearest values to those obtained for a given pair of pixels. As might be expected, this approach is subject to quite large errors. Saturation-recovery spin-echo images are commonly obtained, and the two images must therefore have been obtained with equal echo times (T_E).

Some commercial NMR systems provide more accurate measurements of T_1. The MD800 imager uses alternating saturation-recovery and inversion-recovery sequences with $T_R = 1$ s and $T_I = 200$ ms to calculate T_1 values (Eastwood 1984). Measurements have indicated a reproducibility of 5% (2σ) for solutions having T_1 between 127 and 355 ms (Richards *et al* 1987). At longer T_1, greater variability was observed, as a consequence of the T_R used.

Several approaches have been developed on instruments that do not automatically calculate accurate T_1 values. Hickey *et al* (1986) have reported a technique employing seven saturation-recovery images and six spin-echo images, using an iterative least-squares procedure. They have found good reproducibility in human studies at 0.26 T for T_1 (within about $\pm6.6\%$) and T_2 (within about $\pm4\%$) over two years, but the absolute accuracy has not been reported. The method requires some 50 min to acquire data from one slice (Johnson *et al* 1987b). One method of confirming accuracy is to use the NMR system as a spectrometer (Leach *et al* 1987c, Johnson *et al* 1987a). Johnson *et al* (1987a) have also measured the accuracy of calculation algorithms for a system operating at 0.5 T, using a set of sequences that take about 30 min, with two spin-echo sequences and one inversion-recovery sequence. They obtained accuracies of 7% and 20% for T_1 and T_2, respectively (for $T_1 < 700$ ms and $T_2 < 200$ ms). In animals, using multipoint sequences with 5–10 spin-echo scans and five inversion-recovery scans, taking about 90–120 min, accuracies of 1% and 5% for T_1 and T_2 were achieved. Methods making use of stimulated echoes for simultaneously measuring T_1 and T_2 have recently been proposed (Graumann *et al* 1986). To calculate T_2, two images with equal T_R but differing T_E can be obtained, usually by using a two-echo sequence. This has limited accuracy, and a multi-echo CPMG sequence can be obtained in a practicable clinical acquisition time. By using such a multi-echo sequence giving a large number of images with different echo times, the slope of an exponential can be calculated for each pixel, giving one or more exponential values for the T_2 relaxation time. This approach will clearly be more accurate than the use of just two images to calculate T_2.

The proton density, for those protons giving a measurable signal, can also be calculated approximately by using a spin-echo sequence and values for T_1 and T_2 obtained by the above method. The spin density is

$$\rho = \frac{M_{xy}(2\tau)}{\exp(-2\tau/T_2)[1 - \exp(T_R/T_1)]}. \tag{8.52}$$

In clinical practice, images that are T_1-weighted spin echo (short T_R dependent on field strength, and short T_E, 30 ms or less) or inversion recovery, or T_2-weighted spin echo (long T_R dependent on field strength, longer than 1 s, variable echo time) are usually obtained. These do not provide absolute values of intensity, or quantitative measures of relaxation time, but provide contrast of clinical value. Several examples are included in the next section (§8.6.2).

Diffusion can also be calculated, by using a gradient field superimposed upon the main B_0 field together with a combination of spin-echo sequences having differing values of τ. This is discussed further in §8.6.6.2.

8.6 NUCLEAR MAGNETIC RESONANCE IMAGE ACQUISITION AND RECONSTRUCTION

This section commences by considering the projection reconstruction method of NMR imaging. Although not commonly used now in commercial NMR scanners, this method was used in the early commercial scanners and has much in common with other tomographic imaging methods discussed in this volume (see, for example, Chapters 4, 5, 6, 7 and 11). This is followed by a description of the 2D Fourier-transform (2DFT) method, which is the usual method of image acquisition and reconstruction. The derivation of the signal intensity in NMR images is then presented, indicating how the image can be obtained from the object distribution.

Finally a brief description of other techniques and approaches in NMR imaging is given. A detailed treatment of different NMR imaging methods may be found in Morris (1986).

8.6.1 Projection reconstruction

Many of the NMR imaging methods make use of the property that the resonant frequency of protons is proportional to the applied magnetic field. Thus, if a small magnetic flux gradient (see §8.8.2) along the z axis is added to the main flux density B_0, the resonant frequency of protons will change with z displacement (see figure 8.19). If the applied z gradient is G_z (T m^{-1}), the resonant frequency will change with z displacement as

$$\omega(z) = \gamma B_0 + \gamma z G_z. \tag{8.53}$$

Thus, if a sample is irradiated with a 90° pulse, and the FID occurs in the presence of a z gradient, the signal at each frequency will be a projection of the signal in the plane at the appropriate z location projected onto the z axis. If the FID were then Fourier-transformed, to obtain the frequency distribution, the frequency axis would be equivalent to z displacement. Equally, if the sample were irradiated with a narrow band of frequencies in the presence of the z gradient, only spins in a slice resonating at frequencies within that frequency band would be excited. This process is termed selective excitation, and it is a major component of many imaging pulse sequences. The profile of the selected slice depends upon the shape of the RF envelope and the degree of saturation occurring, a function of T_1 (see §8.6.7.1). In projection reconstruction, the first step is to select a slice, by applying a z gradient if a transaxial slice is required. This is shown in figure 8.20. Having selected the plane of interest, the next task is to resolve x and y displacements within that plane. For a saturation-recovery sequence, if the FID is now read out in the presence of an x gradient, each frequency present in the FID will relate to a particular x displacement. Thus the Fourier transform of the FID will be a projection of the signal intensity in the plane onto the x axis, as shown in figure 8.21. This projection is therefore conceptually similar to those obtained in transmission CT imaging (Chapter 4) or in single-photon emission CT (Chapter 6). If the x and y gradients are combined during read-out of the signal, projections at other angles can be obtained. Thus by cyclically repeating the sequence of selective excitations of the plane of interest in the presence of a z gradient and then reading out the FID in the presence of a combination of x and y gradients that are changed for each measurement, a series of projections of the plane can be built up, allowing an image to be reconstructed by convolution and backprojection (see Chapter 4). Alternatively, the FID can be directly projected into Fourier space, with the FID being multiplied by a suitable one-dimensional window function, or a two-dimensional window function being applied to the two-dimensional space and then a two-dimensional Fourier transform can be carried out to obtain the real-space image.

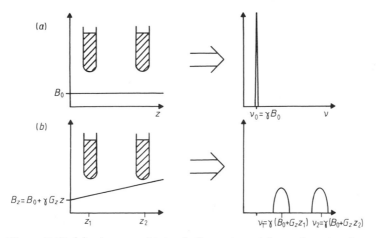

Figure 8.19 (*a*) A FID obtained from two tubes of water in a constant magnetic field B_0 will display only one frequency when Fourier-transformed. (*b*) If the signal is acquired in the presence of a linear gradient G_z, the frequency spectrum will show two frequencies, with a separation proportional to their spatial separation.

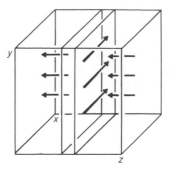

Figure 8.20 Selection of a transaxial slice by application of a selective 90° RF pulse in the presence of a *z* gradient.

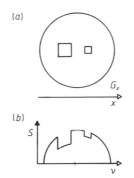

Figure 8.21 (*a*) Following slice selection, the FID is acquired in the presence of an *x* gradient. (*b*) The Fourier transform gives the projection of the signal intensity in the slice onto the *x* axis.

8.6.2 Two-dimensional Fourier imaging

Fourier zeugmatography was introduced by Kumar *et al* (1975a, b). The pulse sequence is shown in figure 8.22. As in projection imaging, a selective excitation is carried out in the presence of a *z* gradient. However, instead of obtaining projections of the spin density at different angles as with projection imaging, the read-out gradient is held constant and the third dimension is obtained by phase encoding. The free induction decay is allowed to evolve under the influence of a *y* gradient applied for a variable period by introducing a phaseshift that is a function of *y* displacement, due to the linear change in precession frequency. After a period equal to the maximum phase-encoding period an *x* gradient is applied, so that the remainder of the FID is frequency-encoded in the *x* direction, and digitisation is performed during this period. A set of FID is obtained, each with an increased phase-encoding period t_y. If the FID are stored in a time space in ascending order of phase-encoding period, the time space can be filled and a two-dimensional Fourier transform carried out to produce the frequency and

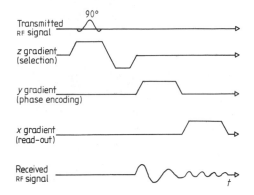

Figure 8.22 In Fourier zeugmatography the FID is acquired in the presence of a read gradient (in this case the x gradient), after a period of phase encoding in the presence of a y gradient, the duration of the phase-encoding period being incremented on each repetition of the sequence.

therefore the spatial distribution in the x and y directions. This is shown in figure 8.23.

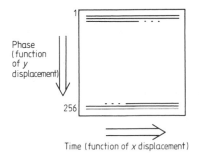

Time (function of x displacement)

Figure 8.23 By carrying out a number of phase-encoding steps (commonly 256), the time, raw data k-space can be filled with the echoes obtained. A two-dimensional Fourier transform then yields the image distribution.

In spin-warp imaging (Edelstein *et al* 1980), slice selection is obtained in the same way as in projection imaging, and use is also made of a read-out gradient, usually the x gradient in transaxial slices, to encode the frequencies in the spin echo. In this case, phase encoding is carried out by increasing the amplitude of the y gradient, rather than by increasing t_y. The phase-encoding period precedes the spin echo, and spin refocusing is obtained with a gradient reversal, rather than a 180° pulse.

Current 2D Fourier-transform imaging techniques are based on this sequence, using variable-strength phase-encoding gradients applied for a fixed time. However, a 180° spin-refocusing pulse is usually used to produce the spin echo. Figure 8.24 shows the raw data sets obtained with the gradient echo sequence and the image obtained following Fourier transformation.

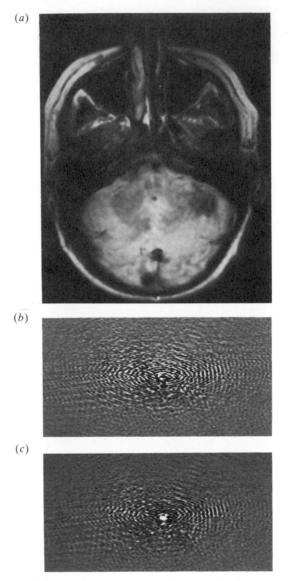

(*a*)

(*b*)

(*c*)

Figure 8.24 (*a*) A transaxial image through the brain of a subject with Hodgkin's disease obtained using a FLASH 90 gradient echo-imaging sequence which uses a 90° flip angle, $T_R = 0.25$ s, $T_E = 12$ ms and a slice width of 7 mm using an extended acquisition matrix of 512 points in the frequency-encoding direction and 256 phase-encoding steps; (*b*) and (*c*) are the corresponding real and imaginary parts of the raw (time domain) data.

This basic sequence can be adapted to a wide range of imaging pulse sequences, including inversion recovery and the parametric imaging methods described in §8.6.6. Figure 8.25 shows a spin-echo sequence.

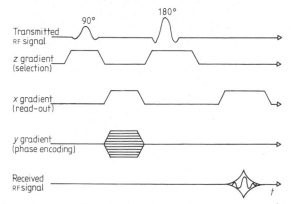

Figure 8.25 A conventional spin-echo pulse sequence used to obtain a single transaxial slice. The time between the initial 90° pulse and the echo is known as the echo time T_E and the time between successive 90° pulses in repetitions of the sequence is the repetition time T_R.

The spin-echo sequence has found wide application in clinical imaging. Figure 8.26 shows brain images obtained in three orthogonal orientations. By appropriate selection of gradients, images at any orientation can be obtained. Surface-coil images of the spine are shown in figure 8.27. Figure 8.28 demonstrates changes in contrast between a proton density weighted image and a T_2-weighted image. They are compared with a short (17 ms) echo sequence, providing increased signal, with and without gadolinium DTPA, and with FLASH 90 and FLASH 50 sequences.

8.6.3 *Derivation of image equations from the spin-density distribution*

Consider the generalised situation of a spin-density distribution $\rho(x, y, z)$ measured in an NMR system having quadrature detection (see §8.8.5) in the presence of three orthogonal gradients. Following a 90° pulse, the signal obtained at time t, $S(t)$, is

$$S(t) = KM_0 \int_x \int_y \int_z \rho(x, y, z) \exp\left(i\gamma \int_0^t [xG_x(t')\right.$$

$$\left. + yG_y(t') + zG_z(t')]\, dt'\right) \exp(-t/T_2^*)\, dx\, dy\, dz. \qquad (8.54)$$

The linear gradients are such that

$$\omega_x = \gamma x G_x \qquad \omega_y = \gamma y G_y \qquad \omega_z = \gamma z G_z$$

and K is a constant depending on the coil geometry. Thus

(a)

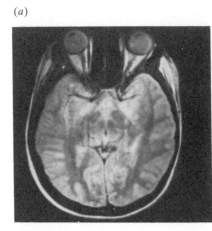

(b)

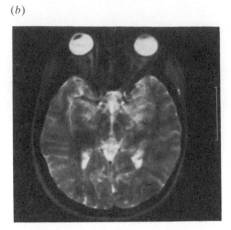

(c)

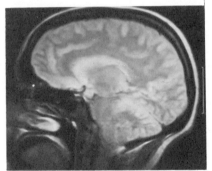

(d)

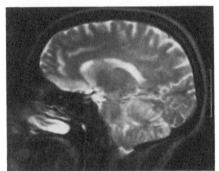

(e)

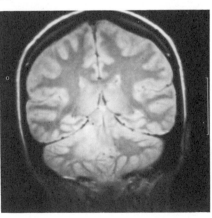

(f)

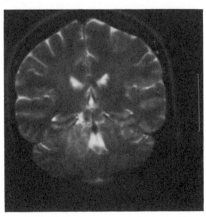

(a)   (b)

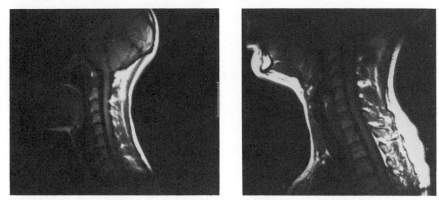

Figure 8.27 Images of the cervical spine and neck in normal subjects obtained using (a) a cervical spine surface coil with $T_R = 0.2$ s, $T_E = 30$ ms and a 10 mm slice thickness, and (b) a Helmholtz surface coil, the transmit coil in both cases being the body coil.

$$S(t) = KM_0 \int_x \int_y \int_z \rho(x, y, z) \exp[i(\omega_x + \omega_y + \omega_z)t] \exp(-t/T_2^*) \, dxdydz.$$
$$(8.55)$$

This has the form of a three-dimensional Fourier transform (see Chapter 12) and thus an inverse three-dimensional Fourier transform can yield the spin distribution.

Most imaging methods reduce the number of gradients to simplify read-out of the signal. So for projection reconstruction, if a transaxial slice is being reconstructed, ω_z will be zero as a z selective excitation will have been carried out. In the simplest case, a projection onto the x axis, ω_y will also equal zero and the signal at time t can be described by

$$S(t) = KM_0 \int_x \rho(x)\exp(i\omega_x t)\exp(-t/T_2^*) \, dx. \qquad (8.56)$$

In the more general case where a combination of x and y gradients are used, then

$$S(t) = KM_0 \int_x \int_y \rho(x, y)\exp[i\gamma(xG_x + yG_y)t]\exp(-t/T_2^*) \, dx \, dy. \qquad (8.57)$$

If

$$r = x \cos \theta + y \sin \theta$$

Figure 8.26 A set of images obtained in three orthogonal planes for a subject with a glioma in the right pons. The lesion is apparent in the lower left quadrant of the coronal sections. Transaxial sections with a double-echo spin-echo sequence with $T_R = 2.7$ s, $T_E = 30$ ms giving a proton density image (a) and with $T_E = 90$ ms providing a T_2-weighted image (b); (c) and (d) show the same sequences applied in the sagittal plane and (e) and (f) show the sequences applied in the coronal plane.

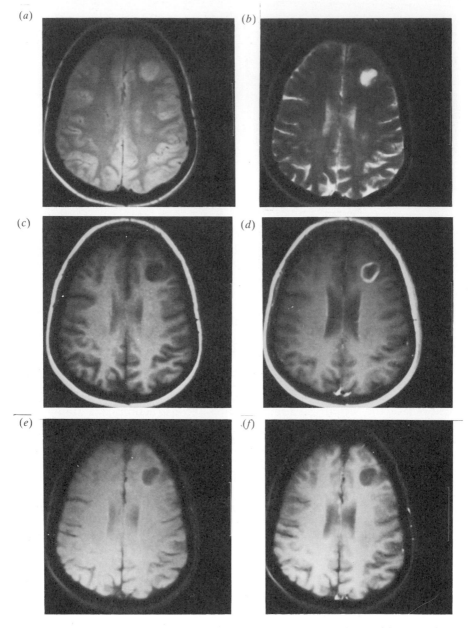

Figure 8.28 Transaxial images through the brain of a patient with secondary deposits from small cell lung cancer showing the effect of gadolinium DTPA and different sequences on lesion contrast. (*a*) and (*b*) are proton density and T_2-weighted images having $T_R = 2.7$ s with $T_E = 30$ ms and 90 ms respectively. (*c*) is a T_1-weighted image with $T_R = 0.5$ s and $T_E = 17$ ms. (*d*) is obtained with the same sequence following administration of gadolinium DTPA, which has been taken up in an active ring of tissue at the periphery of the lesion. (*e*) and (*f*) are FLASH 90 and FLASH 50 gradient echo sequences without gadolinium.

where θ is the angle the composite gradient makes with y, then

$$S(t) = KM_0 \int_x I_1(r) \exp(i\gamma r G_r t)\exp(-t/T_2^*) \, dr \qquad (8.58)$$

where I_1 is the line integral of the spin distribution perpendicular to G_r, the direction of the composite gradient. A one-dimensional Fourier transform of the FID, $S(t)$, therefore gives $I_1(r)$.

To perform a two-dimensional Fourier-transform reconstruction following selective excitation in the z plane, a constant gradient is applied in the x direction for time t_x to frequency-encode the read-out signal. The amplitude of G_y applied for time t_y will vary with each cycle of the pulse sequence,

$$S(t_x, t_y) = KM_0 \int_x \int_y \rho(x, y)\exp[i\gamma(xG_x t_x + yG_y t_y)]\exp(-t/T_2^*) \, dx \, dy. \qquad (8.59)$$

A two-dimensional Fourier transform will then give the spin distribution $\rho(x, y)$. In Fourier zeugmatography, t_y would be varied instead of G_y, which would be constant.

8.6.4 Other methods of nuclear magnetic resonance imaging

8.6.4.1 Point methods. These methods are designed to obtain a signal from only one location in the sample. Compared with plane- and volume-imaging methods, they are therefore of low sensitivity, but they are of value in certain applications.

The sensitive-point method was developed by Hinshaw (1974). This makes use of three orthogonal gradient fields which oscillate in such a way that only one sensitive point is subjected to a non-time-varying gradient. The position of this point can be moved electronically through an object to build up an image. The method can also be used to select the signal from a particular region of interest; for instance, in measuring T_1 or carrying out ^{31}P spectroscopy.

Two other methods depend on modifying the B_0 field such that only a small region is homogeneous. Most of the sample will be in an inhomogeneous field, giving broad spectral components, whereas the signal from the homogeneous region will have a narrow lineshape. Techniques exist to remove the non-homogeneous component to the signal, or a selective pulse can be used to affect only the sample within the homogeneous region. In field-focused nuclear magnetic resonance (FONAR) (Damadian *et al* 1976a,b), this technique has been used for imaging by scanning the patient across the sensitive point. A similar technique, topical magnetic resonance (TMR), has been developed for use with NMR spectroscopy (see §8.7.3), where static-field-profiling coils are used to destroy the homogeneity of the magnetic field other than at a central point (Gordon *et al* 1980).

8.6.4.2 Line-scanning methods. The initial problem in line scanning is how to ensure that a signal is only observed from an individual line. This is technically more difficult than ensuring that the signal comes from a single plane. One method of achieving this is selectively to saturate all but one plane and then excite an orthogonal plane, producing a line of excited magnetisation at the intersection of the two planes. A method (volume-selective excitation) based on two selective 45° pulses applied in the presence of a linear gradient and a 90° broad-band pulse of opposite phase has been developed by Aue *et al* (1984). This method retains the equilibrium magnetisation in the plane perpendicular to an applied gradient, and can be repeated to select a line or an isolated voxel, leading to applications in selective spectroscopy as well as in imaging. Once magnetism in a line has been selected, spatial information along the line can be obtained by using a read-out gradient in the manner described for projection imaging.

Focused selective excitation, proposed by Hutchison (1977), makes use of a different method to define the selected line. A selective 180° pulse in the presence of a z gradient is followed by a selective 90° pulse in the presence of a y gradient, producing spins in the selected line that are 180° out of phase with the signal in the remainder of the sample. A second acquisition is then carried out in the absence of the selective 180° pulse. If the results from the two experiments are then subtracted, the signals from the line will add and the signals from the remainder of the sample will cancel out. The method is, however, particularly sensitive to motion artefacts.

A spin-echo technique that provides both T_1 and T_2 contrast has been suggested by Maudsley (1980). A selective 90° pulse is applied in the presence of a z gradient and this is followed by a selective 180° pulse along the y' axis applied in the presence of a y gradient, which refocuses only those spins lying along the line of intersection of the two planes. The echo can then be read out in the presence of an x gradient to provide displacement in the x direction. The repetition time can be varied to give T_1 contrast and the echo time will provide T_2 contrast. A variant of the sensitive-point method, known as the multiple-sensitive-point method, uses two oscillating field gradients to define the line, read-out along the line occurring in the presence of a linear gradient (Andrew *et al* 1977).

A method of line imaging known as fast-scan imaging has been proposed by Mansfield *et al* (1976). A selective-saturation pulse is applied in the presence of a z field gradient to saturate all the spins in the sample other than those in a plane perpendicular to the z axis. This is followed by a selective-excitation pulse in the presence of a y gradient, which causes only those spins lying in the intersection of the two planes to be rotated into the xy plane. Read-out is again carried out in the presence of an x gradient. By using the same selective-saturation pulse and z gradient for each cycle and varying the selective-excitation pulse, different lines in a plane perpendicular to the z axis can be read out. If T_1 is long, several FID can be sampled before the next application of the selective-saturation pulse.

8.6.4.3 Fast imaging methods. Fast imaging techniques may be considered as falling into two categories. There are those techniques that can be carried out on currently installed commercial systems. These use standard imaging methods as described above, but generally use gradient echoes to reduce the echo time, spoiler gradients to dephase the xy signal and often an adjustable-flip-angle RF pulse rather than a 90° or 180° pulse. Whilst fast compared with the conventional spin-echo and inversion-recovery techniques, they do not enable one to obtain real-time NMR images. The second category covers those techniques that are still largely in a developmental phase, requiring specially constructed hardware optimised for very fast NMR imaging. These latter techniques are often able to produce real-time images.

Fast imaging on conventional equipment   There are a number of fast sequences supplied by the manufacturers of different equipment and many have similar features. Two such sequences are FLASH and FISP.

The pulse sequence for FLASH (fast low-angle single shot) (Haase *et al* 1986) is shown in figure 8.29. It consists of a selective excitation to define the imaging plane with the flip angle selectable to be between 0° and 90°. This is then followed by a phase evolution period to phase-encode one dimension, a switched gradient to induce a gradient echo, and read-out in the presence of an orthogonal gradient. This is then followed by a dephasing gradient to destroy any residual xy component of magnetisation. With this sequence, image contrast can be varied by varying the flip angle and repetition time. This will vary the degree of T_1 suppression, giving a high degree of T_1 contrast. If small angles and long repetition times are used, the T_1 contrast is reduced and T_2^* contrast predominates. Figure 8.30 shows FLASH images having different contrast. Using short repetition times, images can be obtained in times as low as 5 s, allowing the patient to hold their breath during the imaging sequence and thus avoiding artefacts due to respiratory motion.

FISP utilises steady-state free precession and produces contrast that depends on the ratio of T_2 to T_1. The pulse sequence used is shown in figure 8.31 (Oppelt *et al* 1986). This sequence again uses gradient

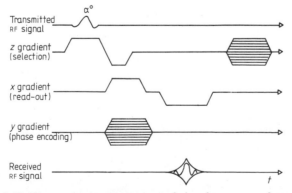

Figure 8.29 The pulse sequence used in the FLASH fast imaging technique. The echo is produced by reversal of the x gradient. A z gradient spoiler pulse is used to dephase residual magnetisation prior to the next repetition of the sequence. T_E, the echo time, is the time between the α^0 pulse and the echo.

(a)

(b)

(c)

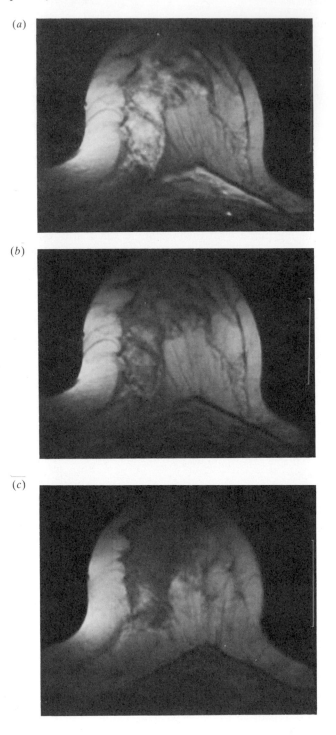

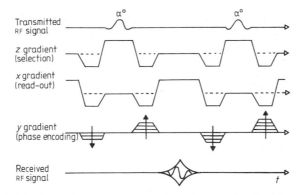

Figure 8.31 The pulse sequence used in the FISP fast imaging technique.

echoes but, rather than destroying the xy magnetisation following the echo, it is refocused with rephasing gradients and rotated back onto the z axis, providing increased image intensity.

Specialised fast imaging methods Although the above methods allow one to obtain images with good contrast and signal-to-noise ratio in much-reduced imaging times, they do not offer the opportunity of obtaining real-time NMR images, which would be particularly of value in cardiology. The most widely applied commercially available method of obtaining images showing the temporal behaviour of the heart are electrocardiograph- (ECG) triggered imaging methods, where either conventional or fast imaging sequences are used to obtain images at different phases of the cardiac cycle. Although these allow the motion of the heart to be visualised and clearly show the movement of blood in the major vessels, allowing motion to be demonstrated in 'ciné mode' apparently as real time, they do not have the time resolution of true real-time images. Also, because they effectively obtain lines of data in the time or k domain at each cycle of the heart, they necessarily include an averaging effect due to the small changes in the pattern of the heart cycle during the course of the study. ECG triggering also prolongs the period of a study. Most of the real-time imaging systems currently under development are based on variants of echo-planar imaging (Mansfield 1977, Mansfield and Pykett 1978). The basic pulse sequence for the two-dimensional form of the measurement is shown in figure 8.32. A selective excitation is applied to define a plane and produce an xy component of magnetisation. This is then dephased and rephased by a strong alternating y gradient. This has the effect of producing a train of

---

Figure 8.30 Gradient echo images of sagittal sections through a right breast containing a nodular area obtained using a breast surface coil with $T_R = 0.25$ s, $T_E = 12$ ms and a slice width of 7 mm. (*a*) FLASH 20, (*b*) FLASH 50, (*c*) FLASH 70, showing the progressive increase in T_1 weighting as the flip angle is increased.

gradient echoes, each of which contains information about the spin density perpendicular to the *y* axis. The *y* gradient is switched in the presence of a much smaller *x* gradient, which also produces gradual dephasing with time, giving information on the profile along the *x* axis. The train of echoes is equivalent to a set of equally spaced delta functions, which has the property of measuring the frequency distribution via a set of discrete frequency samples over the whole sample.

8.6.4.4 Volume imaging. Most clinical imaging examinations require information to be obtained from a volume rather than a single plane. Often this is achieved by the acquisition of a set of planes, which may or may not be contiguous. This is the most common method of acquiring volume data in NMR and, depending on the repetition time and on the number of echoes obtained, different slices can be examined by varying the slice-select pulse during the repetition period. Thus each slice experiences the full repetition time but a number of slices can be examined within this time. This means that there is no time penalty to examining a number of slices rather than one slice, provided that they can be accommodated within the repetition time. As the slice profile is not rectangular, if contiguous slices are to be examined it is normal to use an interlaced sequence so that alternate slices are examined before the intervening slices are measured. This reduces the degree of suppression of adjacent slices due to overlap of the slice profile.

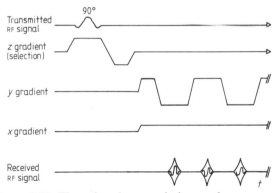

Figure 8.32 The echo-planar technique pulse sequence.

As the slice orientation depends only on the values of the slice gradients chosen, the set of slices can be obtained at any orientation through the body on most commercial instruments, and this is a significant advantage over x-ray CT measurements. As comparison is most often made between NMR and CT, it is useful to compare their behaviour in obtaining volume data. CT scanners (see Chapter 4) obtain slices sequentially and, although the time for each slice may be short, perhaps only a few seconds, if a large number of slices are required the total time is at least the product of the individual scan time and the number of slices. An additional allowance must be made for the reconstruction time and for the power rating of the x-ray tube. With NMR, although the total examination time is longer, there is no penalty in increasing the number of slices within the constraints of the sequence used. This comparative analysis is extended in Chapter 15.

In single-photon emission computed tomography (SPECT) and positron emission tomography (PET) (Chapter 6), data are often gathered from the whole volume of interest. Although in SPECT the planes of interest are normally reconstructed in turn, in PET, particularly with multiwire chambers, the whole volume may be reconstructed as a 3D reconstruction (Clack *et al* 1984). In NMR it is also possible to obtain 3D information directly, and to perform a three-dimensional reconstruction. The most common method for this purpose is the three-dimensional Fourier-transform method, in which the two-dimensional Fourier-transform method is modified by using a non-selective excitation and providing a second period of phase evolution in the presence of a z gradient. If a cube of side n pixels is to be imaged, this means that the total acquisition time will be equal to the product of n^2 and the repetition time rather than the product of n and the repetition time for the two-dimensional Fourier-transform method. The image can then be reconstructed by carrying out a three-dimensional Fourier transform. This dependence of imaging time on n^2 has meant that, until recently, there has been little interest in using 3D Fourier-transform methods for normal imaging. For example, a 256×256 pixel image obtained with a repetition time of 2 s would take about 8.7 min, during which time as many as 20 planes could be acquired. To obtain 20 planes using the same image size by the 3D Fourier-transform method would take approximately 3 h, although, as the whole sample is excited for each repetition, the signal-to-noise ratio would be improved. To obtain a cube of side 256 pixels would take some 36.5 h. However, these methods do become more practicable when used with fast imaging techniques, particularly as they allow thin slices to be obtained quite efficiently without the problems associated with overlapping slice profiles.

On some instruments, the method has now been combined with slice selection to allow each of a number of thick slices to be subdivided into a set of thin phase-encoded slices. The different gradient-selected slices can then be stepped through in the same way as for normal two-dimensional imaging, but very thin slices can be obtained. As an example, with a repetition time of 50 ms for a FLASH sequence, a 256×256 image would take 13 s to acquire, and by using the 3D Fourier-transform method 30 planes could be obtained in 6.5 min. Figure 8.33 shows an image obtained using the 3D FLASH sequence. If thick-slice selection were used, depending on the repetition time, several sets of 30 planes might be obtained in the same total time. In principle, a three-dimensional projection data acquisition and reconstruction method can be used (Shepp 1980), but again the acquisition time is prohibitive and the method has thus produced few applications.

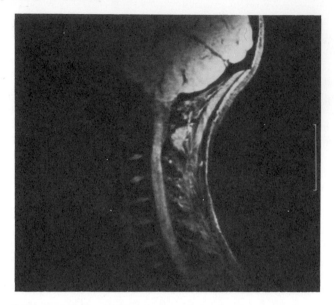

Figure 8.33 A 1.9 mm thick section through the normal cervical spine obtained using a cervical spine coil and a fast 3D FLASH imaging technique, with $T_R = 0.5$ s, $T_E = 20$ ms and a flip angle of 20°.

8.6.5 *Image artefacts*

In common with most imaging techniques, NMR images often contain artefacts. In a well adjusted machine, there are a number of effects that will inevitably give rise to artefacts. There is also a very wide range of effects that can occur due to the failure or poor adjustment of equipment, but these will not be dealt with here, apart from mentioning that it is often instructive to examine the raw data prior to image reconstruction. In projection reconstruction, bad data points present in individual projections in the image space, for example due to short bursts of narrow-band noise, will be propagated into the image as streaks at an angle corresponding to the projection angle. As the linearity of the system and thus the centre of reconstruction is defined via gradients rather than being geometrically defined, misalignment may result in central artefacts and ring artefacts close to the centre. A common phenomenon is Gibb's oscillations or ringing close to high-contrast interfaces in the image, and the magnitude of these will depend on the form of the backprojection filter used. A number of factors will also give rise to blurring in the image, including motion and chemical-shift effects. If the analogue-to-digital converter (ADC) sampling band-width and the frequency filter used in acquiring the projection data are not carefully chosen, information from outside of the projection can be aliased back into the projection.

In spin-warp or 2D Fourier-transform imaging, major artefacts arise due to respiratory motion and to blood flow in the body. Flowing blood usually has a high contrast with respect to surrounding tissue due to the transport of excited magnetism out of the image plane and to other refocusing effects in multislice imaging. This in itself is an advantage, in clearly demarking vessels and also providing a means of assessing blood flow. Figure 8.34 shows the contrast obtained from flowing blood in an EGA gated image. However, the motion of the blood also results in the blood acquiring a different phaseshift compared to the surrounding tissue in the presence of the phase-encoding gradients. This results in signal from the blood being wrongly positioned in the reconstructed image on the basis of the incorrect phase information. Generally fast-flowing blood gives rise to a disturbance across the entire image in the phase-encoding direction aligned with the vessel giving rise to the artefact. In many imaging sequences, the blood has a non-saturated signal compared with the relative saturation of adjacent tissues, and therefore has a very high intensity. Thus the artefact also has a high intensity, and this effect is particularly pronounced at high fields and in fast imaging sequences. The artefact can be reduced in conventional imaging sequences by ECG triggering, although the degree of artefact suppression will depend on the degree to which the ECG trace represents local flow-rate changes. If the orientations of the phase- and frequency-encoding gradients are exchanged, the orientation of the artefact will be rotated through 90°, allowing pathology of interest to be observed. With fast imaging, a more productive approach is to saturate either regions

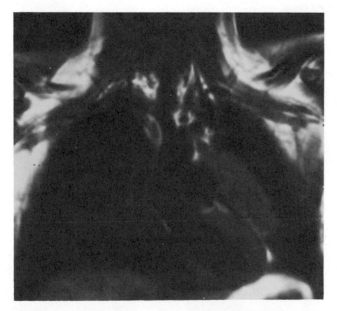

Figure 8.34 An ECG gated coronal section through the chest of a patient with a mediastinal mass from Hodgkin's disease, showing also the high contrast (due in this case to low signal) obtained from flowing blood.

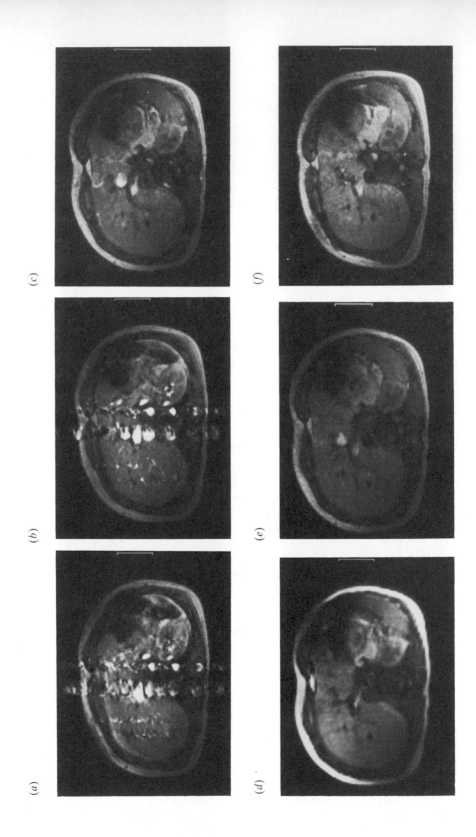

(a)

(b)

(c)

(d)

(e)

(f)

adjacent to the slice, or the whole sample, to reduce the intensity of the signal from flowing blood. Figure 8.35 shows images obtained with and without flow suppression.

Respiratory motion also results in signals being misplaced in the phase-encoding direction, and again the effect is particularly a problem when the moving tissue is of high intensity. The major artefact is due to abdominal and superficial fat, which manifests as ghosts of the body contour or internal fat displaced by a distance related to the period of the motion and the repetition time of the sequence. One commonly used method to reduce this effect is signal averaging (Wood and Henkelman 1985, Axel *et al* 1986), and this is a valuable method for T_1-weighted short-repetition-time sequences. Figure 8.36 shows images with and without averaging. However, for T_2-weighted sequences, imaging times (particularly at high fields) become prohibitively long if four acquisitions are to be obtained. Under these circumstances, respiratory-gated acquisitions are more appropriate. Normally with a gated sequence, the RF and gradient pulses will continue to be delivered, irrespective of whether the data are to be accepted or rejected. This ensures constant repetition times, which are important if relaxation-time measurements are to be performed. Respiratory triggering can also be carried out, but T_R will then be subject to variation, and will be related to the breathing cycle (Lewis *et al* 1986). If fast imaging sequences are used, it is possible to carry out one or perhaps several acquisitions during a breath-holding period (see figure 8.35). If resolution can be sacrificed to some degree, the imaging time can be further reduced by reducing the number of phase-encoding acquisitions to 128 or in some circumstances to 64 steps (see §8.6.7.2). This allows the number of acquisitions in a given time period to be increased.

The chemical-shift artefact occurs in the read-out direction. As fat and water both contain protons that are in different chemical environments and thus resonate at slightly different frequencies, when frequency is

---

Figure 8.35 Images showing the effect of flow suppresion techniques on trans-axial abdominal images from a volunteer. (*a*) FLASH 90, $T_R = 0.08$ s, $T_E = 12$ ms, one acquisition, total time 21 s, with breathing; (*b*) the same sequence but with breath-holding during the acquisition, showing the reduction in breathing motion; artefacts due to blood flow in major vessels are present in both images; (*c*) breath-holding with a flow suppression sequence FL90DSS40 where two non-selective 40° pulses precede the 90° pulse, with a 5 ms delay between the second 40° pulse and the 90° pulse, $T_R = 0.08$ s and 512 points are acquired in the read-out direction with 256 phase-encoding steps, a single acquisition taking a total time of 21 s; (*d*) the same sequence with breath-holding but with $T_R = 0.043$ s, four acquisitions, 128 phase-encoding steps and 256 read-out points, giving a total time of 22 s; (*e*) a FL90D4020 sequence having the initial 40° suppression pulse followed by a 20° suppression pulse, with a 256×256 acquisition but otherwise as in (*c*), giving increased signal but good flow suppression; (*f*) a FL90DSS40 sequence as in (*c*) but with a 256×256 acquisition matrix and a 30 ms delay giving increased signal due to T_1 relaxation during the delay period, which also has the effect of increasing the amplitude of the flow artefact.

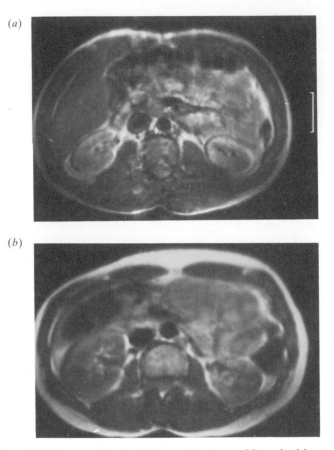

Figure 8.36 Images obtained from volunteers with and without averaging to show the effect of averaging on motion artefacts: (*a*) $T_R = 2.0$ s, $T_E = 30$ ms, one acquisition; and (*b*) $T_R = 0.8$ s, $T_E = 30$ ms with four acquisitions, reducing artefacts due to respiratory motion.

used to define position, water and fat protons from the same location will appear in the image slightly displaced from one another. As the frequency separation of these two components increases with magnetic-field strength, the effect is more pronounced on high-field systems, although it can be reduced by using stronger gradient fields, resulting in a greater frequency bandwidth per pixel. The effect is particularly pronounced where there are large water/fat interfaces, for instance around the bladder, or at interfaces between muscle and fat, as shown in figure 8.37. The effect causes a dark rim on one side of the object where a signal void occurs and a high signal on the other side where the water and fat signals overlap. The artefact can be removed by carrying out chemical-shift imaging to separate the water and fat components (§8.6.6.5). Although there is a great deal of lipid in the brain, it does not normally contribute to proton images, presumably as the protons are

(a)

(b)

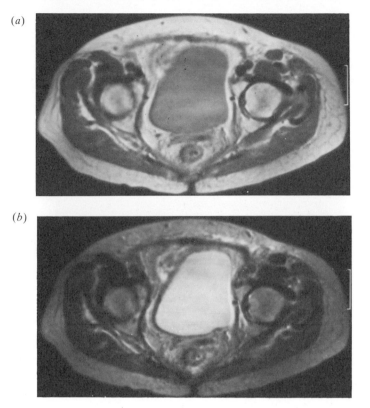

Figure 8.37 Transaxial images through the bladder of a patient with cervical cancer showing the chemical-shift artefact bordering the bladder wall in the frequency-encoding direction obtained using a spin-echo sequence with $T_R = 2.1$ s, a slice width of 10 mm and (a) $T_E = 30$ ms, (b) $T_E = 70$ ms.

more rigidly bound than in fat, and thus the chemical-shift artefact is not a problem in the brain.

A number of other factors can give rise to common image artefacts. Metallic items in the body, even if non-ferromagnetic, can produce some distortions in the local field and may also conduct eddy currents, leading to local distortions and reduction in signal intensity in the image. Distortions in the gradient fields will cause spatial distortions in the image, and these are particularly pronounced close to the gradient coils, and thus at the edges of extended objects. These spatial distortions are a particularly important consideration if NMR data are to be used for radiotherapy treatment planning. Figure 8.38 shows typical gradient distortion in a high-field system. Non-uniformities in the B_1 field will give rise to variations in image intensity throughout the object and, of course, if a surface coil is used as receiver, the image intensity will be non-uniform due to the rapid fall-off in the response of the surface coil. A further effect that may be apparent on images is wrap-round of out-of-plane or out-of-image information. Out-of-plane information may be due to several regions of the body experiencing the same gradient and being excited sufficiently by the B_1 field to produce visible magnetisation in the image. If an image zoom is used to produce a higher-resolution image in the centre of the object, signal from tissue lying outside of the zoomed image may be aliased back into the image due to inappropriate ADC and frequency-filter settings in the read-out direction or due to phase wrap-round in the phase direction. A judicious choice of zoom factor, phase-encoding direction and imaging parameters can usually minimise these problems.

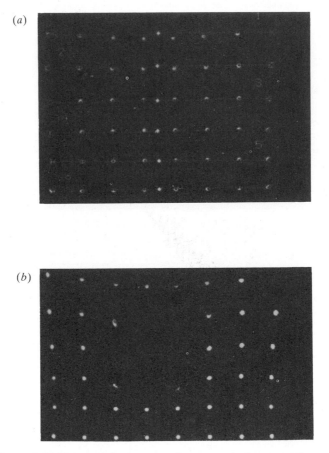

Figure 8.38 Transaxial image of a regularly spaced array of gadolinium-doped water samples showing (*a*) the spatial distortions present in NMR images at the periphery of the image field due to non-linearities in the gradients close to the gradient coils, and (*b*) the distortions resulting from introducing a ferromagnetic object into the imaging field.

8.6.6 Other parametric images

8.6.6.1 Flow imaging. As discussed above, flowing blood is generally displayed with different contrast to static tissue. This contrast will depend on the phase of the cardiac cycle and therefore the speed of blood flow in arteries, and it is possible to produce a magnetic resonance angiograph by using sequences that give different weight to flowing blood but give the same intensity for static tissue. By subtracting the two images obtained either independently or with an interlaced sequence, an image showing only arterial blood can be obtained (Underwood 1987).

Flow can be measured by labelling magnetisation in one slice and then imaging the appearance of this magnetisation in a second slice positioned at some distance from the first slice. A spin-echo sequence can be used where the first slice receives the 90° pulse and the second slice then receives a 180° refocusing pulse, so that only blood that has received both pulses and therefore flowed between the two planes gives a signal. The velocity of the blood may, in principle, be related to the two pulses and the distance between the planes. This method has not, however, produced good quantitative results. A second method is to look at the displacement of blood in the slice by the inflow of unsaturated blood. This can be carried out by first saturating the slice and dephasing the xy signal. If a second 90° pulse is applied, any signal will then have come from blood flowing into the slice, providing that T_1 relaxation is long compared with the time between the pulses.

Phase mapping, where phase-sensitive detection is employed, is one of the more accurate methods of measuring flow. In a spin-echo sequence with phase-encoding gradients on each side of the 180° pulse, static spins will accumulate no net phase change at the centre of the echo. However, moving spins will acquire a phaseshift proportional to the velocity of flow. Depending on the technique adopted, the images may show flow within and through the plane of interest. Several authors (O'Donnell 1985, Le Bihan *et al* 1986) have proposed methods of flow measurement, dependent on phase changes, that may be of value in measuring perfusion.

Where flow in a specific vessel is required, an elegant method is first to excite spins in a plane perpendicular to the vessel and then to refocus those spins in an orthogonal plane containing that vessel. This permits a profile of the flow gradient through the vessel and along the vessel to be obtained. Another method is to utilise the change in effective T_2 relaxation rate resulting from the loss of spin coherence due to the inflow and outflow of magnetisation. This method has been used by Mueller *et al* (1986).

8.6.6.2 Diffusion imaging. From equations (8.50) and (8.51) it will be apparent that the amplitude of a spin echo is in part dependent upon the diffusion coefficient of the spins. By carrying out measurements in the presence of different gradient fields (Stejskal and Tanner 1965), the value of the diffusion coefficient can be calculated, and an image obtained (Bachus *et al* 1987). Imaging measurements of the diffusion coefficient in the yolk of an egg, where perfusion will not make an additional contribution, have also been reported by Taylor and Bushell (1985). *In vivo* it is difficult to separate the contributions from diffusion and perfusion.

8.6.6.3 Susceptibility imaging. Another physical property of the body that can be imaged is magnetic susceptibility. Margosian and Abart (1984) and Cox *et al* (1986) have reported a method of imaging that depends on the phase change resulting from differences in the local magnetic field. Another more direct method of imaging susceptibility effects is to follow a 45° pulse along x' after an appropriate delay by another 45° pulse along x' (Brown 1987). The first pulse will produce a y' component of magnetisation, and the angle through which this rotates in the rotating frame will then depend on the local magnetic field, as influenced by susceptibility. The second 45° pulse will then further rotate the magnetisation, producing a component with a resultant magnitude that depends on the accumulated $x'y'$-plane phase difference. This will therefore give an image where the intensity is modulated by susceptibility effects, T_1 suppression effects and proton density. Figure 8.39 shows images obtained by these two methods. Young *et al* (1987) have demonstrated that susceptibility images show changes at the margins of some brain tumours and in haematomas.

Another important application of susceptibility images is in NMR spectroscopy, where it is necessary to optimise the homogeneity of the local magnetic field, by adjusting the magnet shim-coil settings, prior to measurements on each new sample. Currently this is usually carried out by optimising the FID or linewidth, the success depending to a large part on operator experience and varying considerably from measurement to measurement. It is possible that by imaging and analysing the susceptibility image obtained from a sample, it will prove possible to develop more objectively a highly optimised shimming procedure that depends primarily on a calculation based on the distribution obtained in the susceptibility image. This could considerably improve the reproducibility and quality of results obtained with *in vivo* spectroscopy systems, where these effects are perhaps most pronounced. In spectroscopy, many of the susceptibility effects are due to local interfaces between tissues of different magnetic susceptibility, and it will not be possible to remove all of these effects by adjusting the shim coils. However, it should be possible to determine whether the field homogeneity is as good as can reasonably be achieved.

8.6.6.4 Imaging of other nuclei. Although protons are the easiest nuclei to image and probably the most valuable for clinical purposes, it is also possible to produce images of several other nuclei. A number of measurements of *in vivo* sodium-ion concentration have been reported. The concentration of sodium in the body is very much lower than the concentration of protons, and this leads to a sensitivity that is 3–4 orders of magnitude smaller than the sensitivity of proton imaging. As the

(a)   (b)

(c)   (d)

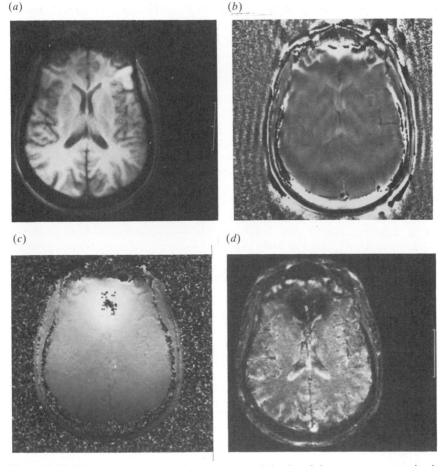

Figure 8.39 Transaxial images through normal brain: (*a*) FLASH 90 magnitude image, (*b*) the corresponding phase image, (*c*) a susceptibility image (Margosian and Abart method) from the same slice, (*d*) an image with proton density modulated by susceptibility obtained with two 45° pulses.

nuclear-spin quantum number of ^{23}Na is $\frac{3}{2}$, it exhibits nuclear quadrupole moments and has very much shorter T_1 and T_2 relaxation times than do protons, within the range 1–60 ms. Imaging sodium is therefore technically far more demanding than imaging protons, and the delays implicit in spin-echo imaging methods result in considerable loss of signal. Nonetheless, useful images have been obtained. Maudsley and Hilal (1984) have reported measurements in the feline brain and canine heart. Perman *et al* (1986) and Turski *et al* (1986) have reported human measurements. The intracellular concentration of sodium is believed to depend strongly on the proper functioning of the cellular ion pump, and sodium appears to concentrate in areas of infarction or in many brain tumours as a result of cellular energy failure, indicating that sodium images may be of considerable value in diagnosis.

^{19}F can be measured with a sensitivity similar to that of measuring protons. However, the concentration of fluorine in the body is very low. If a fluorinated agent can be delivered to the body in sufficient concentration, it can be imaged against a background of zero signal and this is particularly the case, for instance, in fluorinated blood-substitute compounds. This approach may be of particular value in monitoring perfusion or blood flow. The technique has been demonstrated at high field in animals, using ^{19}F-labelled fluorodeoxyglucose (FDG) (Nakada *et al* 1987) and observing intravascular pO$_2$ resulting from the changed T_1 relaxation time of ^{19}F in perfluorocarbon emulsions (Eidelberg *et al* 1987, Fishman *et al* 1987). Images of deuterium distribution have been measured (Ewy *et al* 1987) in the cat brain at 4.7 T. It may in future prove possible to obtain images of other lower-sensitivity nuclei, including ^{31}P, ^{35}Cd, ^{13}C, ^{39}K and ^{17}O, although the most likely applications will be in small-bore systems where higher fields and higher sensitivity can be achieved. Inevitably, such changes will be obtained with a coarse resolution. An increase in pixel dimensions by a factor of 10 will give an increase in sensitivity by a factor of 1000, and this is the most practicable method of compensating for the low sensitivity.

8.6.6.5 Chemical-shift imaging. Chemical-shift imaging is the acquisition of images showing separately the distribution of different metabolites or compounds of the nucleus of interest. This is most commonly performed in fat/water proton imaging, where only two major components are present, and allows chemical-shift artefacts to be corrected, and the differing distribution of water and fat to be mapped. A method proposed by Dixon (1984) uses two measurements having different phases of the 180° RF pulse with respect to the gradients in the spin-echo sequence. An alternative approach used in chemical-shift selective imaging (CHESS) (Rosen *et al* 1984, Haase *et al* 1985) is to modify a normal imaging sequence by selectively saturating one component with a frequency-selective 90° pulse, followed by a spoiler gradient to destroy transverse magnetisation, prior to executing the usual imaging sequence. This has the advantage that a fat or water image can be obtained in only one acquisition, but if the B_0 homogeneity is not high, it can prove difficult to obtain satisfactory suppression over the whole image.

Chemical-shift images have been obtained from other nuclei. For instance, Bendel *et al* (1980) have reported chemical-shift imaging of a phantom containing different ^{31}P metabolites, using projection reconstruction. Such images can also be derived from three- or four-dimensional chemical-shift images (Brown *et al* 1982), described in more detail in §8.7.3.

8.6.6.6 Paramagnetic spin-enhancement agents. Although considerable variation in image contrast can be obtained by varying pulse sequences to exploit the intrinsic proton density, T_1 and T_2 of tissues, the range of contrast available is not always sufficient to demonstrate the pathology of interest. Paramagnetic spin-enhancement agents are being developed to provide a means of increasing tissue contrast by reducing the T_1 relaxation time of tissues (see §8.5.1). Paramagnetic molecules or atoms have unpaired electrons, resulting in a large local magnetic moment. The T_1 relaxation is caused primarily by dipole–dipole interactions, and one atom can affect very many protons, depending upon the tumbling rate of the atom, the closest approach that is possible for the water atoms (this depends on the nature of the compound to which the paramagnetic centre is attached) and the concentration of the paramagnetic species. The different types of relaxation agent include: paramagnetic metal species, such as Gd (often as a chelate), Cr^{3+}, Fe^{3+} and Mn^{2+}; ferromagnetic metal species, such as magnetite (Fe_3O_4), which have a more pronounced effect on T_2 relaxation times; stable free radicals, based on the nitroxide free radical in which the unpaired electrons on the N–O bond are protected from pairing; and molecular oxygen, which has two unpaired electrons, but loses its paramagnetism when bound in oxyhaemoglobin, and is of value when dissolved in perfluorinated blood and plasma substitutes (see §8.6.6.5). Another type of contrast agent is the density substitution agent, where a lipid is used to alter the local water content, for instance in the intestinal tract. Hall and Hogan (1987) have reviewed paramagnetic pharmaceuticals, and a further review was published by Runge *et al* (1983). Where paramagnetic agents can be attached to compounds of sufficient specificity, NMR imaging may combine the functional and localising information available in nuclear medicine together with clear anatomical structure and high resolution. Currently Gd-labelled diethylenetriaminepentaacetic acid (Gd-DTPA) has received most attention and has proven of value in enhancing contrast in cerebral lesions. Figure 8.40 shows images obtained using Gd-DTPA to detect metastatic disease in the brain. Bydder *et al* (1987) have described a number of applications of Gd-DTPA in the central nervous system.

8.6.6.7 NMR microscopy. Recent work has indicated the possibility of carrying out NMR microscopy. The basic principle depends on applying large field gradients, to obtain a much larger frequency change per unit distance, and thus (taking into account any chemical-shift effects) allowing much smaller pixels to be used. Gradients of up to 80 mT cm^{-1} (i.e. 8 T m^{-1}, compared with whole-body imaging systems gradients of typically 6 mT m^{-1}) have been reported (Cho *et al* 1987). This can be achieved for a small object by preparing small gradient coils capable of operating with a large current. Commonly these microscopes have been designed as inserts to whole-body magnets, but, because of the small signal obtained from voxels of side 10–$100 \text{ }\mu\text{m}$, there are advantages to operating at higher fields. Cross-sectional measurements of a rat with a pixel size of $190 \text{ }\mu\text{m}$ and a slice width of 2.5 mm have been reported at 1.5 T (Hedlund *et al* 1986), and images of toad ova at 9.5 T have been obtained with pixels of side $10 \times 13 \text{ }\mu\text{m}$ and slice width $250 \text{ }\mu\text{m}$ (Aguayo *et al* 1986). The size of objects to be measured will depend upon the

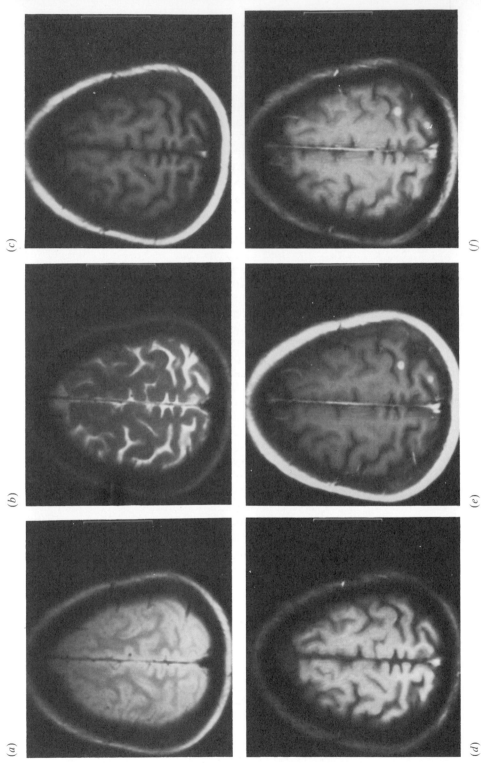

(a)

(b)

(c)

(d)

(e)

(f)

maximum size of gradient coil that can be constructed to give the required spatial resolution, and upon the signal available. Because of these limitations, it will prove difficult to obtain the highest resolution in small regions of human subjects.

8.6.7 Technical considerations in imaging

8.6.7.1 Selective excitation. As discussed above, selective excitation depends upon supplying an excitation pulse of limited bandwidth so that, in the presence of a linear field gradient, only spins over a predetermined slice width are excited. In practice, it is not possible to obtain a slice profile of rectangular cross section. Neglecting T_1-dependent saturation effects, chemical-shift effects and the effects of non-uniform B_1 fields, the slice profile may be obtained by Fourier transforming the time-domain excitation pulse. For instance, if a rectangular RF pulse of length τ is applied, the slice profile will be a sinc function of width $4\pi/\tau$. The bandwidth and therefore the slice width excited is inversely proportional to τ. The slice profile is further affected by the non-linear response of the spin system, which is most pronounced for large RF pulses and has been considered by Hoult (1977, 1979). A slice profile that is a sinc function is not ideal, as it produces considerable excitation in the wings away from the selected slice. Improved slice profiles have been obtained with Gaussian excitation pulses and with a Gaussian excitation pulse modulated by a sinc function (Hutchison *et al* 1978), or with a sinc function excitation pulse, which produces a slice profile that is approximately rectangular.

In practice, the slice profile is further distorted by saturation of the central region of the profile relative to the wings of the profile, leading to a broader slice profile as T_R is reduced or as the local T_1 increases. This effect also has to be taken into account in fast imaging, where it will give rise to different slice profiles for different excitation flip angles. As discussed above, the width of the selected slice will depend upon the width of the exciting RF pulse, as well as upon the function used. However, the accuracy of the slice profile will also depend upon the number of terms in the Fourier series expressing the RF pulse shape, which will depend on the frequency range over which the RF envelope is described, on the duration of each digitisation step and on the dynamic

---

Figure 8.40 Transaxial images through the brain of a patient with secondary deposits from small cell lung cancer, showing the contrast achieved in small metastases with and without Gd-DTPA using different imaging sequences: (*a*) and (*b*) spin-echo sequences with $T_R = 2.7$ s, $T_E = 30$ and 90 ms respectively; (*c*) $T_R = 0.5$ s, $T_E = 17$ ms; (*d*) FLASH 90, all pre-gadolinium; (*e*) $T_R = 0.5$ s, $T_E = 17$ ms and (*f*) FLASH 90, both with gadolinium, showing a small lesion in the lower-right-hand quadrant of the images, indicating that the lesion is better visualised in a shorter time with the aid of gadolinium.

range of the digital-to-analogue converter used. The gradient used should also be determined by the frequency separation of water and fat at the operating frequency.

8.6.7.2 Sampling considerations. In sampling or digitising the FID or echo, the number of pixels in the image is determined by the number of sampling points obtained. The size of each pixel, and hence the resolution in the frequency or read-out direction, depends upon the gradient and the bandwidth per pixel, which is inversely proportional to the duration of each ADC sampling point. The length of each sampling interval is dictated by the sampling theorem, which states that the highest frequency present should be sampled a minimum of twice per cycle (see Chapter 12). If the time signal is undersampled, higher frequencies not completely sampled may alias by being folded back into the low-frequency part of the spectrum. Obviously the frequency range defined depends on the sample extent and the strength of the gradient fields. When resolution is improved by using pixels of smaller dimensions, by increasing the gradient strength, allowing a part of the image to be zoomed, some of the sample often extends outside of the region to be sampled. It is necessary, therefore, both for this purpose and to reduce noise propagation, to analogue-filter the RF signal prior to digitisation. The frequency response of this filter should be adjusted to match the digitisation step used.

The resolution obtained in the frequency direction can be increased by sampling an increased number of points, as there is no time penalty in this operation. The digitised signal can also be conveniently Fourier-transformed during the measurement, thus reducing the final total reconstruction time. In the phase direction, resolution is dictated by the number of phase-encoding steps used, and by halving the number of steps, the total acquisition time can be halved. This results in some loss of resolution, but a part of this can be recovered by zero-packing the raw data set with an equal number of zeroes prior to reconstruction (Bartholdi and Ernst 1973). The technique can also be used to obtain further resolution in the frequency direction. A further time reduction can be achieved by 'half-Fourier' imaging (Margosian 1985, 1987), where the symmetry of the phase data is taken into account by only acquiring the positive phase steps, together with a few negative steps to allow for accurate phase correction. This can also reduce the acquisition time by a factor of approximately 2, but at the expense of signal-to-noise ratio, which is similarly reduced.

8.6.7.3 Signal-to-noise considerations in imaging. A complete description of the signal-to-noise ratio of NMR measurements is complex, depending on factors such as coil design, loading and signal amplification, the RF detection method used (quadrature or single phase), operating field strength and nucleus to be observed, imaging method and pulse sequence used, T_1 and T_2 relaxation times, T_R and T_E, image resolution and slice thickness required. Many of these areas have been considered elsewhere in this chapter and are considered in more depth by a number of authors. Hoult and Richards (1976), Hoult and Lauterbur (1979) and Edelstein *et al* (1986) have considered the system signal-to-noise ratio, with emphasis on the crucial first stage, the detection probe. Using quadrature detection (as is now usual) gives a gain of $\sqrt{2}$ over single-phase detection, and the use of crossed coils (Chen *et al* 1983) can provide a further $\sqrt{2}$ increase in signal-to-noise ratio. In general, signal-to-noise ratio improves with increasing field strength in an approximately linear relationship, but this effect is partially offset by increased T_1, often requiring a longer T_R, and changing contrast between different tissues. The imaging method and pulse sequence have major effects on the image signal-to-noise ratio, which are dependent on tissue relaxation times. These effects have been considered by Brunner and Ernst (1979), Mansfield and Morris (1982) and Morris (1986). Changes in contrast are also important, as ultimately it is the contrast difference between two tissues in the presence of noise that is of interest. This has been considered by Young (1984) for several pulse sequences.

The signal-to-noise ratio (SNR) in 2DFT imaging is related to the image acquisition pixel size. In ECT or x-ray CT, SNR $\propto \sqrt{A}$, where A is the pixel area. For pixels of relative SNR = 5 and side of unit length, if the length of the pixel side is reduced to 0.5, the area is reduced to 0.25 and SNR = 2.5. By averaging four small pixels, the SNR increases to 5, and so no information is lost if the data are acquired with high resolution; it can always be summed to provide large pixels with the SNR that would be obtained if acquiring with that resolution. In NMR, SNR is proportional to the volume, and thus to the area. Data acquired with the same initial SNR for pixels of side 1 unit would have a SNR = 1.25 if acquired with pixels of side 0.5 unit. If the small pixels were then summed to produce pixels of side 1, the SNR would be 2.5, giving a loss of signal compared with acquiring the data with that resolution. This analysis (Edelstein *et al* 1986) assumes that the total imaging time and the total signal-sampling time are held constant. This means that an increase in resolution by a factor of 2 requires twice as many time-domain sampling points, reducing the dwell time per point by a half, and doubling the bandwidth. Thus, it is important to ensure that NMR images are acquired with an adequate SNR.

8.6.7.4 Spin rephasing. A consequence of selective excitation is that spins at different points in the excited slice will accumulate a different phase increment that depends upon their position and upon the local field gradient. This produces an incoherent *xy* magnetisation at the end of the excitation. The method usually adopted to overcome this problem is to reverse the slice-select gradient at the end of the application of the selected pulse. In order to reduce the time period for which the gradient is applied, it is usual practice to increase the amplitude of the reversed gradient and apply it for a correspondingly shorter period. This approach was used by Sutherland and Hutchison (1978).

8.6.7.5 Gradient and stimulated echoes. Most NMR imaging methods make use of spin refocusing via a 180° RF pulse, to produce a spin echo (SE). Edelstein *et al* (1980) proposed a sequence where the spins were refocused by allowing dephasing in a gradient, and then reversing it for the same period, to produce refocusing. However, as the spin direction is not reversed in the presence of B_0 field inhomogeneities, the dephasing due to this will not be reversed, and spin-echo intensity will fall off as T_2^*, rather than as T_2. By increasing the strength of the reversed gradient, the rephasing time can be reduced, producing a more rapid echo, which has proved of value in fast imaging sequences such as FLASH (Haase *et al* 1986).

Stimulated echoes were originally reported by Hahn (1950) and have recently been proposed for use in imaging experiments (Frahm *et al* 1985). The basic pulse sequence is

$$90°—t_1—90°—t_2—90°—t_3(\text{STE})$$

where STE is the stimulated echo, and this is shown in figure 8.41. The first 90° (x') pulse rotates longitudinal magnetisation into the xy plane. This then dephases, depending on local magnetic fields. The second 90° (x') pulse then rotates the (phase-encoded) y' component onto the z axis. If the spins were equally distributed in the xy plane, this would amount to half of the magnetism. The remaining magnetisation (the x' component) refocuses at t_1 after the second 90° pulse to produce a spin echo (SE). The third 90° pulse then rotates the 'stored' z component into the xy plane and this produces the stimulated echo (STE) at a time $t_3 = t_1$ after the third pulse. During 'storage', the phase-encoded z magnetisation decays with the spin–lattice T_1 relaxation time, offering the possibility of measuring T_1 without the long T_R otherwise required. A further two secondary echoes are produced as spin echoes, excited by the third pulse, which mirrors the primary spin echo and the FID after the first pulse. Dependent upon the pulse flip angle, and number of pulses, additional echoes can be produced. These sequences are likely to produce a variety of new techniques for improved contrast, and will also have spectroscopic applications.

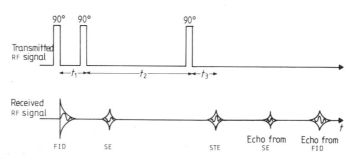

Figure 8.41 The production of a stimulated echo by a sequence of three 90° pulses showing the other signals also produced during the sequence.

8.7 SPATIALLY LOCALISED SPECTROSCOPY

Until recently, NMR spectroscopy was applied only to studies of small samples or to small animals. In the former, there was little need to localise the source of the signal as the sample was usually homogeneous; in the latter, adequate localisation could usually be obtained either by using a surface coil to examine muscle, or, if particular organs or tumours were to be examined, by growing tumours superficially so that they could be examined without contamination from other tissues, or by implanting the coil via a surgical procedure. Thus, although there were some applications where localisation techniques would have been advantageous prior to the advent of large-bore high-homogeneity magnets, and some localisation techniques pre-dated these instruments, the major drive for spatial localisation has occurred since large-bore magnets suitable for spectroscopy have become available and interest has grown in the non-invasive studies of metabolism *in vivo*. Indeed, much of the research with the early whole-body systems has been concerned with the study of energy metabolism, using ^{31}P NMR, and many of these studies have been performed on muscle, where again the use of a surface coil often provides sufficient localisation.

As this book is concerned primarily with imaging, this section will provide only the necessary background material to explain NMR spectroscopy, and will concentrate on localisation methods, a number of which are similar to methods used in NMR imaging measurements.

8.7.1 Chemical-shift spectroscopy

The relationship between resonant frequency and applied magnetic field for a particular isotope was discussed in §8.3. An isolated atom is subject only to the externally imposed magnetic field. In a molecule, however, atoms are shielded by the electron cloud of the molecule, which produces a small additional magnetic field that is characteristic of the atom's position in a given molecule. This results in a chemical shift (δ) with respect to a fixed reference frequency (v_{ref}), given by

$$\delta \text{ (ppm)} = \frac{v_{ref} - v_{sample}}{v_{ref}} \times 10^6 \qquad (8.60)$$

where v_{sample} is the Larmor precession frequency of the atom of interest in a particular molecule and δ is expressed in terms of parts per million (ppm), which produces a value for chemical shift that is independent of frequency. Figure 8.42 shows a ^{31}P spectrum obtained from human muscle in a 1.5 T Siemens Magnetom, showing inorganic phosphate (Pi), phosphocreatine (PCr) and nucleotide triphosphates (principally adenosine triphosphate (ATP)), together with the chemical forms of the compounds. The separation of characteristic peaks, expressed in terms of frequency, increases with the strength of the applied magnetic field. The dispersion of chemical shifts for different isotopes varies greatly, and table 8.1 shows typical chemical-shift ranges for a number of isotopes.

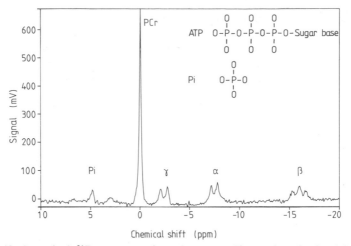

Figure 8.42 A typical ^{31}P spectrum from human calf muscle, obtained in 4 min with a 5 cm diameter coil. The data have not been smoothed. The phosphocreatine (PCr) peak, the inorganic phosphate (Pi) peak and the γ, α and β peaks due primarily to adenosine triphosphate (ATP) can be clearly seen. The chemical formulae for ATP and Pi are also shown. The γ peak may also include a contribution from β adenosine diphosphate (ADP) nuclei, and the α peak may include contributions from α ADP nuclei together with NAD and NADP. These three peaks are sometimes referred to as nucleotide triphosphates (NTP). In this spectrum, the spin–spin splitting of the γ and α ATP peaks into doublets and the β ATP peak into a triplet can be seen.

The chemical shift for a particular molecule is due to the sum of shielding effects from local diamagnetism, diamagnetism and paramagnetism of neighbouring atoms, and intra-atomic currents. In many molecules, the diamagnetic effects dominate. These are due to small electron currents induced perpendicular to B_0, which in turn generate a small induction B_{opp} that opposes B_0. The other effects may generate inductions that either oppose or reinforce B_0. The total induction experienced by the nucleus, B_{eff}, will then be the difference between B_0 and the sum of these shielding inductions $B_{shielding}$, i.e.

$$B_{eff} = B_0 \pm B_{shielding}. \tag{8.61}$$

The chemical-shift characteristics of atoms in many molecules for a number of nuclei have been tabulated and in many cases calculated. Typical chemical-shift ranges are included in table 8.1. Several factors will give rise to a shift in peak position from its characteristic position, including local pH, the solvent used, in some circumstances the temperature, and the presence of paramagnetic or ferromagnetic agents. Another important feature of some NMR spectra is spin–spin splitting, which causes a single peak to be split into a multiplet as a result of the interaction between the magnetic moment of that nucleus with another neighbouring spin also having a magnetic moment (see figure 8.42). The spin–spin coupling constant J describes the magnitude of the splitting. Spin–spin or J coupling does not require an external magnetic field and is therefore independent of the applied field. The coupling constant is defined in hertz (Hz). If rapid exchange of some of the atoms concerned occurs, this can lead to expected splitting not being observed or to the multiplets being broadened. The splitting can be removed by the technique of double irradiation, where a second strong RF field is used to saturate the resonance of the nuclei that are causing the splitting, which reduces the multiplet to a singlet. This is sometimes known as spin decoupling, and may be either homonuclear for like nuclei or heteronuclear for unlike nuclei.

As discussed in §8.3.4, the linewidth of a singlet resonance is inversely related to the T_2 relaxation time,

$$\Delta v_0 = 1/\pi T_2^* \tag{8.62}$$

where Δv_0 is the full width at half-maximum of the spectral line and T_2^* is the effective T_2 relaxation time including the natural T_2 and the effect of magnetic-field inhomogeneities. Usually the effects of magnetic-field inhomogeneities are minimised, and in small prepared

Table 8.1 Properties of NMR nuclei of medical interest.

Nucleus	Spin	Gyromagnetic ratio, $\gamma/2\pi$ (MHz T^{-1})	Natural abundance (%)
^{1}H	$\frac{1}{2}$	42.57	100
^{2}H	1	6.54	0.015
^{13}C	$\frac{1}{2}$	10.71	1.108
^{14}N	1	3.08	99.63
^{15}N	$\frac{1}{2}$	−4.31	0.37
^{17}O	$\frac{5}{2}$	−5.77	0.037
^{19}F	$\frac{1}{2}$	40.05	100
^{23}Na	$\frac{3}{2}$	11.26	100
^{31}P	$\frac{1}{2}$	17.23	100

Nucleus	Typical concentrations *in vivo* (mmol l^{-1})	Relative sensitivity per nucleus at constant field, corrected for natural abundance	Approximate chemical-shift range (ppm)
^{1}H	10^5	1	10
^{2}H	10^5	1.4×10^{-6}	–
^{13}C	10	1.8×10^{-4}	250
^{14}N	10	1.0×10^{-3}	–
^{15}N	10	3.8×10^{-6}	1000
^{17}O	50	1.2×10^{-5}	650
^{19}F	–	8.3×10^{-1}	500–1000
^{23}Na	80	9.3×10^{-2}	–
^{31}P	10	6.6×10^{-2}	700

samples the expression can be reduced to

$$\Delta v_0 = 1/\pi T_2 \qquad (8.63)$$

where the T_2 of the molecule dominates the expression for the linewidth. In large samples, for instance in *in vivo* studies, the effects of local susceptibility often dominate and ensure that the local field inhomogeneity is still an important component.

In general, in NMR spectroscopy, it is desirable to operate at as high a field as possible, to increase the frequency separation of different spectral lines and to maximise the signal-to-noise ratio available. In order fully to realise the benefits of the higher field, it is necessary to retain or improve on the field homogeneity at higher fields. Although this is practicable in small samples, *in vivo* the local susceptibility effects may prevent a significant gain in resolution at high fields.

As a result of the effect of the magnetic-field inhomogeneity on the linewidth, it is necessary to make the magnetic field as uniform as possible. The B_0 field homogeneity is usually adjusted by varying currents in a number of shim coils, which superimpose small high-order magnetic fields on B_0, proportional to z, x, y, z^2, xz^2, etc. These adjustments will normally be carried out for each measurement, dependent on the location of the region of interest and the effects of the sample on the field distribution. They are in addition to the normal optimisation of the field at installation via static iron shims and possibly a superconducting shim set. Account should be taken of the symmetry of the sample, which will also affect the field distribution and the ability to compensate easily for any distortions produced. In whole-body magnets, the subject is, of course, large, and will have a pronounced effect on the magnetic-field distribution.

Sample susceptibility is an important effect and, in certain cases, where tissues of very different composition lie close together, this may give rise to large changes in local susceptibility that cannot be compensated for and will therefore dominate the linewidth in the spectrum obtained. These effects may be particularly pronounced in experimental tumours in animals, where the tumour causes a large deviation from cylindrical symmetry, or in tumours that have large necrotic regions. As discussed in §8.6.6.3, the use of susceptibility images may help to optimise shimming in these cases. Currently, shimming is usually carried out by optimising the length of the free induction decay of protons in the sample, either by observing the free induction decay on an oscilloscope or display and optimising the shape and maximising the length of the T_2^* decay, or by observing the Fourier-transformed spectra and minimising the linewidth. Some systems provide a figure of merit as an aid to shimming. Automatic shimming methods, together with the ability to store shim settings in the computer and to load shim settings appropriate to the region being studied, may render shimming more objective and reliable in the future.

In order to excite an appropriate range of spectral lines at differing frequencies, it is necessary to use a pulse with sufficient bandwidth to cover the chemical-shift range. For an RF pulse of length τ, the frequency bandwidth is $\Delta v(\text{Hz}) = 1/\tau$. In receiving the NMR signal, it is important again to ensure that the equipment is appropriately set up. As with imaging, a hardware filter is applied to limit the bandwidth of the received signal and therefore to prevent noise aliasing back into the spectrum. These filters should cover an appropriate bandwidth, and the sampling rate of the ADC should be adjusted to sample the complete bandwidth. If the dwell time (d) of the ADC is equal to the time between sampling points, the maximum frequency sampled with quadrature detection is

$$v_{\max}(\text{Hz}) = 1/d \qquad (8.64)$$

and the number of spectral points will again be limited to the number of sampling points. Thus, if one were to sample for a total of 500 ms with 1024 samples, $d = 500/1024$ ms, and therefore $v_{\max} = 1/d$, which equals 2 kHz. This should be the maximum frequency present in the FID or else data will be undersampled and aliasing will occur. The frequency resolution is another important factor, which is equal to the acquired bandwidth divided by the number of sample points. This resolution should be adequate to discriminate between spectral lines of interest. The FID is often smoothed or multiplied by a decaying exponential to reduce noise prior to the Fourier transformation.

8.7.2 *Human in vivo applications of* NMR *spectroscopy*

Although NMR spectroscopy measurements in human subjects have only become practicable over the last few years, the field is developing extremely rapidly, with a variety of methods being transferred from the more complex techniques practised in high-resolution spectroscopy. Measurements in living tissue have been possible for over 10 years, and a wide range of experimental studies of ^{31}P, ^{1}H, ^{13}C and ^{19}F are now in progress in animals. Many of these techniques will in time be transferred to human studies. It is not the place of this section to discuss the *in vivo* applications of NMR in detail, and the reader is referred to Gadian (1982) and Stephens (1987).

The majority of human studies to date have been performed using ^{31}P NMR spectroscopy, with particular interest in investigating defects in energy metabolism. ^{31}P NMR provides a valuable means of monitoring the cellular energy balance due to the use of high-energy phosphate bonds to store energy. A typical ^{31}P spectrum obtained at 1.5 T from human brain is shown in figure 8.43. Detailed peak assignments are given in the figure caption. The main features of these spectra are the three peaks due to: nucleotide triphosphates, predominantly adenosine triphosphate (ATP), which is the major source of energy in the body; phosphocreatine (PCr), which provides a readily accessible short-term supply of energy, useful for meeting short-term high demands for energy, for instance, in muscle contraction, but not present in liver; and inorganic phosphate (Pi), which is the product of the breakdown of ATP to produce adenosine diphosphate (ADP) and energy. By measuring the chemical shift of the Pi peak with respect to a reference frequency (commonly PCr is used although the resonant frequency of water can also be used as a reference), the local pH can be determined. For instance, in the absence of an adequate oxygen supply, the Krebs cycle cannot provide ATP as a source of energy and it must be derived less efficiently via anaerobic glycolysis. This produces lactic acid as an end-product, which gives rise to more acidic conditions and therefore to a fall in pH. This is reflected by a shift of the Pi peak towards the PCr peak. If adequate resolution is available, it may be possible to see a separation of the Pi peak into components from intracellular and extracellular compartments. *In vivo* NMR spectra may represent a large population of differing cell types under varying degrees of anoxia, and the Pi peak may be broadened due to this distribution.

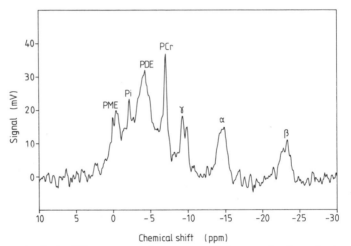

Figure 8.43 A ^{31}P spectrum from normal human brain, obtained using the FROGS technique to suppress signal from superficial tissues. In addition to the peaks described earlier, peaks due to PME, including phosphorycholine, glucose-6-phosphate, glycerol-3-phosphate and AMP and to PDE including GPE and GPC are evident. These are underlain by a broad, short T_2 component due to phospholipids. This broad component is asymmetrical, and this effect, as well as the broadening of other spectral lines in the brain, may be the result of relaxation due to chemical-shift anisotropy.

The PME peak is representative of phosphomonoesters, such as phosphocholine, and also of sugar phosphates, such as glucose-6-phosphate and glycerol-3-phosphate, and adenosine monophosphate (AMP). Some of the constituents of the PME peak acts as precursors to lipid production. The phosphodiester (PDE) peak is due to compounds such as glycerolphosphorylethanolamine (GPE) and glycerolphosphorylcholine (GPC), and some of these substances appear to provide evidence of lipid breakdown. By quantitating the areas under these different peaks and taking account of factors such as T_1-suppression effects, and the bandwidth of the irradiating pulse and the analogue filter that may affect the relative heights of different peaks, it is possible to observe relative changes in the concentration of these metabolites during the course of an investigation. In well controlled laboratory systems, the absolute concentrations of different compounds may also be assessed. As will be apparent from ^{31}P spectra, different organs present character-istic spectra indicative of their differing metabolism. The liver's PME and PDE peaks are related to its role in producing and storing glucose. In the brain, the high PME and PDE peaks may be related to the higher lipid content of the brain.

Phosphorus spectroscopy studies *in vivo* have been used to investigate the effect of exercise on metabolism (Taylor *et al* 1983) and metabolic myopathies such as McArdle's syndrome (Ross *et al* 1981) and enzyme deficiencies (Radda 1984). The method has also proved valuable in the study of mitochondrial myopathies (Radda *et al* 1982) and in studies of muscular weakness and dystrophies (Arnold *et al* 1984). An important application has been in the study of birth asphyxia in premature babies (Reynolds *et al* 1987, Cady *et al* 1983).

There is also considerable interest in the use of NMR spectroscopy for monitoring tumour metabolism as a method of measuring response to therapy, and possibly as an aid to diagnosis or to predict the appropriate therapy. Figure 8.44 shows ^{31}P spectra from a number of different human tumours measured *in vivo* together with fluorine spectra showing the catabolism of the chemotherapeutic drug 5FU in the human liver, and figure 8.45 shows changes in spectra from a breast tumour during the course of therapy. These spectra were obtained from a whole-body magnet operating at 1.5 T. Compared with normal tissue spectra, the PME and PDE peaks are more pronounced, as often is the Pi peak. In general, however, the tumours do not appear to be markedly acidic. Spectra from a number of tumours have also recently been reported by Oberhaensli *et al* (1986).

Proton spectroscopy in human studies is at a very early stage, but promises, in the longer term, to prove valuable. It is technically more complex due to the smaller chemical-shift range and, therefore, higher resolution requirements for proton spectroscopy and due to the domi-nating effect of the water resonance and to a lesser extent of the lipid resonance, which require suppression sequences such as the 1331 sequ-ence (Hore 1983) or other spectral editing techniques. Another valuable nucleus is ^{13}C, which, although having low natural abundance and low NMR sensitivity, can still be seen in a number of molecules at natural abundances. Owing to the cost of isotopically enriched carbon, it is unlikely to prove practicable to use enriched ^{13}C-labelled compounds in human studies. However, the large chemical-shift range and the ability to detect a number of hydrocarbon compounds of interest will continue to support study of this nucleus. Decoupling techniques are also possible in both carbon and hydrogen spectroscopy, but these techniques require substantial RF power, and care is required to ensure that such sequences do not exceed RF power deposition guidelines (see §8.9.1). ^{19}F has proved of interest, particularly in the absence of a significant natural

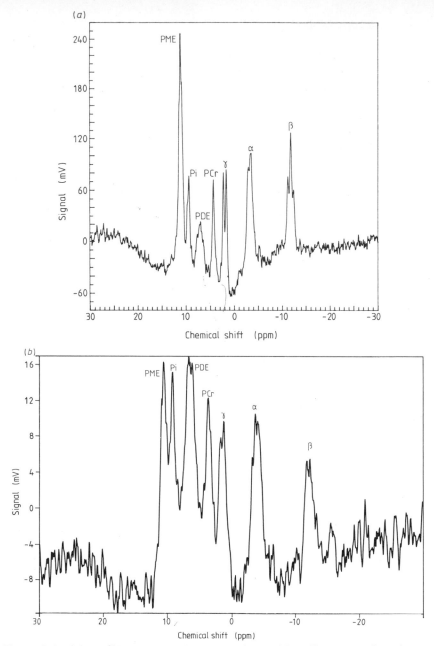

Figure 8.44 (*a*) A ^{31}P spectrum from a patient with a liver secondary from a carcinoid, using the FROGS localisation technique (§8.7.3). (*b*) A ^{31}P spectrum from a patient with an advanced primary breast tumour. In these spectra large PME and PDE peaks can be seen which are typical of tumour spectra. These are believed to be associated with precursors to cell building (PME) and degradation products resulting from cell breakdown (PDE). The large inorganic phosphate peak may result from anaerobic glycolysis or from cellular breakdown. The separation between the PCr and the Pi peak can give a measure of local pH. The rolling baseline in these two spectra is due to a 1800 μs delay having been

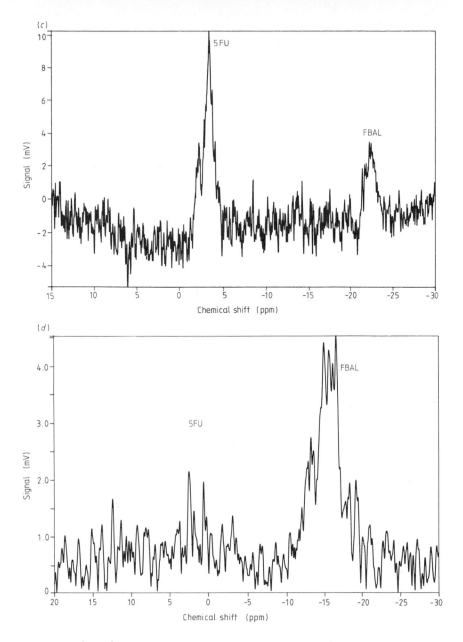

Figure 8.44 (*cont.*)

used between the end of the read pulse and the start of acquisition of the FID. The PCr signal may include a small component due to neighbourhood muscle. (*c*) and (*d*) show fluorine spectra measured in the liver of a patient and obtained as part of ~60 min series of spectra following the appearance of metabolites after the administration of the chemotherapy agent 5-fluorouracil (5FU) by (respectively) intravenous and intraperitoneal delivery. The 5FU peak can be seen and the product of the catabolic pathway α-fluoro-β-alanine (FBAL), this peak also containing a contribution from the intermediate product α-fluoroureidopropionic acid (FUPA).

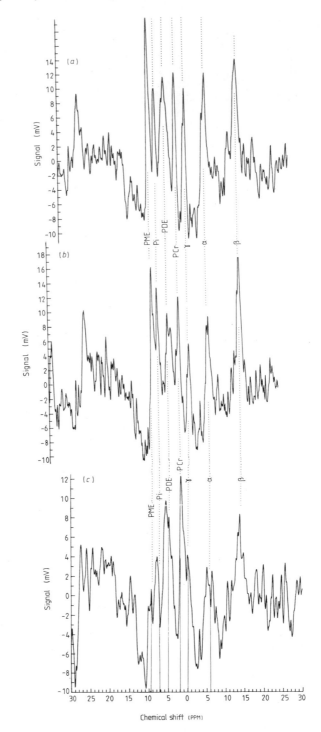

Chemical shift (PPM)

fluorine signal from the body. Several chemotherapeutic drugs contain fluorine, and studies of the metabolism of 5-fluorouracil (5FU) have been reported (Wolf *et al* 1987). A more recent comparison of intravenous and intraperitoneal delivery of 5FU is shown in figure 8.44. It has the advantage that the isotope ^{19}F is 100% naturally abundant (i.e. the only form of fluorine existing naturally) and also has a high NMR sensitivity.

8.7.3 *Spatially localised NMR spectroscopy techniques*

The importance of localisation techniques in spectroscopy has been discussed briefly above. The development of these techniques is an area of current research and development. Localisation is required to ensure that a signal comes from an organ or part of an organ or tissue where the metabolism is of interest either because of its function or because of a particular stimulus that is being applied. This is particularly the case in studies of abnormal pathology, where one wishes to obtain a measurement characteristic of that pathology rather than of the normal tissue that may surround it. The problem is particularly compounded if one is measuring a tissue other than muscle, such as tumour, where the signal may be considerably lower than that produced from muscle. A small amount of surrounding muscle can cause a high degree of contamination in a tumour spectrum.

Ideally, one would wish to identify a region of any shape on a proton image and then obtain a spectrum that came only from that defined volume and was subject to no distortions as a consequence of the localisation method used. In practice, particularly for phosphorus spectroscopy, none of the techniques reported to date meet this objective. In measurements of tumours, the interpretation of results is further complicated by the heterogeneity of the tissue, as most tumours contain a range of cells and cell types, often in different stages of development and with variable oxygen supply. An important advantage of a localisation technique is the ability to provide simultaneously signals from more than one localised region. This is likely to be important both where a reference spectrum is required and where changes within a tissue or organ are of interest.

Localisation techniques can conveniently be divided into those that rely on variations in the B_1 field and those that rely on varying the B_0 field. In addition, there are some techniques that combine both of these approaches. Aue (1986) has published a recent review of localisation techniques. The simplest technique, and for some superficial tissues, the best technique, is simply to use a surface coil (Ackermann *et al* 1980).

Figure 8.45 A series of three ^{31}P spectra measured in a primary breast tumour in a patient undergoing MMM chemotherapy (mitomycin, mitozantrine and methotrexate). (*a*) was obtained prior to the start of therapy, (*b*) and (*c*) were obtained during the course of the therapy, during which the tumour responded to treatment. This is consistent with the fall in the PME peak and the increase in the PDE peak. There is also an overall loss of signal during the period. The increase in PCr is due to an increasing component being detected from chest-wall muscle as the tumour shrank.

A surface coil in its simplest form is one or more loops of wire wound to obtain signal only from a localised region of the sample, in which the localisation is obtained by the rapid fall-off of the B_1 field from the coil. More complex resonator structures have now also been developed, and the topic of surface coils will be discussed further in the next section.

In proton imaging, surface coils are generally used to receive only and must therefore be decoupled from the transmitter coil, often by crossed diodes, to avoid large currents being induced in the surface coil and, in turn, inducing local RF heating in the subject and possibly also damaging the receiver electronics. For spectroscopy, surface coils may also be used only to receive, either because a large coil is to be used to transmit a homogeneous B_1 excitation, or because the receive surface coil is surrounded by a larger transmit surface coil, which, because of its larger size, gives a more homogeneous B_1 excitation in the region of interest. However, a higher degree of spatial localisation is achieved if the surface coil is used as transmitter and receiver, and this is the common mode for spectroscopy. Where appropriate, the coil can be designed to operate simultaneously at the resonant frequency of several nuclei, and figure 8.46 shows a proton image obtained from a double-tuned coil (Leach *et al* 1986) with the surface coil acting as receiver only, and an image obtained using the phosphorus channel with the coil acting as transmitter and receiver.

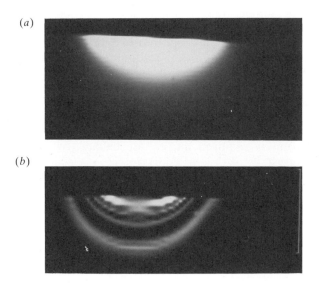

(*a*)

(*b*)

Figure 8.46 A proton image was obtained using a double-tuned coil in receive mode only at 1.5 T, the body coil acting as transmitter, and (*b*) a proton image obtained at 0.5 T using the phosphorus channel of the same coil operating in transmit and receive mode. The non-homogeneous B_1 profile can be seen, with the dark bands corresponding to flip angles that are integer multiples of π.

The phosphorus image was obtained using a phantom containing triethyl phosphate doped with gadolinium and a modified fast imaging technique that uses a non-selective pulse and thus provides a projection image (Leach *et al* 1987a). Although this is a simple method of obtaining localisation, figure 8.46(*b*) shows the highly non-uniform nature of the B_1 field, leading to a complex bowl-shaped region being excited with a variable flip angle. This means that, if the repetition time of the sequence used is not longer than five times the longest T_1 of the metabolites being measured, the contribution of some metabolites to the spectrum will be partially suppressed, varying their peak height with respect to the peak heights of samples having shorter T_1 relaxation times. The degree of this suppression will vary spatially, depending on the local flip angle delivered by the B_1 field, and the overall spectrum obtained will be an integration over the degree of suppression and the local metabolite concentration at each point. As metabolites do exhibit a considerable range of T_1 values, this is obviously an important factor in interpreting spectra obtained from surface-coil examinations. The sensitivity of the surface coil in receiving signal also falls off rapidly with distance from the coil, and the spectrum obtained with the surface coil is therefore strongly weighted by the spatial variation in coil sensitivity, and is a product of this with the degree of excitation at different depths.

In general, simple surface coils are sensitive over a hemisphere defined by the radius of the coil, although this response can be biased by varying the transmitter voltage delivered to the coil. Figure 8.47 shows the variation in integrated sensitivity with depth for a 4 cm surface coil for a range of transmitter voltages.

A number of localisation techniques that are particularly suitable for use with surface coils rely on varying the B_1 field. One approach has been to limit the range of pulse angles from which signal is acquired to a narrow range of angles, for example, $90° \pm 10°$. This uses depth and refocusing pulse sequences (Bendall and Gordon 1983). This approach overcomes some of the problems of non-uniform excitation with surface coils, but the region excited is still a complex curved surface. As the method relies on refocusing the signal, a T_2 dependence is also introduced.

Rotating-frame spectroscopy (Cox and Styles 1980) has been used in a number of clinical applications, particularly at Oxford. A set of acquisitions are obtained from free induction decays produced by pulses with differing values of the applied pulse length (or pulse amplitude), which varies the local flip angle experienced at a given position in the object. A Fourier transformation can be carried out on these data to produce a set of spectra each from a region of constant B_1 field. This again gives curved regions of complex shape, but, by using a larger transmitter coil and a small receiver coil (Styles *et al* 1985), the regions approximate to planes parallel to the plane of the coil. If the variation in flip angle required is obtained by varying pulse length, this can also have an effect on the bandwidth of the signals excited, and may lead to distortions at the extremes of the frequency spectra obtained. The method does,

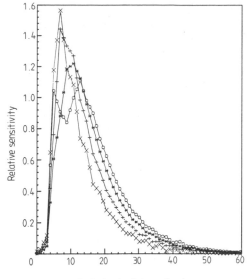

Figure 8.47 The variation of integral sensitivity with depth of a 4 cm diameter surface coil for a number of transmitter voltages, measured using a doped triethyl phosphate phantom. ×, 10 V; +, 15 V; *, 20 V; ○, 25 V.

however, allow a signal to be obtained from a number of planes simultaneously. Hoult (1980) has proposed a method of rotating-frame selective excitation (RFSE) where the transmitter coil is designed to provide a B_1 gradient, which is the rotating-frame analogue of the linear gradient in the B_0 field that is usual in the laboratory frame. The B_1 field is phase-modulated to provide an effective B_2 field precessing about B_1 in the rotating frame. Spatial selectivity using this method has recently been shown in phantoms (Hedges and Hoult 1987). The technique avoids the need to refocus the spins, which is necessary following B_0 selective excitation.

A large class of techniques modify the B_0 field by the addition of gradient fields. The method of topical magnetic resonance (Gordon *et al* 1980) has already been described (§8.6.4.1). A set of static non-linear magnetic-field gradients profile the B_0 field, to limit the B_0 field to a predetermined volume. This provides a spectrum that is the sum of a well resolved spectrum from the highly homogeneous region and a very broad inhomogeneously broadened contribution from the remainder of the sample. This broad component can be subtracted by convolution differencing. Although the technique has been used successfully, it suffers from the limitations that the sensitive region can be moved only over a limited spatial displacement, that a limited range of volumes can be measured, and that the edges of the region are not well defined.

A number of localisation methods make use of selective-excitation or selective-suppression techniques. In these methods, selection of the region to be excited or suppressed is achieved by applying a selective-excitation pulse in the presence of a gradient field in the same way as would be carried out for selected excitation in imaging. In depth-resolved surface-coil spectroscopy (DRESS) (Bottomley *et al* 1984), selective excitation is carried out using a surface coil in the presence of a gradient oriented along the axis of the surface coil. Thus a slice would be excited parallel to the surface coil, the shape of the slice depending on the power transmitted to the surface coil and the size of the surface coil. In this case, although the plane is relatively well defined, the lateral extent of the region is less well defined, and there will be some variation of flip angle and sensitivity within the region. As acquisition of the spectrum is carried out shortly after a selective excitation, the method can also be affected by eddy currents (see §8.8.2). To reduce the effect of eddy currents, the signal is acquired after a wait period and will consequently be T_2-weighted. The method has been extended in slice-interleaved (SLIT) DRESS to provide multislice sampling within the repetition time of the sequence (Bottomley *et al* 1985). In common with other methods of selective excitation, the method also suffers from a chemical-shift artefact in the position of different metabolites, which depends on the gradient strength employed and on the variation in resonance frequency of the different metabolites being observed.

Fast-rotating-gradient spectroscopy (FROGS) has been proposed as a method of suppressing unwanted signal, particularly from superficial regions close to the surface coil (Sauter *et al* 1987). In this technique, a number of selective RF pulses are applied in the presence of an x or y gradient, to saturate the magnetisation in the selected slice. Because the B_1 field is non-uniform, it is necessary to cycle a number of pulses to achieve the required degree of saturation. The sequence used has applied a number of different θ, 2θ pulses, each pulse being followed by a spoiler gradient. Following this saturation and after an adjustable wait time to minimise eddy-current effects, a normal non-selective sampling pulse is applied followed by the immediate collection of the signal as in a normal spectroscopy experiment. The technique has been optimised for use with small surface coils and applied clinically (Sharp *et al* 1987a), and has been further modified so that it can be used at any angle irrespective of surface-coil orientation. Initially, the technique was of most use for large tumours with overlying muscle, where the signal from muscle could be suppressed. Figures 8.48 and 8.49 give examples of the use of the technique for a diabetic liver and for a leiomyosarcoma in the abdomen. The large degree of suppression of signal from the superficial layers can be seen. In order to improve the usefulness of the technique, it has been further modified to allow several slices at different orientations to be suppressed during the same sequence, as is shown in the phosphorus images in figure 8.50 (Sharp *et al* 1987b). The technique is less sensitive to eddy currents than is the case for DRESS, as the pulse obtaining the signal is non-selective. The degree of suppression obtained depends on the delay before the readout pulse, as this provides a period of time for T_1 relaxation to occur. Again, the saturated regions will be affected by chemical-shift effects. It may prove possible with an

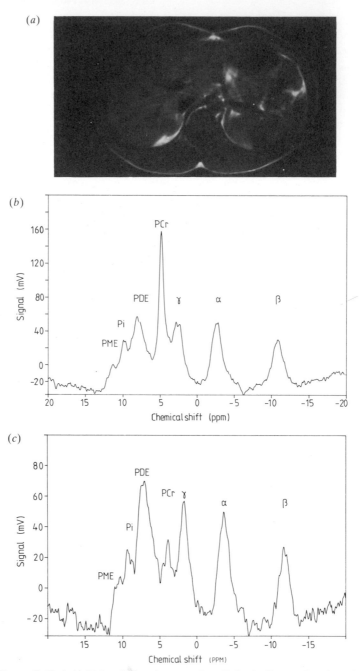

Figure 8.48 (*a*) A localising image of a diabetic liver, showing the location of an 8 cm diameter surface coil (marked by doped proton markers), (*b*) a ^{31}P spectrum from the liver with no FROGS suppression and (*c*) with FROGS suppression showing the reduction in amplitude of the PCr peak, which is not present in liver.

appropriate RF envelope to saturate two parallel regions with the same pulse, and by applying this technique at two orientations to define a rod-shaped region of rectangular cross section. Other suppression techniques have been proposed by Haase (1986) and by Doddrell *et al* (1986).

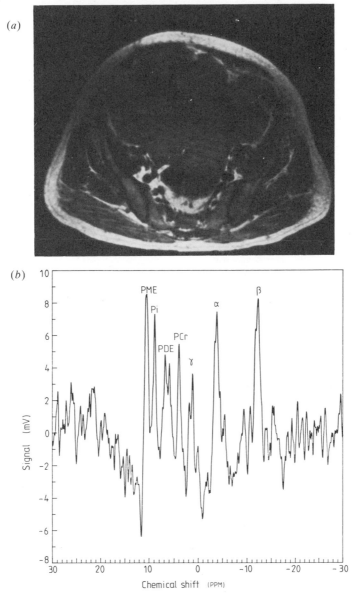

Figure 8.49 (*a*) A proton image of a patient showing a large leiomyosarcoma and the position of the surface coil; (*b*) the ^{31}P spectrum obtained using the FROGS technique to suppress muscle signal from the superficial region.

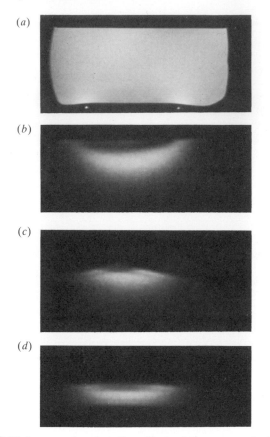

Figure 8.50 Images showing the effect of the FROGS technique. (*a*) A proton image of a phantom containing doped triethyl phosphate showing the coil position, and a set of phosphorus projection images of the coil response (with the geometry inverted) (*b*) showing the coil response with no suppression; (*c*) with two angled slices simultaneously suppressed and (*d*) with two parallel slices simultaneously suppressed to provide increased localisation.

Volume-selective excitation (VSE) (Aue *et al* 1984) uses a combination of selective 45° pulses and non-selective 90° pulses applied along each Cartesian coordinate to excite a volume of interest. The method has been shown to work on small-bore systems, but the technical requirements are such that it may be difficult to implement successfully on clinical systems. A second method in which magnetism is prepared by a series of selective 180° pulses prior to a measurement and read-out pulse is image-selected *in vivo* spectroscopy (ISIS) (Ordridge *et al* 1986). A set of spectra are obtained having received none, one or more orthogonal selective inversion pulses prior to the sampling pulse. When the spectra obtained are summed in an appropriate manner, the signal from points in the sample other than at the intersection of the three orthogonally selected slices is cancelled out, with only the contribution from the central voxel adding. There is no loss of signal-to-noise ratio in this method as the signal from the central volume contributes to each of the spectra obtained. By using a selective pulse shape such as a hyperbolic-secant pulse, where an inversion is always obtained provided that the power deposited is above a threshold value (Silver *et al* 1984, 1985), the method can be used with a coil giving a non-homogeneous excitation. For both of these methods, the selective excitation can give rise to chemical-shift artefacts. If gradients are increased to reduce the chemical-shift artefact, shorter RF pulses are needed to give increased bandwidth, and thus larger-amplitude RF pulses are required to excite the same volume. This can cause difficulties in supplying sufficient RF power. With minor modifications, ISIS can be used to select a number of volumes simultaneously. As magnetisation is prepared prior to the sample-and-read pulse, there is no T_2 dependence and the sequence is relatively resistant to eddy-current effects.

A number of other single-voxel localisation methods have been proposed, including SPARS (Luyten *et al* 1986), PRESS (Bottomley 1984) and STEVE (Frahm *et al* 1986). The localisation of the SPARS technique suffers from poor off-axis performance due to the number of different spin populations produced. The results are T_2-weighted. PRESS achieves good localisation, but again the results are T_2-weighted. STEVE makes use of stimulated echoes, providing data that arc both T_1- and T_2-weighted. Another approach to spatially selective excitation has been reported by Bottomley *et al* (1987). In this case, a two-dimensional spatially selective NMR pulse (δ pulse) is applied in the presence of two orthogonal time-dependent magnetic-field gradients. This two-dimensional selective excitation has been used in conjunction with DRESS to control the size of the sensitive disc defined by the DRESS technique. It also has obvious applications with other localisation techniques. A similar approach has been reported where a two-dimensional ISIS selective experiment has been combined with rotating-frame spectroscopy to achieve a similar result (Segebarth *et al* 1987).

A different approach to the use of gradients in localised spectroscopy is the application of Fourier imaging techniques (discussed in §8.6.4.4) in spectroscopy. Brown *et al* (1982) and Maudsley *et al* (1983) have reported an extension of 3D Fourier-transform imaging to provide localised spectroscopy. In these techniques, a non-selective excitation is applied followed by phase encoding in the required number of orthogonal directions, with an appropriate resolution. Chemical shift then occupies a further dimension and, to obtain a three-dimensional set of spectra, a four-dimensional Fourier transform must be performed on the final data set. The time requirement for this approach, as discussed earlier, is demanding, and thus it can only be used with a low resolution in each spatial dimension. With ^{31}P spectroscopy, sensitivity considerations would in any case preclude voxels smaller than a few cubic centimetres. Pykett and Rosen (1983) have also reported a similar method, and Haselgrove *et al* (1983) have reported a one-dimensional phase-encoding technique.

Recently, a four-dimensional Fourier-transform technique has been applied to clinical studies in a whole-body magnet. In this case, a three-dimensional array of resolution $8 \times 8 \times 16$ has been reported (Bryant *et al* 1987). These approaches have the advantages of no chemical-shift artefacts, as phase encoding is used, and the acquisition of a large number of voxels, which gives the spatial distribution of changes in the spectra. The small pixels also display high spectral resolution due to the reduced contribution of field inhomogeneities within the small voxel. This is the most promising approach to spatial localisation currently available. However, due to the phase-encoding period, the spectra are T_2-weighted and distortions due to eddy currents can present problems. Owing to the small number of Fourier terms, truncation artefacts occur but can be reduced by filtering with a cosine function in the three spatial directions. This also, however, reduces the spatial resolution of the technique, and each pixel may contain an approximately 30% contribution from outside of the defined volume. An important consideration with all of these techniques is the effect that motion may have on the localised spectra. The four-dimensional Fourier-transform technique described above was employed in conjunction with three-dimensional field-mapping techniques (Young *et al* 1987), which is important if adjacent voxels are to be summed to improve signal-to-noise ratio.

Many of these techniques of localised spectroscopy provide a region of interest that is based on a rectilinear coordinate system, and therefore define a rectangular volume of interest. This often does not conform well with the irregular boundaries of tumours. A method of correcting for spectral contamination, particularly from muscle, phosphorus-image muscle-subtracted spectroscopy (PIMSS) (Leach *et al* 1987b), is currently under development. The method depends on identifying regions of contamination on proton images, obtaining a reference spectrum from the contaminating tissue, and then, by using maps of surface-coil response, calculating the proportion of sensitivity due to the contaminating tissue and subtracting that proportion of signal from each spectral peak in the spectrum from the region of interest. The method can be used without a localisation method if a separate reference spectrum is obtained, and in this case it may be useful, for instance, in subtracting small contributions to breast spectra from muscle in the chest wall. If the approach is used with a method such as ISIS, which provides signal from two regions, a separate spectroscopy measurement of the contaminating tissue is not required.

Localisation techniques are important to improve quantitation in *in vivo* spectroscopy, as it is essential to ensure that the spectrum comes from the region of interest. A variety of other factors, however, affect the ability to quantitate the spectra obtained. These include: the absence of any internal standard for phosphorus spectroscopy; differences in coil loading; differences between coil response for different coils and transmitter powers; the variable sensitivity with depth of many spectroscopy techniques; a variable degree of suppression and in some cases also variable T_2 dependence; the possibility of patient movement during the study; and the difficulty of separating adjacent spectral peaks and also compensating for changes in linewidth due to differences in shimming. In addition, there are a number of instrumental factors that may affect reproducibility and quantitative measurements. It is possible to some extent to monitor changes in Q by measuring the RF power required to obtain an inversion in a calibration sample at a fixed point in the coil. If a uniform excitation can be carried out, the problems due to varying pulse angle through the sample can be minimised. A map of the coil response that is representative of the product of the B_1 field and sensitivity in tissue may allow some correction to be made for variations in B_1 field and coil sensitivity. Spectral fitting and analysis routines are also required to allow changes in B_0 homogeneity, which may of course vary with position in the sample, to be taken into account. A susceptibility map may be of help here. Quantitation in *in vivo* NMR spectroscopy is currently at a very early stage, and this is an area in which developments are likely to occur.

8.8 INSTRUMENTATION

The development of hardware for NMR imaging and spectroscopy is a very rapidly moving field and reference should be made to the recent literature, conference reports or patents. This section will give an overview of the major components of the NMR system. Figure 8.51 shows a block diagram of an idealised NMR imaging and spectroscopy system. It can be conveniently divided into the magnet, the gradient system and coils, the RF system including the transmitter and receiver coils, the shim coils, the controlling processor with appropriate peripherals, and the displays and software appropriate for both imaging and spectroscopy.

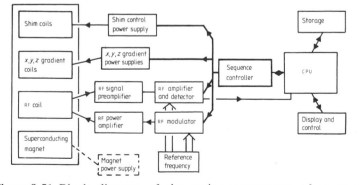

Figure 8.51 Block diagram of the major components of an NMR imaging and spectroscopy system.

8.8.1 The magnet

The magnet is the major component in an NMR system. Most systems utilise either a resistive magnet or a superconducting magnet. Other types of magnet such as permanent magnets or electromagnets have been designed and are available for some commercial equipment. At low flux densities, 0.3 T or below, resistive magnets are used. These require a continuous power supply and generate a significant amount of heat, often being air- or water-cooled. An important consideration is the time taken for thermal stability to occur. Higher field strengths are normally obtained by using cryogenic superconducting systems. For clinical imaging, a large bore, preferably of 1 m diameter, is required. For imaging, the flux density may range from 0.05 to 2 T, the homogeneity requirement typically being some 20 ppm over the central 50 cm. If spectroscopy is to be carried out on the instrument, a homogeneity in the central 10 cm of 0.1 ppm is required, this being a much more exacting requirement. For spectroscopy, a flux density of 1.5 T or greater is required. For imaging, there are advantages for many applications in using lower fields, particularly as motion artefacts appear to be less pronounced and the repetition times to achieve the same contrast tend to increase with field strength, increasing the length of the examination period. However, higher signal-to-noise ratio is obtained at higher field strengths, and this is particularly valuable in the brain, where motion is in any case less pronounced. With the advent of a variety of fast imaging techniques, where the signal-to-noise ratio achieved is particularly important, there may be further advantages in higher-field machines. In the superconducting magnets produced to date, the long axis of the patient generally lies in the direction of B_0. This is generally also the case with resistive systems, and this orientation requires a saddle-shaped receiver coil, which has a lower signal-to-noise ratio (Hoult and Richards 1976) than the equivalent solenoidal coil. A number of low-field resistive magnets have been designed where the subject lies between the two centre coils of the magnet and thus perpendicular to the B_0 field direction (Hutchison *et al* 1980, Mansfield *et al* 1978), and this permits solenoidal coils to be used, giving a higher signal-to-noise ratio. This approach is also possible with permanent and iron-cored electromagnets.

Most of the NMR systems in clinical use now use superconducting magnets. These magnets typically have a clear room-temperature bore of 1 m diameter and their windings constructed using superconducting niobium–titanium alloy strands contained within copper wires. In order to maintain the superconducting state, it is necessary to immerse the magnet windings in liquid helium at 4.3 K. Thus, considerable effort must be directed at minimising the heat losses from the helium vessel. The magnet is normally constructed with a number of different evacuated spaces filled with insulation and helium-gas-cooled heat shields. Generally a liquid-nitrogen chamber surrounds the helium vessel. A typical superconducting magnet is shown diagrammatically in figure 8.52. The construction of the magnet is designed to minimise heat loss by reducing the number of conduction, convection or radiation paths. It is possible to obtain systems in which a refrigeration head that uses helium gas as a refrigerant is attached to the magnet. This both reduces the helium boil-off rate and obviates the need for liquid nitrogen. The capital investment is high and, in terms of reduced running costs, can be difficult to justify. However, an important additional consideration is that the considerable amount of time required for periodic replenishment with nitrogen and helium is very much reduced, and this can be a considerable advantage. An important aspect of magnet design is the degree to which eddy currents are produced within the magnet, and careful design of the magnet can reduce the amplitude of these eddy currents. Magnet design is considered further by Hanley (1984) and Morris (1986).

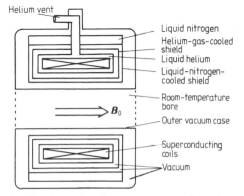

Figure 8.52 A diagram showing a cross section through a whole-body superconducting magnet.

With the advent of materials that superconduct at higher temperatures, and with the theoretical possibility of operation at room temperature, there is obviously likely to be a change in the design of superconducting magnets, and in the long term a considerable reduction in their cost. Currently, most high-temperature superconducting materials are extremely brittle and are also crystalline, with high conductivity only being achieved within the crystal. Significant development is therefore required before a material that has satisfactory superconducting properties and can withstand the considerable forces within a magnet will be available. However, given the widespread application of such materials, research activity in this field will be considerable.

There have recently been important developments in superconducting magnet design. A number of manufacturers are producing magnets that are capable of being ramped in field strengths in a period of 15 min or less. This has advantages for mobile NMR centres where magnets need to be transported on trailers, but is also important for those installations that may wish to perform imaging at several field strengths, depending on the region of interest, or wish to carry out spectroscopy as well as imaging. The shielding of the magnetic field is another important consideration. As will be discussed below, the extent of the magnetic field is a major consideration in siting and designing facilities for NMR. It is most convenient to reduce the field close to the magnet, and a number of approaches to shielding the magnetic field now exist. These include erecting a soft-iron room about the magnet and fitting very large steel yolks to the magnet, which may consist of four very large tie-bars and end-pieces or a larger number of bars positioned around the circumference of the crysotat. An alternative approach, developed by Oxford Magnet Technology, is to use additional superconducting coils to generate a shielding magnetic field, which significantly reduces the floor loading compared with steel self-shielded magnets. Magnets have been shown to be rampable with some types of self-shielding, although sometimes it is necessary slightly to overfield the magnet and then reduce the field strength to the desired field to take account of hysteresis effects. With care and suitable adjustment, it has proven possible to obtain high central homogeneities with the self-shielding structures.

Recently, several manufacturers have produced prototype whole-body high-field magnets operating at approximately 4 T (figure 8.53). The system produced by Siemens (Vetter *et al* 1987) has a room-temperature bore of 1.25 m and the magnetic field can be ramped rapidly. The major current area of interest for these large magnets is in their application in *in vivo* spectroscopy.

Figure 8.53 A 4 T whole-body superconducting magnet produced by Siemens. The magnet has a room-temperature bore of 1.25 m diameter.

8.8.2 Magnetic-field gradients

The magnetic-field gradients are an essential part of an imaging system and are of considerable value in localised spectroscopy. Three sets of gradient coils generate magnetic fields directed along the z axis but varying linearly in magnitude in the x, y and z directions. Thus, they always add to or subtract from the main B_0 field. The way in which these gradients are used depends upon the application. Most imaging applications require only static gradients that can be switched on and off as required, with only the overall amplitude being changed. Some fast imaging methods require the gradients to be switched very rapidly, and here it can be advantageous if the gradient coils form part of a tuned circuit. For other applications, there are requirements that the gradients be time-varying in a predetermined way. An important consideration in gradient systems is the maximum gradient amplitude, which affects the degree of chemical shift experienced in selective excitation and in the read-out direction. This is a particularly important consideration in high-field systems, and gradients of strength $3\text{--}6\,\text{mT}\,\text{m}^{-1}$ are common; recently, gradients of $12\,\text{mT}\,\text{m}^{-1}$ have become available. Another important consideration is the time taken for the gradients to reach their maximum value when switched on, and to return to zero when switched off, which can be a limitation in some fast imaging and spectroscopic applications. Image currents and eddy currents in the magnet can also present problems, the effect depending on the amplitude of the gradient pulses and the proximity of conducting material to the gradients. Whilst the effects are often now minimised by building the room-temperature bore of superconducting magnets of a material such as glass-reinforced plastic, shielded gradient designs have also been proposed (Mansfield and Chapman 1986, Bowtell and Mansfield 1987) to overcome or minimise these problems.

Gradient power supplies must be capable of driving very high currents and providing rapid switching. Gradients are generally derived from digital-to-analogue converters driven by a microprocessor or by the controlling central processor unit. Often a fairly well defined instruction set is incorporated, which will provide the basic requirements of the system, but may not provide all of the flexibility required for experimental work, particularly if one wishes to control the time variation of the gradient waveform. There are considerable advantages in being able to switch between different gradients for the phase-encoding direction, for example. If motion artefacts are present, this can allow the operator to rotate them through 90° on the image, thus avoiding disturbance to pathology of interest.

8.8.3 Radiofrequency system

For a fixed-frequency system, the RF source can conveniently be obtained from quartz-crystal oscillators, but if a variable-frequency system is to be used for multinuclear applications, a frequency synthesiser would be required. The RF system needs to provide non-selective and selective RF pulses of predetermined bandwidth, and generally these are obtained by gating or modulation. This pulse processing is carried out prior to power amplification of the signal. On versatile systems, the modulation function is generally derived from a digital-to-analogue converter. There are considerable advantages in using single-sideband modulation of the RF signal, which reduces the RF power requirements of the system. The power amplifier must amplify the RF signal to provide considerable power, and RF systems capable of delivering up to some 17 kW are now in use. For a system to be used for a number of nuclei, the power amplifier must be capable of performing over the necessary frequency range, and this may be achieved either by using a tuned power amplifier or by using a linear power amplifier. If one wishes to switch rapidly between nuclei, there would be substantial advantages in using a broad-band RF power amplifier. On some instruments, the maximum power available from the RF system will be limited by attenuators to prevent too great a power being delivered to small coils.

8.8.4 Transmit-and-receive coil

The NMR coil is a critical component of the NMR system. Some probes are responsible for both transmitting and receiving RF signal, whereas others are responsible only for transmitting or only for receiving signal. The optimum requirements for these two functions differ, and the use as a transmitter will be dealt with first. For imaging, a transmitter coil giving a homogeneous B_1 field is generally a primary requirement. This is often achieved by using a pair of saddle coils or by an arrangement such as a birdcage coil (Hayes *et al* 1985) at higher frequencies. The coil will normally form part of a tuned circuit, which must be matched to the RF transmitter output impedance of $50\,\Omega$ to prevent power being reflected back to the transmitter. It is important that the coil should not have too high a Q (quality factor), otherwise it will continue to ring for an extended period following termination of the RF pulse. This conflicts with the requirement of a high Q for a receiver coil. It may therefore be necessary to include an active switch that provides for low Q during transmission and high Q during reception. The capacitors should ideally be of high Q and should be capable of withstanding high voltages.

Apart from the requirement for a homogeneous B_1 field, particularly for imaging applications, the receiver system needs to be well coupled to the sample and to have as high a signal-to-noise figure as is possible. The design here, therefore, is far more crucial. A high coil Q is required to give a high signal-to-noise ratio, and a coil that is closely coupled to the sample provides the maximum received signal-to-noise ratio. Another important consideration, which to some extent conflicts with the requirement for high Q, is the bandwidth required of the coil. This is particularly so in echo-planar systems, where a very high bandwidth is required.

Many successful coils are of relative simple design, often requiring only a number of turns of wire. However, at higher fields, the requirements are more demanding, and self-capacitance can present a particular problem. This is most readily overcome by deliberately introducing distributed capacitance into the coil, which is then in series with the stray capacitance and reduces the overall capacitance of the coil. In some applications, it has proven possible to wind particularly closely coupled coils, for instance for head imaging.

Where separate transmit and receive coils are used, it is important that the two coils are decoupled to prevent large signals being induced in the receiving coil, which is tuned to the same resonant frequency, and thus potentially damaging the receiving equipment and also possibly leading to a large local power deposition in the subject. The easiest way to arrange this decoupling is to ensure that the RF fields from the two coils are orthogonal. It is, of course, also important that the RF fields generated by these coils are orthogonal to the B_0 field, and so the axes of these coils generally lie in the xy plane. In imaging, most of the surface coils act as receivers only, although the body and often the head coils may act as transmitters and receivers. For *in vivo* spectroscopy, simple surface coils are often used, or double coils with the larger-diameter concentric outer coil giving a more homogeneous transmit field and the centre smaller-diameter coil being used as the receiver. As discussed above, however, there are considerable advantages in achieving a homogeneous excitation for spectroscopy. Where the same coil is used for both transmitter and receiver, it is necessary to provide protection for the receiver circuitry whilst the transmitter is transmitting to the coil. This is conveniently achieved by using crossed diodes with a quarter-wavelength coaxial line.

8.8.5 Receiver and detection system

This provides the high degree of amplification required followed by a frequency filter, generally a quadrature detection system and a pair of analogue-to-digital converters. In the detector system, the RF signal, generally in the range 20–80 MHz, is converted to an audiofrequency signal, typically 2–10 kHz, by using a phase-sensitive detector to measure the phase variations of the RF signal with respect to a reference signal. This corresponds to the component of magnetisation in the $x'y'$ plane in the direction determined by the phase of the reference signal. In single-phase detection, only one reference frequency is used; in quadrature detection, two reference signals in quadrature (i.e. with a 90° phaseshift) are used, allowing an improvement in signal-to-noise ratio of $\sqrt{2}$ (see §8.6.7.3), and permitting spins having a positive phase difference to be distinguished from those with a negative phase difference. High gain and low noise are important considerations in the amplification system, with the noise level of the preamplifier being particularly crucial. It is also important that the total gain of the system can be varied so that the gain can be matched to the ADC range. Quadrature detection is common now in NMR systems and ensures that the detection frequency is at the centre of the receiver frequency band. In accordance with sampling theory, the maximum frequency that is digitised is equal to $1/(2\tau)$, where τ is the ADC sampling interval. It is therefore important to ensure that no frequencies greater than the maximum frequency digitised are present, and this is achieved by using an appropriate in-line filter, thus preventing the aliasing of noise from outside the bandwidth of interest. This bandwidth must, of course, be adjusted to the analysis problem in hand; for instance, if high field gradients are used, a higher bandwidth is required, and in spectroscopy the bandwidth must be matched to the chemical-shift bandwidth of interest.

8.8.6 Computer and peripherals

The central processor unit (CPU) of a modern NMR system is generally a powerful minicomputer. It provides the channel of communication with the operator and controls the functions of the equipment outlined above. It also provides storage and archival facilities, display facilities and often is interfaced to fast processing equipment such as an array processor (see also Chapter 14). In order to reduce overheads on the central processor, the task of controlling the gradients, the RF system and some of the processing equipment is often given to a microprocessor unit, which controls the different operations of these units in accordance with an appropriate time pattern, as will be apparent from some of the pulse sequences shown in this chapter. This therefore often incorporates a pulse controller or programmer. These pulse sequences are generally down-loaded from the memory of the CPU. In a flexible system, the CPU will provide the capability for programming these pulse sequences and also, for instance, for calculating and down-loading particular RF waveforms. The digitised signal is acquired and normally transferred to memory or disc prior to Fourier transformation. Transformation is normally carried out on an array processor.

The transformed images are then stored on a large disc, and systems for imaging commonly have discs as large as 500 Mbytes with a memory of 2–3 Mbytes. A smaller removable disc is of value in many systems, as it allows random access to images that are to be frequently used. A method of archiving is important, and magnetic tape is the most usual method. However, given the problems of finding storage space for tape in NMR installations, owing to the presence of the magnetic field, optical discs have particular advantages in this environment, and in addition they also allow random access to the image store. The CPU will normally have the function of maintaining a directory of patient studies and will operate the software necessary for manipulating images on the viewing consoles. Modern systems will often support more than one viewing system and support their simultaneous operation as well as the operations of controlling the NMR system and carrying out data processing.

8.8.7 Siting considerations

Siting an NMR system raises a number of problems not normally encountered in the design of a clinical installation. Not only is the magnetic assembly itself often massive, and if self-shielded it may weigh up to 32 tons, a non-shielded superconducting system often weighing of the order of 6–8 tons, the magnetic field produced by the magnet also raises a number of demanding considerations. Much equipment common in hospitals and research institutes is highly sensitive to the presence of magnetic fields. These include most devices in which charged particles are accelerated over a distance. This includes most radioactive counting equipment utilising photomultiplier tubes, much x-ray equipment where image intensifier systems are particularly strongly affected, electron microscopes, linear accelerators for radiotherapy treatment, some CT scanners, and gamma cameras. If these devices are static, the effects are sometimes less important or can be corrected for. However, where components move, for instance in rotating gamma-camera systems, it is difficult to correct for the effects if local mu-metal shielding is not sufficient, and siting of NMR units therefore has to be carefully arranged. Normally, if the magnetic field from the magnet in the area at risk is of the order of the Earth's magnetic field, then the effect due to the magnetic field is likely to be within the adjustment range of the equipment. However, should a ramping magnet be used, this may place more demanding criteria on siting, as a single or very infrequent recalibration of equipment will not necessarily then suffice. Most manufacturers will provide field maps of their magnets in varying configurations, and these can be used in planning the location for a magnet. A site-planning guide has been published by Oxford Magnet Technology (1983). Obviously self-shielded magnets will have very much reduced external fields compared with an unshielded magnet. It should be borne in mind that local iron structures may vary the local effect of the additional magnetic field created by the NMR system.

Whilst the magnet will certainly have an effect on its surrounding environment, the effect of the environment on the magnetic field and the operation of the NMR system also has to be considered. Here the influences of a number of different effects require attention. The principal effects are those due to ferrous materials, and these can be divided into moving and non-moving steel. The effect of static ferrous objects can usually be allowed for provided that they are not too massive or too close to the magnet, by shimming the magnet after installation. This can be achieved as described above with steel shims or the shim coils supplied with the magnet. It should be noted here that the demands of a system to be used for spectroscopy are much higher than those of an imaging system, and this is particularly so when one considers the second class of objects, moving ferrous objects. These need to be very carefully controlled where they may have an effect on the system, as it is not possible to correct for them where changes may occur during the course of the measurement. The magnet needs to be sited far from items like lifts, trains and lorries. Close to the magnet, care must be taken to ensure that small moving steel objects will not have an adverse effect on homogeneity. Other factors that must be taken into consideration include the presence of power cables close to the magnet, particularly those at lower voltages where higher currents may be present, generating a 50 Hz magnetic field. Access must also be provided for cryogen replenishment in a superconducting system, and it is prudent to allow for the possible replacement of the magnet itself.

Safety is an important part of the design of the facility, and this will be dealt with in more detail below. An important factor is to ensure that persons wearing cardiac pacemakers cannot be exposed to a field in excess of guidelines or regulations. This is generally taken to be a flux density of 0.5 mT. In designing a facility, it is also important to ensure that access to the magnet room can be carefully controlled and inadvertent access of unauthorised personnel outside normal working hours can be prevented. It is also advisable to provide some means of screening personnel for the presence of metallic items, and of checking whether tools that, for instance, may need to be used by maintenance staff in the magnet room are non-ferromagnetic. Provision also needs to be made for staff and visitors to deposit items sensitive to a magnetic field, such as credit cards and watches, in a safe place prior to approaching close to the magnet. It is also convenient to check at this point whether they have pacemakers or other delicate equipment such as hearing aids, which may be affected by the magnet. These factors are dealt with in more detail in the section on general safety below.

8.9 NUCLEAR MAGNETIC RESONANCE SAFETY

There are two areas of NMR safety to consider in operating an NMR facility. First, there is the biological effect of the NMR exposure on staff and patients in terms of effects in tissue or organs or the body as a whole, which may give rise to either acute short-term effects or long-term effects. The second category is the prevention of accidents that may result from the potential hazard of the high magnetic field and the presence of cryogens.

8.9.1 *Biological effects and hazards of* NMR

NMR equipment and measurements involve exposure to three different types of magnetic and electromagnetic field. These are the main B_0 or static magnetic field, the alternating magnetic fields produced by the gradients, which have a variable time dependence and may not be present in some spectroscopy equipment, and the radiofrequency field produced from the transmitting RF coils. Both patients and, to a lesser degree, staff and maintenance engineers are exposed to these different fields. At the exposure levels encountered in current NMR measurements, no evidence of any deleterious long-term or short-term effects have been reported, and nor have mechanisms been proposed by which such damage could occur. However, at much higher values of RF and alternating magnetic fields, there are possible mechanisms that may lead to damage. Several authors have considered the physiological effects or long-term hazards that might arise from NMR exposure (Budinger 1979, Saunders and Smith 1984), and guidelines have been produced by the NRPB (1983) in the UK and by the FDA (1982) in the USA.

Clinical NMR systems have static magnetic fields with flux densities that operate in the range 20 mT to 2 T. Most clinical systems operate between 0.3 and 1.5 T. A large number of studies have been carried out directed at providing evidence of physiological or mutagenic effects from exposure to static magnetic fields; the evidence obtained is contradictory and no conclusive corroborated evidence of effects is available (Budinger 1981).

One effect that is evident in magnets is the potential difference generated by charges flowing in blood in the presence of a magnetic field. This can be observed on an ECG as an additional flow-potential peak. Although this induced potential can be observed, it appears to have no effect on the heart rate. Saunders (1982) and Saunders and Orr (1983) have calculated the peak flow potential generated on the aorta wall, assuming a peak blood velocity of $0.63 \, \mathrm{m \, s^{-1}}$ and an aortic diameter of $0.025 \, \mathrm{m}$, to be $16 \, \mathrm{mV \, T^{-1}}$. Thus, $2.5 \, \mathrm{T}$ would create flow potentials of about $40 \, \mathrm{mV}$. This is the depolarisation threshold for individual cardiac muscle fibres. This potential would, of course, be spread across very many individual cells, and thus for each cell it would be well below the cell depolarisation threshold. The NRPB have therefore recommended that the static magnetic field should not exceed $2.5 \, \mathrm{T}$ to the whole or a substantial portion of the body. They have also recommended that staff operating the equipment should not be exposed for prolonged periods to more than $0.02 \, \mathrm{T}$ to the whole body or $0.2 \, \mathrm{T}$ to the arms or hands. For short periods totalling less than 15 min, these limits may be increased to $0.2 \, \mathrm{T}$ for the whole body and $2 \, \mathrm{T}$ to the arms and hands. A number of such short exposures in any day is permissible provided reasonable intervals occur between examinations.

Time-varying magnetic fields, caused by the switched gradients on imaging systems, will induce currents in conductive pathways in the body. Saunders and Orr (1983) have estimated the tissue current density at which a number of effects in man occur. These have generally been derived by supplying a current flowing between electrodes placed on the skin surface. Phosphenes, which are the sensations of flashes of light, can be induced with a tissue current density of the order of $1-10 \, \mathrm{A \, m^{-2}}$ at 60 Hz. This range of current densities also includes the thresholds for the voluntary release or let-go of a grip contact, thoracic tetanisation (which inhibits breathing) and ventricular fibrillation, with cardiac arrest and inhibition of respiratory centres being thought to occur at greater currents than those producing ventricular fibrillation. The thresholds for these effects are very dependent on both frequency and pulse lengths. Saunders and Orr (1983) have also calculated that the current density per tesla per second produced in inductive loops in peripheral head and trunk tissue will be of the order of $10 \, \mathrm{mA \, m^{-2}}$, which is very much less than the threshold values described above.

Based on this evidence, the NRPB recommended that, for periods of magnetic flux density change exceeding 10 ms, exposures should be restricted to less than $20 \, \mathrm{T \, s^{-1}}$ (RMS) for all persons. Because the effects tend to fall as frequency increases, they have recommended that, for periods of change less than 10 ms, the expression $(\mathrm{d}B/\mathrm{d}t)^2 t$ should be less than 4, where $\mathrm{d}B/\mathrm{d}t$ is the RMS value of the rate of change of magnetic flux density in any part of the body in tesla per second, and t is the duration of the change of magnetic field in seconds. For sinusoidally varying magnetic fields or other continuously varying periodic fields, the duration of the change can be considered to be the half-period of the waveform. For conventionally oriented systems, the z gradient is the most important gradient to consider.

It should be noted that in the considerable number of patients and volunteers scanned to date, no *acute* ill effects have been noted resulting from exposure to the alternating gradient fields. No studies monitoring long-term effects from NMR, where epidemiological studies of large populations are required, have been reported, and such studies are technically difficult to design. However, a number of studies investigating effects resulting from exposure to low-frequency electromagnetic fields have been carried out. These exposures include an electric component as well as the magnetic component, and operate at frequencies of 50–60 Hz, which approach the alternating gradient switching frequencies met in NMR studies. In studies of residentially exposed populations, no conclusive link between leukaemia and exposure to low-frequency electromagnetic fields has been established (McDowall 1986). In electrical workers, an increased incidence of leukaemia has been observed in a number of studies (McDowall 1983, Howe and Lindsay 1983), although this may be due to factors other than the low-frequency electromagnetic field.

Power deposition from RF radiation is well known to produce local heating if sufficient power is absorbed in tissue. As is apparent from studies of RF hyperthermia, most tissues in the body have a high capacity to dissipate deposited heat. However, certain tissues, such as the eye and the testes, have a low blood flow, and so cannot readily dissipate heat. The lens of the eye and the testes are therefore particularly susceptible tissues. Exposure limits have been drawn up on the basis that any significant rise in the temperature of the sensitive tissues of the body should be avoided. Acceptable exposures should not result in a rise of body temperature of more than 1 °C as shown by skin and rectal temperature or more than 1 °C in any mass of tissue not exceeding 1 g in the body. This may be ensured by limiting the mean specific absorption rate in the whole body to 0.4 W kg^{-1} and by limiting the specific rate in any mass of tissue not exceeding 1 g to 4 W kg^{-1}. This may be compared with an additional heat load to the body equivalent to the basal metabolic rate that would result in a temperature rise of about 1 °C.

The RF power limit is perhaps the most important limit in the practical operation of an NMR facility. If a receiver coil is not properly decoupled from the transmitter coil, a large local RF field-focusing effect can occur, giving much increased local power deposition. It is also important to note that RF power deposition in loops of wire inadvertently placed on the body, for instance due to ECG leads, can also give rise to local heating in the cables. Whilst the manufacturers take particular precautions to ensure that the pulse sequences and equipment they supply will not deliver too great an RF power to tissues, by means of fuses in transmitter coils, limitations in the power applied by the hardware, and calculations of body weight related to the power deposition for particular sequences, it is possible, especially in spectroscopy, to design sequences that exceed the guidelines. This is particularly so where a small surface coil is used as a transmitter. The field produced is non-homogeneous and falls off rapidly, leading to a relatively high dose close to the coil. With new pulse sequences and new techniques, it is necessary to calculate the power that will be deposited in the subject to ensure that it does not exceed these guidelines. The potential risks of NMR exposure are placed in context with those of other imaging modalities in Chapter 15.

8.9.2 Safety considerations in NMR facilities

As discussed above, cardiac pacemakers may be adversely affected by strong magnetic fields as well as by alternating magnetic fields. Most modern pacemakers usually run in a demand mode, only providing a stimulus when the heart is not functioning correctly. In order to prevent failure in the presence of RF interference, the units are usually designed to operate in automatic pacing mode in the presence of electromagnetic fields of sufficiently high strength. Often they also have a provision to be switched to this mode by switching an internal reed switch with a small hand-held magnet. Thus, normally, anyone with a cardiac pacemaker approaching the magnet would only suffer the effect of their pacemaker switching to automatic pacing mode and producing a higher-than-normal pulse rate. However, if the electromagnetic interference should mimic the profile of a naturally detectable cardiac signal, this could falsely inhibit the pacemaker or cause false synchronisation should the pacemaker have failed to switch to automatic mode. There is also some possibility that induced voltages in the pacemaker leads could cause direct stimulation of the heart, although this is not currently considered to be a risk with present imaging sequences. Thus, it is necessary to indicate that the hazard exists by placing appropriate signs around an NMR facility. Patients and visitors must be screened for the presence of cardiac pacemakers, and unscreened persons must be restricted from having access to areas where the field strength exceeds 0.5 mT.

A major hazard in operating an NMR facility is the risk of ferromagnetic materials inadvertently being brought close to the magnet and then acting as projectiles in the presence of the magnetic field. This is a problem with both large ferromagnetic objects, which may partially crush anyone caught between them and the magnet, and smaller items such as scissors or small tools, which will travel with considerable velocity close to the magnet and can again inflict serious damage on staff close to the magnet or to patients within the magnet. Particular care therefore has to be taken to exclude all ferromagnetic objects from the vicinity of the magnet and to devise methods of ensuring that such objects cannot be introduced. This is particularly a problem as staff become more familiar with the system and may therefore be less alert to the potential hazards. These considerations also affect equipment required for maintenance of the system, replenishment of cryogens and resuscitation. Perhaps one of the major risks is to staff, for instance cleaning staff, entering the building outside of normal working hours with conventional cleaning equipment, and not having been familiarised with the risks. Thus, it is necessary to take precautions to prevent staff, service personnel and patients entering the magnet room during normal operation with any ferromagnetic items, and also to prevent any unauthorised persons entering the magnet room when the facility is unattended. This may perhaps be best done by ensuring that the magnet facility is locked when the building is unattended, with a key that is not available to normal domestic personnel. During normal working hours, anyone entering the magnet room should be screened with, for instance, a metal-detector archway, with provision also to check tools that are supposedly non-ferromagnetic for ferromagnetic content. This can best be done with a strong magnet. The problems are most severe with a large high-field superconducting magnet, but precautions must also be taken with lower-field resistive magnets.

With a superconducting magnet, there is also the risk that a major incident would necessitate an emergency shutdown of the magnet, requiring the magnetic field to be quenched, with a release of the energy stored in the magnetic field as heat by boiling off much of the liquid helium. This can also obviously involve considerable expense and may involve some risk of damaging the superconducting coils of the magnet. During an emergency quench, the field is typically reduced to 50% in 10 s and 99% in 30 s. There has been some concern that the rate of collapse during a quench could cause field induced currents *in vivo* sufficient to cause cardiac arrest. However, this problem was evaluated in a recent deliberate emergency quench of a magnet system (Bore *et al* 1985); no adverse effects due to the quench were observed.

It is particularly important that all staff working close to the magnet and also all patients are screened for the presence of ferromagnetic plates in the skull, post-operative clips or other implanted metal. New *et al* (1983) have studied a variety of metal surgical implants and found that, of these, a number suffered sufficient forces and torques to produce a risk of haemorrhage or injury by displacement. Evidently great care is necessary in screening patients for the presence of such objects, and unless steps have been taken to ensure that the patient has been operated on with a protocol that permits NMR scanning, in other words that non-ferromagnetic clips were used as part of that protocol, very great care must be exercised in deciding whether it is safe to examine that patient in an NMR system. Care also has to be taken to exclude patients with shrapnel injuries or fragments of steel imbedded after industrial accidents, as they may also be at risk. Local heating can also occur in some non-ferromagnetic implants, and patients should be warned to be aware of this possibility so that the examination can be stopped immediately if local heating occurs. Some non-ferromagnetic implants will cause local image distortions. The best method of ascertaining any risks is a detailed discussion with each patient, with the use of a form indicating the possible operations or experiences that may give rise to hazards, followed by an x-ray examination if there is any doubt.

The use of cryogens also presents particular hazards in superconducting facilities. If handled improperly, liquid helium and nitrogen can cause severe burning and soft tissues are rendered brittle if they approach these low temperatures. Thus, great care is required during filling the cryostat, particularly where there is any risk of the filling lines fracturing. Most facilities and magnets are designed to vent exhaust cryogen gases from the room, usually by means of a large-diameter vent pipe that is connected not only to the main vents from the magnet but also to rupture discs, so that, in the event of a quench or any other large release of gases, these will not enter the magnet room but will be released to the atmosphere. Alternatively (Bore and Timms 1984), the magnet room can be designed so that the low-density helium released during a quench would not occupy the lower 2 m of the room. In these circumstances, it is also necessary to provide the room with windows or panels designed to open on overpressure, and this would not be practicable in an RF screened room. If there is a possibility that, due to large spillages or leaks of cryogen, a significant proportion of oxygen in the atmosphere of the room could be displaced, some warning system should be incorporated in the design.

The response to fire alarms in an NMR building also presents particular problems, as the Fire Brigade are usually clad in a considerable quantity of steel. Thus, should they need to enter the magnet room close to a high-field magnet, it is necessary for them to remove all steel, or for the magnet to be quenched. Thus it may be advisable in designing a facility to include provision external to the room for quenching the magnet. However, it is also important to ensure that the Fire Brigade are familiar with the risks that they would encounter on entering the magnet room, and are also well aware of the costs and implications of quenching the magnet. The situation is best handled by having a member of staff familiar with the system on call, but it is also necessary to provide for emergencies where no expert member of staff is available.

REFERENCES

ABRAGAM A 1983 *Principles of Nuclear Magnetism* (Oxford: Oxford University Press)
ACKERMANN J J H, GROVE T H, WANG G G, GADIAN D G and RADDA G K 1980 Mapping of metabolites in whole animals by ^{31}P NMR using surface coils *Nature* **283** 167–70
AGUAYO J B, BLACKBAND S J, SCHOENIGER J, MATTINGLY M A and HINTERMANN M 1986 Nuclear magnetic resonance imaging of a single cell *Nature* **322** 190–1
ANDREW E R 1958 *Nuclear Magnetic Resonance* (Cambridge: Cambridge University Press)
ANDREW E R, BOTTOMLEY P A, HINSHAW W S, HOLLAND G N, MOORE W S and SIMAROJ C 1977 NMR images by the multiple-sensitive-point method: application to larger biological systems *Phys. Med. Biol.* **22** 971–4
ARNOLD D L, BORE P J, RADDA G K, STYLES P and TAYLOR D J 1984

Excessive intracellular acidosis of skeletal muscle on exercise with a patient in a post viral exhaustion/fatigue syndrome *Lancet* i 1367–9

AUE W P 1986 Localization methods for *in vivo* nuclear magnetic resonance spectroscopy *Rev. Magn. Reson. Med.* **1** 21–72

AUE W P, MÜLLER S, CROSS T A and SEELIG J 1984 Volume-selective excitation. A novel approach to topical NMR *J. Magn. Reson.* **56** 350–4

AXEL L, SUMMERS R M, KRESSEL H Y and CHARLES C 1986 Respiratory effects in two-dimensional Fourier-transform MR imaging *Radiology* **160** 795–801

BACHUS R, MUELLER E, KOENIG H, BRAECKLE G, WEBER H and REINHARDT E R 1987 Functional imaging using NMR *Functional Studies Using NMR* ed V R McCready, M O Leach and P J Ell (London: Springer) pp43–60

BARTHOLDI E and ERNST R R 1973 Fourier spectroscopy and the causality principle *J. Magn. Reson.* **11** 9–19

BENDALL M R and GORDON R E 1983 Depth and refocussing pulses designed for multi-purpose NMR with surface coils *J. Magn. Reson.* **53** 365–85

BENDEL P, LAI E M and LAUTERBUR P C 1980 ^{31}P spectroscopic zeugmatography of phosphorus metabolites *J. Magn. Reson.* **38** 343–56

BLEANEY B I and BLEANEY B 1976 *Electricity and Magnetism* 3rd edn (Oxford: Oxford University Press)

BLOCH F, HANSEN W W and PACKARD M E 1946 Nuclear induction *Phys. Rev.* **69** 127

BLOEMBERGER N 1957 Proton relaxation times in paramagnetic solutions *J. Chem. Phys.* **27** 572–3

BORE P, GALLOWAY G, STYLES P, RADDA G, FLYNN G and PITTS P 1985 Are quenches dangerous? *Proc. 4th Annu. Meet. Society of Magnetic Resonance in Medicine, London* (Berkeley, CA: SMRM) pp914–15

BORE P J and TIMMS W E 1984 The installation of high-field NMR equipment in a hospital environment *Magn. Reson. Med.* **1** 387–95

BOTTOMLEY P A 1984 Selective-volume method for performing localized NMR spectroscopy *US Patent* 434688

BOTTOMLEY P A, FOSTER T B and DARROW R T 1984 Depth-resolved surface-coil spectroscopy (DRESS) for *in-vivo* ^{1}H, ^{31}P, ^{13}C NMR *J. Magn. Reson.* **59** 338–42

BOTTOMLEY P A, HARDY C J and LOWE W M 1987 2D spatially selective excitations with a single NMR (ρ) pulse: PROGRESS in 3D localized ^{31}P spectroscopy and surface-coil imaging *Proc. 6th Annu. Meet. Society of Magnetic Resonance in Medicine, New York* (Berkeley, CA: SMRM) p133

BOTTOMLEY P A, SMITH L S, LEUE W M and CHARLES C 1985 Slice-interleaved depth-resolved surface-coil spectroscopy (SLIT-DRESS) for rapid ^{31}P NMR imaging *in vivo Proc. 4th Annu. Meet. Society of Magnetic Resonance in Medicine, London* (Berkeley, CA: SMRM) pp946–7

BOWTELL R and MANSFIELD P 1987 An integrated screened gradient coil for magnets with transverse-field geometry *Proc. 6th Annu. Meet. Society of Magnetic Resonance in Medicine, New York* (Berkeley, CA: SMRM) p180

BROWN T R 1987 Personal communication

BROWN T R, KINCAIL B M and UGURBIL K 1982 NMR chemical-shift imaging in three dimensions *Proc. Natl Acad. Sci. USA* **79** 3523–6

BRUNNER P and ERNST R R 1979 Sensitivity and performance time in NMR imaging *J. Magn. Reson.* **33** 83–106

BRYANT D J, BAILES D R, BYDDER G M, CASE M A, COLLINS A G, COX I J, HALL A S, HARMAN R R, KHENIA S, MCARTHUR P, ROSS B D and YOUNG I R 1987 *In-vivo* four-dimensional Fourier-transform ^{31}P spectroscopy with correction for static-field inhomogeneity and tissue susceptibility by ¹H field mapping *Proc. 6th Annu. Meet. Society of Magnetic Resonance in Medicine, New York* (Berkeley, CA: SMRM) p156

BUDINGER T F 1979 Thresholds for physiological effects due to RF and magnetic fields used in NMR imaging *IEEE Trans. Nucl. Sci.* **NS-26** 2821–5

—— 1981 Nuclear magnetic resonance (NMR) *in-vivo* studies: known thresholds for health effects *J. Comput. Assist. Tomogr.* **5** 800–11

BURCHAM W E 1973 *Nuclear Physics, An Introduction* (London: Longman)

BYDDER G M, NIENDORF H P and YOUNG I R 1987 Clinical use of intravenous gadolinium-DTPA in magnetic resonance imaging of the central nervous system *Functional Studies Using NMR* ed V R McCready, M O Leach and P J Ell (London: Springer) pp129–45

CADY E B, COSTELLO A M DE L, DAWSON M J, DELPHY D T, REYNOLDS E O R, TOFTS P S and WILKIE D R 1983 Non-invasive investigation of cerebral metabolism in newborn infants by ³¹P NMR *Lancet* **i** 1059–62

CARR H Y and PURCELL E M 1954 Effects of diffusion on free precession in nuclear magnetic resonance experiments *Phys. Rev.* **94** 630–8

CHEN C N, HOULT D I and SANK V J 1983 Quadrature detection coils—a further √2 improvement in sensitivity *J. Magn. Reson.* **54** 324–7

CHO Z H, AHN C B, JUH S C, LEE H G, LEE S, YI J H and JO J M 1987 Some experiences on a 4 μm NMR microscopy *Proc. 6th Annu. Meet. Society of Magnetic Resonance in Medicine, New York* (Berkeley, CA: SMRM) p233

CLACK R, TOWNSEND D and JEAVONS A 1984 Increased sensitivity and field of view for a rotating positron camera *Phys. Med. Biol.* **29** 1421–31

COX I J, BYDDER G M, GADIAN D G, YOUNG I R, PROCTOR E , WILLIAMS S R and HART I 1986 The effect of magnetic susceptibility variations in NMR imaging and NMR spectroscopy *in-vivo J. Magn. Reson.* **70** 163–8

COX S J and STYLES P 1980 Towards biochemical imaging *J. Magn. Reson.* **40** 209–12

DAMADIAN R, GOLDSMITH M and MINKOFF L 1977 NMR in cancer: XVI. FONAR image of the live human body *Physiol. Chem. Phys.* **9** 97–100

DAMADIAN R, MINKOFF L, GOLDSMITH M, STANFORD M and KOUTCHER J 1976a Tumour imaging in a live animal by field focussing NMR (FONAR) *Physiol. Chem. Phys.* **8** 61–5

—— 1976b Field-focussing nuclear magnetic resonance (FONAR) visualization of a tumour in a live animal *Science* **194** 1430–2

DIXON T 1984 Simple proton spectroscopic imaging *Radiology* **153** 189–94

DODDRELL D M, BULSING J M, GALLOWAY G J, BROOKS W M, FIELD J, IRVING M G and BADDESLEY H 1986 Discrete isolation from gradient-governed elimination of resonances. DIGGER, a new technique for *in-vivo* volume-selected NMR spectroscopy *J. Magn. Reson.* **70** 319–26

DWEK R A 1973 *Nuclear Magnetic Resonance (NMR) in Biochemistry: Applications to Enzyme Systems* (Oxford: Clarendon)

EASTWOOD L 1984 Nuclear magnetic resonance proton imaging *Technical Advances in Biomedical Physics* ed P P Dendy, W D Ernst and A Sengun (The Hague: Nijhoff) pp377–410

EDELSTEIN W A, GLOVER G H, HARDY C J and REDINGTON R W 1986 The intrinsic signal-to-noise ratio in NMR imaging *Magn. Reson. Med.* **3** 604–18

EDELSTEIN W A, HUTCHISON J M S, JOHNSON G and REDPATH T 1980 Spin-warp NMR imaging and applications to human whole-body imaging *Phys. Med. Biol.* **25** 751–6

EIDELBERG D, JOHNSON G, BARNES D, TOFTS P and McDONALD W I 1987 ¹⁹F imaging of cerebral intravascular pO₂ *Proc. 6th Annu. Meet. Society of*

Magnetic Resonance in Medicine, New York (Berkeley, CA: SMRM) p240

Ewy C S, Ackerman J J H and Balaban R S 1987 Deuterium NMR imaging of the cat brain *in-situ Proc. 6th Annu. Meet. Society of Magnetic Resonance in Medicine, New York* (Berkeley, CA: SMRM) p238

Farrer T C and Becker E D 1971 *Pulse and Fourier-Transform NMR: Introduction to Theory and Methods* (New York: Academic Press)

FDA (Food and Drug Administration) 1982 *US Bureau of Radiological Health of the FDA Report* BRM (HFX-460) FDA

Fishman J E, Carolin M J, Joseph P M, Mukherji B and Sloviter H A 1987 Tumour blood oxygenation using ^{19}F magnetic resonance imaging *Proc. 6th Annu. Meet. Society of Magnetic Resonance in Medicine, New York* (Berkeley, CA: SMRM) p242

Frahm J, Merboldt K D and Hanicke W 1986 Localised proton spectroscopy. New steps using stimulated echoes *Proc. 5th Annu. Meet. Society of Magnetic Resonance in Medicine, Montreal* Works in progress (Berkeley, CA: SMRM) pp158–9

Frahm J, Merboldt K D, Hanicke W and Haase A 1985 Stimulated echo imaging *J. Magn. Reson.* **64** 81–93

Fukushima E and Roeder S B W 1981 *Experimental Pulse NMR: A Nuts and Bolts Approach* (Reading, MA: Addison-Wesley)

Gadian D G 1982 *Nuclear Magnetic Resonance and Its Applications to Living Systems* (Oxford: Clarendon)

Gordon R E 1985 Magnets, molecules and medicine *Phys. Med. Biol.* **30** 741–70

Gordon R E, Henley P E, Shaw D, Gadian D G, Styles P, Bore P J and Chan L 1980 Localisation of metabolites in animals using ^{31}P topical magnetic resonance *Nature* **287** 736–8

Graumann R, Fischer H and Oppelt A 1986 A new pulse sequence for determining T_1 and T_2 simultaneously *Med. Phys.* **13** 644–7

Gunther H 1980 *NMR Spectroscopy: An Introduction* (Chichester: Wiley)

Haase A 1986 Localisation of unaffected spins in NMR imaging and spectroscopy (LOCUS spectroscopy) *Magn. Reson. Med.* **3** 963–9

Haase A, Frahm J, Hanicke W and Matthaei D 1985 ^{1}H NMR chemical-shift selective imaging (CHESS) *Phys. Med. Biol.* **30** 341–4

Haase A, Frahm J, Matthaei W, Hanicke W and Merboldt K D 1986 FLASH imaging. Rapid NMR imaging, using low flip-angle pulses *J. Magn. Reson.* **67** 258–66

Hahn E L 1950 Spin echoes *Phys. Rev.* **80** 580–94

Hall L D and Hogan P G 1987 Paramagnetic pharmaceuticals for functional studies *Functional Studies Using NMR* ed V R McCready, M O Leach and P J Ell (London: Springer) pp107–27

Hanley P 1984 Magnets for medical applications of NMR *Br. Med. Bull.* **40** 125–31

Haselgrove J C, Subramanian V H, Leigh J S, Gyulai L and Chance B 1983 *In-vivo* one-dimensional imaging of phosphorus metabolites by phosphorus-31 nuclear magnetic resonance *Science* **220** 1170–3

Hayes C E, Edelstein W A, Schenck J F, Mueller O M and Eash M 1985 An efficient highly-homogeneous radiofrequency coil for whole body imaging at 1.5 T *J. Magn. Reson.* **63** 622–8

Hedges L K and Hoult D I 1987 Rotating-frame selective excitation: an approach to spatially-localised spectroscopy *Proc. 6th Annu. Meet. Society of Magnetic Resonance in Medicine, New York* (Berkeley, CA: SMRM) p134

HEDLUND L W, JOHNSON G A, KARIS J P and EFFMANN E L 1986 MR 'microscopy' of the rat thorax *J. Comput. Assist. Tomogr.* **10** 948–52

HICKEY D S, CHECKLEY D, ASPDEN R M, NAUGHTON A, JENKINS J P R and ISHERWOOD I 1986 A method for the clinical measurement of relaxation times in magnetic resonance imaging *Br. J. Radiol.* **59** 565–76

HINSHAW W S 1974 Spin mapping: the application of moving gradients to NMR *Phys. Lett.* **48A** 87–8

HORE P J 1983 Solvent suppression in Fourier-transform nuclear magnetic resonance *J. Magn. Reson.* **55** 283–300

HOULT D I 1977 Zeugmatography: a criticism of the concept of a selective pulse in the presence of a field gradient *J. Magn. Reson.* **26** 165–7

—— 1979 The solution of the Bloch equations in the presence of a varying B_1 field: an approach to selective pulse analysis *J. Magn. Reson.* **35** 69–86

—— 1980 NMR imaging: rotating-frame selective pulses *J. Magn. Reson.* **38** 369–74

HOULT D I and LAUTERBUR P C 1979 The sensitivity of the zeugmatographic experiment involving human samples *J. Magn. Reson.* **34** 425–33

HOULT D I and RICHARDS R E 1976 The signal-to-noise ratio of the nuclear magnetic resonance experiment *J. Magn. Reson.* **24** 71–85

HOWE G R and LINDSAY J P 1983 A follow-up study of a ten-percent sample of the Canadian labour force. 1. Cancer mortality in males 1965–73 *J. Natl. Cancer Inst.* **70** 37–44

HUTCHISON J M S 1977 Imaging by nuclear magnetic resonance *Medical Images: Formation, Perception and Measurement* Proc. 7th L H Gray Conf. 1976, ed G A Hay (New York: Wiley) pp135–41

HUTCHISON J M S, EDELSTEIN W A and JOHNSON G 1980 A whole-body NMR imaging machine *J. Phys. E: Sci. Instrum.* **13** 947–55

HUTCHISON J M S, SUTHERLAND R J and MALLARD J R 1978 NMR imaging: image recovery under magnetic fields with large non-uniformities *J. Phys. E: Sci. Instrum.* **11** 217–21

JAMES T L 1975 *Nuclear Magnetic Resonance in Biochemistry: Principles and Applications* (New York: Academic Press)

JOHNSON G, ORMEROD I E C, BARNES D, TOFTS P S and McMANUS D 1987a Accuracy and precision in the measurement of relaxation times from nuclear magnetic resonance images *Br. J. Radiol.* **60** 143–53

JOHNSON R J, JENKINS J P R, ISHERWOOD I, JAMES R D and SCHOFIELD P F 1987b Quantitative magnetic resonance imaging in rectal carcinoma *Br. J. Radiol.* **60** 761–4

KIBBLE T W B 1966 Rotating frames *Classical Mechanics* (New York: McGraw-Hill)

KITTEL C 1966 *Introduction to Solid State Physics* 3rd edn (New York: Wiley)

KUMAR A, WELTI D and ERNST R R 1975a NMR Fourier zeugmatography *J. Magn. Reson.* **18** 69–83

—— 1975b Imaging of macroscopic objects by NMR Fourier zeugmatography *Naturwissenschaften* **62** 34

LAUTERBUR P C 1973 Image formation by induced local interactions: examples employing nuclear magnetic resonance *Nature* **242** 190–1

LEACH M O 1987 The physical basis of NMR studies measuring physiological function and metabolism *Functional Studies Using NMR* ed V R McCready, M O Leach and P J Ell (London: Springer) pp15–42

LEACH M O, HIND A J, SAUTER R, REQUARDT H and WEBER H 1986 Design and use of a dual-frequency surface coil providing proton images for improved localisation in ^{31}P spectroscopy of small lesions *Med. Phys.* **13** 510–13

LEACH M O, HIND A J, SHARP J C, COLLINS D, McCREADY V R, BABICH J and HAMMERSLEY P A G 1987a ^{31}P imaging as a method of calibrating surface-coil response *Proc. 6th Annu. Meet. Society of Magnetic Resonance in Medicine, New York* (Berkeley, CA: SMRM) p937

LEACH M O, McCREADY V R, HIND A J, SHARP J C and COLLINS D 1987b Phosphorus image muscle subtracted spectroscopy (PIMSS) *Proc. 6th Annu. Meet. Society of Magnetic Resonance in Medicine, New York* (Berkeley, CA: SMRM) p1050

LEACH M O, SHARP J C, GOWLAND P A and HIND A J 1987c The use of a whole-body imaging and spectroscopy system as a spectrometer for relaxation-time measurements *Proc. 6th Annu. Meet. Society of Magnetic Resonance in Medicine, New York* (Berkeley, CA: SMRM) p877

LE BIHAN D, BRETON E, LALLEMAND D, GRENIER P, CABANIS E and LAVAL-JEANTET M 1986 MR mapping of intravoxel incoherent motions: applications to diffusion and perfusion in neurological disorders *Radiology* **161** 401–7

LERSKI R A 1985 *Physical Principles and Clinical Applications of Nuclear Magnetic Resonance* (London: Institute of Physical Sciences in Medicine)

LEWIS C E, PRATO S, DROST D J and NICHOLSON R L 1986 Comparison of respiratory triggering and gating techniques for the removal of respiratory artefacts in MR imaging *Radiology* **160** 803–10

LUYTEN P R, MARIEN A J H, SIJTSMA B and DEN HOLLANDER J A 1986 Solvent-suppressed spatially-resolved spectroscopy. An approach to high resolution NMR in a whole-body MR system *J. Magn. Reson.* **67** 148–55

McCREADY V R, LEACH M O and ELL P J (eds) 1987 *Functional Studies Using NMR* (London: Springer)

McDOWALL M E 1983 Leukacmia mortality in electrical workers in England and Wales *Lancet* **i** 246

—— 1986 Mortality of persons resident in the vicinity of electricity transmission facilities *Br. J. Cancer* **53** 271–9

MANSFIELD P 1977 Multi-planar image formation using NMR spin echoes *J. Phys. C: Solid State Phys.* **10** L55–8

MANSFIELD P and CHAPMAN B 1986 Active magnetic screening of gradient coils in NMR imaging *J. Magn. Reson.* **66** 573–6

MANSFIELD P and GRANNELL P K 1973 NMR diffraction in solids *J. Phys. C: Solid State Phys.* **6** L422–6

MANSFIELD P, MAUDSLEY A A and BAINES T 1976 Fast-scan proton density imaging by NMR *J. Phys. E: Sci. Instrum.* **9** 271–8

MANSFIELD P and MORRIS P G 1982 *NMR Imaging in Biomedicine* (London: Academic Press)

MANSFIELD P and PYKETT I L 1978 Biological and medical imaging by NMR *J. Magn. Reson.* **29** 355–73

MANSFIELD P, PYKETT I L, MORRIS P G and COUPLAND R E 1978 Human whole-body line-scan imaging by NMR *Br. J. Radiol.* **51** 921–2

MARGOSIAN P 1985 Faster MR imaging with half the data *Proc. 4th Annu. Meet. Society of Magnetic Resonance in Medicine, London* (Berkeley, CA: SMRM) pp1024–5

—— 1987 MR images from a quarter of the data: combination of half-Fourier methods with a linear recursive data extrapolation *Proc. 6th Annu. Meet. Society of Magnetic Resonance in Medicine, New York* (Berkeley, CA: SMRM) p375

MARGOSIAN P and ABART J 1984 Rapid measurement of magnetic field inhomogeneities using imaging techniques *Proc. 3rd Annu. Meet. Society of Magnetic Resonance in Medicine, New York* (Berkeley, CA: SMRM) pp495–6

MAUDSLEY A A 1980 Multiple-line-scanning spin density imaging *J. Magn. Reson.* **41** 112–26

MAUDSLEY A A and HILAL S K 1984 Biological aspects of sodium-23 imaging *Br. Med. Bull.* **40** 165–6

MAUDSLEY A A, HILAL S K, PERMAN W H and SIMON H E 1983 Spatially-resolved high-resolution spectroscopy by 'four-dimensional' NMR *J. Magn. Reson.* **51** 147–52

MEIBOOM S and GILL D 1958 Modified spin-echo method for measuring nuclear relaxation times *Rev. Sci. Instrum.* **29** 688–91

MOON R B and RICHARDS J H 1973 Determination of intracellular pH by ^{31}P magnetic resonance *J. Biol. Chem.* **248** 7276–8

MORRIS P G 1986 *Nuclear Magnetic Resonance Imaging in Medicine and Biology* (Oxford: Clarendon)

MUELLER E, DEIMLING M and REINHARDT E R 1986 Quantification of pulsatile flow in MRI by an analysis of T_2 changes in ECG-gated multiecho experiments *Magn. Reson. Med.* **3** 331–5

NAKADA T, KWEE I L, CORD P J, MATWIYOFF N A, GRIFFEY B V and GRIFFEY R M 1987 ^{19}F NMR metabolic imaging of 3-fluoro-3-deoxy-*d*-glucose *Proc. 6th Annu. Meet. Society of Magnetic Resonance in Medicine, New York* (Berkeley, CA: SMRM) p239

NEW P J, ROSEN B R, BRADY T J, BUONANNO F S, KISTLER J P, BURT C T, HINSHAW W S, NEWHOUSE J M, POHOST G M and TAVERAS J M 1983 Potential hazards and artefacts of ferromagnetic and non-ferromagnetic surgical and dental materials and devices in nuclear magnetic resonance imaging *Radiology* **147** 139–48

NRPB (NATIONAL RADIOLOGICAL PROTECTION BOARD) 1983 Revised guidance on acceptable limits of exposure during nuclear magnetic resonance clinical imaging (NRPB *ad hoc* advisory group on NMR clinical imaging) *Br. J. Radiol.* **56** 974–7

OBERHAENSLI R D, HILTON-JONES D, BORE P J, HANDS L J, RAMPLING R P and RADDA G K 1986 Biochemical investigations of human tumours with phosphorus-31 magnetic resonance spectroscopy *Lancet* **ii** 8–11

O'DONNELL M 1985 NMR blood-flow imaging using multiecho, phase-contrast sequences *Med. Phys.* **12** 59–64

OPPELT A, GRAUMANN R, BARFUSS H, FISCHER H, HARTL W and SCHAJOR W 1986 FISP—a new fast MRI sequence *Electromedica* **54** 15–18

ORDRIDGE R J, CONNELLY A and LOMAN J A B 1986 Image-selected *in-vivo* spectroscopy (ISIS). A new technique for spatially-selective NMR spectroscopy *J. Magn. Reson.* **66** 283–94

OXFORD MAGNET TECHNOLOGY 1983 *Magnets in Clinical Use: Site-Planning Guide* (Oxford: Oxford Magnet Technology)

PARTAIN C L, JAMES A E, ROLLO F D and PRICE R R 1983 *Nuclear Magnetic Resonance (NMR) Imaging* (Philadelphia: W B Saunders)

PERMAN W H, TURSKI P A, HOUSTON L W, GLOVER G H and HAYES C E 1986 Methodology of *in-vivo* human-sodium MR imaging at 1.5 T *Radiology* **160** 811–20

PURCELL E M, TORREY H C and POUND R V 1946 Resonance absorption by nuclear magnetic moments in a solid *Phys. Rev.* **69** 37–8

PYKETT I L 1982 NMR imaging in medicine *Sci. Am.* **246** 78–88

Pykett I L and Rosen B R 1983 Nuclear magnetic resonance: *in-vivo* proton chemical-shift imaging *Radiology* **149** 197–201

Radda G K 1984 Clinical studies by ^{31}P NMR spectroscopy *Proc. 3rd Annu. Meet. Society of Magnetic Resonance in Medicine, New York* (Berkeley, CA: SMRM) pp605–8

Radda G K, Bore P J, Gadian D G, Styles P, Taylor D and Morgan-Hughes J 1982 ^{31}P NMR examination of two patients with NADH CoQ reductase deficiency *Nature* **295** 608–9

Reynolds E R, Azzopardi D, Cady E B, Delphy D T, Wyatt J S, Hamilton P A and Hope P L 1987 Magnetic resonance spectroscopy for the investigation of perinatal hypoxic-ischaemic brain injury *Proc. 6th Annu. Meet. Society of Magnetic Resonance in Medicine, New York* (Berkeley, CA: SMRM) P3

Richards M A, Gregory W M, Webb J A W, Jewell S E and Reznek R H 1987 Reproducibility of spin–lattice relaxation time (T_1) measurement using an 0.08 tesla MD800 magnetic resonance imager *Br. J. Radiol.* **60** 241–4

Rosen B R, Wedeen V J and Brady T J 1984 Selective-saturation NMR imaging *J. Comput. Tomogr.* **8** 813–18

Ross B D, Radda G K, Gadian D G, Rocker G, Esiri M and Falconer-Smith J 1981 Examination of a case of suspected McArdle's syndrome by ^{31}P NMR *New Engl. J. Med.* **304** 1338–42

Runge V M, Clanton J A, Lukehart C M, Partain C L and James A E 1983 Paramagnetic agents for contrast-enhanced NMR imaging: a review *Am. J. Roentgenol.* **141** 1209–15

Saunders R D 1982 The biological hazards of NMR *Proc. 1st Int. Symp. NMR Imaging, Bowmen Gray School of Medicine, Winston-Salem, NC* pp65–71

Saunders R D and Orr J S 1983 Biologic effects of NMR *Nuclear Magnetic Resonance Imaging* ed C L Partain, A E James, F D Rollo and R R Price (Philadelphia: W B Saunders)

Saunders R D and Smith H 1984 Safety aspects of NMR clinical imaging *Br. Med. Bull.* **40** 148–54

Sauter R, Mueller S, Klose U, Beckmann N and Weber H 1987 Selective saturation: a simple technique for volume-selective *in-vivo* ^{31}P spectroscopy *Eur. Workshop on Magnetic Resonance in Medicine, London* Book of Abstracts, pp82–3

Segebarth C, Luyten P and den Hollender J 1987 Depth selection of single surface coil ^{31}P MR spectra using a combination of B_1 and B_0 selection techniques *Proc. 6th Annu. Meet. Society of Magnetic Resonance in Medicine, New York* (Berkeley, CA: SMRM) pp29–30

Sharp J C, Leach M O, Hind A J, McCready V R, Weber H and Sauter R 1987a A ^{31}P spectroscopy surface coil localisation technique: FROGS *Eur. Workshop on Magnetic Resonance in Medicine, London* Book of Abstracts, pp31–2

—— 1987b FROGS: a surface coil ^{31}P spectroscopy localisation technique *Proc. 6th Annu. Meet. Society of Magnetic Resonance in Medicine, New York* (Berkeley, CA: SMRM) 600

Shaw D 1976 *Fourier Transform NMR Spectroscopy* (Amsterdam: Elsevier)

Shepp L A 1980 Computerized tomography and nuclear magnetic resonance *J. Comput. Assist. Tomogr.* **4** –107

Silver M S, Joseph R I and Hoult D I 1984 Highly selective $\pi/2$ and π pulse generation *J. Magn. Reson.* **59** 347–51

—— 1985 Selective spin inversion in nuclear magnetic resonance and coherent optics through an exact solution of the Bloch–Riccati equations *Phys. Rev.* A **31** 2753–5

SLICHTER C P 1978 *Principles of Magnetic Resonance* (Berlin: Springer)

SOLOMON I 1955 Relaxation processes in a system of two spins *Phys. Rev.* **99** 559–65

STEJSKAL E O and TANNER J E 1965 Spin diffusion measurements: spin echoes in the presence of a time-dependent field gradient *J. Chem. Phys.* **42** 288–92

STEPHENS A N 1987 NMR spectroscopy: application to metabolic research *Functional Studies Using NMR* ed V R McCready, M O Leach and P J Ell (London: Springer) pp61–84

STYLES P, SCOTT C A and RADDA G K 1985 A method for localising high resolution NMR spectra from human subjects *Magn. Reson. Med.* **2** 402–9

SUTHERLAND R J and HUTCHISON J M S 1978 Three-dimensional NMR imaging using selective excitation *J. Phys. E: Sci. Instrum.* **11** 79–83

TAYLOR D G and BUSHELL M C 1985 The spatial mapping of translational diffusion coefficients by the NMR imaging technique *Phys. Med. Biol.* **30** 345–9

TAYLOR D J, BORE P J, STYLES P, GADIAN D G and RADDA G K 1983 Bioenergetics of intact human muscle: a ^{31}P NMR study *Mol. Biol. Med.* **1** 77–94

TURSKI P A, PERMAN W H, HALD J K, HOUSTON L W, STROTHER C M and SACKETT J F 1986 Clinical and experimental vasogenic edema; *in-vivo* sodium MR imaging *Radiology* **160** 821–5

UNDERWOOD S R 1987 Functional studies of the cardiovascular system using magnetic resonance imaging *Functional Studies Using NMR* ed V R McCready, M O Leach and P J Ell (London: Springer)

VETTER J, SIEBOLD H and SOLDNER L 1987 A 4 T superconducting whole-body magnet for MR-imaging and spectroscopy *Proc. 6th Annu. Meet. Society of Magnetic Resonance in Medicine, New York* (Berkeley, CA: SMRM) p181

WOLF W, ALBRIGHT M, SILVER M S, WEBER H, REICHARDT U and SAUER R 1987 Fluorine-19 NMR spectroscopic studies of the metabolism of 5-fluorouracil in the liver of patients undergoing chemotherapy *Magn. Reson. Imaging* **5** 165–9

WOOD M L and HENKELMAN R M 1985 MR image artefacts from periodic motion *Med. Phys.* **12** 143–51

YOUNG I R 1984 Considerations affecting signal and contrast in NMR imaging *Br. Med. Bull.* **40** 139–47

YOUNG I R, KHENCA S, THOMAS D G T, DAVIS C H, GADIAN D G, COX I J, ROSS B D and BYDDER G M 1987 Clinical magnetic susceptibility mapping of the brain *J. Comput. Assist. Tomogr.* **11** 2–6

CHAPTER 9

PHYSICAL ASPECTS OF INFRARED IMAGING

C H JONES

9.1 INTRODUCTION

Infrared (IR) radiation may be used to image surface vascularity, to transilluminate soft tissue and to image surface temperature distributions. This chapter is concerned principally with IR thermal imaging, but for clarity these other techniques will be described briefly.

9.2 INFRARED PHOTOGRAPHY

Clinical IR photography depends upon the spectral transmission and reflection characteristics of tissue and blood in the wavelength range 0.7–0.9 μm. The technique requires the subject to be photographed in diffuse, shadowless illumination with a lens filter to absorb radiation of wavelengths less than 0.7 μm. Penetration and reflection are maximal in the red end of the visible spectrum, where the radiation penetrates the superficial layers of skin and tissue up to a depth of 2.5 mm and is then reflected out again. The spectral transmission of radiation through blood depends upon the relative concentration of reduced and oxygenated haemoglobin: the latter transmits radiation of wavelengths greater than 0.6 μm, whereas reduced haemoglobin transmits radiation significantly when the wavelength is greater than 0.7 μm. These properties allow images to be recorded on IR-sensitive film and will show blood vessels (veins) that lie within 2.5 mm of the skin. IR film can be replaced by a silicon-target vidicon tube (see also Chapter 11) to allow real-time viewing of vascular patterns (Jones 1982).

9.3 TRANSILLUMINATION

This is a very old method of imaging, which is based on the finding that it is sometimes possible to identify pathological tissue by the shadow it casts when the surrounding tissues are transilluminated by an intense beam of light. Ohlsson *et al* (1980) have combined transillumination and photography for imaging the female breast by using specially designed equipment consisting of a variable-intensity light and a camera with IR colour film. Transmission of light through the breast depends upon tissue constituency and blood supply: veins carrying reduced blood or tumours with increased blood supply cast dark shadows. Transmission images may also be recorded with IR-sensitive vidicon cameras (Watmough 1982) and multispectral scanners. In the latter technique, the breast is illuminated with light of two alternating spectral characteristics, which produce an image containing information in the red and IR parts of the spectrum.

Transillumination may be used for the visualisation and measurement of the oxygenation state of brain and muscle in the neonate. In this case, the light source is a xenon arc lamp and the detector is an image intensifier. At wavelengths between 0.7 and 0.9 μm, the fall-off in transmitted light intensity is between 10- and 30-fold per centimetre of tissue, depending on wavelength, tissue type and the relative sizes of source and collector (Arridge *et al* 1986). The physics of transillumination imaging is discussed in more detail in Chapter 11.

9.4 INFRARED IMAGING

Thermal imaging is dependent upon the detection of radiant emission from the skin surface. A typical temperature-measuring device consists of a system for collecting radiation from a well defined field of view and a detector that transduces the radiation focused onto it into an electrical signal. The build-up of the thermal image is achieved by means of either an optical scanning system or a pyroelectric vidicon television (TV) tube; a monitor supported by an image-processing unit displays the image. Thermal detectors, such as thermocouples and semiconductor bolometers, that rely upon the temperature rise in the absorbing element are less sensitive than photon detectors and have longer time constants. Photon detectors respond rapidly to radiation flux changes with time constants of a few microseconds but detect only a small proportion of the total radiation emitted from the object's surface. The measurement of surface temperature by infrared imaging techniques requires knowledge of the emissive properties of the surface being examined over the range of wavelengths to which the detector is sensitive.

9.4.1 Thermal radiation

The Stefan–Boltzmann law for total radiation emitted from a perfectly black body is given by

$$R(T) = \sigma A T^4 \tag{9.1}$$

where $R(T)$ is the total power radiated into a hemisphere, σ is the Stefan–Boltzmann constant (5.67×10^{-8} W m^{-2} K^{-4}), A is the effective radiating area of the body and T is the absolute temperature of the radiating surface. For a thermal black body, the emissivity (ε) is unity. In practice, a thermal body is not only radiating energy but is also being irradiated by surrounding sources: it is the net radiant exchange that is important. For surfaces that are not full radiators

$$R(T) = \varepsilon(T)\sigma A T^4$$

and the rate of heat loss between two surfaces at T_1 and T_2 is

$$W = A\sigma[\varepsilon(T_1)T_1^4 - \varepsilon(T_2)T_2^4]. \tag{9.2}$$

This radiant heat loss forms the basis of thermal imaging. Although $\varepsilon(T)$ is a slowly varying function of T, at body temperatures it can be considered to be constant.

The spectral distribution of emitted radiation from a surface depends upon its temperature and is represented accurately by Planck's formula. Skin temperature is normally maintained within the range 25–35 °C, and over this range emission occurs between 2 and 50 μm with maximum emission around 10 μm. Steketee (1973a) has measured the emissivity of living tissue between 1 and 14 μm at normal incidence: for white skin, black skin and burnt skin he found that the emissivity was independent of wavelength and equal to 0.98 ± 0.01. The emissivity of a surface is affected principally by surface structure, and consequently surface contaminants (such as talcum powder and cosmetics) can alter the emissive properties of skin and affect the apparent surface temperature.

If the surface is at T_0 and has an emissivity ε_0, the radiant power emitted from the surface will be proportional to $\varepsilon_0 T_0^4$. If T_b is the temperature of a black body emitting the same radiant power, then

$$T_b^4 = \varepsilon_0 T_0^4 \qquad \text{and} \qquad \varepsilon_0 = (T_b/T_0)^{1/4} \tag{9.3}$$

so a 1% change in emissivity will cause an apparent reduction in T of about 0.75 °C. In the case of photon detectors, emissivity changes affect temperature measurements in a complex way, which depends upon the spectral sensitivity of the detector and the temperature of the skin surface.

Electromagnetic theory for electrical insulators shows that maximum emission always occurs at normal incidence to a skin surface. Watmough *et al* (1970) and Clark (1976) have predicted the variation of emissivity with angle of view for a smooth surface and have shown that significant errors occur if view angles of greater than 60° are used. Clinically, this is not a problem except, maybe, in female-breast examinations. Sometimes it is helpful to use a plane mirror or reflector for viewing the scene to be imaged. In these situations, allowance must be made for the reflectivity of the mirror and the viewing angle. Conventional glass-fronted mirrors are unsuitable for IR imaging, and it is customary to use front-surface silvered mirrors or, alternatively, aluminised Mylar. For an object temperature of 32 °C at viewing angles up to 45°, polished copper and silver reflectors will produce an apparent temperature reduction of 0.2 °C, whereas aluminium and aluminised Mylar will show an apparent temperature reduction of 0.5 °C.

9.4.2 Photon detectors

Most thermal imaging systems employ photon detectors: these are semiconductor-type devices in which photon absorption results in the freeing of bound electrons or charge carriers in proportion to the intensity of the incident radiation. Indium antimonide (InSb) and cadmium mercury telluride (CdHgTe) (often abbreviated to CMT) have been used most frequently in medical imaging, although lead telluride (PbTe) is also being used. Figure 9.1 illustrates the spectral detectivities D^* of these detectors as well as typical thermal detectors, some of which were used in early imagers. Photon detectors may be used in either photoconductive or photovoltaic modes. In the photoconductive mode, the conductivity change due to incident photon flux is used to cause a corresponding voltage change across a resistor, which in turn is amplified. The photovoltaic effect is an internal photoeffect in which the action of photons within a semiconductor produces a voltage that can be detected without the need for bias supply or load resistor. Although photoconductive detectors are easier to couple to the required preamplifiers, the heat generated by bias current can cause cooling problems. When many detector elements are used close together, photovoltaic detectors are to be preferred. Spectral response and detectivity depend on the mode of operation: in the case of InSb and CMT detectors, both modes have been used for medical imaging. To make the detector sensitive enough to resolve small temperature differences and to overcome thermal noise within the detector, cryogenic cooling to −196 °C is used. This is achieved with liquid nitrogen, by Joule–Thomson cooling using air at high pressure or by a Stirling-cycle cooling engine.

The parameters by which IR detectors are usually specified are responsivity, noise, detectivity, cut-off wavelength and time constant. Responsivity (R) is the ratio of output voltage to radiant input power, expressed in volts per watt. Because the output voltage due to incident IR radiation is a very small fraction (about 10^{-5}) of the DC bias voltage across the detector, the responsivity is measured by exposing the detector to chopped radiation from a calibrated source (frequently 500 K), and measuring the alternating voltage component at the chopping frequency. Responsivities (R, 500 K) are typically 10^4–10^5 V W^{-1} for CMT detectors operated at −196 °C. Noise, together with responsivity, determines a detector's ability to detect small input signals: this is specified in volts per hertz at one or a number of frequencies or as a noise spectrum. Detectivity (D) is the reciprocal of NEP, the noise equivalent power, which is the RMS value of the sinusoidally modulated radiant power falling upon a detector that will give rise to an RMS signal voltage (V_s) equal to the RMS noise voltage (V_n) from the detector,

$$D = 1/\text{NEP}. \qquad (9.4)$$

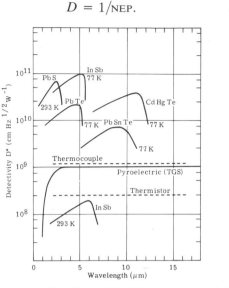

Figure 9.1 Variation of D^*: detectivity of IR detector with wavelength for various detectors. (From Jones and Carnochan (1986) courtesy of Research Studies Press Ltd.)

For many photon detectors, the NEP is directly proportional to the square root of the area of the detector, and it becomes appropriate to use a normalised detectivity D^* given by

$$D^* = DA_d^{1/2}(\Delta f)^{1/2} = \frac{V_s}{V_n} \frac{[A_d(\Delta f)]^{1/2}}{W} \qquad (9.5)$$

where A_d is the area of the detector, Δf is the frequency bandwidth of the measuring system and W is the radiation power incident on the detector (RMS value in watts). As D^* varies with wavelength of the radiation and the frequency at which the noise is measured, these are stated in parentheses: $D^*(5\ \mu\text{m}, 800\ \text{Hz}, 1)$ means that the specified value was measured at $5\ \mu\text{m}$ wavelength with the noise measured at 800 Hz and that the measurement has been normalised to unit bandwidth. D^* may also be measured at a specific black-body temperature, and this is used in place of the wavelength. As a figure of merit, D^* enables a theoretical maximum detectivity to be calculated that would apply when performance is limited only by noise due to fluctuation of the background radiation. The condition of operation of a detector when limiting noise is solely due to incident photon fluctuations has been termed BLIP (for background-limited infrared photoconductor). The specific detectivity of a detector operating under the BLIP condition (D^*_{BLIP}) varies with the temperature of the background, the solid angle subtended by the background at the detector and the long-wavelength cut-off of the detector. This latter parameter, denoted λ_{co}, is the longer of the two wavelengths at which the responsivity of the detector is down to half its maximum. The detector time constant Γ is the time between incident radiation being cut off and the output of the detector falling by 63%; typical time constants range from a fraction of a microsecond for CMT detectors to a few microseconds for InSb detectors.

9.4.3 Imaging systems

The detector forms part of an imaging system whose performance is usually specified in terms of temperature resolution, angular resolution and field of view. Temperature resolution is a measure of the smallest temperature difference in the scene that the imager can resolve. It depends on the efficiency of the optical system, the responsivity and noise of the detector, and the signal-to-noise ratio of the signal-processing circuitry. Temperature resolution can be expressed in two ways: noise equivalent temperature difference (NETD), which is the temperature difference for which the signal-to-noise ratio at the input to the display is unity; and minimum resolvable temperature difference (MRTD), which is the smallest temperature difference that is discernible on the display. Most medical thermography has been carried out with systems that have MRTD between 0.1 and 0.3 K. Angular resolution is typically 1–3 mrad but can be as small as 0.5 mrad.

In practice, the imaging system comprises many complex components, each of which can affect image quality. A comprehensive analysis of the factors influencing the design and construction of thermal imaging systems has been given by Wolfe and Zissis (1978) and Lawson (1979). Table 9.1 summarises the principal parameters for medical thermography.

In an imaging system, the scene is viewed by an optical system capable of transmitting and focusing IR radiation. A high value of IR refractive index is advantageous in lens design, but materials that have high refractive indices tend to have low transmittances. For example, for 2 μm radiation, germanium has a refractive index of about 4 and a transmittance of 47%. This high reflective loss may be eliminated by antireflection coatings, which raise the transmittance to as high as 95–97% for a given wavelength interval. Germanium and silicon are used frequently for IR optical components. Scanning systems might employ configurations of lenses, rotating prisms, rocking mirrors or rotating multisided mirror drums. The precise design is dependent on commercial features and the functional purpose of the imager. The bulk of medical imaging has been carried out with single-element detectors, which have the advantage of simplicity, both electronically and mechanically. They also lend themselves better to the production of quantitative data, since non-uniformity of the elements in an array of multiple detectors can lead to errors of temperature measurement. Figures 9.2 and 9.3 illustrate the compactness of the AGEMA Thermovision 870 Camera, which has been designed to operate at room temperatures. The advantage of replacing a single detector by a multi-element array of n

similar detectors is that the signal increases in proportion to n whereas noise increases in proportion to $n^{1/2}$. Figure 9.4 shows different scanning methods with detectors, most of which have been used for medical thermography. The higher scan speeds obtainable with array detectors make these instruments important for many investigations. In a clinical context, high-resolution real-time imaging allows precise focusing on the skin surface and continuous observation of thermal changes so that transient or dynamic studies can be made on patients.

Table 9.1 Details of thermographic imaging systems.

Component/parameter	Range of values
Detector	
Single element	InSb (3–5.5 μm), 77 K
	CdHgTe (8–15 μm), 77 K
	CdHgTe (3–5 μm), 195 K
	PbTe (3–5 μm), 77 K
	PbSnTe (8–12 μm), 77 K
Linear arrays	× 10 InSb
	SPRITE CdHgTe (CMT)
	× 512 PbTe
Matrix	6 × 8 InSb
	1× 8 SPRITE CMT
Pyroelectric	TGS
	Vidicon TGS, DTGS (room temp.)
Spatial resolution	0.5–3 mrad
Number of horizontal lines	90–625
Elements per line	100–600
NETD	0.05–0.4 °C at 30 °C
MRTD	0.1–0.5 °C at 30 °C
Number of frames per second	0.5–50
Field of view	5° × 5° to 60° × 40°
Temperature ranges	1–50 °C

Figure 9.2 AGEMA Thermovision 870 Camera with germanium lens and CMT detector. (Courtesy of AGEMA Infrared Systems Ltd.)

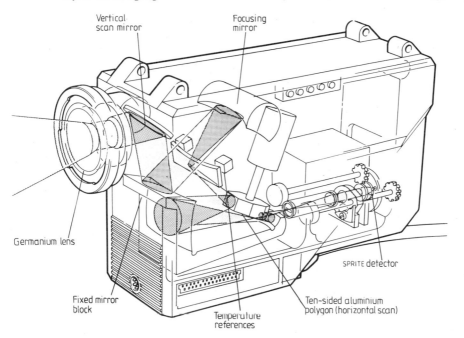

Figure 9.3 Schematic of AGEMA Thermovision 870 Camera. (Courtesy of AGEMA Infrared Systems Ltd.)

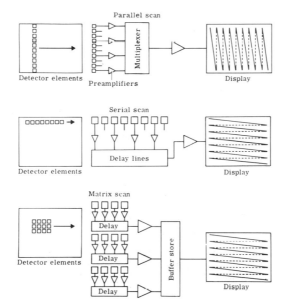

Figure 9.4 Schematic of typical scan-display interfaces. (From Jones (1987) after Lawson (1979).)

A significant development in imaging technology has been the SPRITE (signal processing in the element) detector. The SPRITE CMT detector carries out the delay and add functions within the element, so that a single detector replaces a linear array of detectors. In a conventional in-line array, the signal from each IR-sensitive element is preamplified and then added to the signal that is generated in the adjacent element. In the SPRITE, the individual elements are replaced by a single IR strip mounted on a sapphire substrate. It requires only one amplifier channel and has optimum gain at high speeds. An eight-element SPRITE detector is equivalent in performance to an array of at least 64 discrete elements but requires far fewer connections. By arranging the detectors in a stack, outputs can be stored in parallel in-line registers and serially combined to a TV-compatible display rate (Mullard 1982) (figure 9.5).

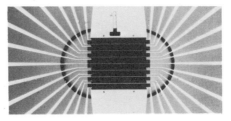

Figure 9.5 Eight-element SPRITE CMT array. (Courtesy of Mullard Ltd.)

9.4.4 Pyroelectric imaging systems

The pyroelectric effect is exhibited by certain ferromagnetic crystals such as barium titanate and triglycine sulphate (TGS). When exposed to a change in radiance, these materials behave as capacitors on which electrical charge appears. The magnitude of the effect depends on the rate of temperature change in the detector, and so the sensor does not respond to a steady flux of radiation. Pyroelectric detection has been developed as a cheaper alternative to photon-detector-based systems. Although pyroelectric detectors have been incorporated into systems that employ mechanical scanning devices to build up a thermal image, most interest is being shown in the development of pyroelectric vidicon camera tubes for clinical investigations. The scene is panned or modulated by a rotating disc and the IR radiation enters the vidicon tube by means of a germanium IR-transmitting lens (8–14 μm), which focuses the image of the thermal scene onto a thin disc of TGS pyroelectric material (20 mm diameter, 30 μm thick). On the front of the TGS target there is an electrically conducting layer of material chosen to be a good absorber of thermal radiation. The target is scanned in a TV raster by the electron beam of the vidicon tube and the image displayed on a TV monitor (Watton *et al* 1974).

At a temperature of 30 °C and a chopping frequency of 25 Hz, an irradiance of $20\,W\,m^{-2}$ produces a signal current of 5 or 12 nA in a 14 μm target depending upon whether it is TGS or deuterated TGS. The spatial resolution of a target is limited by thermal diffusion within the detector itself; target reticulation is being researched as a method of circumventing this, but the method is costly to implement (Goss *et al* 1984).

Pyroelectric vidicon cameras are being manufactured for specific industrial and commercial purposes such as security, fire-fighting investigations and viewing in smoke and darkness. Image quality is inferior to photon-detector scanning systems and, generally, systems are not able to record clinical temperature patterns with an MRTD of better than 0.3–0.4 °C. Nevertheless, pyroelectric vidicon systems have several merits: they do not require cryogenic cooling, they are relatively inexpensive, and they have the advantages of compactness and compatibility with standard video cameras so that full overlay of video and thermal images is possible. Figure 9.6 shows the basic components of a pyroelectric vidicon IR camera tube.

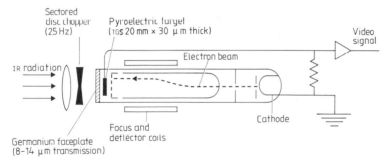

Figure 9.6 Schematic diagram of a pyroelectric vidicon camera.

9.4.5 *Temperature measurement*

Medical investigations require quantification of the thermal scene. Some systems have adjustable, built-in, temperature reference sources, and the detector output signal is compared with that from the reference source at a known temperature. Generally, it is more reliable to use an independent temperature reference located in the field of view. The image display of typical thermographic equipment is equipped with an isotherm function by which a signal is superimposed upon the grey-tone picture so that all surface areas with the same temperature are presented as saturated white (or in a specific colour if the image is displayed in colour mode). The isotherm can be adjusted as required within the temperature range of the thermogram (figure 9.7).

Temperature measurement of the skin by an IR camera must allow for the following:

(*a*) radiation emitted by the skin as a result of its surface temperature (T_0) and its emissivity (ε);

(*b*) radiation reflected by the skin into the measuring system as a result of the ambient temperature of the surroundings (T_a), which will be proportional to the reflectance (ρ) of the skin; and

(*c*) radiation transmitted through the skin as a result of the temperature (T_d) at a depth within the body of the object.

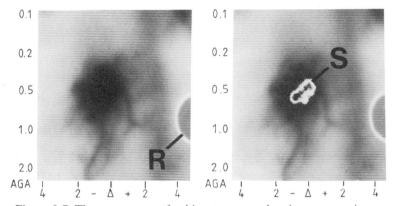

Figure 9.7 Thermograms of skin tumour showing a maximum temperature of 34.9 °C indicated by isotherm S. The reference temperature R is 32 °C.

In practice, the skin is non-transparent for the wavelengths detected, and the transmitted contribution can be neglected. The signal S resulting from a small quantity of the total radiation emitted by an object irradiating the detector can be written as

$$S = \varepsilon f(T_0) + \rho f(T_a) \tag{9.6}$$

where $\varepsilon f(T_0)$ is the emitted radiation, which is a function of the object's surface temperature, and $\rho f(T_a)$ is the reflected radiation, which is a function of the ambient temperature. Since $\rho = 1 - \varepsilon$, then

$$S = \varepsilon f(T_0) + (1 - \varepsilon)f(T_a). \tag{9.7}$$

The camera signal versus black-body temperature characteristic curve

$$S_b = f(T_b) \tag{9.8}$$

depends upon the detector and system optics but is usually exponential in form. Skin radiates with a spectral distribution similar to a black body and, if ε is the surface emissivity, the corresponding characteristic response curve for the detector will be

$$S_0 = \varepsilon f(T_0). \tag{9.9}$$

Calibration curves are usually provided by manufacturers for various aperture settings in isotherm units (or signal voltages). Since skin emissivity is very high, the calibration curve for unit emissivity may be used to correlate isotherm-level differences with temperature-magnitude differences. The isotherm facility is used to measure the temperature of an area relative to the temperature of a reference standard: the isotherm can be adjusted first to match the reference and then to match the area of interest. With the aid of the appropriate calibration curve, the difference between these two measurements can be used to read off the unknown temperature.

9.4.6 Clinical thermography: physiological factors

As a homeotherm, man maintains a stable deep-body temperature within relatively narrow limits (37 ± 1 °C), even though the environmental temperature may fluctuate widely. The minimal metabolic rate for resting man is about $45 \, \mathrm{W \, m^{-2}}$, which for the average surface area of $1.8 \, \mathrm{m^2}$ gives a rate of 81 W per man (Mount 1979). Metabolic rates increase with activity, with a consequential temperature rise. Furthermore, the core temperature is not constant throughout the day: it shows a 24 h variation, with maximum body temperature occurring in the late afternoon and minimum temperature at 4.00 a.m., the difference being about 0.7 °C. Blood perfusion has an important role in maintaining deep-body temperature, even though the environmental temperature might alter. The rate of blood flow in the skin is the factor that has the chief influence on the internal conductance of the body: the higher the blood flow and the conductance, then the greater is the rate of transfer of metabolic heat from the tissues to the skin for a given temperature difference. For these reasons, thermal imaging is best carried out in a constant ambient temperature. Blood flowing through veins coursing just below the skin plays an important part in controlling heat transfer. The thermal gradient from within the patient to the skin surface covers a large range, and gradients of $0.05-0.5 \, \mathrm{°C \, mm^{-1}}$ have been measured. It might appear that, when there is a large thermal gradient, IR temperature-measuring techniques could be inaccurate because of the effect of radiation that emanates from deeper tissues being transmitted through the skin. Steketee (1973b) has shown that this is not the case for IR radiation of wavelength greater than $1.2 \, \mu\mathrm{m}$.

The surface temperature distribution of a person depends upon age, sex and obesity as well as on metabolism and topography. Superimposed on a general pattern will be a highly specific pattern due to the thermal effects of warm blood flowing through subcutaneous venous networks. Surface temperature gradients will be affected by these vessels providing they are within 7 mm or so of the skin surface (Jones and Draper 1970); the deeper the vessel is, the smaller the surface temperature gradient and the wider the half-height of the thermal profile.

Temperature increments associated with clinical conditions are typically 2–3 °C in magnitude (table 9.2), and prominent patterns due to the effect of underlying vasculature have profiles with widths at half-height of 1–3 cm and gradients of 1–2.5 °C cm^{-1} (figure 9.8). To image these profiles requires equipment of high resolution. Figure 9.9, which shows the distribution of tiny droplets of perspiration over sweat pores on the surface of a finger, illustrates the image quality that high-resolution systems are now capable of producing.

Table 9.2 Details of clinical investigations.

Thermographic investigation	Typical temperature changes, ΔT (°C)	Notes
Malignant disease		
Breast	1.5–4	Tumours with large ΔT tend
Malignant melanoma	2–4	to have a poor prognosis
Soft-tissue tumours	2–5	Vascular patterns are
Bone tumours	2–5	anarchic
Skin lesions	0.5–3	Skin infiltration causes increased ΔT
Vascular studies		
DVT	>1.5	Vascular patterns are significant
Scrotal varicocele	>1(>32)	Examination under stress
	>1.5(32)	Presence of varicocele indicated by ΔT; effect on infertility greater if $T >$ 32 °C
Locomotor-disease studies		
Rheumatic diseases	2	Thermographic index useful for monitoring thermal changes
Trauma		
Burns	1.5–3	Higher-degree burn has
Pain	2, also −1	larger drop in ΔT
Frostbite		Pain is sometimes associated with hypothermia
Therapy studies		
Radiotherapy	2	Radiotherapy skin reactions
Chemotherapy	~ 1.5	increase ΔT
Hormone therapy	~ 1.5	

There are two main methods of thermographic investigation:

(*a*) Examinations made in a constant ambient temperature (within 19–26 °C, depending upon the investigation); this type of study requires patient equilibrium for about 15 min prior to data collection.

(*b*) Examinations on patients undergoing thermal stress; usually this involves the application of a hot or cold load onto an area of the patient's skin or getting the patient to place a hand or foot into a bath of cold (or hot) water for a short time, after which thermal patterns are recorded sequentially.

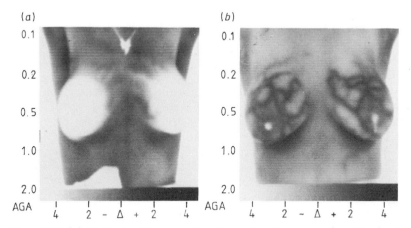

Figure 9.8 (*a*) Patient with avascular thermal pattern over breasts due to mammoplasty. (*b*) Patient with prominent vascular pattern associated with lactation.

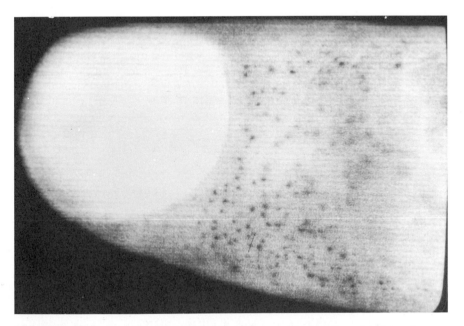

Figure 9.9 High-resolution thermogram of finger showing tiny droplets of perspiration over sweat pores. Image recorded with Barr and Stroud IR18 Mark 2 Camera. (Courtesy of Barr and Stroud).

Most studies are based on the observation of temperature differences (or temperature changes from a known baseline) rather than the measurement of temperature magnitudes. A knowledge of the normal baseline pattern is a fundamental requirement.

9.4.7 Clinical thermography: applications

Medical thermography has a large bibliography (Jones 1983, Ring and Phillips 1984), and the technique has been used to investigate a wide variety of clinical conditions. Most important among these are the following, and we now consider briefly each of these in turn:

(*a*) The assessment of inflammatory conditions such as rheumatoid arthritis.

(*b*) Vascular-disorder studies including
      (i) the assessment of deep vein thrombosis,
      (ii) the localisation of varicosities,
      (iii) the investigation of vascular disturbance syndromes, and
      (iv) the assessment of arterial disease.

(*c*) Metabolic studies.

(*d*) The assessment of pain and trauma.

(*e*) Oncological investigations.

9.4.7.1 The assessment of inflammatory conditions.
Arthritis is frequently a chronic inflammatory lesion that results in overperfusion of tissue and a consequential increase in skin temperature. Thermography is able to distinguish between deep-seated inflammation and more cutaneous involvement. Furthermore, it is useful for evaluating and monitoring the effects of drug and physical therapy. By standardising conditions and cooling peripheral joints so that the skin is within a specific temperature range (26–32 °C for lower limbs and 28–34 °C for the joints of upper limbs), it is possible to quantify the thermal pattern in the form of a thermographic index on a scale from 1.0 to 6.0, in which healthy subjects are usually found to be less than 2.5 and inflammatory joints are raised to 6.0 (Collins *et al* 1974).

9.4.7.2 Investigations of vascular disorders

Deep vein thrombosis (DVT)   Phlebography is the most reliable means of detecting the presence of deep venous thrombosis, but it is a time-consuming and labour-intensive technique. An increase in limb temperature is one of the clinical signs of DVT, and thermography can be used to image the suspected limb. The thermal activity is thought to be caused by the local release of vasoactive chemicals associated with the formation of the venous thrombosis, which cause an increase in the resting blood flow (Cooke 1978). The thermal test is based on the observation of delayed cooling of the affected limb: the patient is examined in a resting position and the limbs cooled, usually with a fan. Recent calf-vein thrombosis produces a diffuse increase in temperature of about 2 °C.

The localisation of varicosities The localisation of incompetent perforating veins in the leg prior to surgery may be found thermographically. The limb is first cooled with a wet towel and fan and the veins are
drained by raising the leg, to an angle of 30–40° with the patient lying
supine. A tourniquet is applied around the upper third of the thigh to
occlude superficial veins and the patient then stands and exercises the
leg. The sites of incompetent perforating veins are identified by areas of
'rapid rewarming' below the level of the tourniquet (Patil *et al* 1970).

Varicosities occurring in the scrotum can alter testicular temperatures.
It is known that the testes are normally kept at about 5 °C lower than
core-body temperature: if this normal temperature differential is abo-
lished, spermatogenesis is depressed. The incidence of subfertility in
man is high and the cause is often not known, but the ligation of
varicocele in subfertile men has been shown to result in improvements
in fertility, with pregnancies in up to 55% of wives (Dubin and Amelar
1977). The presence of a varicocele is not always clinically evident. Even
after surgical ligation, residual dilated veins can affect testicular
temperature adversely. Thermography has been used extensively to
locate varicocele and to enable objective assessment of the efficacy of
surgery. In a cool environment of 20 °C, the surface temperature of the
normal scrotum is 32 °C, whereas a varicocele can increase this to
34–35 °C (Vlaisavljevic 1984).

Vascular disturbance syndromes Patients with Raynaud's disease and
associated disorders have cold and poorly perfused extremities. The
viability of therapeutic intervention will depend upon whether the
vasculature has the potential for increased blood flow. This may be
assessed by subjecting the hands (or feet) to hot or cold stress tests and
comparing the thermal recovery of the skin temperature with that of
healthy limbs (Clark and Goff 1986) (figure 9.10).

The assessment of arterial disease Peripheral arterial disease can
cause ischaemia, necessitating amputation of an affected limb. Tissue
viability of the limb has to be assessed to determine the optimal level of
amputation; this will depend upon skin blood flow, perfusion pressure
and arterial pressure gradients. Skin-temperature studies that show
hypothermal patterns can indicate the presence of arterial stenosis
(Spence *et al* 1981). The non-invasive nature of thermal imaging allows
the method to be used even on patients who are in extreme pain.

9.4.7.3 Metabolic studies. Skin temperature is influenced by the proximity of the skin and superficial tissues to the body core and the effects of
subcutaneous heat production, blood perfusion and the thermal properties of the tissues themselves. Subcutaneous fat modifies surface-
temperature patterns, as does muscular exercise. The complex interplay
of these factors limits the role of infrared imaging in metabolic investigations to the study of the most superficial parts of the body surface. For
example, in the case of new-born infants, it has been postulated that the
tissue over the nape of the neck and interscapular region consists of

brown adipose tissue, which plays an important role in heat production. Thermal imaging has been used to study this heat distribution directly after birth (Rylander 1972). The presence of brown adipose tissue in man has also been investigated by this means: it has been observed that metabolic stimulation by adrenaline (ephedrine) produces an increase in skin temperature in the neck and upper back.

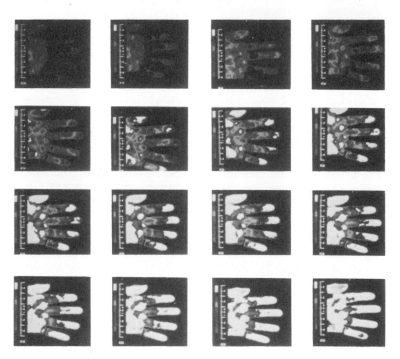

Figure 9.10 Sequence of thermograms taken at 20 s intervals after mild cold stress in a healthy hand. The sequence is from left to right along each row from top to bottom. The palm first warms to the base of the fingers; this is followed by warming of the finger tips due to the opening of the arteriovenous anastomoses. (Courtesy of Dr R P Clark and M R Goff.)

9.4.7.4 The assessment of pain and trauma. The localisation of temperature changes due to spinal-root-compression syndromes, impaired sympathetic function in peripheral nerve injuries and chronic-pain syndromes depends largely upon the finding that in healthy subjects thermal patterns are symmetrical. Asymmetrical heat production at dermatomes and myotomes can be identified thermographically (Wexler 1979). Temperature changes are probably related to reflex sympathetic vasoconstriction within affected extremity dermatomes and to metabolic changes or muscular spasm in corresponding paraspinal myotomes.

Thermography can be used to assess tissue damage caused by a burn or frostbite. The treatment of a burn depends upon the depth of injury and the surface area affected. Whereas a first-degree burn shows a skin erythema, a third-degree burn is deeper and shows a complete absence of circulation. Identification of a second-degree burn is sometimes difficult, and temperature measurements are used to assist with this assessment. Third-degree burns have been found to be on average 3 °C colder than surrounding normal skin. In conditions of chronic stress (such as bed sores, poorly fitting prosthetic devices), thermal imaging can be used to assess irritated tissue prior to frank breakdown.

9.4.7.5 Oncological investigations. Thermal imaging has been used as an adjunct in the diagnosis of malignant disease, to assess tumour prognosis and to monitor the efficacy of therapy. Malignant tumours tend to be warmer than benign tumours due to increased metabolism and, more importantly, due to vascular changes surrounding the tumour. It has been observed that surface-temperature patterns are accentuated and temperature differences increased by cooling the skin surface in an ambient of 20 °C. This procedure reduces blood flow in the skin and subcutaneous tissues and, since blood flow through tumour vasculature is less well controlled than through normal vasculature, the effects of cooling the skin surface are less effective over the tumour than over normal tissue. Imaging has been used to determine the extent of skin lesions and to differentiate between benign and malignant pigmented lesions. The method has been used by many investigators as an aid in the diagnosis of malignant breast disease, but it lacks sensitivity and specificity. It has been advocated as a means of identifying and screening 'high-risk groups' and for selecting patients for further investigation by mammography (Gautherie and Gros 1980). Breast tumours that cause large temperature changes (more than 2.5 °C) tend to have a poor prognosis (Jones *et al* 1975).

The treatment of malignant disease by radiotherapy, chemotherapy or hormone therapy can be monitored thermographically. Serial temperature measurements indicating temperature drops of 1 °C or more are usually consistent with tumour regression.

9.5 LIQUID-CRYSTAL THERMOGRAPHY

Thermochromic liquid crystals are a class of compound that exhibit colour–temperature sensitivity. They are used encapsulated in pseudo-solid powders and incorporated in a thin film with a black background to protect the crystals from chemical and biological contamination. The reflective properties of cholesteryl nonanoate and chiral nematic liquid crystals are temperature-dependent: when viewed on a black background, the scattering effects within the material give rise to iridescent colours, the dominant wavelength being influenced by small changes in temperature. The absolute temperature range of response of each plastic film depends upon the liquid-crystal constituents, but is typically 3 °C. When used clinically, the plate should be placed in uniform contact with the skin surface, but this is not always possible. The response time varies according to plate thickness (which ranges from 0.06 to 0.3 mm) and is typically 20–40 s. Although the method provides an inexpensive method of recording temperature distributions, it suffers from the major disadvantage that it can alter the temperature pattern it is being used to measure.

9.6 MICROWAVE THERMOGRAPHY

Emission of IR radiation from tissue at 30 °C is most copious at wavelengths around 10 μm; the intensity of radiation at this wavelength is greater by about 10^8 than that of 10 cm wavelength radiation emitted from the same source. However, suitably designed microwave radiometers placed in contact with the skin surface can be used to detect this radiation. Since body tissue is partially transparent to this radiation, the method can be used to estimate body temperatures at depths of a few centimetres. Surface probes sensitive to 1.3 GHz (23 cm wavelength) and 3.3 GHz (9.1 cm wavelength) radiation have been used experimentally for this purpose (Myers *et al* 1979, Land *et al* 1986). Thermal imaging has been achieved with 30 and 68 GHz radiations, but at these frequencies only radiation originating in the most superficial layers of body tissue is detected. Microwave computed tomography is discussed in Chapter 10.

REFERENCES

ARRIDGE S R, COPE M, VAN DER ZEE P, HILLSON P J and DELPY D T 1986 Near infrared transillumination as a method of visualization and measurement of the oxygenation state of brain and muscle in newborn infants *Recent Developments in Medical and Physiological Imaging* Suppl. to *J. Med. Eng. Technol.* ed R P Clark and M R Goff (London: Taylor and Francis) pp24–31

CLARK J A 1976 Effects of surface emissivity and viewing angle on errors in thermography *Acta Thermogr.* **1** 138–41

CLARK R P and GOFF M R 1986 Dynamic thermography in vasospastic diseases *Recent Developments in Medical and Physiological Imaging* Suppl. to *J. Med. Eng. Technol.* ed R P Clark and M R Goff (London: Taylor and Francis) pp95–101

COLLINS A J, RING E F J, COSH J A and BACON P A 1974 Quantitation of thermography in arthritis using multi-isotherm analysis. 1. The thermographic index *Ann. Rheum. Dis.* **33** 113–15

COOKE E D 1978 *The Fundamentals of Thermographic Diagnosis of Deep Vein Thrombosis* Suppl. 1 to *Acta Thermogr.* (Padova: Bertoncello Artigrafiche)

DUBIN L and AMELAR R 1977 Varicocelectomy: 986 cases in a twelve-year study *Urology* **10** 446–9

GAUTHERIE M and GROS C M 1980 Breast thermography and cancer risk prediction *Cancer* **45** 51–6

GOSS A J, NIXON R D, WATTON R and WREATHALL W M 1984 Infrared television using the pyroelectric vidicon EEV thermal television *SPIE 28th Annual Technical Symp., San Diego* (Chelmsford: English Electric Valve Co Ltd) pp1–6

JONES C H 1982 Review article: methods of breast imaging *Phys. Med. Biol.* **27** (4) 463–99

—— 1983 Thermal imaging *Imaging With Non-Ionising Radiations* ed D F Jackson (Glasgow: Surrey University Press) pp151–216

—— 1987 Medical thermography *IEE Proc. A* **134** (2) 225–35

JONES C H and CARNOCHAN P 1986 Infrared thermography and liquid crystal plate thermography *Physical Techniques in Clinical Hypothermia* ed J W Hand and R J James (Letchworth: Research Studies Press) pp 507–47

JONES C H and DRAPER J W 1970 A comparison of infrared photography and thermography in the detection of mammary carcinoma *Br. J. Radiol.* **43** 507–16

JONES C H, GREENING W P, DAVEY J B, McKINNA J A and GREEVES V J 1975 Thermography of the female breast: a five year study in relation to the detection and prognosis of cancer *Br. J. Radiol.* **48** 532–8

LAND D V, FRASER S M and SHAW R D 1986 A review of clinical experience of microwave thermography *Recent Advances in Medical and Physiological Imaging* Suppl. to *J. Med. Eng. Technol.* ed R P Clark and M R Goff (London: Taylor and Francis) pp109–13

LAWSON W D 1979 Thermal imaging *Electronic Imaging* ed T P McLean and P Schagen (New York: Academic Press) pp325–64

MOUNT L E 1979 *Adaptation to Thermal Environment—Man and His Productive Animals* (London: Edward Arnold) p146

MULLARD 1982 *Electronic Components and Applications* vol 4, no. 4, Mullard Technical Publ. M82-0099

MYERS P C, SADOWSKY N L and BARRETT A H 1979 Microwave thermography: principles, methods and clinical applications *J. Microwave Power* **14** 105–13

OHLSSON B, GUNDERSEN J and NILSSON D M 1980 Diaphanography: a method for evaluation of the female breast *World J. Surg.* **4** 701–5

PATIL K D, WILLIAMS J R and LLOYD-WILLIAMS K 1970 Localization of incompetent perforating veins in the leg *Br. Med. J.* **1** 195–7

RING E F J and PHILLIPS B (ed) 1984 *Recent Advances in Medical Thermology* (New York: Plenum)

RYLANDER E 1972 Age dependent reactions of rectal and skin temperatures of infants during exposure to cold *Acta Paediatr. Scand.* **61** 597–605

SPENCE V A, WALKER W F, TROUP I M and MURDOCH G 1981 Amputation of the ischemic limb: selection of the optimum site by thermography *Angiology* **32** (3) 155–69

STEKETEE J 1973a Spectral emissivity of skin and pericardium *Phys. Med. Biol.* **18** 686–94

—— 1973b The effect of transmission on temperature measurements of human skin *Phys. Med. Biol.* **18** 726–9

VLAISAVLJEVIC V 1984 Thermographic characteristics of the scrotum in the infertile male *Recent Advances in Medical Thermology* ed E F J Ring and B Phillips (New York: Plenum) pp415–20

WATMOUGH D J 1982 Light torch for transillumination of female breast tissues

Br. J. Radiol. **55** 142–6

WATMOUGH D J, FOWLER P W and OLIVER R 1970 The thermal scanning of a curved isothermal surface: implications for clinical thermography *Phys. Med. Biol.* **15** (1) 1–8

WATTON R, SMITH C, HARPER B and WREATHALL W M 1974 Performance of the pyroelectric vidicon for thermal imaging in the 8–14 micron band *IEEE Trans. Electron Devices* **ED-21** (8) 462–9

WEXLER C E 1979 Lumbar, thoracic and cervical thermography *J. Neurol. Orthop. Surg.* **1** 37–41

WOLFE W L and ZISSIS O O (ed) 1978 *The Infrared Handbook* (Washington, DC: Naval Research Department of the Navy)

CHAPTER 10

IMAGING OF TISSUE ELECTRICAL IMPEDANCE

S WEBB

10.1 THE ELECTRICAL BEHAVIOUR OF TISSUE

It is an exciting possibility to consider making images of the spatial variation of the electrical properties of biological tissue. Biological tissue exhibits at least two important passive electrical properties. First, it comprises free charge carriers and may thus be considered an electrical conductor. It would be expected that electrical conductivity is a characteristic property of different tissues and that images of electrical conductivity may resolve structure and even be indicative of pathology. Secondly, tissue also contains bound charges leading to dielectric effects, and it might also be possible to form an image of relative electrical permittivity.

Both the electrical conduction current and displacement current, which arise when a potential gradient is applied to tissue, are frequency-dependent. The conduction current is, however, only very weakly dependent on frequency and may be thought of as essentially constant over six decades between 10 Hz and 10 MHz. The displacement current exhibits strong frequency dependence. Below 1 kHz, the conduction current is some 3.5 orders of magnitude greater than the displacement current, and even at 100 kHz the conduction current is still 2 orders of magnitude greater for soft biological tissue. It is therefore reasonable to neglect dielectric effects when considering measurements of tissue impedance at frequencies at or below 100 kHz. If an image of a biological structure could be made at several frequencies, then it would be expected that different structures would be seen at different frequencies in view of the complex impedance. At very high frequencies, however (more than about 200 kHz), stray wiring capacitance becomes a serious experimental problem. The concept of producing spatial images of electrical conductivity (or the inverse, resistivity or specific impedance) is attractive but extremely difficult, and there are currently no commercial machines based on these principles. There is, however, a body of research that has led to the development of a few prototype systems (Brown 1983).

When low-frequency currents are applied to the skin via a pair of electrodes, a threshold of sensation may be found, which increases with frequency by some two decades as frequency varies between 10 Hz and 10 kHz. The dominant electrical mechanism for sensation also changes with frequency, and there are three identifiable causes of sensation. At low frequency, the sensation arises because of local electrolysis. The body is electrically most sensitive at around 50 Hz, explaining the potential danger of domestic electric shock. At middle frequencies electrolysis appears to be reversible, and neural stimulation is the dominant mechanism. At much higher frequencies, in the range 10–100 kHz, the dominant biological mechanism is heating of the tissue. In view of the inherent delay for a nerve to respond to stimulus as a result of the low propagation speed of an impulse along a nerve, alternating currents, of some tens of kilohertz may be used to probe the body without any danger to heart, nerves and muscles.

For electrical conductivity to be a useful property to attempt to image, it is necessary to know whether the conductivity of biological material is linear with electric field strength, and there is experimental evidence to support linearity to within 5% in the region 100 Hz to 100 kHz. There is a large body of experimental data on the electrical conductivity of biological tissue, but there are many discrepancies in the reported literature, which is not surprising when one considers the difficulty of making measurements *in vivo*. Much of this difficulty arises from electrode contact impedance. It is believed that, at around 50 kHz, cell membranes are insulators and the electric current flows around the cell boundaries. To give a sense of reference we note, for example, that, at room temperature, blood has a resistivity of 1.5 Ω m, liver 3–6 Ω m, neural tissue 5.8 Ω m (brain), 2.8 Ω m (grey matter), 6.8 Ω m (white matter), bone 40–150 Ω m and skeletal muscle 1–23 Ω m (anisotropic resistivity). For comparison, the resistivity of sea water is less than 1 Ω m. These values arise from the different concentrations of electrolytes in the individual organs and tissues. The value for liver depends on the blood content. A very detailed table of resistivities for biological tissue is provided in the review by Barber and Brown (1984), including full references to the sources of these and other measurements. It should then be possible to differentiate tissues *in vivo*, and indeed it is noteworthy that electrical conductivity can vary very greatly between two tissues whose x-ray linear attenuation coefficients may be somewhat similar and whose appearance in x-ray CT would then be more difficult to discriminate. For example, table 10.1 shows how five biological tissues compare. Muscle and blood, for example, have the same x-ray linear attenuation coefficient at x-ray energies typical of CT scanners, but differ by almost a factor of 2 in resistivity. The ratio of bone to muscle is about 50 for resistivity but less than 2 for linear attenuation coefficient. This is a further example of the general truism (see introduction) that new imaging modalities tend to complement rather than replace existing ones, since they generally rely on different physical mechanisms.

In view of the ratios of resistivity of blood, fat and muscle, it is necessary to remember that, when the volume of blood changes in the region being imaged, the gross resistivity of blood-filled tissues also changes. It is possible to evaluate the magnitude of this effect in some cases by noting the (possibly anisotropic) expansion of (say) a limb, whose volume will be greater during systole than during diastole, and using the elementary formula $R = \rho L/A$ relating the resistance R to the length L, area A and resistivity ρ of a cylinder. Such expansion leads to non-trivial changes in electrical resistance. The resistivity of body fluids *in vitro* can be readily obtained using laboratory conductivity cells, but measurements *in vivo* are considerably more difficult to perform.

Table 10.1 Typical values of resistivity and x-ray linear attenuation coefficient for five biological tissues. (From Barber *et al* (1983).)

Tissue	Resistivity (Ω m)	X-ray attenuation coefficient (m^{-1})
Bone	150	35.0
Muscle	3.0	20.4
Blood	1.6	20.4
Fat	15.0	18.5
Cerebrospinal fluid	0.65	20.0

10.2 TISSUE IMPEDANCE IMAGING

The major difficulty to be overcome in constructing maps of tissue impedance is that caused by the divergence of the electrical field lines when a potential is applied between two electrodes attached to the tissue. Although the curvilinear field lines can be predicted for uniform tissue, for the real case of unknown inhomogeneous tissue even this becomes impossible. The measurement made at a recording electrode does not relate directly to the bioelectrical properties along a straight line to the source electrode. It is simply not true that, if a potential gradient is applied and an exit-current measurement made, the summed electrical resistance along the path between the electrodes is the ratio of the potential difference to the current. If this difficulty did not arise, one might imagine making measurements of the straight-line integral resistivity in many directions and employing identical reconstruction techniques to those used in x-ray computed tomography (CT) and single-photon emission computed tomography (SPECT) (see Chapters 4 and 6). If we consider for the moment the electrical analogue of a planar x-radiograph, two experimental possibilities arise. Either one could constrain the electrical field lines to become straight by applying guard potentials to the (small) recording electrodes, or one could, with unguarded electrodes, attempt to backproject along curved isopotentials to cope with the complicated field pattern. The latter is more difficult. Considering the former, one could imagine applying a large pair of electrodes to a slab of biological tissue and making measurements of exit current. In such an arrangement, the field lines are relatively straight (for uniform tissue) away from the periphery of the electrodes, and only bend at the edges. If one now imagined that the recording electrode were divided up into a matrix of small recording electrodes, the electric field lines remain relatively straight between each and the driving electrode (for uniform tissue), and straight-line integral resistivity measurements may be made. In this experimental arrangement of essentially guarded electrodes, the straight-line assumption holds exactly only if the tissue is uniform, and for electrically inhomogeneous material this becomes only a first-order approximation. Henderson and Webster (1978) have reported measurements, using an impedance camera, of integral thoracic impedance by such a technique but, in addition to poor spatial resolution, image distortion was inevitable due to electrical inhomogeneity (figure 10.1).

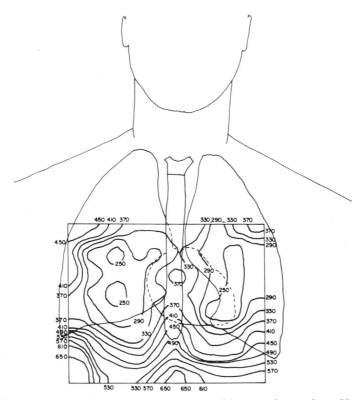

Figure 10.1 Isoadmittance contour map of human thorax from Henderson and Webster's impedance camera. (Reprinted from Henderson and Webster (1978). ©1978 IEEE.)

These experimental arrangements strongly suggest a method of impedance computed tomography by which rotating the electrode array would generate straight-line integral resistivity measurements at different orientations. From these projections, reconstruction within parallel geometry is possible. This has been attempted, but the results are unsatisfactory. The reason lies in the susceptibility of the technique to errors in the data that arise from the finite electrode impedance itself, which distorts the measurement of exit current. Against this background, it is difficult not to conclude that both planar imaging and CT with guard-electrode systems and measurement of exit current are unlikely to be useful techniques in clinical practice. Additionally, Bates *et al* (1980) provide a convincing argument to show that conventional CT approaches to impedance imaging are subject to fundamental limitations, which make the solutions non-unique. Although the method is just possible for certain unrealistic distributions of conductivity, in general the method seems unlikely to succeed.

A more viable image reconstruction technique, known as applied potential tomography (APT), and extensively developed in Sheffield, UK (Barber and Brown 1984), relies for its physical basis on a departure from the measurement of exit current in favour of the measurement of the potential distribution arising from an impressed current. This technique is also referred to as electrical impedance imaging, conductivity imaging, impedance CT and electrical impedance tomography. The technique is as follows.

When a current is applied to two points on the surface of a conducting biological material, a potential distribution is established within the material. If this is sampled by making measurements of peripheral potential, the ratio of such measurements to what would be the expected peripheral potentials if the medium were uniform can be backprojected along the (uniform material) curved isopotentials to create an image. Repeating the procedure for different locations of the stimulating electrodes leads to a form of computed tomography. Imagine a circular planar distribution of electrical conductivity (or equivalently an irregularly shaped biological specimen immersed in a circular saline bath) around which are arranged N electrodes. When current is applied across a pair of electrodes at each end of a diagonal of the system, $(N - 3)$ measurements of peripheral potential gradient may be made. (No measurements may be made connecting to the driven electrodes.) If for the moment we imagine the electrical conductivity to be uniform, then an exact solution to Laplace's equation for the electrical potential distribution exists, predicting the peripheral potential gradients between the electrodes. Indeed, for symmetrically placed electrodes and assuming a cylindrical object for measurement, the electrodes may be diagrammatically joined by curved equipotentials defining bands of electrical conductivity within the circle (see figure 10.2). The departure of the measured peripheral potential gradients from these calculated gradients may then be interpreted as arising from departures of the integral conductivity within these bands from uniform values as a result of the spatial dependence of electrical conductivity. Where the band conductivity between equipotentials is lower, the peripheral potential gradient will be higher, and vice versa. This explains the philosophy behind backprojection of the measured-to-calculated peripheral potential gradient along curved equipotentials, which should then generate the relative distribution of electrical conductivity, in the limit of many backprojections for every pair of peripheral electrodes. The system of backprojection along such curved isopotentials is difficult to handle mathematically. The techniques described in Chapter 4 for x-ray CT and SPECT do not apply. In practical realisations of this scheme, it has proved simpler to *measure* the electrode potentials for a uniform circle of conductivity by putting uniform saline solution in the field of view. This generates the required isopotentials experimentally rather than by calculation. Alternatively, dual-frequency data collection enables the construction of static images without the need for measurements or calculations for uniform conductivity (Griffiths and Ahmed 1987).

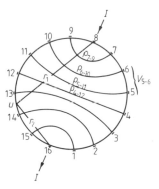

Figure 10.2 The isopotentials for a circular conducting system in which current is applied across diagonally opposite electrodes. (From Brown (1983).)

If there are N electrodes, there are $N(N-1)/2$ possible pairs through which the current may be driven, and for each of these configurations one might naively think N measurements can be made, leading to a total of $N^2(N-1)/2$ measurements. These are not all independent, however, and considerations of reciprocity lead to the result that there are only $N(N-3)/2$ independent measurements. We see that for a typical experimental arrangement of 16 electrodes, only 104 independent measurements can be made, and one could argue that the reconstructed pixel values are only independent if the matrix size is 10×10 or smaller. Note that, in the practical realisation, the stimulation is not restricted to diagonally opposed electrodes, but instead is applied to adjacent electrodes. Use of this configuration appears to provide the best resolution, although not the highest sensitivity (Barber and Brown 1986). The isopotentials corresponding to this experimental arrangement are shown in figure 10.3, and the shaded area is that 'receiving' backprojected data from the measurement across electrodes 5 and 6.

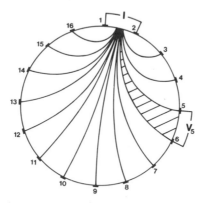

Figure 10.3 The isopotentials of a practical APT system for uniform conductivity. (From Brown (1986).)

There are several factors that work in favour of this method of image reconstruction. The measurement of peripheral potential is inherently accurate to within 1% over a dynamic range of 40 dB. No electrode guarding is necessary and the measurements are not affected by the electrode contact impedance, which is much higher than the tissue impedance. We recall that the latter was the major impediment to image reconstruction using measured exit currents. A guard ring above and below the slice of interest can constrain the currents to the slice of interest (Price 1979).

The reconstructed image, being a backprojection, is blurred (as is the case for all CT backprojections), and it would be desirable to deconvolve the point response function of the system (see Chapter 12). Using a detailed analysis of the reconstruction algorithm, Barber and Brown (1986, 1987) show that the point-source response function is not spatially invariant. It proves possible, however, to convert the filtration problem to a position-independent one by a coordinate transformation. After deconvolution, the reverse transformation is applied to the result. Additionally, the method is only accurate to first order because the experimental data are backprojected along the equipotentials corresponding to a uniform distribution of conductivity rather than the true non-uniform distribution, giving rise to the measured peripheral potentials, and in which the isopotentials are somewhat different from in the uniform case. Iterative reconstruction schemes have been devised in which the first-order isopotentials are recomputed from the first-order backprojected distribution of conductivity and used to recompute difference projections prior to re-backprojection along the new first-order isopotentials.

Distinguishing features of applied potential tomography are the possibility of very rapid data collection (about 0.5 s or less per slice), modest computing requirements, relatively low cost and no known significant hazards. Against these may be set the limitations described of poor spatial resolution and technical difficulties arising from the oversimplified assumptions concerning the distribution of the isopotentials. Brown *et al* (1985) have estimated that the best resolution that applied potential tomography might expect to achieve is of the order 1.5% of the reconstructed field diameter (with 128 electrodes), but presently the experimental limit has been 10% of the reconstructed-field diameter. When the number of drive electrodes is increased, the practical problems of data collection rise, and Brown and Seagar (1985) have provided a detailed analysis of the factors involved. The data collection may be serial or parallel. The currents must, of course, be applied serially, but the recording electrodes can be read in parallel providing an arrangement employing one amplifier for each electrode was used. In the system described by Brown (1983), the data collection is in fact serial but still takes less than a second.

Figure 10.4 shows an experimental 2D reconstruction of data from this 16-electrode system for a human forearm in which the bones and major blood vessels are resolved. It seems that at present the best that APT is able to achieve is the resolution of gross anatomy and a ranking of such in terms of the relative conductivity.

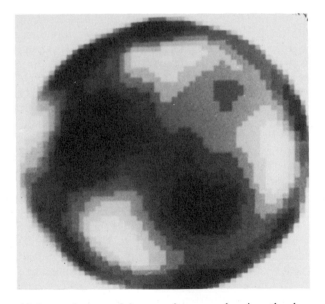

Figure 10.4 APT image of human forearm showing the bones and blood vessels. (From Brown *et al* (1985).)

10.3 SUGGESTED CLINICAL APPLICATIONS OF APPLIED POTENTIAL TOMOGRAPHY

It must be stated at the outset that APT is not currently at the same state of development as most of the imaging modalities described in this volume. Commercial equipment is *not* available and *in vivo* imaging has only been performed in the laboratory at several centres to demonstrate the potential of APT in a number of novel clinical applications. In this section, we therefore report on those proposed applications for which there is some limited evidence of clinical viability. In fairness to the few workers in this field, the images must be judged with some generosity, mindful of the physical limitations of tissue impedance imaging and the present lack of the impetus that would arise if there were interest from a manufacturer and widespread clinical trials. At the time of writing, the most recently expressed clinical view of the usefulness of APT has been provided by Dawids (1987).

First, let us note the extraordinary sensitivity of the method to changes in water content. Figure 10.5 shows a 2D reconstructed plane through a grapefruit (whose integral resistivity is similar to that of the human head) into which just 1 ml of tap water had been injected. The 16-electrode system used to make this image has a resolution limited to one-tenth of the reconstructed field diameter, but 1% changes in resistivity may be seen (Brown 1986). Thus it is apparent that, whereas APT can never be expected to compete with x-ray CT and nuclear magnetic resonance (NMR) for high-resolution anatomical imaging, it may offer a new route to rapid dynamic measurement of physiological change. In this respect, it is more reasonable to make comparisons between what is offered and measured by APT and the imaging of physiological function using labelled pharmaceuticals and a gamma camera. The most likely first clinical implementations of APT will be dynamic rather than static studies. With typical data collection times of about 0.1 s, changes in impedance during the cardiac cycle should be amenable to imaging. During dynamic imaging, static structures do not appear in the images, although they may contribute to image artefacts (Brown *et al* 1985). For example, figure 10.6 shows the appearance of normal lungs from APT in which the distribution of conductivity before and after inspiration were compared. The data collection time was 0.1 s. Inflated lungs are more resistive (23 Ω m compared to 7 Ω m deflated) due to current flow around the aerated alveoli. It should be possible to image pulmonary oedema (fluid in the lungs) in this way, since this would appear as a region of reduced resistivity. The symptom is a consequence of many diseases and, although pulmonary oedema is usually readily diagnosed by clinical and simple radiological means (chest x-ray), any technique that allowed quantitation of the amount of water present would be useful, for example, in toxicological studies. It may also be interesting to look at differences in resistivity between simple pulmonary oedema, i.e. a fluid overload, and oedema associated with, for example, pulmonary infection. In particular, this would be useful if it allowed a more definite discrimination between pulmonary oedema and some of the unusual pulmonary infections that can appear very similar on a chest radiograph (Cherryman 1987). Normal lung comprises 80% air and only 5% fluid (Brown *et al* 1985), and it has been estimated that volumes of fluid as small as 10 ml should be able to be imaged. There is a possibility of using APT to follow the course of treatment in established cases of pulmonary oedema. Changes in pulmonary resistivity during inspiration for six normal volunteers have been mapped using APT techniques by Harris *et al* (1987). An impedance index derived from the images was shown to vary linearly with the volume of air inspired up to 5 l if the subject sits upright. The function of right and left lungs may be observed separately. In particular, it was demonstrated that for a normal volunteer lying in the lateral decubitus position (on his/her right-hand side), the upper (left) lung increases in impedance more rapidly than the right for a low inspiration volume, possibly due to a reduction in the compliance of the lower lung under gravity. Harris *et al* (1987) also provide images of a patient with emphysematous bulla in the right lung, the defect being conspicuous by non-uniform changes in impedance over the affected lung. The differential impedance in the lung between inspiration and expiration has been used via impedance plethysmography to provide a gating signal for a gamma camera in order to reduce respiratory artefacts by collecting data during only part of the respiratory cycle (Heller *et al* 1984). This is claimed to be more appropriate than spirometry for extremely ill patients.

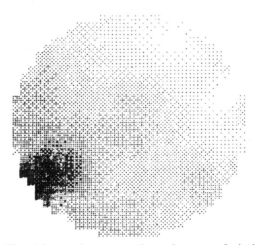

Figure 10.5 The image of a 2D section of a grapefruit into which 1 ml of tap water has been injected. (From Brown (1986).)

Experiments have also been performed to demonstrate gastric transit by rapid dynamic imaging of the passage of a drink. It has been argued (Brown 1986) that the method might replace mechanically invasive and unpleasant procedures by which dyes are transmitted to the stomach or radiologically invasive procedures by which technetium-labelled meals may be dynamically imaged by a gamma camera. The possibility is intriguing. A detailed comparison of electrical APT and radionuclide studies of gastric emptying is included in a recent review of APT by Brown *et al* (1986), and Mangall *et al* (1987) have performed detailed studies that validate the technique.

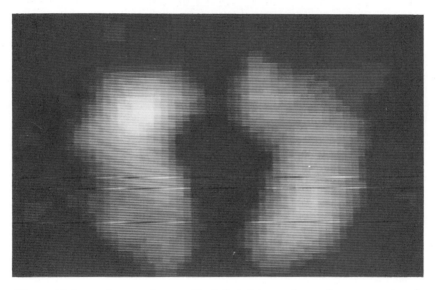

Figure 10.6 APT image of normally inflated lungs (no pulmonary oedema). (From Brown (1986).)

The most recently suggested (and demonstrated) application of APT is to measuring cardiac-related changes in resistivity within the cardiac–pulmonary system due to the movement of blood (Barber and Brown 1987). Dynamic movies have been constructed showing the movement of blood between the right ventricle of the heart and the lungs, with subsequent drainage into the left ventricle. It should be possible to detect pulmonary embolism, which would present as a lung region with good ventilation but poor perfusion. Eyuboglu *et al* (1987) have pointed to the necessity to perform signal averaging, time locked to the electrocardiograph (ECG), to allow cancellation of the respiratory effect if the heart rate is not close to the breathing rate. This is to discriminate between cardiac-related changes in electrical impedance ($0.2\ \Omega$ for a healthy adult) and respiratory changes in impedance ($2\ \Omega\ l^{-1}$), both of which vary the electrical resistivity of the thorax.

So far uninvestigated suggestions are that APT might be a useful monitor of tissue viability, since well defined changes in resistivity occur when tissue dies. This could have some relevance to the monitoring of tumour development during and after therapy. Since blood has a lower resistivity than many body tissues, the presence of increased blood flow might also be amenable to imaging by APT. Blood-flow changes are also known to be sequelae of hyperthermia and, if it were not for this complication, an intriguing possibility of using APT to measure temperature during hyperthermia might arise since resistivity is, of course, sensitively temperature-dependent. The temperature coefficient of the impedance of muscle tissue is approximately 2% $°C^{-1}$, making the technique more sensitive than (say) x-ray CT, where the temperature coefficient of the x-ray attenuation coefficient is only 0.04% $°C^{-1}$. The measurement of temperature in this context is notoriously difficult (Hand 1985, Carnochan *et al* 1986). Some *in vitro* and *in vivo* laboratory experiments have been reported (Conway *et al* 1985). In the *in vitro* study, the heating and cooling curves (temperature versus time) recorded by APT and a thermocouple closely correlated when an agar phantom was heated by microwaves. *In vivo* the scapula region of a volunteer was heated by 7 °C, monitored by APT. Once more, the somewhat coarse spatial resolution of APT would presently preclude its use for monitoring the small-scale temperature gradients in tumours being treated by hyperthermia.

The measurement of intraventricular haemorrhage (IVH) in the neonatal head has also been attempted. This application is potentially very important since IVH is an important cause of death and handicap in premature babies. It is almost always within the first few days of life, bleeding into the cerebral ventricles displacing the cerebrospinal fluid that normally fills them. Tarassenko and Rolfe (1984) have proposed an APT system of imaging whose principal advantage is that it provides for continuous monitoring with little disturbance to the normal care of the infant. Murphy *et al* (1987) have presented clinical images of IVH in the neonate. The APT imaging system at Oxford is also used to study gastric emptying and pulmonary dysfunction in the premature new-born. The pattern of gastric emptying in the older infant with congenital heart disease, gastro-oesophageal reflux or pyloric stenosis may also be studied by APT, avoiding methodological and ethical problems in applying techniques such as dye dilution and scintigraphy.

It has also been suggested that APT could be used to study neural changes from afferent stimuli and perhaps even identify the focus of an epileptic attack (Brown *et al* 1985). Another untested but intriguing possibility is to exploit the different resistivities of fat and muscle (5:1 ratio) to aid nutritional studies by measurement of lean:fat ratios and in particular those within a short timescale.

It has also been possible to make a hand-held linear probe of electrodes and investigate the distribution of resistivity to a depth half that of the length of the probe. The technique has increased sensitivity over ring-system APT and is, in this respect, the APT equivalent of surface coils in NMR imaging (see Chapter 8). Powell *et al* (1987) have constructed an array and used it to make preliminary images of the chest wall. Currently they conclude that the design of the optimal array should depend on the clinical task envisaged, and this development is still awaited. It has been pointed out that the imaging principle is similar to that used for electrolocation in the mildly electric South American knife fish and the African elephant nose fish.

Jossinet *et al* (1981) developed a system using a ring of $N = 128$ electrodes, which might be placed around the breast, making (*sic*) 8128 independent measurements. The conductivity of the normal and malignant breast has been reported to differ by a factor of 3 (Nyboer 1970). There has been a great deal of interest in new imaging modalities that might offer non-invasive and potentially harmless investigation of breast disease. The implications are very important and APT is an attractive possibility. The necessary requirements of a breast imaging method to rival mammography have already been discussed in Chapter 2 (see also Chapter 11). Skidmore *et al* (1987) have designed and constructed a battery-operated, 16-electrode ring, operating at 90 kHz, which completely encompasses the breast and makes contact with gel. The breast is manipulated so that good contact is made with each electrode. The ring encases the electronics, reducing the interference difficulties of having the electrodes remote from the electronics. Data have been recorded, but to date no clinical images have been reconstructed.

It has also been noted that there is a large difference in microwave absorption between fatty and non-fatty substances and, since breast tissue has a large component of fatty tissue whilst carcinoma tends to be non-fatty, microwave computed tomography is another potential candidate for breast examination. Microwave CT should produce no long-lasting biological damage. A laboratory microwave CT scanner has already been reported (Rao *et al* 1980) utilising 10.5 GHz, 10 mW microwaves and a superheterodyne detector capable of detecting power levels to 10^{-11} mW. The scanner used first-generation rotate–translate geometry (see Chapter 4). It is important to provide good impedance matching with the sample, and ethylene glycol, with an attenuation coefficient of 4.3 cm^{-1}, has been suggested for this purpose. Figure 10.7 shows an image of a frankfurter obtained from 36 equispaced projections in $0 \to \pi$ with a detector sampling of 1.6 mm. The spatial resolution is of the order of 2 cm. At present, microwave CT also remains a laboratory tool whose clinical importance has yet to be explored.

APT is a new and exciting development. It is probably apparent that most of the clinical studies to date and cited here have emerged from just a few centres that are pioneering the work (Sheffield, Oxford, Cardiff and Bristol). At other centres, equipment is still at the developmental stage, with image processing being simulated. Some centres, on their own admission, are experiencing less success with *in vivo* imaging. In the spirit of the opening paragraph of this section, it is concluded that currently APT shows much clinical promise, can rival existing alternative techniques for selected applications, but retains significant disadvantages that probably destine its arrival in the clinic to be a slow process rather than an acclaimed overnight innovation.

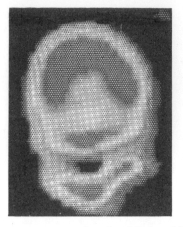

Figure 10.7 Microwave CT scan of a frankfurter (diameter 1.8 cm) and a test tube of ethylene glycol in polyethyleneglycol, separated by a centre-to-centre distance of 3.1 cm. The frankfurter is clearly imaged as its water content offers a large contrast to the polyethyleneglycol coupling medium, but the ethylene glycol is poorly imaged as its characteristics are mid-way between those of water and polyethyleneglycol. (From Rao *et al* (1980).)

REFERENCES

BARBER D C and BROWN B H 1984 Applied potential tomography *J. Phys. E: Sci. Instrum.* **17** 723–33
—— 1986 Recent developments in applied potential tomography—APT *Information Processing in Medical Imaging* ed S L Bacharach (The Hague: Nijhoff)
—— 1987 Construction of electrical resistivity images for medical diagnosis *SPIE Conf., Long Beach* Paper 767 04
BARBER D C, BROWN B H and FREESTON I L 1983 Imaging spatial distributions of resistivity using applied potential tomography *Electron. Lett.* **19** 933–5
BATES R H T, McKINNON G C and SEAGAR A D 1980 A limitation on systems for imaging electrical conductivity distributions *IEEE Trans. Biomed. Eng.* **BME-27** (7) 418–20
BROWN B H 1983 Tissue impedance methods *Imaging With Non-Ionising Radiations* ed D F Jackson (Guildford: Surrey University Press)
—— 1986 Applied potential tomography *Phys. Bull.* **37** 109–12
BROWN B H, BARBER D C, HARRIS N and SEAGAR A D 1986 Applied potential tomography: a review of possible clinical applications of dynamic impedance imaging *Recent Developments in Medical and Physiological Imaging* Suppl. to *J. Med. Eng. Technol.* ed R P Clark and M R Goff (London: Taylor and Francis) pp8–15
BROWN B H, BARBER D C and SEAGAR A D 1985 Applied potential tomography: possible clinical applications *Clin. Phys. Physiol. Meas.* **6** (2) 109–21
BROWN B H and SEAGAR A D 1985 Applied potential tomography—data collection problems *Proc. IEE Conf. on Electric and Magnetic Fields in Medicine and Biology, London* Conf. Publ. no. 257, pp79–82

CARNOCHAN P, DICKINSON R J and JOINER M C 1986 The practical use of thermocouples for temperature measurement in clinical hyperthermia *Int. J. Hyperthermia* **2** 1–19

CHERRYMAN G 1987 Personal communication

CONWAY J, HAWLEY M S, SEAGAR A D, BROWN B H and BARBER D C 1985 Applied potential tomography (APT) for noninvasive thermal imaging during hyperthermia treatment *Electron. Lett.* **21** 836–8

DAWIDS S G 1987 Evaluation of applied potential tomography: a clinician's view *Clin. Phys. Physiol. Meas.* **8** Suppl. A 175–80

EYUBOGLU B M, BROWN B H, BARBER D C and SEAGAR A D 1987 Localisation of cardiac related impedance changes in the thorax *Clin. Phys. Physiol. Meas.* **8** Suppl. A 167–73

GRIFFITHS H and AHMED A 1987 A dual frequency applied potential tomography technique: computer simulations *Clin. Phys. Physiol. Meas.* **8** Suppl. A 103–7

HAND J W 1985 Thermometry in hyperthermia *Hyperthermic Oncology 1984* ed J Overgaard (London: Taylor and Francis) pp299–308

HARRIS N D, SUGGET A J, BARBER D C and BROWN B H 1987 Applications of applied potential tomography (APT) in respiratory medicine *Clin. Phys. Physiol. Meas.* **8** Suppl. A 155–65

HELLER S L, SCHARF S C, HARDAFF R and BLAUFOX M D 1984 Cinematic display of respiratory organ motion with impedance techniques *J. Nucl. Med.* **25** (10) 1127–31

HENDERSON R P and WEBSTER J G 1978 An impedance camera for spatially specific measurements of the thorax *IEEE Trans. Biomed. Eng.* **BME-25** 250–3

JOSSINET J J, FOURCADE C and SCHMITT M 1981 A study for breast imaging with a circular array of impedance electrodes *Proc. 5th Int. Conf. on Electrical Bioimpedance, Tokyo* pp83–6

MANGALL Y F, BAXTER A J, AVILL R, BIRD N C, BROWN B H, BARBER D C, SEAGAR A D, JOHNSON A G and READ N W 1987 Applied potential tomography: a new non-invasive technique for assessing gastric function *Clin. Phys. Physiol. Meas.* **8** Suppl. A 119–29

MURPHY D, BURTON P, COOMBS R, TARASSENKO L and ROLFE P 1987 Impedance imaging in the newborn *Clin. Phys. Physiol. Meas.* **8** Suppl. A 131–40

NYBOER J 1970 Electrorheometric properties of tissues and fluids *Int. Conf. on Bioelectrical Impedance* ed S E Markovich (*Ann. NY Acad. Sci.* **170** (2))

POWELL H M, BARBER D C and FREESTON I L 1987 Impedance imaging using linear electrode arrays *Clin. Phys. Physiol. Meas.* **8** Suppl. A 109–18

PRICE L R 1979 Imaging of the electrical conductivity and permittivity inside a patient: a new CT technique *Proc. Soc. Photo-Opt. Instrum. Eng.* **206** 115–419 (Recent and Future Developments in Medical Imaging II)

RAO P S, SANTOSH K and GREGG E C 1980 Computed tomography with microwaves *Radiology* **135** 769–70

SKIDMORE R, EVANS J M, JENKINS D and WELLS P N T 1987 A data collection system for gathering electrical impedance measurements from the human breast *Clin. Phys. Physiol. Meas.* **8** Suppl. A 99–102

TARASSENKO L and ROLFE P 1984 Electrical impedance tomography—a new method to image the head continuously in the newborn *Proc. BES 6th Nordic Meet., Aberdeen* Paper CIG3:5

CHAPTER 11

IMAGING BY DIAPHANOGRAPHY

S WEBB

11.1 CLINICAL APPLICATIONS

Skin and subcutaneous tissue are partially transparent in the visible and near-infrared regions of the spectrum. An interest has developed in establishing whether the observed pattern of transillumination through moderately thick biological tissues might be indicative of pathology, and a body of evidence has accrued to support the importance of this developing area. The technique has potential application in the examination of the female breast, and has been successful in the differential diagnosis of cystic (benign) lesions and malignant tumours. Cystic (fluid-filled) lesions show up as regions of increased brightness if the fluid they contain is not blood, whereas dark regions in the transilluminated image are strongly indicative of malignant tumours. Fibroadenosis gives rise to an intense cherry-red colour. The same differential diagnosis between solid tumours and cystic lesions can be performed for the scrotum. Other potential applications are neonatal brain studies.

The role of diaphanography for the breast is, however, questionable and must be evaluated against competing procedures such as mammography (Chapter 2), ultrasound (Chapter 7) and biopsy. Diaphanography with good equipment is fairly new, so far only tested in a limited number of centres and with a variety of technical forms, and, although promising, is still without a consensus of favourable opinion. In §11.2 the physical basis of transillumination is reviewed. To set the scene, however, it is noted that there have been relatively few clinical studies addressing the problem of comparison with other modalities, and these studies come to varying conclusions. Hussey et al (1981) found the sensitivity of diaphanography was 30% lower than for mammography, and the false-negative rate was three times higher. In a recent large clinical trial involving 1000 women, a comparison of mammography and diaphanography was performed (Marshall et al 1984). This study showed a higher false-positive rate with mammography, but all patients with true-positive light scans had positive mammograms. On the other hand, a clinical comparison involving 1476 women reported by Drexler *et al* (1985) concluded that mammography was significantly more sensitive than either physical examination or diaphanography (by approximately a factor of 2), and that diaphanography does not satisfy the criteria of a screening procedure. A 'Diagnostic and Therapy Technology Assessment' (1984) remained cautious with regard to the comparative assessments performed to date. The enormous research effort in mammography has perhaps eclipsed the progress of diaphanography. The major criticism levelled against x-ray mammography, however, as a routine screening tool, is the potential hazard of the cumulative radiation dose (Mole 1978) (see Chapter 15), and in this respect diaphanography offers a harmless alternative. In symptomatic women with palpable disease, it offers the possibility to distinguish between malignant and cystic tumours, which in general cannot be determined by mammography.

The poor spatial resolution associated with transillumination imaging makes it a poor competitor, however, if the clinical interest is detection of microstructural disease in an asymptomatic population. On the other hand, in mass screening, 10% of cancers occur within a year of supposedly negative mammography (known as interval cancers), and in this respect diaphanography offers a useful adjunct to personal monitoring by palpation or possibly hazardous mammography, increasing the sensitivity of mass screening. When an abnormality is detected by transillumination, the patient may proceed to mammography or biopsy for confirmation. The limitations of mammography in the case of the young dense breast is further reason for interest in diaphanography. It is perhaps unfortunate that (inevitably) comparisons of imaging modalities are generally made with patients showing signs or symptoms of disease, and in this sense may not be representative of relative efficacy in a screening study. For example, in the study by Marshall *et al* (1984) all women had been referred for clinical symptoms related to the breast, baseline studies, presurgical examination or follow-up of the contralateral breast after mastectomy. It should also not be ignored that this study showed that, in response to a questionnaire, 95% of women stated a preference for diaphanography if subsequent examinations were needed.

It should also be remembered that the physical basis of diaphanography is quite different from its competitors, and it is yielding information on a different biological property whose importance is as yet only partly understood. There is thus good reason to continue to develop these techniques and to study further the mechanics of the biological attenuation and scatter of visible and infrared photons.

11.2 PHYSICAL BASIS OF TRANSILLUMINATION

The percentage of transmitted light varies with the wavelength of the radiation, the type of tissue through which it passes and the thickness of the tissue. Reliable information on the optical properties of tissue is scarce. Cartwright (1930) demonstrated that cheek, 5 mm thick, transmitted 14% of the incident 860 nm photons. At 1150 nm, the percentage increased to 20%. Other biological tissue exhibits a similar wavelength dependence of the light transmission curve. Beyond 1400 nm, the presence of water within the tissues renders them strongly absorbing. Other workers have studied the percentage transmission of very thin specimens; their interest was more in selecting the wavelength that was optimum for skin heating. There are also some data (Watmough 1982b) on the transmission properties of sections of material taken at operation from lesions within the breast, but there seems to be no clear distinction between the transmission properties of neoplastic and benign lesions *in vitro*. This laboratory evidence has considerable significance. The transmission curves for a range of samples of varying thickness and representing normal tissue, carcinoma and fibroadenosis are similar (figure 11.1). This implies that simple variations in transmission coefficient alone do *not* explain the observed features in clinical diaphanography. These curves show the characteristic absorption bands of oxyhaemoglobin at 576, 542 and 412 nm pointing towards a mechanism involving the blood pool.

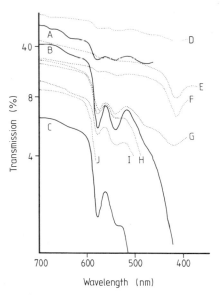

Figure 11.1 Transmission curves for post-operative breast specimens: curves A–C, carcinoma; E, pork fat; G, normal breast; all others, fibroadenosis. The samples were of varying thickness: A, 5 mm; B, 2 mm; C, 10 mm; D–J, 2 mm. (From Watmough (1982b).)

There is laboratory evidence to suggest that the number of red blood cells (erythrocytes) per unit volume is the major factor leading to the brightness and coloration of the transilluminated image. Whole blood, for example, is almost totally absorbing, and spectrophotometric measurements of blood at a range of dilutions from 1:2000 towards whole blood show a range of colours including brown, yellow and straw colour, as seen in breast examinations, the colours also depending on the illuminating spectrum. Red-cell densities in essence provide a range of colour filters and these experiments also suggest that the red-cell density in normal breast is about 1:10. This gives a hint at the mechanism by which malignant lesions (which are often associated with an abnormally active local blood supply) appear dark in transillumination. Their advanced neovascular front is indicative of increased absorption. The theory is supported by ultrasound Doppler measurements (see Chapter 7) in which neovascularisation around malignant lesions leads to characteristic signals. The characteristic increased brightness associated with the presence of cysts is also explicable on this basis. The absorption is reduced because blood-containing tissues are displaced by the fluid-filled cavity. If during examination a patient's own blood is injected into the cyst it reverts to a characteristic distinctly outlined dark region (Wallberg 1985).

The basis for breast diaphanography is the differential transmission of optical radiation, and mathematical models have been developed to describe the interaction of the electromagnetic wave with biological tissue (Ertefai and Profio 1985). Photons are both scattered and absorbed in biological tissue such that diffuse transmittance results. Scattering arises from variations of refraction in the microscopic and macroscopic structure, whilst absorption is due to electronic transitions and excitations associated with atomic vibrational and rotational modes. Correspondingly, Beer's law

$$I = I_0 \exp(-\mu_a x)$$

relating incident intensity I_0 to exit intensity I, absorption coefficient μ_a and thickness x, and appropriate to a transparent medium, fails, and may be replaced by a solution to the steady-state diffusion equation

$$D \frac{d^2 I}{dx^2} - \mu_a I = 0 \tag{11.1}$$

with diffusion coefficient

$$D = \frac{1}{3[\mu_s(1 - w) + \mu_a]}. \tag{11.2}$$

Here w is the average of the cosine of the scattering angle and μ_s is the scattering coefficient. For isotropic scattering ($w = 0$), believed to hold for breast tissue (see §11.3), the solution is

$$I = \frac{4I_0}{1 + 2D\alpha} \exp(-\alpha x) \tag{11.3}$$

with

$$\alpha = (\mu_a/D)^{1/2} = 1/L \tag{11.4}$$

where L is called the diffusion length or depth of penetration. A different model, known as the Kubelka–Munk model, leads to a very similar conclusion that light traversing tissue obeys a simple exponential law with the rate of attenuation governed by absorption and scattering coefficients. Such coefficients have been experimentally determined for breast tissue (Ertefai and Profio 1985) and used in a model-based calculation of the expected contrast in diaphanography. It may be noted that when $\mu_s = 0$, equation (11.3) reverts to Beer's law (except for constants), illustrating the similarity with the laws of x-ray attenuation.

11.3 EXPERIMENTAL ARRANGEMENTS

Diaphanography may be performed with the patient on a seat, capable of rotation, and applying an infrared torch to the underside of the pendulous breast, with the patient leaning slightly forwards. The examining physician sits in front of the patient in a darkened room and an intensity of radiation is selected to match the thickness of tissue. The inspection is a visual search for areas of increased or decreased brightness. Dark areas are possibly malignant, although blood-filled cysts contribute to the differential diagnosis. Superficial veins also show up black, as do the sites of previous biopsies and needle aspirations, which appear dense black. For this reason, diaphanography should preferably precede biopsy where both techniques are called for. Other potential generators of false positives are haematoma, chronic or acute mastitis, sclerosing adenosis, fibroadenoma and papillomatosis. Figure 11.2 shows the typical appearance of a breast cyst in diaphanography, and figure 11.3 shows the appearance of infiltrating ductal carcinoma. After visual inspection, images may be recorded on photographic film using a xenon flash tube with a high transient light output. It is noteworthy that Cutler (1929) reported the first attempts at transillumination some 60 years ago, using white light to which the breast acts as an optical filter, appearing yellow or red depending on the relative amounts of fat and fibroglandular tissue. It was only in the late 1970s, however, that the first commercially available equipment came into use (Sinus Medical Equipment AB, Stockholm, Sweden). It proves to be extremely important to image the breast from at least five orientations (craniocaudal, lateral, medial, inferolateral and inferomedial), since most tumours are generally observed in only one view. Since film exposure may be critical, it is usual to record three images at different exposures at each orientation. It is also very important that observers have received lengthy and adequate training in the interpretation of diaphanograms. The nature of this technique leads to two disadvantages. First, development of infrared photographs is very time-consuming (up to 1 h) and, secondly, because of the fluctuations in the number of patients presenting for diaphanography, there may be a delay of up to one week before a whole film has been used. In the meantime, it may be important to make an urgent clinical decision on the management of the patient. Additionally, if experimental errors occurred, the patient would have to be recalled.

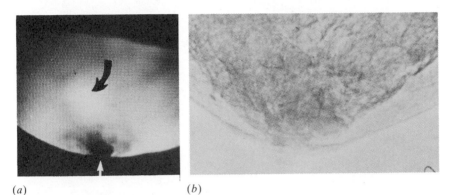

(*a*)   (*b*)

Figure 11.2 (*a*) Craniocaudal view showing a lighter area (curved arrow) above the nipple (small arrow), representing a cyst. (*b*) The corresponding xeromammogram. (From Marshall *et al* (1984).)

Recently, diaphanography using the Sinus Medical Equipment system has been extensively documented in a series of papers from the Huddinge University Hospital, Huddinge, Sweden, and the Royal Institute of Technology, Stockholm, Sweden (Wallberg 1985, Wallberg *et al* 1985a, b, c, d). Regarding experimental arrangements, their experiments demonstrated no preferential distance between 10 and 25 cm for the camera from the film. By comparing imaging with and without polarising filters, it was also shown that the breast is optically isotropic, dispelling the view that tension in breast tissue caused by a tumour might create a state of optical anisotropy. These studies indicated that the diameter of the light probe did not influence image contrast, but this is not the view held by Watmough (1982a), who demonstrated (with different equipment) the necessity for a variable diaphragm by which the source size could be reduced for beneficially imaging small lesions in the breast and to reduce the scatter, which might otherwise confuse the depression of transilluminated intensity due to the presence of the tumour. The Swedish study employed a computer-controlled image scanner OSIRIS to translate images stored on film to a linear diode array. By using moving mirrors, the array was made to sweep the entire image and the digital diaphanogram was recorded by computer (Sperry Univac V77-600). By taking profiles across the image, the tumour extent was recorded.

A new torch for diaphanography incorporating a variable diaghragm has been reported by Watmough (1982d). Its efficiency for demonstrating small lesions was tested using a phantom comprising front and back

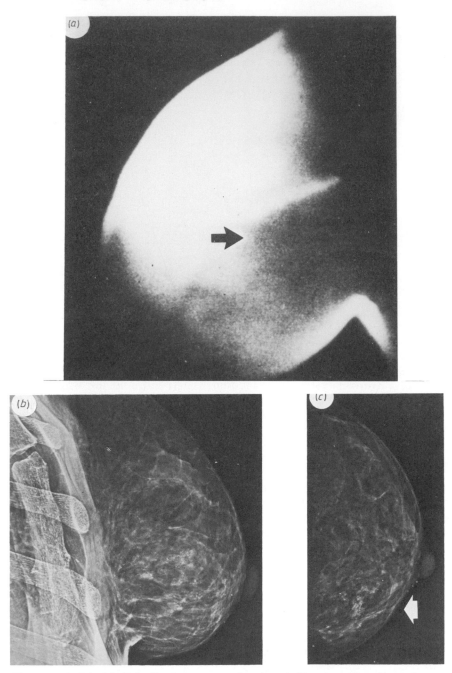

Figure 11.3 (*a*) Medial diaphanogram showing infiltrating ductal carcinoma (arrow). (*b*) Mammogram, craniocaudal view: this shows malignant calcifications scattered throughout almost entire lower half of breast, although no discrete masses are showing. (*c*) Mammogram, mediolateral view: the arrow shows a cluster of malignant calcifications. (From Drexler *et al* (1985).)

sheets of opal Perspex between which simulated absorbing lesions could be placed at various depths. The thickness of the sheets was adjusted to give a background transilluminatcd intensity of about 1%, which is believed to be the approximate transilluminated intensity through the normal breast at the appropriate frequency. The torch was superior in resolving 3 mm 'lesions' and demonstrating the importance of the variable diaphragm. This work also warns that, above a certain level, increased light output from the torch may lead to reduced contrast. With regard to phantoms, Drexler *et al* (1985) have made use of a milk-filled rubber balloon with simulated lesions, and detectability was correlated with depth of lesion. A tumour was just detectable at a depth slightly more than twice its size.

To offset the disadvantages associated with the use of film, tele-diaphanography has been developed by a number of workers (Watmough 1983a, b, Drexler *et al* 1985), offering the possibility of real-time diaphanography. The colour infrared film is replaced by a television detector. At 800 nm, there are very few television camera detectors that are sensitive enough. Two possibilities appear to be the chalnicon and the silicon vidicon. The latter is sensitive up to 1000–1100 nm, and at 800 nm its absolute spectral sensitivity (about $0.3 \, \mu A \, \mu W^{-1}$) is two orders of magnitude more than the plumbicon, whose sensitivity has fallen dramatically between its peak, at about 500 nm, and 800 nm. It is sufficient to use a tungsten filament lamp, dispensing with the xenon flash tube used in film imaging, as the radiative power of the lamp can be matched to the television detector. The radiated power output as a function of wavelength depends on the temperature of the filament and at about 3200 K peaks at about 800 nm. For permanent records, a video tape recorder is used. The role of telediaphanography has yet to be established, and, in general, the relative merits of diaphanography and other breast imaging tcchniques remain disputed at this time. The most recent report of preliminary clinical results of using computerised telediaphanography has been published by Bundred *et al* (1987). In a study of 129 patients examined to distinguish palpable carcinoma from benign breast lesions, a sensitivity of 94% was reported. The technique was most useful for the group of women aged over 50 and for large lesions, but it was considered that the benefit for young women with dense dysplastic breasts was yet to be established.

Very recently a system has been proposed for performing transillu-mination computed tomography, with the aim of improving contrast (Jackson *et al* 1987). A rotate–translate (see Chapter 4) assembly has been constructed by which an optical beam generator and optical collimator were scanned linearly by 12.8 cm and in angle by 180° in 127 rotational steps. Transillumination profiles were recorded by computer and reconstructed via a convolution and backprojection algorithm. It was demonstrated that the signal contrast in projections is greatest when the regions of high absorption are close to the illuminated surface, indicating the need to consider attenuation correction in the reconstruction algorithm. Cross-sectional images of a breast phantom have been obtained, but the system has not yet been optimised for clinical utility.

Finally, with another context, we note that near-infrared transillumination has been used as a method of measuring the state of oxygenation of brain and muscle in new-born infants (Arridge *et al* 1986). Cerebral cellular hypoxia in the new-born is a major contributor to adverse neurodevelopment, particularly in small premature infants. Equipment has been developed to produce digital images of the transillumination with 658 ± 50 nm photons, and its use demonstrated in the head and also with the human palm. It has been estimated that the resolution may be poor for tissue thicknesses greater than a few centimetres but that deconvolution techniques (see Chapter 12) may improve the imaging.

It is clear that diaphanography has a role to play, but the weight of this role in relation to other imaging techniques is currently uncertain. The next few years could witness this being clarified.

REFERENCES

ARRIDGE S R, COPE M, VAN DER ZEE P, HALLSON P J and DELPY D T 1986 Near infrared transillumination as a method of visualisation and measurement of oxygenation state of brain and muscle in newborn infants *Recent Developments in Medical and Physiological Imaging* Suppl. to *J. Med. Eng. Technol.* ed R P Clark and M R Goff (London: Taylor and Francis) pp24–31

BUNDRED N, LEVACK P, WATMOUGH D J and WATMOUGH J A 1987 Preliminary results using computerised telediaphanography for investigating breast disease *Br. J. Hosp. Med.* **37** (1) 70–1

CARTWRIGHT C H 1930 Infra-red transmission of the flesh *J. Opt. Soc. Am.* **20** 81–4

CUTLER M 1929 Transillumination as an aid to diagnosis of breast lesions *Surg. Gynaecol. Obstet.* **48** 721–7

DIAGNOSTIC AND THERAPEUTIC TECHNOLOGY ASSESSMENT 1984 Diaphanography (transillumination of the breast) for cancer screening *J. Am. Med. Assoc.* **251** (14) 1902

DREXLER B, DAVIS J L and SCHOFIELD G 1985 Diaphanography in the diagnosis of breast cancer *Radiology* **157** 41–4

ERTEFAI S and PROFIO A E 1985 Spectral transmittance and contrast in breast diaphanography *Med. Phys.* **12** (4) 393–400

HUSSEY J K, MacDONALD A F, NICOLS D M and WATMOUGH D J 1981 Diaphanography—a comparison with mammography and thermography *Br. J. Radiol.* **54** 163

JACKSON P C, STEVENS P H, SMITH J H, KEAR D, KEY H and WELLS P N T 1987 The development of a system for transillumination computed tomography *Br. J. Radiol.* **60** 375–80

MARSHALL V, WILLIAMS D C and SMITH K D 1984 Diaphanography as a means of detecting breast cancer *Radiology* **150** 339–43

MOLE R H 1978 The sensitivity of the human breast to cancer induction by ionising radiation *Br. J. Radiol.* **51** 401–5

WALLBERG H 1985 Diaphanography in various breast disorders: clinical and experimental observations *Acta Radiol. Diagn.* **26** (3) 271–6

WALLBERG H, ALVERYD A, BERGVALL U, NASIELL K, SUNDELIN P and

TROELL S 1985a Diaphanography in breast carcinoma *Acta Radiol. Diagn.* **26** (1) 33–44

WALLBERG H, ALVERYD A and CARLSSON K 1985b Physical interpretation of diaphanograms using the computer-controlled image scanner OSIRIS *Acta Radiol. Diagn.* **26** (4) 417–24

—— 1985c Breast carcinoma and benign breast lesions; diaphanography and quantitative evaluation using the computer controlled image scanner OSIRIS *Acta Radiol. Diagn.* **26** (5) 535–41

WALLBERG H, ALVERYD A, NASIELL K, SUNDELIN P, BERGVALL U and TROELL S 1985d Diaphanography in benign breast disorders *Acta Radiol. – Diagn.* **26** (2) 129–36

WATMOUGH D J 1982a A light torch for the transillumination of female breast tissues *Br. J. Radiol.* **55** 142–6

—— 1982b Diaphanography; mechanism responsible for the images *Acta Radiol. Oncol.* **21** (1) 11–15

—— 1983a Diaphanography *Imaging With Non-Ionising Radiations* ed D F Jackson (Guildford: Surrey University Press) pp217–25

—— 1983b Transillumination of breast tissues: factors governing optimal imaging of lesions *Radiology* **147** 89–92

CHAPTER 12

THE MATHEMATICS OF IMAGE FORMATION AND IMAGE PROCESSING

S WEBB

12.1 THE CONCEPT OF OBJECT AND IMAGE

Why is it interesting or necessary to study the mathematics of image formation, and can it really be possible that the many imaging modalities that we have met so far can be represented by the same set of equations? In this chapter, we shall see that, in general, it is indeed possible to write down a small set of equations that are sufficiently general to be applicable to many imaging situations. This property endows the equations of image formation with a certain majesty, and they become powerful tools for coming to grips with the underlying principles of imaging. As such, they also provide the basis for understanding why and how images are imperfect, and they suggest techniques for image processing. In this chapter, the words 'image processing' are used to imply the *processing* of images for the removal of degradations that have arisen during their formation. (Elsewhere in this book, these same words have often been used more generally to imply the *creation* of images from some form of raw or recorded data. For example, the theory of reconstruction of tomographic images from projection data has been referred to as image processing in earlier chapters.)

Here we describe as the *object* that property which is distributed in multidimensional space and which we require to measure. We describe as the *image* that measurement (also, in general, multidimensional) which has been made and which is regarded as best representing the object distribution. It has become somewhat conventional to represent the object mathematically by the symbol f and the image by the symbol g. We can write straightaway that a perfect imaging system is one in which, at all locations in the space, $f = g$. Almost without exception, no medical imaging system achieves this and, in general, it is necessary to conduct experiments to parametrise the relationship between f and g.

In general, both f and g are three-dimensional when they describe the distribution of some time-stationary property *in vivo*. However, we have already seen that many medical imaging modalities are tomographic and that the complete three-dimensional distribution of a property is substituted by a contiguous sequence of two-dimensional descriptors. This achievement of tomography has not only revolutionised our understanding of the true three-dimensional distribution of some property, removing artefactual interference or cross talk, but has also led to a most useful simplification of the mathematical description of imaging. Hence, in this chapter, we shall restrict the mathematics to two spatial dimensions, with the understanding that this will be applicable both to distributions that are genuinely two-dimensional and to tomographic images. We shall make use of the terms 'image plane' and 'object plane' freely, implying total containment of information within these planes.

To perceive the beautiful generality of the mathematics that follows, some examples are given here to illustrate the conceptual statement in the previous paragraph. Stepping aside from medicine for one moment, we might imagine the simplest optical imaging method of taking a photograph (image plane) of a scene painted on a two-dimensional canvas (object plane). Introducing a coordinate scheme, the painting is $f(\alpha, \beta)$ and the photograph is $g(x, y)$. In this example, both f and g might be thought of as having the same dimensions, since they are both the spatial distribution of colour or light intensity in a plane. The photograph was taken by quanta of light reflected from the painting through the lens system of the camera and onto the photographic film. Such image formation is, of course, degraded by distortions and the detector response, and hence f is not equal to g. If we had a description of the degradation, image processing might be performed to compensate for the degradation in image formation, as we shall see from what will follow.

As a second example, let us recall that, from a series of two-dimensional images recorded by a rotating gamma camera at sequential orientations around a patient, it is possible to generate a series of two-dimensional images g of the distribution f of the radiopharmaceutical *in vivo* by the process of single-photon emission computed tomography (SPECT) (Chapter 6). In this example, the diagnostician is led to believe that the images being viewed are actually at the location of interest in the patient. There is a suspension of disbelief, of course, because what is being observed (the image g) is a 2D distribution of reconstructed pixel values, whereas the clinician is thinking of this as a 2D distribution of the uptake of activity within the patient. The image g is thus a *representation* of the object f, which has different dimensions but is in more or less the same spatial location. The aim of SPECT is, of course, that g be related to f in a linear quantitative way. Interestingly, the gamma camera, which was the means of recording the data, does not feature at all in the mathematical description relating image to object! Indeed, it is quite incidental that the data used to reconstruct SPECT tomograms are themselves images. We recall, for instance, that the recorded data for a multiwire proportional chamber (MWPC) positron camera (see Chapter 6) were no more than a string of coordinate values, images only appearing after backprojection.

It is thus apparent that we should remove ourselves from the restriction of thinking that images of an object are created at a different place from the source distribution and with the same dimensions as the object itself. Indeed, it is more appropriate to think of the image and object as being physically coincident, sometimes with differing dimensions, but related to each other by the processes characterising the imaging modality.

12.2 THE RELATIONSHIP BETWEEN OBJECT AND IMAGE

Against this background, we proceed to consider two-dimensional distributions, knowing these to incorporate tomographic images. We shall represent the object distribution by $f(\alpha, \beta)$, entirely contained within the object plane, and the image distribution by $g(x, y)$, entirely contained within the image plane. In general, there is no perfect 1:1 correspondence between the information at a particular location (α', β') and the corresponding location (x', y'). 'Information' can, in principle, disperse *to all* image locations *from each* object location. For any useful imaging modality, however, the principal contribution to each (α', β') will come from a particular single location (x', y'). Other neighbouring points will contribute smaller amounts of information, the contribution decreasing rapidly away from the principal contributing location (x', y'). This is known as the neighbourhood principle, and simply recognises that the image of some point in object space may be dependent on the object point and on points in an infinite neighbourhood surrounding the object point. Later we shall see that this concept can be described quantitatively by use of the point source response function (PSRF), which will be narrow for a 'good' imaging system in which the above neighbourhood is confined to a small (certainly not infinite!) region. By way of illustration, imagine a gamma camera fitted with a parallel-hole collimator viewing a point source of activity in a scattering medium. One particular image pixel (that corresponding to the direct line-of-sight view) will record maximum counts, but adjacent pixels will register some recorded events, with the number of events decreasing rapidly with distance from the maximum-valued pixel. This is known as the point process of image formation. Conversely, imagine reversing all the photon paths from a single camera pixel back towards a plane parallel to the face of the camera. Most paths will intersect the direct line-of-sight object pixel but paths will also branch out to a myriad of other pixels because of imperfect collimation and photon scatter.

What is this physical link between object and image space? Conceptually, it is the transport of information. The image plane is built up from a knowledge of what is in the object plane brought by whatever are the appropriate 'coding messengers' of the imaging modality. Thus a photograph is created by optical photon transport. A distribution of acoustic scattering centres is mapped to a B-scan brightness image by scattered longitudinal ultrasound waves. A gamma camera records a measure of the uptake of a radiopharmaceutical by counting the emitted γ-ray photons. A transmission x-radiograph is a brightness image related to the linear attenuation of x-ray photons that traverse the object space to form the image. A thermogram is a map of pixel intensity corresponding to an *in vivo* temperature distribution, the messengers carrying the information being infrared photons. Because the smallest unit of radiant energy transport is non-negative (implying that there is either something to measure or there is not), then the image distribution must also be non-negative. Mathematically we can write

$$f(\alpha, \beta) \geq 0 \qquad \text{and} \qquad g(x, y) \geq 0. \qquad (12.1)$$

Measuring the total information in an object and image distribution is also physically meaningful, and so it is reasonable to assume that f and g are integrable functions, since integration corresponds to measuring totals.

Now postulate a function $h(x, y, \alpha, \beta)$ that describes the spatial dependence of the point process. For a point process in which the object is only non-zero at (α', β'), the recorded image is

$$g'(x, y) = h(x, y, \alpha', \beta', f'(\alpha', \beta')). \qquad (12.2)$$

Here the express dependence on the magnitude of the point-object signal has been made clear by including f as a fifth argument of the function h. Now imagine a second signal at the same location giving rise to recorded image

$$g''(x, y) = h(x, y, \alpha', \beta', f''(\alpha', \beta')). \qquad (12.3)$$

Since the superposition principle states that radiant energies are additive

$$g'(x, y) + g''(x, y)$$
$$= h(x, y, \alpha', \beta', f'(\alpha', \beta')) + h(x, y, \alpha', \beta', f''(\alpha', \beta')). \qquad (12.4)$$

Equation (12.4) is non-linear superposition (since equation (12.2) is non-linear), i.e. additive components in the object plane do not lead to additive measurements in the image plane. If the imaging system is linear, then equation (12.2) becomes

$$g'(x, y) = h(x, y, \alpha', \beta')f(\alpha', \beta') \qquad (12.5)$$

and equation (12.4) becomes

$$g'(x, y) + g''(x, y) = h(x, y, \alpha', \beta')[f'(\alpha', \beta') + f''(\alpha', \beta')]. (12.6)$$

From this, we see that additive components in the object plane lead to additive measurements in the image plane via a single use of the transformation function h. Mathematically, this is such an important simplification that, as we shall see, linearity is often assumed to first order even when it is not strictly true.

It is now possible to invoke the concept of a distribution as the superposition of a finite set of points to extend the ideas above to generate the general equations that link the spaces. For a non-linear imaging system

$$g(x, y) = \int\int h(x, y, \alpha, \beta, f(\alpha, \beta)) \, d\alpha \, d\beta \qquad (12.7)$$

and for a linear system

$$g(x, y) = \int\int h(x, y, \alpha, \beta)f(\alpha, \beta) \, d\alpha \, d\beta. \qquad (12.8)$$

The function h, which we have been using to relate f to g, is called the point source response function (PSRF). In equations (12.7) and (12.8) the PSRF h is a function of all four spatial coordinates and is consequently referred to as a space-variant PSRF (or SVPSRF). These are the most general descriptions of imaging that it is possible to write down.

If, however, the imaging system is such that the point process is the same for all locations of the point in the object plane, then the system is said to be spatially invariant and the PSRF is called a space-invariant PSRF (or SIPSRF). In these special circumstances, the function h depends only on the *difference* coordinates $(x - \alpha, y - \beta)$, because the point source response function depends only on relative distance between points in the object plane and points in the image plane. Another description is to refer to this situation as isoplanatic imaging. This concept is quite separate from any consideration of linearity, and so we may write down that for a space-invariant imaging system that is non-linear

$$g(x, y) = \int\int h(x - \alpha, y - \beta, f(\alpha, \beta)) \, d\alpha \, d\beta \qquad (12.9)$$

whereas for a linear space-invariant imaging system

$$g(x, y) = \int\int h(x - \alpha, y - \beta) f(\alpha, \beta) \, d\alpha \, d\beta. \qquad (12.10)$$

This is recognised as the familiar *convolution integral*. An image is the convolution of the object distribution with the point source response function. It is this function h that carries the information from object to image space and embodies all the geometric 'infidelities' of the imaging modality. It is therefore not surprising that attempting to remove these degradations is often referred to as *deconvolution*. Strictly, this only applies to image processing when the modality is space-invariant, although the word 'deconvolution' has slipped into common use as a general substitute for the words 'image processing'. This casual use will not be made here. We shall return at length to the subject of deconvolution later in this chapter.

A final simplification of the general image equations arises if the behaviour in orthogonal directions is unconnected. By this is meant that the 2D PSRF can be constructed by multiplying together two 1D PSRF. For a space-variant system

$$h(x, y, \alpha, \beta) = h'(x, \alpha)h''(y, \beta) \qquad (12.11)$$

and for a space-invariant modality

$$h(x, y, \alpha, \beta) = h'(x - \alpha)h''(y - \beta). \qquad (12.12)$$

This property is known as separability. Finally, then, we give the imaging equation for a linear, space-invariant, separable modality as

$$g(x, y) = \int h'(x - \alpha)f(\alpha, \beta) \, d\alpha \int h''(y - \beta)f(\alpha, \beta) \, d\beta \quad (12.13)$$

which is the product of two orthogonally separate convolution integrals.

12.3 THE GENERAL IMAGE PROCESSING PROBLEM

Against this background, it is easy to appreciate that by a suspension of disbelief, when we observe an image, we imagine we are looking at an object distribution. That, after all, was the rationale for creating the image. We are not! We are inspecting an image that is a *fair representation* of the object distribution. It has already been seen that this may not even have the same dimensions (i.e. be the same physical quantity) as the object in which we are interested. Built into the measured image g are all the characteristics and imaging imperfections of the modality in question. To what we might call first order, this does not matter. If the instrument has been constructed to a high specification, we would expect that 'fair' means 'good' and that a more or less 1:1 spatial correspondence exists between the spaces, and that the measured quantity is more or less linear with the object distribution that it represents. However, when we become a little more critical, we lose this suspension of disbelief, and notice the imperfections introduced by the imaging process. In some cases, what we are inspecting clearly cannot be real, and we want to rid the image of these imperfections. Using a simple analogy, if an optical photograph shows a familiar face with four eyes, we can be pretty sure the camera or object has moved during the exposure! Generally, degradations introduced in medical imaging are unfortunately more subtle and consequently more difficult to remove. For example, a SPECT tomogram taken through a small point source will produce an image whose width is much larger than the source. We immediately recognise that degradations are often synonymous with loss of resolution.

This state of affairs should not be interpreted that 'something has gone wrong'. Imaging degradations are present because of the underlying physical laws governing the image formation process. *They have to be there* and it would be 'wrong' if they were not. It is for this reason that considerable theoretical and experimental effort is expended to describe properly the physical processes of image formation. In attempting image processing, one is thus using the known performance of a system (usually its point response) to cheat nature and produce an image that the system could not otherwise possibly generate!

Let us take a naive approach to the problem of image processing and also concentrate on a linear space-invariant model. We have already found that, in this case, a convolution integral (equation (12.10)) links the object f to the image g. In simpler notation, equation (12.10) may be written:

$$g(x, y) = h(x, y) * f(x, y) \qquad (12.14)$$

where $*$ represents convolution. Notice that in equation (12.14) x and y are used simply to represent whatever are the local 2D coordinates in the appropriate space. When we rewrite equation (12.10) in Fourier space, the convolution becomes a simple multiplication by the familiar theorem. Using upper-case letters to represent Fourier variables, we can write

$$G(u, v) = H(u, v)F(u, v) \qquad (12.15)$$

where (u, v) are spatial frequencies corresponding to the x and y directions.

(*Aside*. The remainder of this chapter makes extensive use of Fourier transforms. To follow the arguments it is, however, only necessary to be familiar with the *concept* of the Fourier transform (as defined below) and with the theorem that states that convolutions in one space correspond to multiplications in the reciprocal space. Readers who wish to implement the techniques described would, of course, require a greater understanding. The theory necessary to follow the rest of this chapter is summarised in the appendix (§12.9), and the book *Radiological Imaging* by Barrett and Swindell (1981) is a good starting point for understanding the Fourier transform.)

Before considering the implications of equation (12.15), let us step aside to review what is meant by the Fourier representation of the object and the image. The linking equations are

$$F(u, v) = \int \int f(x, y) \exp[-2\pi i(ux + vy)] \, dx \, dy \qquad (12.16)$$

and

$$f(x, y) = \int \int F(u, v) \exp[+2\pi i(ux + vy)] \, du \, dv. \qquad (12.17)$$

Equation (12.16) shows how the object f may be split up into its component spatial frequencies, and equation (12.17) shows how these components can be recombined to yield again the original object f. The Fourier representation F contains the same information as f, but in a different form. The same set of statements can be made about the image g and its Fourier representation G. In verbal terms, we can now interpret the information in the Fourier representation. An image that is 'sharp' contains spatial frequencies that are higher than those in a 'more blurred' image. We recognise sharpness as relating to resolution, and so we would expect that the description of the image in spatial-frequency terms contains information pertaining to the available resolution.

More importantly, returning to equation (12.15), it is clear that the 2D Fourier transform H of the PSRF h gives the fraction of the component of the object distribution at spatial frequency (u, v) that is transferred to the image distribution at the same spatial frequency. H regulates the transfer of information at each spatial frequency and is often called the modulation transfer function (MTF) (see also Chapter 2). If the imaging system introduced no loss of spatial-frequency information (i.e. $H = 1$ for all spatial frequencies up to infinity), then this would imply that $G = F$ and, in turn, $g = f$ by definition. That is, the image would be a perfect representation of the object. Remembering that the Fourier transform of unity is a delta function, this corresponds to an infinitely narrow PSRF, which, in turn, means a precise 1:1 correspondence between object space and image space. We recognise the impossibility of this situation in practice. For all imaging systems, the PSRF has a finite width, and this corresponds to a falling-off in magnitude of the modulation transfer function with increasing spatial frequency. One can define mathematically the resolution of a system in terms of either the width of the PSRF at (say) half-maximum or correspondingly the width of the MTF at half-maximum, there being a reciprocal relationship between the two.

It is thus clear that, for all real imaging systems, there is a loss of spatial-frequency information at high spatial frequencies, this being parametrised by the MTF. What are the implications for image processing? *They are dire!* Returning to equation (12.15), we might naively imagine that we can recover the object distribution from the measured image by direct deconvolution. Inverting equation (12.15) gives

$$F(u, v) = G(u, v)/H(u, v)$$

and hence

$$f(x, y) = \int \int [G(u, v)/H(u, v)] \exp[+2\pi i(ux + vy)] \, du \, dv. \quad (12.18)$$

Directly invoking equation (12.18) is a foolish procedure. The reason is that the MTF H will decrease in magnitude with increasing spatial frequency, and there will exist a cut-off frequency beyond which it has zero magnitude. Even before this cut-off is reached, the magnitude will become small. Hence, it is clear that, because at high spatial frequencies the divisor of G becomes vanishingly small or zero, the integral becomes dominated by the larger terms generated at these frequencies. Moreover, it is precisely at such high spatial frequencies that the image (and hence G) becomes dominated by noise. Direct deconvolution thus leads to unacceptable noise amplification. An example from SPECT imaging of the liver is shown later in the chapter. We shall return to this difficulty later in this chapter when a resolution of the problem is proposed. Before that, however, we need to give more thought to the nature of the image formation process and its mathematical representation.

To conclude this section, let us recall that the experimental technique of imaging a point source can yield the MTF. That is, when $f = 1$ we have directly from equation (12.15) that $G = H$ or $g = h$. The image measured *is* the PSRF and its 2D Fourier transform *is* the MTF. The PSRF of many imaging systems in nuclear medicine, for example, is a Gaussian function, and since the Fourier transform of a Gaussian is another Gaussian, then so also is the MTF. High spatial frequencies are lost and the system behaves as a low-pass filter.

12.4 DISCRETE FOURIER REPRESENTATION AND THE MODELS FOR IMAGING SYSTEMS

Returning to equation (12.7), we have

$$g(x, y) = \int \int h(x, y, \alpha, \beta, f(\alpha, \beta)) \, d\alpha \, d\beta.$$

This can be regarded as an operator equation in which the function H $\{ \ \}$ operates on the object to give the image, i.e.

$$g = H\{f\}. \tag{12.19}$$

Here $H\{ \ \}$ implies a real-space operation on whatever is within the braces $\{ \ \}$. In the special circumstances that the imaging system is linear and spatially invariant, we have (equation (12.10))

$$g(x, y) = \int \int h(x - \alpha, y - \beta) f(\alpha, \beta) \, d\alpha \, d\beta.$$

Equation (12.10) is a Fredholm integral with a two-dimensional kernel, and a model of this type is referred to as 'continuous–continuous', in that both the object and the image space are depicted as a continuous distribution of values. More realistically, the image plane is usually discrete, comprising a matrix of sensors that sample the image. In medical imaging, images are usually made and stored as discrete matrices of numbers in picture elements (pixels). Also, within the conceptual generalisation introduced in §12.1, the tomographic image, which is thought of as being physically located within the object distribution (patient), is also generally discrete or digital. Under these circumstances, the Fredholm integral becomes

$$g_{i,j} = \int \int h_{i,j}(\alpha, \beta) f(\alpha, \beta) \, d\alpha \, d\beta. \tag{12.20}$$

The PSRF depends on the discrete image-space variables i and j and on the continuous object-space variables α and β. This is referred to as the 'continuous–discrete' model. *It is the closest representation of what really occurs.*

All biological systems are, of course, continuous on the spatial scale that we are considering, rather than discrete, and yet routinely we inspect images (which are discrete) that purport to represent objects, which by inference are therefore being regarded as if they were discrete. We have become very familiar with this, the greatest suspension of disbelief. Not only are we not really looking at the distribution of the parameter of interest—we are looking at its image—we are also imagining the representation to be discrete. Nevertheless, this is extremely convenient and is, in practice, what corresponds with our requirements to store digital images and display them as matrices. If we therefore imagine the object to comprise digital pixels also, we can write down the 'discrete–discrete' model of imaging as

$$g_{i,j} = \sum_{k=1}^{N} \sum_{l=1}^{N} h_{i,j,k,l} f_{k,l}. \qquad (12.21)$$

This is unreal, but is customary practice in many medical imaging modalities. Equation (12.21) is familiar as a matrix multiplication in which the object f and image g are $N \times N$ 2D matrices and the PSRF is an $N^2 \times N^2$ 2D matrix. In view of our expectation that imaging systems will have 'good' resolution and that, to first order, there is a more or less 1:1 correspondence between locations in object and image spaces, the matrix h will be very sparse, i.e. the majority of terms will be zero. It turns out that this can be a useful property in image processing. Imagine, however, the enormous task of storing the digital form of h. If $N = 256$, as is common, then h has close to half a million terms. It is also easy to see the enormous computational complexity of attempting digital deconvolution by real-space matrix multiplication techniques! We already know that direct deconvolution will have other physical disadvantages. The size problem alone is a very important incentive to attempt image processing in the Fourier-space, rather than the real-space, domain.

Let us therefore return to the Fourier representation and discuss the representation when the object and image are regarded as discrete matrices. We have already seen that the advantage of Fourier-space representation is that the behaviour of each frequency component in object space can be traced through to image space via the MTF. All convolutions disappear and are replaced by simpler multiplications. The MTF carries within it the resolution of the system in a manner that is easier to understand than the somewhat arbitrary real-space definitions of resolution.

For a discrete distribution f, the integrals in equation (12.16) become replaced by discrete summations, and we have

$$F(u, v) = (1/N) \sum_{x=0}^{N-1} \sum_{y=0}^{N-1} f(x, y) \exp[-2\pi i(ux + vy)/N] \qquad (12.22)$$

where it is implicit that x, y, u and v are discrete variables representing sample points in the object and image space.

Evaluating equation (12.22) on a digital computer is today relatively simple even for large matrix sizes N. The computation is separable because of the form of the exponential and an 'intermediate transform' can first be taken in the x direction to generate $F(u, y)$. This can then be transformed by a series of 1D transforms in the orthogonal y direction to yield $F(u, v)$. Thus an N^2 2D transform breaks down into $2N$ 1D transforms. The development of the Cooley–Tukey algorithm (which has become loosely known as the fast Fourier transform or FFT) has revolutionised the computation of Fourier transforms, which today no longer increase linearly with increasing size N but rather increase in computational time required roughly as $N^2 \ln N$. Many different forms of the algorithm exist and each user has a favourite.

Reversing equation (12.22) gives

$$f(x, y) = (1/N) \sum_{u=0}^{N-1} \sum_{v=0}^{N-1} F(u, v) \exp[+2\pi i(ux + vy)/N] \quad (12.23)$$

and shows how a discrete distribution is constructed from its discrete Fourier representation. From this equation, it is very easy to introduce the idea of digital image processing. Imagine that some function $T(u, v)$ is inserted on the right-hand side to give

$$\hat{f}(x, y) = (1/N) \sum_{u=0}^{N-1} \sum_{v=0}^{N-1} F(u, v) T(u, v) \exp[+2\pi i(ux + vy)/N]. \quad (12.24)$$

The function acts as a frequency modulator or filter, since it multiplies into the distribution F. By choosing different forms of T, a set of filtered images $\hat{f}$ can be generated from f by first operating equation (12.22) followed by equation (12.24) (see also §6.5). Indeed, direct deconvolution is just a special form of equation (12.24) when the function T is chosen to be $1/H$ and the operation is performed on G, the resulting equation

$$\hat{f}(x, y) = (1/N) \sum_{u=0}^{N-1} \sum_{v=0}^{N-1} [G(u, v)/H(u, v)] \exp[+2\pi i(ux + vy)/N] (12.25)$$

being the discrete version of equation (12.18). Once again beware the dangers!

Before leaving the consideration of imaging models, let us briefly return to the form of the PSRF. If the imaging system were perfect, then, in the 'discrete–discrete' notation, we have

$$h_{i,j,k,l} = \delta(i - k, j - l) \quad (12.26)$$

and hence, from equation (12.21), $g_{i,j} = f_{i,j}$ and the image perfectly maps the object. When this is not so (which is always), the degradation is coded into h, but the nearer diagonal (or more sparse) is h, the better is the representation. In general, the degradation is not known *a priori*, being a characteristic of the instrument and possibly also a function of the imaging conditions. At worst, it may be object-dependent. Hence, in general, experimental work has to establish the form of the PSRF. For medical imaging modalities, this is often possible by imaging a point source. The theory described, however, would equally apply, say, to an optical telescope, and in this case isoplanatic areas of the sky would be required to contain isolated point sources of optical emission in order to determine the PSRF. The theory presented also applies to processing the images from TV cameras carried by space platforms, and in this case known degradations have often been accepted in the design of such instruments knowing that they can be largely removed after transmission to Earth of the imperfect images. The reason this is done is to save on transmission bandwidth.

12.5 THE GENERAL THEORY OF IMAGE RESTORATION

Imaging can be considered as an operation, and in operator format we have written (equation (12.19))

$$g = H\{f\}.$$

The image restoration problem is then to find the inverse operator S such that the operation

$$S\{H\{f\}\} = f \qquad (12.27)$$

recovers f or at least a best estimate $\hat{f}$ of f. Several difficulties may arise:

(*a*) It may be that S does not exist. In this case, the problem is said to be singular and the problem is, of course, insoluble.

(*b*) If S exists, it may not be unique. Two different inverse operators could yield the same result. This is not really a problem except that it disguises the true nature of the degradation.

(*c*) Even if S exists and is unique, it may be ill conditioned. By ill conditioning we mean that a trivial perturbation in g could lead to non-trivial perturbations in the estimate of f, i.e.

$$S\{g + \varepsilon\} = f + Y \qquad \text{where} \qquad Y \gg \varepsilon. \qquad (12.28)$$

In practice, most medical imaging systems conform to the third situation, and non-trivial perturbations (noise) in the image g give artefactual signals in the estimate of f which swamp the true signal.

In reality, it has been wrong to represent the imaging process by equation (12.8), which gives the false impression that object and image space are connected only by geometrical transforms. In practice, images are contaminated by a variety of noise processes; these differ from modality to modality and have been described in earlier chapters. That being the case, the correct mathematical description of imaging for a linear system is

$$g(x, y) = \int \int h(x, y, \alpha, \beta) f(\alpha, \beta) \, d\alpha \, d\beta + n(x, y) \qquad (12.29)$$

where $n(x, y)$ represents the distribution of noise in the image. There is *no unique solution* to this equation. The reason is as follows. For every element in the ensemble of the noise process and for every corresponding image element, there is an element for f (always assuming S exists, is unique and is well conditioned). Since, for every realisation of the noise process, the restored element f is different, it makes no sense to talk about *the* restored image. The fact must be faced that image restoration depends on finding the 'best'-fitting solution to the imaging equation with some *a priori* constraints. We are now in a position to quantitate noise amplification simply in mathematical terms.

As a further simplification, let us invoke lexicographic stacking to portray the functions f and g as 1D vector strings of data and the operator $H\{\ \}$ (in the sense of equation (12.19)) as a 2D stacked matrix. In order to make this distinction, the 1D vectors will be set in bold italic type (e.g. g) and the 2D matrix will be set in bold sans serif type (e.g. H). Equation (12.29) now becomes

$$g = \mathsf{H}f + n. \tag{12.30}$$

Restated, the problem is that from a measured image g there is an infinity of solutions for f if the noise term n is non-zero (as it always is). Suppose we operate on both sides of equation (12.30) with the operator H^{-1}, which is the inverse operator; we obtain

$$\mathsf{H}^{-1}g = \mathsf{H}^{-1}\mathsf{H}f + \mathsf{H}^{-1}n. \tag{12.31}$$

If we define a measure of the true object $(\hat{f})$ as $\mathsf{H}^{-1}g$ then we find

$$\hat{f} = f + \mathsf{H}^{-1}n. \tag{12.32}$$

This equation may be interpreted that the processed image data, which give a measure of the true object, are the sum of the true object and a term that is the noise amplification. If the operator H is singular, clearly we cannot even proceed this far, but if it is 'merely' ill conditioned, then the second term in equation (12.32) will dominate the first and effectively invalidate the technique. The degree of ill conditioning is sometimes represented by an 'imaging Reynolds number' given by the square root of the ratio of the largest to the smallest eigenvalue of the composite matrix operator $\tilde{\mathsf{H}}\mathsf{H}$ (here the superscript tilde $\sim$ represents transpose).

It is then clear that it may not be too wise to perform direct deconvolution of the measured data. Some more mathematics gives an expression for the degradation of signal-to-noise ratio in direct deconvolution. Representing the Euclidean norm of a vector by $|g|$ $(= \Sigma_i g_i^2)$ from equation (12.30)

$$|g| = |\mathsf{H}f| + |n| \tag{12.33}$$

but by the conservation of energy implied by the theorem of non-negativity discussed earlier, by which information cannot become 'lost' in being transferred from object to image space,

$$|\mathsf{H}f| = |f| \tag{12.34}$$

and so the signal-to-noise ratio in the observed image g is

$$(S/N)_{\text{im}} = |f|/|n|. \tag{12.35}$$

Now performing the same operations on equation (12.32) we have

$$|\hat{f}| = |f| + |\mathsf{H}^{-1}n| \tag{12.36}$$

and so the signal-to-noise ratio in the estimate of the object is

$$(S/N)_{\text{ob}} = |f|/|\mathsf{H}^{-1}n|. \tag{12.37}$$

Now although H is energy-preserving, there is no reason why H^{-1} should be, and we thus have

$$(S/N)_{\text{ob}} = |\mathsf{H}^{-1}|^{-1}(S/N)_{\text{im}}. \tag{12.38}$$

Now the 'magnification factor' $|\mathsf{H}^{-1}|$ could be of many orders of magnitude, the noise amplification destroying the aims of the image processing. Hence we may make the statement that *for all real medical imaging systems, which are necessarily ill conditioned, it is usual to seek some solution to the image processing problem that optimises some feature*. We shall now consider such optimisations.

12.5.1 Least-squares image processing

Let us suppose that the image formation process may be represented by equation (12.30). Now seek a solution in which the norm of the noise is minimised. That is, the solution is in this sense 'least squares'. The norm of the noise vector may be represented by the outer product $\widetilde{n} \cdot n$. From equation (12.30)

$$\widetilde{n} \cdot n = (\widetilde{g - H\hat{f}})(g - H\hat{f}) \tag{12.39}$$

where $\hat{f}$ is now the best estimate, under these conditions, of the object. All we have to do now is solve for this estimate. Differentiating with respect to the elements f_k of $\hat{f}$, putting $d_k = \partial \hat{f}/\partial f_k$ and setting $\partial(\widetilde{n} \cdot n)/\partial \hat{f} = 0$, we obtain

$$0 = d_k \widetilde{H} g - \widetilde{g} H d_k + d_k \widetilde{H} H \hat{f} + \hat{f} \widetilde{H} H d_k. \tag{12.40}$$

This equation is satisfied by

$$\widetilde{H} H \hat{f} = \widetilde{H} g \tag{12.41}$$

because this, plus its transpose, sum to zero in equation (12.40). In turn, this reduces to the simpler form

$$\hat{f} = (\widetilde{H} H)^{-1} \widetilde{H} g$$

or

$$\hat{f} = H^{-1} g. \tag{12.42}$$

This is surprising! The result seems to be advocating precisely what we have been counselling against, namely direct deconvolution. Equation (12.42) is just the same as if we had struck out the noise term as if it did not exist. It is just the direct reversal of equation (12.19). What we are being told is that a least-squares solution minimising the norm of n is equivalent to just this. As such, we have unfortunately been once again thwarted. The result is of very little use in view of the previous discussion. Not all is lost, however, as the method is suggestive of one further refinement, which will prove to be most useful.

12.5.2 Constrained deconvolution

Suppose now we choose f such that some linear operator acting on f, namely $Q(f)$, is minimised, subject to the constraint that the norm of the noise is fixed. This is a Lagrangian problem, which is equivalent to minimising the function

$$|Qf| + \Gamma|g - Hf| \tag{12.43}$$

where Γ is a Lagrange multiplier. Writing out the Euclidean norm in full as the outer product of a vector with its transpose, we need to minimise

$$\hat{f}\,\widetilde{\mathbf{Q}}\mathbf{Q}f + \Gamma\widetilde{(g - \mathbf{H}\hat{f})}(g - \mathbf{H}\hat{f}). \tag{12.44}$$

Differentiating with respect to $\hat{f}$ and setting the derivative to zero, we obtain

$$d_k\widetilde{\mathbf{Q}}\mathbf{Q}\hat{f} + \hat{f}\,\widetilde{\mathbf{Q}}\mathbf{Q}d_k - \Gamma d_k\widetilde{\mathbf{H}}g - \Gamma\widetilde{g}\mathbf{H}d_k$$
$$+ \Gamma d_k\widetilde{\mathbf{H}}\mathbf{H}\hat{f} + \Gamma\hat{f}\widetilde{\mathbf{H}}\mathbf{H}d_k = 0 \tag{12.45}$$

with d_k as in §12.5.1. This equation is satisfied by

$$\widetilde{\mathbf{Q}}\mathbf{Q}\hat{f} - \Gamma\widetilde{\mathbf{H}}g + \Gamma\widetilde{\mathbf{H}}\mathbf{H}\hat{f} = 0 \tag{12.46}$$

since this, plus its transpose, sum to zero to give equation (12.45). Simplifying

$$(\widetilde{\mathbf{Q}}\mathbf{Q} + \Gamma\widetilde{\mathbf{H}}\mathbf{H})\hat{f} = \Gamma\widetilde{\mathbf{H}}g \tag{12.47}$$

or

$$\hat{f} = (\tau\widetilde{\mathbf{Q}}\mathbf{Q} + \widetilde{\mathbf{H}}\mathbf{H})^{-1}\widetilde{\mathbf{H}}g \tag{12.48}$$

with $\tau = \Gamma^{-1}$. This is the most general solution to the image processing problem because it contains the free operator $\mathbf{Q}$. In order to implement the technique, it is necessary to consider some specific forms for $\mathbf{Q}$. First, let us recover a trivial result by setting $\tau = 0$. We then obtain

$$\hat{f} = (\widetilde{\mathbf{H}}\mathbf{H})^{-1}\widetilde{\mathbf{H}}g \tag{12.49}$$

or (remembering to reverse terms in taking the inverse of the product)

$$\hat{f} = \mathbf{H}^{-1}g. \tag{12.50}$$

This is the familiar least-squares solution (or direct deconvolution) appearing again. This is not surprising because setting τ to zero was equivalent to letting the minimisation of the Euclidean norm of the noise term dominate the Lagrange minimisation. In mathematical terms, expression (12.44) is under these conditions reduced to expression (12.39).

Now let the operator $\mathbf{Q}$ be the identity operator $\mathbf{I}$. Equation (12.48) becomes

$$\hat{f} = (\tau\mathbf{I} + \widetilde{\mathbf{H}}\mathbf{H})^{-1}\widetilde{\mathbf{H}}g. \tag{12.51}$$

Choosing this form of $\mathbf{Q}$ is such that the restored picture itself has a minimum norm because in equation (12.43) $|f|$ has been minimised. Equation (12.51) is a very important result for image processing and has been described by several names including 'pseudo-inverse filter', 'constrained deconvolution' and 'maximum-entropy deconvolution'. Essentially, the reason why this equation leads to successful image processing is that, for those frequencies where the modulation transfer function becomes vanishingly small or even zero, the term $\tau\mathbf{I}$ dominates the denominator in equation (12.51) and avoids noise amplification. It is therefore clear that the value of τ, which remains a free parameter in general, has to be chosen taking due account of the shape of the decline of the modulation transfer function with increasing frequency. When τ itself is considered to be frequency-dependent, the filter is a Wiener filter. If the modulation transfer function is real (corresponding to a symmetric PSRF), then the outer product can be replaced by the modulus and the transpose by the complex conjugate (denoted by an asterisk), giving the neater form for the equation as

$$\hat{f} = \frac{\mathbf{H}^* g}{\tau + |\mathbf{H}^2|}. \tag{12.52}$$

12.5.3 Maximum-entropy deconvolution

The entropy in a scene f can be defined as

$$-f \ln f \tag{12.53}$$

by analogy with entropy in information theory or statistical mechanics. If we maximise this expression, this is equivalent to yielding a solution to the image processing problem which makes the fewest presuppositions concerning the form of the processed result. The analysis is similar to that detailed in the previous two sections, except that it is a little more complicated, and here the result only is given that

$$\hat{f} = \exp[-1 - 2\Gamma\widetilde{\mathbf{H}}(g - \mathbf{H}\hat{f})]. \tag{12.54}$$

This is a transcendental equation because $\hat{f}$ appears on both sides of the equation. Suppose, however, the result is expanded to just the first term in the Taylor series for the exponential. Then we obtain

$$\hat{f} = 1 - 2\Gamma\widetilde{\mathbf{H}}(g - \mathbf{H}\hat{f}) + 1. \tag{12.55}$$

Rearranging the result and cancelling terms gives

$$(\mathbf{I} - 2\Gamma\widetilde{\mathbf{H}}\mathbf{H})\hat{f} = -2\Gamma\widetilde{\mathbf{H}}g \tag{12.56}$$

or

$$\hat{f} = (\tau\mathbf{I} + \widetilde{\mathbf{H}}\mathbf{H})^{-1}\widetilde{\mathbf{H}}g \tag{12.57}$$

with τ replacing $-1/2\Gamma$.

Equation (12.57) is none other than the equation for constrained deconvolution (12.51) and it is now clear from where its other name derives.

Necessarily, the preceding discussion is a somewhat simplified summary of some of the mathematics of image formation and image processing, presented to illustrate the essential concepts rather than provide a comprehensive review. The subject is treated in depth in the book by Andrews and Hunt (1977) and readers are referred there for further reading.

12.6 IMAGE SAMPLING

In §12.4 the concept of discretely sampled space was introduced for both object and image. The formal imaging equation was interpreted as a matrix multiplication and the implications for the Fourier-space representation were considered. It was appreciated that a form of self-deception was in operation when an observer studied a discrete representation of an image and interpreted it as the object distribution of some continuously distributed property. In this section, we further discuss the relationship between a continuous function and its sampled form. The description is given in terms of image space but the same considerations apply in the domain of the object. Let us also confine the discussion to images that are sampled uniformly at spacing δx and δy and let us require that the image being sampled is band-limited at Ω_x and Ω_y. By band limiting we mean that the image does not comprise any spatial frequencies greater than Ω_x and Ω_y. Sampling the continuous image g to form the discrete sampled image g_s is equivalent to multiplication by a 2D Dirac delta function or SHAH function $d(x, y)$, i.e.

$$g_s(i\delta x, j\delta y) = \sum_i \sum_j g(x, y)\delta(x - i\delta x, y - j\delta y). \quad (12.58)$$

Let us now inspect the implications for Fourier space and we will discover that this leads to a very important statement concerning the relationship between the sampling intervals and the band-limiting frequencies. Equation (12.16) gave the Fourier transform $G(u, v)$ of g and with the same notation we write $G_s(u, v)$ as the Fourier transform of the sampled image g_s. Let us write $D(u, v)$ as the Fourier transform of the 2D Dirac sampling comb $d(x, y)$. This also turns out to be another 2D Dirac comb in frequency space, namely

$$D(u, v) = (\delta x \delta y)^{-1} \sum_i \sum_j \delta(u - i/\delta x, v - j/\delta y). \quad (12.59)$$

Equation (12.58) is a multiplicative expression in real space, i.e.

$$g_s(i\delta x, j\delta y) = g(x, y)d(x, y) \quad (12.60)$$

and hence the corresponding Fourier-space relationship involves a convolution

$$G_s(u, v) = G(u, v)*D(u, v) \quad (12.61)$$

or in full

$$G_s(u, v) = (\delta x\ \delta y)^{-1} \sum_i \sum_j G(u - i/\delta x, v - j/\delta y). \quad (12.62)$$

Equation (12.62) implies that the transform $G(u, v)$ is *replicated* on an infinite regularly spaced grid. Wherever a value $G(u', v')$ occurs for the continuous transform, it reappears at all locations spaced apart from (u', v') by $(\delta x^{-1}, \delta y^{-1})$. This is best appreciated diagrammatically and in figure 12.1 is shown a figurative form of the transform $G(u, v)$ of the continuous image g. In figure 12.2 is shown the corresponding transform $G_s(u, v)$ of the sampled image g_s. This diagram enables a clear interpretation of the relationship between the sampling interval δx and the cut-off frequency Ω_x and the corresponding orthogonal pair δ_y and

Ω_y. It is required that

$$2\Omega_x \leq (\delta x)^{-1} \qquad \text{and} \qquad 2\Omega_y \leq (\delta y)^{-1}. \qquad (12.63)$$

If these conditions are not satisfied, then the sampled function $G_s(u, v)$ will self-interfere and contributions at a particular frequency may arise from a sample and its neighbour. Such overlapping in frequency space is called aliasing, and it is an unwanted phenomenon. Rewriting equation (12.63) we arrive at the rule that must govern unaliased sampling of an image, namely

$$\delta x \leq 1/2\Omega_x \qquad \text{and} \qquad \delta_y \leq 1/2\Omega_y. \qquad (12.64)$$

These are known as the Nyquist sampling criteria.

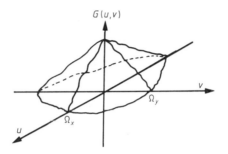

Figure 12.1 Schematic 2D Fourier transform $G(u, v)$ of continuous image g.

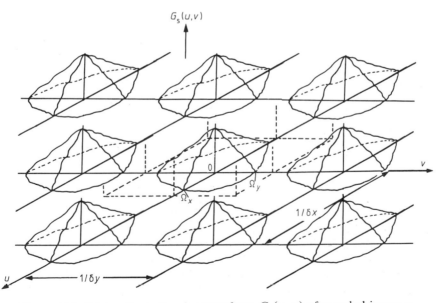

Figure 12.2 Schematic 2D Fourier transform $G_s(u, v)$ of sampled image g_s.

Hence, for a given image, band-limited at Ω_x and Ω_y, there is a maximum size limitation on the sampling intervals in real space. There is, of course, no minimum limit except that determined by the physical limitations of the technique. As the sampling intervals are made smaller and smaller, so the 'Fourier ghost images' 'move' away from the central transform. Of course, in the limit as the sampling intervals become vanishingly small, all the 'ghost transforms' have congregated at infinity! This is then entirely consistent with the statement that a continuous function has only a single-valued transform. It is as if the situation of a variable sampling interval corresponds to the Fourier images being on a rubber sheet that is being continually stretched.

This view of the relationship between a continuous image and its sampled form generates a technique for recovery of a continuous distribution from a sampled image. If it can be arranged that a 'box' isolates just the central part of the Fourier transform and sets to zero all the replicated part of the transform, then inverse-transforming the function so generated back to real space gives a continuous image. Such a Fourier box filter is the 'zero–one' filter shown as broken lines (base only shown) in figure 12.2. This filter (let us call it $P(u, v)$) is multiplied in Fourier space. Thus the operation is equivalent in real space to convolving the sampled distribution with the Fourier transform $p(x, y)$ of $P(u, v)$, i.e.

$$g(x, y) = g_s(i\delta x, j\delta y) * p(i\delta x, j\delta y). \qquad (12.65)$$

The real-space transform of a 'zero–one' box function is a 2D sinc function (see appendix (§12.9.4))

$$p(i\delta x, j\delta y) = \text{sinc}[2\Omega_x(x - i\delta x)]\, \text{sinc}[2\Omega_y(y - j\delta y)]. \qquad (12.66)$$

The operation of recovery in real space can be thought of as an interpolation. This is also useful if the image that one requires to view does not contain pixels of the right size to match the visual acuity of the eye (see Chapter 13). In short, inspection of an image may be obscured by the obvious presence of large pixels. The theory presented here can be used to redisplay an image at a finer resolution than that at which it could be acquired. Note that it is not a unique interpolation because there are a number of ways in which one could choose the box function in frequency space. Also, unless the convolution were taken to an infinite number of terms, equation (12.65) never quite generates the correct interpolated result. Additionally, our discussion has glossed over the fact that images are not normally sampled at a finite number of *points* but are rather binned into finite-size *pixels*. In this sense, the sampling is not truly a 2D Dirac comb but a comb in which the 'teeth' have a finite size and in fact fill up all the space with only infinitesimally thin 'inter-teeth' gaps. The corresponding full analysis would be more complex and for this reason the simpler analysis has been presented.

In conclusion, the above analysis has demonstrated the Shannon sampling theorem that a band-limited function is fully specified by samples spaced at intervals not exceeding $1/2\Omega$.

12.7 TWO EXAMPLES OF IMAGE PROCESSING FROM MODERN CLINICAL PRACTICE

12.7.1 An example from single-photon emission computed tomography

In this section we discuss an example of constrained deconvolution of SPECT liver tomograms, which will add more substance to the earlier theoretical discussions. Some additional observations will be made in terms of further difficulties that may arise during the evaluation of medical images created by these techniques. It is necessary to realise that evaluation of medical image processing is equally as important as its development and application. Such evaluation is difficult in that there are no universally accepted criteria for image assessment. Many authors present their results in hearsay terms showing a few striking examples, and it is almost certain that there is no universally optimum processing technique that is independent of the imaging modality and the particular clinical image being processed. Partly to avoid this problem, in a recent study that we performed (Webb *et al* 1985), specific numerical criteria were used in order to quantify the improvements consequent on image processing. In a second study (Webb 1985), the features that image processing was hoping to reveal were known to be present from confirming data from other imaging modalities. Possibly the best way of assessing the results of image processing is to perform observer perception experiments, but work in this direction is not so far advanced.

Figure 12.3 shows a single tomographic slice through a liver phantom filled with a solution of $^{99}Tc^m$-pertechnetate in water at the level of a Perspex sphere of diameter 2.5 cm, which represents a cold spot. The SPECT (see Chapter 6) tomogram was reconstructed, by a filtered backprojection technique utilising a Hanning window rolling to zero at the Nyquist frequency, from 64 projections equally spaced in 0–2π. Projection data were recorded on a GE 400T gamma camera fitted with a H2503BC low-energy general-purpose collimator. To quantitate performance numerically, the contrast of the cold spot was defined as $C = 1 - a$, where a is the ratio of the minimum pixel value in the cold spot to the largest mean of pixel values in circular annuli surrounding this minimum. In figure 12.3 this contrast is 39%. The image 'mottle' was characterised by taking the ratio of the standard deviation (σ) of pixel values to the mean (m) in a region of interest representing uniform uptake of activity. In figure 12.3, σ/m was 14%.

The aim of processing this image was to improve the contrast of the cold spot whilst not adversely affecting image mottle. Although it is always possible to argue in favour of different quantitative schemes, it is the trend in C and σ/m rather than their absolute values that is of interest. If the imaging modality had been perfect, then one would have expected $C = 100\%$. The smaller value of C that arises in practice is due to the 'infilling effect' of the finite-sized PSRF. Hence it makes sense to attempt a constrained deconvolution using equation (12.52). In figure 12.4 is shown the behaviour of contrast and mottle in processed forms of figure 12.3 as the free parameter τ was varied from 10^{-1} to 10^7 in decade steps. Figure 12.5 shows the processed tomogram corresponding to $\tau = 10^{-1}$ and figure 12.6 shows the processed tomogram corresponding to $\tau = 10^5$.

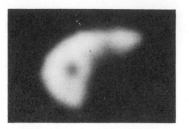

Figure 12.3 SPECT liver phantom tomogram, showing 39% tumour contrast and 14% normal tissue mottle. (From Webb *et al* (1985).)

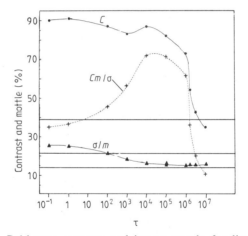

Figure 12.4 Cold-spot contrast and image mottle for liver phantom SPECT after digital image restoration (●, cold-source contrast C; ▲, mottle (σ/m) in a region of interest comprising 98 pixels in the uniform part of the liver; also shown (+) is the ratio contrast/mottle (Cm/σ)). Trends have been indicated by curves. The contrast, mottle and ratio for the unprocessed tomogram are shown as horizontal lines. (After Webb *et al* (1985).)

These results may be interpreted as follows. For low values of τ, the deconvolution is close to unconstrained and noise amplification occurs. Thus, although the contrast in figure 12.5 is 90%, there is 26% mottle and the amplified noise clearly makes the image next to useless. At high values of τ, the τ term dominates the denominator of equation (12.52) at all spatial frequencies, and so the processed image is just a scaled version of the unprocessed image. At intermediate values of τ (such as $\tau = 10^5$), there is significant improvement in contrast (83% in figure 12.6) with only 15% mottle. In figure 12.4 it is clear that there is a peak in the ratio of contrast to mottle for these intermediate values of τ. The reason follows from an inspection of the histogram of values of the modulus of the MTF in which 72% of the values of $|H^2|$ were less than or equal to 10^5.

Figure 12.5 Deconvolved SPECT liver tomogram constrained by maximum-entropy filter with $\tau = 0.1$, showing 90% tumour contrast, but unacceptable image mottle (26%). (From Webb *et al* (1985).)

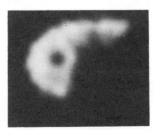

Figure 12.6 Deconvolved SPECT liver phantom tomogram constrained by maximum-entropy filter with $\tau = 10^5$, showing 83% tumour contrast and 15% image mottle. (From Webb *et al* (1985).)

This example has been chosen to illustrate the technique for an object with a well defined large cold region. It is, of course, quite reasonable to argue that in such cases image processing is not really necessary since the space-occupying defect can be visualised on the unprocessed image. In figure 12.7 is shown an example of a SPECT liver tomogram where the SPECT image was equivocal. A lesion was known to be present from confirmatory CT data. The unprocessed image has both contrast and mottle evaluated at 10%. Figure 12.8 shows the image processed using equation (12.52) and $\tau = 5 \times 10^7$, a value found to be optimum for these data. In figure 12.8 the lesion contrast has increased to 43% with no significant increase in mottle. The visual perception of the lesion is clearly greater. This example is one of several from a clinical study involving several patients, and further details can be found in Webb *et al* (1985) and Webb (1985).

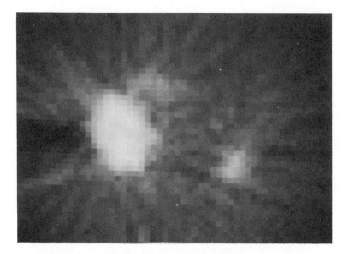

Figure 12.7 A liver SPECT tomogram with a suspected tumour. (From Webb (1985).)

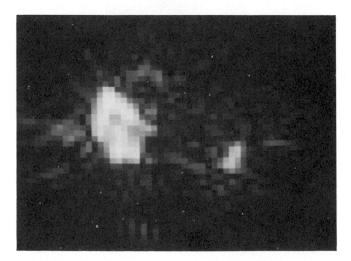

Figure 12.8 The tomogram of figure 12.7 after PSRF deconvolution, showing improved contrast. (From Webb (1985).)

12.7.2 An example from positron emission tomography

Inherent in the above example of processing image data created by SPECT was the assumption that the 'only' degradation in the image was due to the point source response function, which blurs the data in the section of interest. The nature of SPECT is such that information from adjacent tomographic slices is largely rejected by the data collection technique.

This might also be considered the case for PET data from ring systems with collimation, but it is certainly not the case for PET tomograms reconstructed from data taken with MWPC detectors (see Chapter 6). The data reconstruction technique for the latter entails backprojecting the recorded data along the lines determined by the gamma interaction coordinates in the detectors. Hence the backprojected reconstruction, which comprises a set of planes parallel to the faces of the detectors (for a stationary detector system), is inherently blurred by the three-dimensional point source response function of the system, and information spills over from one tomographic plane to its neighbours as an additional phenomenon to the in-plane blurring. For this reason, it is necessary to deconvolve the three-dimensional response function from the backprojected images. In this section, the 'tricks' associated with the deconvolution of backprojected data taken with the prototype Royal Marsden Hospital Positron Camera (described in Chapter 6) will be briefly reviewed. Further details are to be found in Webb *et al* (1984).

The three-dimensional deconvolution equation used was

$$F(u) = \frac{G(u)W(u)}{H_t(u)[1 + \Gamma(2\pi|u|)^4/H_t^2(u)]} \quad (12.67)$$

where *F*, *H* and *G* are, as before, the (now 3D) Fourier transforms of the object, point source response function and image data, respectively, *u* is the 3D spatial frequency (which could be represented in orthogonal coordinates by (u, v, w)), $|u|$ is the modulus of the 3D frequency *u* (note that $|u| \neq u$), Γ is a free Lagrange constant and the subscript 't' indicates a truncated MTF *H*, which will be explained shortly. Finally, *W* is a 3D Hanning window in frequency space given by

$$W(u) = \tfrac{1}{8}[1 + \cos(\pi u/u_N)][1 + \cos(\pi v/v_N)][1 + \cos(\pi w/w_N)] \quad (12.68)$$

and $u_N = (u_N, v_N, w_N)$ is the 3D Nyquist frequency. In equation (12.67) there are three explicit methods of 'tuning' the deconvolution to avoid noise amplification. First, the window *W* rolls off the high-frequency components of the transform *G* of the image, attenuating the components at frequencies where noise may dominate. Secondly, we have introduced the possibility of an amplitude threshold on the 3D MTF *H*. The implementation involves setting *H* to some lower threshold value at all frequencies for which the amplitude falls below this value. This specification is quite different from the modulation introduced by *W*, being amplitude- not frequency-dependent. The motivation is two-fold. In addition to those high frequencies at which the image may be dominated by noise (and at which the 3D MTF may become very small), there is also attenuation introduced at those (low) frequencies for which the 3D MTF H vanishes in view of the limited stereoscopic angle of the camera. The lower bound is determined by (as before) inspecting the histogram of H.

The third method of 'tuning' for controlling the behaviour of the deconvolution involves setting the Lagrange parameter Γ. The control is exercised in two ways, since from equation (12.67) it is clear that the second term in the square brackets can become large if either the frequency is large (at which H will almost certainly be small) and/or if H becomes small (which can also occur at some low spatial frequencies).

The several tuning methods can if desired be 'switched on or off' independently, and this was fully investigated by Webb *et al* (1984). Here, for illustrative purposes, we show in figure 12.9 some of the backprojected data planes corresponding to two lines of activity at 45° to each other, in parallel planes, and separated by 8 cm in a tank of water. The cross talk between these planes is only too apparent! In figure 12.10, in order to illustrate the point that unconstrained deconvolution is next to useless, we see the result of setting $\Gamma = 0$ and switching off the window and amplitude thresholding on H. The noise in the deconvolved images swamps the signal. In figure 12.11 is shown the result of carefully selecting the amplitude cut-off on the 3D MTF, frequency modulating using the window W and (since these are sufficient) switching off the other tuning by setting $\Gamma = 0$. The result is a striking improvement, showing the virtual elimination of cross talk. Each line is imaged clearly with little extra contamination from noise.

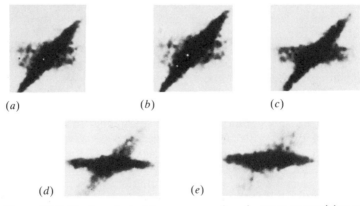

(a)   (b)   (c)

(d)   (e)

Figure 12.9 Backprojected tomograms of a phantom comprising two line sources separated by 8 cm in a scattering medium (water): the diagonal source is in plane (b); the horizontal source is in plane (d); plane (c) is mid-way between (b) and (d); plane (a) is 4 cm from (b) on the side remote from (d); plane (e) is 4 cm from (d) on the side remote from (b). (From Webb *et al* (1984).)

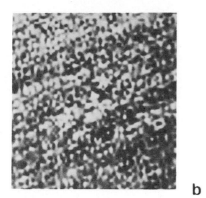

b　　　　　　　　　　　　　　　　**d**

Figure 12.10 Planes (*b*) and (*d*) of figure 12.9 after unconstrained deconvolution. (From Webb *et al* (1984).)

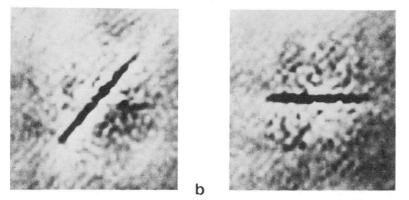

b　　　　　　　　　　　　　　　　**d**

Figure 12.11 Planes (*b*) and (*d*) of figure 12.9 after constrained deconvolution. (From Webb *et al* (1984).)

This technique has been used clinically and greatly clarifies the diagnostic usefulness of the images. Indeed, it is fair to say that, without image deconvolution, the tomographic data are too confused to be of much use.

12.8 ITERATIVE IMAGE PROCESSING

For a linear space-invariant image processing system, it has been seen that the image is related to the object distribution via the convolution integral (equation (12.10))

$$g(x, y) = \int \int h(x - \alpha, y - \beta) f(\alpha, \beta) \, d\alpha \, d\beta$$

and several methods of inverting this equation have been discussed. A neat iterative solution exists, due to Iinuma and Nagai (1967), which, for a finite number of iterations, falls short of deconvolution but in the infinite limit approaches unconstrained deconvolution. The advantage lies in the user's ability to terminate iteration interactively before noise amplification becomes too great. To illustrate the method, consider the one-dimensional form of equation (12.10)

$$g(x) = \int h(x - \alpha)f(\alpha)\,d\alpha \qquad (12.69)$$

or equivalently

$$g(x) = \int h(\phi)f(x - \phi)\,d\phi. \qquad (12.70)$$

First, make a 'zeroth approximation' to the object distribution, f_0, which is simply the observed image distribution g. When this is substituted for f on the RHS of equation (12.70), a zeroth-order approximation to the image (or pseudo-image), g_0, is generated

$$g_0(x) = \int h(\phi)f_0(x - \phi)\,d\phi. \qquad (12.71)$$

The first-order iteration for the object distribution f_1 is then defined as the sum of the zeroth-order f_0 and the difference between the measurement g and the pseudo-image g_0, i.e.

$$f_1(x) = f_0(x) + \left(g(x) - \int h(\phi)f_0(x - \phi)\,d\phi\right). \qquad (12.72)$$

By continuing the same reasoning up to the nth order

$$f_n(x) = f_{n-1}(x) + \left(g(x) - \int h(\phi)f_{n-1}(x - \phi)\,d\phi\right). \qquad (12.73)$$

The underlying reasoning supporting this particular iterative scheme is as follows. If the imaging modality were noiseless, the term in large parentheses in equation (12.73) would gradually tend to zero as n increases and $f_n(x)$ would tend to the true object distribution $f(x)$, which is the inversion of equation (12.70). This is consequent on realising that the nth-order pseudo-image

$$g_n(x) = \int h(\phi)f_{n-1}(x - \phi)\,d\phi.$$

gradually tends to $g(x)$ as n increases.

In practice, noise in the imaging modality precludes proceeding to high orders of iteration and the scheme is concluded when the value of $|g_n(x) - g(x)|$ falls below some prearranged value. Now generalising for a 2D imaging system gives

$$f_n(x, y) = f_{n-1}(x, y) + \left(g(x, y) - \iint h(\phi, \theta)f_{n-1}(x - \phi, y - \theta)\,d\phi\,d\theta\right). \qquad (12.74)$$

12.8.1 *Spatial frequency analysis of the iterative method*

An analysis of this method in Fourier space gives further insight into why it succeeds and why in the limit as n tends to infinity it approximates unconstrained deconvolution. The first iteration is given by equation (12.72) and the second yields

$$f_2(x) = f_1(x) + \left(g(x) - \int h(\phi)f_1(x - \phi) \, d\phi\right). \tag{12.75}$$

Rewriting equation (12.72) in Fourier space and recalling that a real-space convolution becomes a multiplication in frequency space, we obtain

$$F_1(u) = F_0(u) + [G(u) - H(u)F_0(u)] \tag{12.76}$$

and since $f_0 = g$, $F_0 = G$ and we may group terms to obtain

$$F_1(u) = F_0(u)K_1(u) \qquad \text{where} \qquad K_1(u) = 2 - H(u). \tag{12.77}$$

Doing the same for equation (12.75) we find

$$F_2(u) = F_1(u) + [F_0(u) - H(u)F_1(u)] = F_0(u)K_2(u)$$

$$\text{where} \qquad K_2(u) = 3 - 3H(u) + H^2(u). \tag{12.78}$$

If this conversion to Fourier space is continued up to the nth iteration

$$F_n(u) = F_0(u)K_n(u) \tag{12.79}$$

where

$$K_n(u) = 1 + [1 - H(u)] + [1 - H(u)]^2 + \ldots$$
$$+ [1 - H(u)]^{n-1} + [1 - H(u)]^n. \tag{12.80}$$

Equation (12.80) is a geometric series with first term 1 and common ratio $1 - H(u)$. Its sum, to the nth term is thus

$$K_n(u) = \frac{1 - [1 - H(u)]^n}{1 - [1 - H(u)]} = \frac{1}{H(u)} \{1 - [1 - H(u)]^n\}. \tag{12.81}$$

Combining equations (12.79) and (12.81), remembering $F_0 = G$, we get

$$F_n(u) = \frac{G(u)}{H(u)} \{1 - [1 - H(u)]^n\}. \tag{12.82}$$

As the normalised MTF is always less than 1, as $n \to \infty$, the term in braces in (12.82) tends to 1 and equation (12.82) reduces to pure unconstrained deconvolution (equation (12.18)). For any other finite n, equation (12.82) represents a truncated deconvolution. The analogy is further clarified by noting that unconstrained deconvolution involves the multiplication of the transform of the image, $G = F_0$, by the function $1/H(u)$. Now

$$1/H(u) = [1 - S(u)]^{-1} \qquad \text{with} \qquad S(u) = 1 - H(u) \tag{12.83}$$

and expanding binomially to an infinite number of terms

$$1/H(u) = 1 + S(u) + S^2(u) + S^3(u) + \ldots + S^n(u) + \ldots \text{(to } \infty\text{)}.$$
$$(12.84)$$

Thus unconstrained deconvolution is

$$F_\infty(u) = F_0(u)K_\infty(u) \qquad (12.85)$$

where

$$K_\infty(u) = 1 + [1 - H(u)] + [1 - H(u)]^2 + \ldots$$
$$+ [1 - H(u)]^n + \ldots \text{(to } \infty\text{)}. \qquad (12.86)$$

We now observe the important difference between the pairs of equations (12.79) and (12.80) and equations (12.85) and (12.86). In the former, the series has a finite number n of terms, whilst the latter has an infinite number of terms. Hence finite iteration corresponds to *truncating* the deconvolution. It is thus possible to perform image processing that can approach *arbitrarily* close to deconvolution without reaching the limit and incurring the penalties described earlier in the chapter.

Inspecting equations (12.79) and (12.80) more closely, it is clear they represent a frequency-dependent modulation with the precise dependence governed by the number of iterations and the shape of the modulation transfer function. K_n achieves its largest value $(n + 1)$ at and above those frequencies for which $H(u) = 0$. For any finite number of iterations $K_n(u) < H(u)^{-1}$ at *all* frequencies u. It is for precisely this reason that noise amplification otherwise associated with small values of the modulation transfer function is not so severe. There is, however, a computational penalty associated with the implementation in real space, namely that n iterations necessitate n discrete convolutions (see equation (12.73)) and convolution is a time-consuming computational process. This technique was originally applied by Iinuma and Nagai (1967) to improving the images of radioactive distributions from a rectilinear scanner, and further suggestions were made for overcoming this computational problem. The method is included here because it is completely general and could be applied to images from other medical imaging modalities. The discussion also nicely completes our review of the physical limitations of processing images whose point source response function is linear and space-invariant.

12.9 APPENDIX

12.9.1 The Fourier transform

In this appendix the equations that define the Fourier transform are introduced and the convolution theorem, which is vital to understanding the basis of image deconvolution, is proved.

It is shown in elementary mathematical texts that, provided a function $f(x)$ has certain basic properties, it can be expanded in a trigonometric series

$$f(x) = a_0 + \sum_{n=1}^{\infty} a_n \cos(nx) + \sum_{n=1}^{\infty} b_n \sin(nx). \qquad (12.87)$$

For this to be possible, $f(x)$ must be periodic. The coefficients of the series are unique and the series itself is differentiable. If the function $f(x)$ is odd, then the series has only sine terms, and if the function $f(x)$ is even, then the series has only cosine terms. It is not true that all trigonometric series are Fourier series and some functions have finite rather than infinitely extending Fourier series. As an example, the function

$$f(x) = \begin{cases} 0 & \text{for } -\pi \leqslant x < 0 \\ 0.5 & \text{when } x = 0 \\ 1 & \text{for } 0 < x \leqslant \pi \end{cases} \qquad (12.88)$$

can be represented by the infinite series:

$$f(x) = \tfrac{1}{2} + (2/\pi) \{ \sin x + \tfrac{1}{3} \sin(3x) + \tfrac{1}{5} \sin(5x)$$
$$+ [(1 - (-1)^n)/2n] \sin(nx) + \dots \}. \qquad (12.89)$$

Now making use of the complex exponential $\exp(i\phi) = \cos\phi + i\sin\phi$ where $i = \sqrt{(-1)}$, a more general Fourier series may be defined as

$$f(x) = \sum_{n=-\infty}^{\infty} z_n \exp(2\pi i n x / X)$$

where

$$z_n = (1/X) \int_X f(x) \exp(-2\pi i n x / X). \qquad (12.90)$$

By expanding the complex exponential, it is easy to relate the coefficients of this series to those of the trigonometric series.

It is now a small step to develop the Fourier transform. First *define* the function $F(u)$ as

$$F(u) = \int f(x') \exp(-2\pi i u x') \, dx'. \qquad (12.91)$$

Now evaluate the integral

$$I = \int F(u) \exp(2\pi i u x) \, du.$$

Substituting from equation (12.91) for $F(u)$ gives

$$I = \int\int f(x') \exp[2\pi i u (x - x')] \, du \, dx'$$
$$= \int f(x') \, dx' \int \exp[2\pi i u (x - x')] \, du$$
$$= \int f(x') \delta(x - x') \, dx'$$

by the definition of the delta function:

$$\delta(x - x') = \int \exp[2\pi i u (x - x')] \, du.$$

Hence

$$I = \int F(u) \exp(2\pi i u x) \, du = f(x). \qquad (12.92)$$

Equations (12.91) and (12.92) define the one-dimensional Fourier pair. Equations (12.16) and (12.17) are the two-dimensional extension and can be similarly derived in view of the orthogonal separability of the Fourier transform.

12.9.2 The convolution theorem

It is now a simple matter to develop the convolution theorem. Consider a one-dimensional version of equation (12.14)

$$g(x) = \int h(x - x')f(x') \, dx'.$$

Taking the one-dimensional Fourier transform (via equation (12.91))

$$G(u) = \int g(x) \exp(-2\pi i u x) \, dx \qquad (12.93)$$

and substituting from the equation above for $g(x)$ gives

$$G(u) = \int \int h(x - x')f(x') \exp(-2\pi i u x) \, dx \, dx'$$

$$= \int \int h(x'')f(x') \exp[-2\pi i u(x' + x'')] \, dx' \, dx''$$

$$= \int h(x'') \exp(-2\pi i u x'') \, dx'' \int f(x') \exp(-2\pi i u x') \, dx'$$

$$= H(u)F(u). \qquad (12.94)$$

This is the one-dimensional form of equation (12.15). That is, a convolution in real space has become a multiplication in Fourier space. It is an easy matter to prove the converse that a convolution in Fourier space becomes a multiplication in real space.

12.9.3 The autocorrelation function and the power spectrum

We show that the Fourier transform of the autocorrelation function is the power spectrum of the image. The autocorrelation function is

$$c(X) = (1/J) \int f^*(x)f(x + X) \, dx \qquad (12.95)$$

for a real-space shift X, where the asterisk represents complex conjugate and where

$$J = \int [f(x)]^2 \, dx.$$

The Fourier transform of the autocorrelation function is $C(u)$ where

$$C(u) = \int c(X) \exp(-2\pi i u X) \, dX$$

$$= (1/J) \int \int f^*(x)f(x + X) \exp(-2\pi i u X) \, dX \, dx$$

$$= (1/J) \int f^*(x) \exp(2\pi iux) \, dx \int f(x + X) \exp[-2\pi iu(x + X)] \, dX$$

$$= F^*(u)F(u)/\int [F(u)]^2 \, du \qquad (12.96)$$

where we have used that in the second integral $(x + X)$ is a dummy and the denominator J can be replaced by the corresponding Fourier-space integral by Parseval's theorem that

$$\int [f(x)]^2 \, dx = \int [F(u)]^2 \, du.$$

Equation (12.96) is the normalised power spectrum and the proof is complete.

12.9.4 The interpolation function

In §12.6 on image sampling it was stated that the zero–one box function that isolates the central order part of the discrete Fourier transform of a sampled image is a sinc function. We prove this here for the one-dimensional case as an example of the calculation of a Fourier transform.

A one-dimensional box function is specified by

$$f(x) = \begin{cases} b & \text{for } |x| \le a/2 \\ 0 & \text{elsewhere.} \end{cases} \qquad (12.97)$$

From equation (12.91)

$$F(u) = \int_{-\infty}^{\infty} f(x) \exp(-2\pi iux) \, dx$$

$$= \int_{-a/2}^{a/2} b \exp(-2\pi iux) \, dx$$

$$= b[\exp(-2\pi iux)]_{-a/2}^{a/2}/(-2\pi iu)$$

$$= b \, [-2i \sin(\pi au)]/(-2i\pi u)$$

$$= ab \sin(\pi au)/\pi au$$

$$= ab \, \text{sinc}(au). \qquad (12.98)$$

The generalisation to the 2D box function follows the same line of argument.

REFERENCES

ANDREWS H C and HUNT B R 1977 *Digital Image Restoration* (Englewood Cliffs, NJ: Prentice-Hall)
BARRETT H H and SWINDELL W 1981 *Radiological Imaging: The Theory of Image Formation, Detection and Processing* vol 1 (New York: Academic Press)
IINUMA T A and NAGAI T 1967 Image restoration in radioisotope imaging systems *Phys. Med. Biol.* **12** 501–9

WEBB S 1985 Comparison of data-processing techniques for the improvement of contrast in SPECT liver tomograms *Phys. Med. Biol.* **30** 1077–86

WEBB S, LONG A P, OTT R J, LEACH M O and FLOWER M A 1985 Constrained deconvolution of SPECT liver tomograms by direct digital image restoration *Med. Phys.* **12** 53–8

WEBB S, OTT R J, BATEMAN J E, FLESHER A C, FLOWER M A, LEACH M O, MARSDEN P, KHAN O and McCREADY V R 1984 Tumour localisation in oncology using positron emitting radiopharmaceuticals and a multiwire proportional chamber positron camera; techniques for 3D deconvolution *Nucl. Instrum. Meth. Phys. Res.* **221** 233–41

CHAPTER 13

PERCEPTION AND INTERPRETATION OF IMAGES

C R HILL

13.1 INTRODUCTION

Medical imaging occupies a position part way between science and art. Both activities are attempts to transmit to the eye and brain of observers some more or less abstract *impression* of an object of interest and, in so doing, somehow to influence their state of mind. The pure artist will be aiming for an emotional interpretation of the impression, whilst the medical imager will want to use it to attain an objective judgment about the object, perhaps concerned with the relationships or abnormalities of its anatomy. Thus, even the most technically remarkable of medical images might reasonably be likened to a Picasso, but should never be thought of as the real thing. Figure 13.1 illustrates this point. In fact, as outlined below, part of the function of a well designed medical imaging system is to extract certain features from the real object and present them to observers in a form that is well matched to their particular perceptual faculties. The conceptual and mathematical relationships between the object and image spaces and frames of reference have already been discussed in Chapter 12. In that chapter, the image in space was considered the physical end-point, and no attempt was made to examine how such an image conveys its information to the brain of the observer. In this chapter we proceed to examine this latter question.

Whether in art or science, the ultimate goal of the medical image maker can only be reached through the operation of the particular properties of two sequential channels: the visual faculties of the observers and their mental processes. Both of these are remarkably powerful, and far from fully understood, but both have important characteristics and limitations. The main purpose of this chapter is to give a brief account of the properties and limitations of the human visual perception process, which constitutes a final physical stage, and as

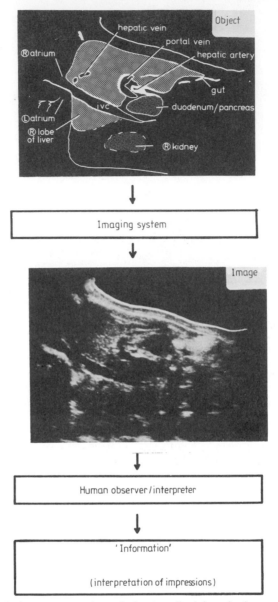

Figure 13.1 A representation of the 'medical imaging' process. An 'object' (in this case, a thin section through the body) has a highly complex (and time-variant) physical–chemical structure. An imaging system transfers (generally by a distorted and non-linear process) a very limited subset of the original data (modified by noise and other artefacts) to form a light amplitude (and/or colour) modulation pattern in 'image space'. This new data set is in turn further filtered (spatiotemporally) by an observer and its contents interpreted in the context of his/her experience and expectations, to provide some information about the hitherto unknown (or partly known) object. (Anatomical section and image reproduced by permission from Cosgrove and McCready (1982).)

such is describable by its own 'optical' or 'modulation transfer function', in an imaging system. The mental or interpretative faculties of an observer will only be touched on briefly.

All images are to some extent noisy, in the sense that there are statistical errors in the image data. For many forms of imaging, the magnitude of such noise is small in relation to the true data ('signal') and the perception process is not appreciably affected. Almost all forms of medical imaging procedures, however, are subject to a specific constraint on the magnitude of the signal: that it is proportional to the radiation exposure of, and hence to some actual or hypothetical damage to, a patient. This is a central consideration in almost all medical imaging and related procedures and is taken up more fully in Chapter 15. A mathematical treatment of the relationship between signal-to-noise ratio and dose was derived in §2.4.2 in the context of diagnostic radiology with x-rays. Many medical imaging procedures are thus characterised by low signal-to-noise ratios, and the ability of the human eye and brain to function effectively under such conditions comes to be of crucial importance. Some relevant aspects of visual performance and also of problems of image interpretation under low S/N conditions will be discussed.

The scope of this chapter is necessarily limited and it is not possible to deal, for example, with the important topic of the perception of colour. However, this and other detailed aspects of the subject are well covered in more specialised texts. A stimulating introduction to methods of image analysis and the essential characteristics of human vision is given in a book by Pearson (1975), which is oriented particularly to the engineering of systems for communication of image data. More detailed and comprehensive treatments of the visual perception process have been provided by Cornsweet (1970) and by Haber and Hershenson (1973), whilst engineering aspects, and particularly the implications of noise in imaging systems, are dealt with by Biberman (1973) and Overington (1976). Part of the material of the present chapter is drawn from an earlier review by the author (Hill 1986).

13.2 THE EYE AND BRAIN AS A STAGE IN AN IMAGING SYSTEM

Elsewhere in this book we discuss the quantitative description of imaging systems that is necessary to document their behaviour in handling image data. In this context, the eye–brain combination constitutes the final component of such a system, and it will be useful here to recall some of the relevant quantitative measures of spatial transfer characteristics.

The classical, and apparently most straightforward, approach to expressing the spatial properties of an imaging system, in transforming a stimulus in an object space to a signal in the corresponding image space, is to determine the spatial distribution of the image signal that results from a point-object stimulus. Such a distribution is termed a point spread function (PSF), and a related function, corresponding to an infinitely thin line object, is the line spread function (LSF). The spatial properties of the PSF were discussed in §12.2, where it was shown that, when the imaging system is space- and time-invariant and linear, the object is convolved with the PSF to form the image distribution.

In dealing with any but the simplest imaging systems, the above formulation proves unsatisfactory, particularly since the procedure for computing the combined effect, on the total imaging process, of the set of PSF contributed by each stage of the process entails mathematically a series of convolutions for linear space-invariant systems. Thus, it becomes both simpler mathematically and computationally, and more intuitively enlightening, to deal with the situation in the frequency domain (see §§2.4.1 and 12.3). The unidimensional Fourier transform of the LSF is the *optical transfer function* (OTF), which is in general a complex quantity. whose modulus is termed the *modulation transfer function* (MTF). The MTF for an imaging system, or for any stage of it, is thus the ratio of the amplitudes of the imaged and original set of spatial sine waves, corresponding to the object that is being imaged, plotted as a function of sine-wave frequency. The MTF for a system made up of a series of stages is now the product of the MTF of the individual stages.

This approach to the analysis of imaging systems has proved to be very powerful, and is dealt with in detail in other texts, such as that by Pearson (1975). It is important to note, however, that its applicability in an exact sense is limited by a number of important conditions, some of which are not met in certain medical imaging systems. Included in these restrictions are that the processes should be *linear* and *spacetime-invariant* (i.e. the MTF should not vary either with time or over the surface being imaged), and also that they should be *non-negative*. This latter restriction, which implies that an image function should not possess negative values, can be met in imaging systems using incoherent radiation (e.g. x-rays or radioisotope γ emissions (see Metz and Doi 1979)) but not in those using coherent radiation, which is the case in most ultrasonic systems.

Analytically, a fundamental difference between the use of OTF and MTF is that only the former recognises and preserves phase information. As discussed in the context of Chapter 7, this proves, in practice, to be a vital consideration in the understanding of coherent-radiation imaging systems.

As a preliminary to describing, in a suitably quantitative fashion, the behaviour of human vision, it is necessary to remind ourselves of the conventions used in measuring and reporting the display amplitude of a particular element of an image. This is termed the *luminance* (L) of the image element and is defined (in an observer-independent manner) in terms of the rate of emission, per unit area of the image, of visible light of specified spectral range and shape. Luminance is normally expressed in units of candela per square metre ($cd\,m^{-2}$), although some older texts use the millilambert ($1\,mL = 3.183\,cd\,m^{-2}$). The term *brightness* (B), whilst physically equivalent to luminance, is by definition observer-dependent (Pearson 1975). Some examples of typical luminance values found in images and generally in the environment are given in table 13.1.

Table 13.1 Typical luminance levels.

	Luminance (cd m^{-2})
White paper in sunlight	3×10^4
Highlights of bright CRT display[a]	1000
Comfortable reading	30
Dark region of low-level CRT display[a]	0.1
White paper in moonlight	0.03

[a] In any one CRT display the ratio of maximum to minimum luminance is seldom more than 100:1, and is typically less than this.

An immediate point to note from this table is the very wide (more than 10^6 and probably 10^8 in luminance, or approximately 80 dB) *dynamic range* of the visual system. This is the range between those luminance values that are so low as to be indistinguishable from noise background and those beyond which any increase in luminance no longer results in an increased perceived response.

13.3 SPATIAL AND CONTRAST RESOLUTION

Both the sensitivity and resolution characteristics of vision are closely related to the structure of the human retina, in which the sensitive elements include both 'cones' and the relatively more sensitive 'rods' (which constitute highly miniaturised—with dimensions in micrometres—photomultipliers that provide useful electrical output in response to the input of a single photon). Many of the remarkably powerful properties of the human and mammalian visual faculties are related to the manner in which the outputs of these sets of detector elements can be combined, according to need, to achieve what is in effect a programmable, hard-wired preprocessing system, capable of functions such as edge detection, movement detection and S/N enhancement. Detailed treatment of this subject is beyond the scope of this chapter. It is worth recalling that the mammalian visual system, which evolved essentially as a response mechanism for survival (animals requiring to detect a potential predator), is driven by individual photons. As discussed in §13.2, the mechanism has evolved to be responsive even to very dwindling fluxes of photons. The energy of a single photon is only able to disturb a single atom or molecule. Since a nerve pulse involves the movement of millions of atoms or ions, the visual system demands to be a highly efficient photomultiplier, whose mechanism is still imprecisely understood.

The central region of the retina, the fovea, which subtends (at the focal plane of the lens) an angle of between 1° and 2°, is lined almost entirely with closely packed cones, whilst more peripheral regions include both cones and rods. Within the fovea, the spacing of cones is sufficiently close (about $10'\mu$m) to enable grating resolution to be achieved up to about 60 cycles/degree. The behaviour at lower spatial frequencies, for two display luminances (see table 13.1), is illustrated in figure 13.2.

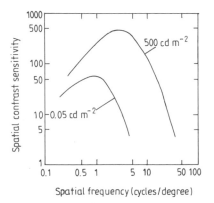

Spatial frequency (cycles/degree)

Figure 13.2 Typical contrast sensitivity of the eye for sine-wave gratings. Evidently the perception of fine detail is dependent on luminance level; this has practical implications, for example, on the choice between positive and negative modulation for the display of particular types of image. (After Pearson (1975).)

The existence of the peak in sensitivity (at about 1 and 3 cycles/degree for the two luminance levels chosen here) illustrates the remarkable and important fact that the human eye–brain is specifically adapted for the perception of sharp boundaries. It is for this reason that processes such as edge enhancement, grey-level quantisation and colour mapping can prove to be effective features in the design of an imaging system. The spectral response of the visual system peaks near the peak of the Sun's radiation and shifts towards the blue at twilight (lower light levels) in order to match the shifting spectral content of light scattered from the sky.

Other practical implications of the behaviour illustrated in figure 13.2 are that there will be a limit to the degree of fine detail that can be perceived (even if it is present in an image), that very gradual boundaries (e.g. a diffusely infiltrating border of a tumour) may easily be missed unless processing measures are taken to enhance them, and that spatial frequencies roughly in the 1–5 cycles/degree range will be maximally perceived. This latter feature may clearly be beneficial or the reverse according to whether the structure and magnification of the image are such that the detail in question is anatomical or artefactual (e.g. due to a raster pattern or even to coherent speckle; see below).

There is a further important practical message here: that one should be careful, in viewing images and in designing image display arrangements, to ensure a viewing geometry that will enable optimum perception of important image detail whilst maximally suppressing the perception of noise and other artefacts. Practical appreciation of this point may be gained by viewing the scan in figure 13.1 at a range of viewing distances and also with a magnifying lens.

The value of about 60 cycles/degree, quoted above for optically achievable grating resolution, represents a limit set by the anatomical structure of the foveal region of the retina and only holds for high levels of illumination and low levels of image noise. Thus, degradation in acuity to below this anatomical limit can be seen as arising in two separate ways: from the statistics of the visual averaging process that becomes necessary in the absence of adequate illumination, and from the limitations imposed by image noise. These two factors will be considered here in turn, together with their relevance to the question, which is of central importance in several branches of medical imaging, of contrast resolution—the ability to discriminate between neighbouring regions of differing image brightness.

The definition and measurement of contrast resolution, even in the absence of significant image noise, has been a matter for considerable research and, for a detailed account of the various factors involved, reference should be made to one of the specialised texts, e.g. chapter 5 of Haber and Hershenson (1973). Generally, experiments have been carried out in which observers have been presented with a large screen, illuminated in two adjacent segments at different uniform luminance levels $(L, L + \Delta L)$, and have been tested for their ability to detect a difference in those levels. Contrast resolution threshold $\Delta L/L$ is then determined from the difference in luminance ΔL that is just perceivable (reported in 50% of observations) at a given luminance, L. The ratio $\Delta L/L$, termed the Weber ratio, varies considerably with the level of light falling on the retina, in the manner indicated in figure 13.3. As the data in this figure indicate, the human eye is capable under ideal conditions (bright illumination and a sharp boundary between two semi-infinite object areas) of discriminating between grey levels separated by as little as 1%. In practical situations, performance is commonly much reduced as a result of four particular factors: the use of suboptimal illuminance, the absence of sharp boundaries (the significance of which has been discussed above in relation to figure 13.2), the limited size of the target area for discrimination, and the presence of image noise and 'clutter'. The significance of these last two factors will now be considered.

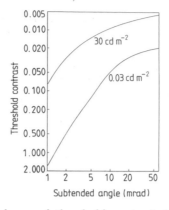

Figure 13.3 Dependence of threshold contrast $\Delta L/L$ (the 'Weber ratio') on size of an observed circular disc object, for two levels of background luminance, with zero noise and a 6 s viewing time. The effect of added noise and/or of shorter viewing time will generally be to increase threshold contrast relative to the levels indicated here. (From Blackwell (1946).)

In a major series of experiments, Blackwell (1946) demonstrated the manner in which the ability of a human observer to detect the presence of a circular target against a contrasting background depends on the degree of contrast, the level of illumination and the angular size of the target. For this purpose contrast C is defined as $C = (B_s - B_0)/B_0$, where B_0 is the brightness of the background and B_s is the brightness of the stimulus (target), and $B_s > B_0$; for $B_s < B_0$, $C = (B_0 - B_s)/B_0$. This dependence is summarised in figure 13.3. From these data, it will be seen in particular that, for a given brightness level, there is an inverse relationship between the linear size of a target and the degree of contrast necessary for its discrimination. Although this is sometimes stated to be an inverse *linear* relationship, such an approximation is clearly only valid over a limited range of target size.

The influence of the noise content of an image on its perceptibility has been the subject of a great deal of study, particularly in relation to electro-optical imaging and photographic grain noise. An often-quoted result is that to detect a single pixel differing from its N neighbours by a contrast C ($C = 0.01$ means 1% contrast), the total number of photons that must be detected is k^2N/C^2 where $k \approx 5$ (Rose 1973). For example, to detect a 1% single-pixel contrast in a 256×256 digital image would require the detection of 16.38×10^9 photons. In a typical 64×64 gamma-camera image with 10^6 detected photons, the smallest detectable contrast on this basis would be 32%, although this would reduce if a smaller value of k were adopted corresponding to decreasing the confidence in the detected contrast. It should, however, be remembered that a signal-to-noise value of 5 is not always required if the object to be detected is not simply a single pixel and has some correlation, and also if it is acceptable not to be completely certain that the detected event is not spurious (see also Webb 1987).

In the present context it is convenient to extend the concept of noise to include that of 'clutter', a term that has the more general connotation of an unwanted signal. A distinction between noise and clutter is that noise is generally incoherent in nature, whereas clutter may exhibit a degree of coherence in relation to the wanted signal. Examples of such coherent 'clutter' in medical images are the reconstruction artefacts sometimes seen in x-ray CT and NMR images (cf figure 5.7) and the reverberation artefacts in some ultrasound images (see Chapter 7).

Apart from the general level of noise, a factor of major importance is its effective spatial-frequency distribution. Quantitatively, this dependence can be investigated by measuring the degree of modulation that is necessary for visual detectability in a sine-wave bar pattern (raster) as a function of spatial frequency. In the target recognition field, this is sometimes referred to as the demand modulation function (DMF) or 'noise required modulation' (NRM).

The motive for introducing this concept is to provide a somewhat quantitative basis for predicting the effect, on an overall imaging-and-perception process, of noise or clutter having particular spatial-frequency characteristics. A practical example where this concept might usefully be applied in medical imaging is the problem of coping with the phenomenon of coherent speckle in ultrasound images (Chapter 7). Here the designer has some control over the spatial frequency of the resultant noise, and may indeed ultimately be able to eliminate its ill effects. Figure 13.4 illustrates graphically, for this example, how the relationship between the system MTF and the noise-induced degradation

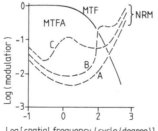

Log [spatial frequency (cycle/degree)]

Figure 13.4 Illustrative relationship between the modulation transfer function (MTF) of an imaging system and the noise required modulation (NRM) functions for visual perception of image detail under various noise conditions: curve A, low noise; B and C, noise or clutter with spatial spectra having peaks at about 10 and 1 cycles/degree, respectively. Note that curve A is an alternative representation of the behaviour illustrated in figure 13.2. The area between the MTF and appropriate NRM curves is termed the modulation transfer function area (MTFA) and is considered to be a useful measure of overall performance (Biberman 1973). The examples given here might arise practically in ultrasonic imaging with coherent speckle noise (Chapter 7) and indicate the advantage (increase in MTFA) of shifting the noise spectrum upwards (C to B) towards frequencies where perceptual ability and MTF are falling off.

of visual perception (expressed as noise required modulation) can lead to a quantitative predictor of the relative system performance for different designs.

In experiments in which varying amounts of grain noise were deliberately added to optical images, it was found that an observer, if given the choice, will tend to adjust the viewing magnification in a manner that keeps constant the spatial-frequency spectrum of the perceived image. This again emphasises the importance, in practical system design, of either providing an optimum set magnification or of allowing for operator selection. A useful discussion of the application of the MTF concept to the problem of image noise is given by Halmshaw (1981).

The concept of operator performance is clearly important in relation to medical imaging, since an image here is generally produced in order to assist in a particular task of detection or recognition of an abnormality. In this connection, therefore, it is of interest to note that, in optical imagery, three stages of perception of an object are formally categorised: as 'detection' (i.e. a decision is made as to whether some, as yet unspecified, abnormality is present), 'recognition' (i.e. features such as size and shape of an abnormality are quantified) and 'identification' (i.e. decisions are made as to likely disease patterns that correspond to the recognised, detected abnormalities). Furthermore, these different degrees of perception have been related empirically to the detectability, using the same imaging conditions and degree of modulation, of bar patterns of given spatial frequency. Thus it is found (in work on non-medical imaging applications) that 'detection', 'recognition' and 'identification' occur for target angular widths $\alpha \simeq 1/v$, $4/v$ and $6.5/v$, respectively, where v is the period of the highest detectable spatial frequency (cycles per unit angle). An alternative way of framing this statement would appear to be (if, for simplicity, one assumes that two image samples can be taken from each modulation period—the so-called Nyquist rate) that the linear dimensions of targets that are 'detectable', 'recognisable' and 'identifiable' will be 2, 8 and 13 resolution cell widths, respectively. This somewhat simplistic statement should, however, be qualified by noting that, in practice, performance of such perceptual tasks will be influenced by a number of other factors, including the time available for the task and also the *a priori* expectation of finding a particular target in a particular region of a 'scene'. It would be of interest to investigate the possible extension of the above approach for relating operator performance to imaging system parameters in some particular areas of medical imaging. In particular, it may have relevance to questions of the type: 'What is the smallest size of tumour that we can expect to see with the XYZ equipment?'

13.4 PERCEPTION OF MOVING IMAGES

Useful information about human anatomy and pathology can be derived from studies of movement, or variation in time. It is unfortunate in this respect that a number of medical imaging modalities are severely restricted, through the combined effects of noise and the requirement to avoid excessive radiation damage, in the rate at which image data can be generated, but dynamic or 'real-time' imaging is nevertheless possible in some situations, notably with ultrasound but also in some x-ray procedures.

It is therefore important to be aware of the time-related factors that affect human visual perception. Discussion here will be limited to the behaviour of the immediate visual process, but it should also be borne in mind that the processes of pattern and feature recognition that take place in the higher levels of the brain will in general be rate-limited and therefore time-dependent.

It appears that the human visual process has evolved in a manner that is responsive to specific perception of movement. Quantitatively, this can be measured and expressed, analogously to the spatial response, as a temporal-frequency response or 'flicker sensitivity' (Pearson 1975). If a small, uniform-luminance source is caused to fluctuate sinusoidally in luminance about a mean value L, the resulting stimulus will be

$$L + \Delta L \cos{(\pi f t)}$$

where ΔL is the peak amplitude of the fluctuation and f is its frequency. If the value of ΔL that produces a threshold sensation of flicker is determined experimentally as a function of f, a *flicker sensitivity* or *temporal contrast sensitivity* can be derived as the ratio $L/\Delta L$.

The typical response of the human eye under representative display viewing conditions is illustrated in figure 13.5. From this, it will be seen that, at low luminance levels, the eye behaves as though it were integrating with a time constant of around 0.2 s and is maximally sensitive to static objects. By contrast, at higher luminance, the system discriminates quite strongly against relatively static aspects of a scene and is maximally sensitive at frequencies around 8 Hz. The criteria for flicker-free viewing will evidently entail that operation must be on the negative slope of the response curve and that relative brightness fluctuation must be below some value set by the flicker frequency. The

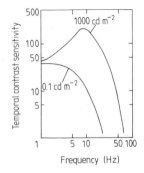

Figure 13.5 Typical flicker sensitivity of the eye for two values of retinal illuminance. (After Pearson (1975).)

limitation will be somewhat less stringent at moderate than at high luminance. The time constant of the human visual system appears to be a consequence again of evolution for survival. Since signals from potential predators are arriving in discrete quanta, there is a dilemma between the need to integrate for sufficiently long that the image is not quantum-limited (and possibly useless) and the need to react quickly for survival! The human nervous/muscular response time is about 0.1 s and the visual system has evolved to match this, with a maximal sensitivity around 8 Hz at high illumination.

13.5 QUANTITATIVE MEASURES OF INVESTIGATIVE PERFORMANCE

Medical imaging is, of course, only one subdivision of mankind's endeavour to 'investigate' his environment and, as such, it shows a fairly general need to have quantitative measures of how well the investigation is proceeding. Certainly, in the medical imaging context, it is worth giving some thought to the nature and meaning of 'investigation'. Normally one will be trying to answer a fairly specific problem (e.g. does the patient have breast cancer?); one will already have expectations as to the result (e.g. from knowledge of a palpable lump, the patient's age and family history), and one may never know the 'true' answer and therefore whether one's interpretation of the images was correct (e.g. because the patient does not go to surgery, or biopsy is unsatisfactory, etc). Thus, useful measures of investigative performance will have to be based on the *incremental improvement* in knowledge achieved and on the comparison with the currently best-available approximation to a true answer (sometimes referred to as a 'gold standard'). It will be clear also that we are dealing with a process of *decision making*.

Several different kinds of decision are involved in the total procedure of image interpretation. The first is that of 'detection': a decision as to whether an abnormality is present. Beyond this, however, there is 'localisation' (where is the abnormality?) and 'classification' (what sort of abnormality is it?). Of these, the detection process has been most fully discussed, and seems to be best understood, although both localisation and classification can usefully be considered as modifications of the detection process.

It is clearly important to be able to assess the quality of diagnostic decisions that result from a particular imaging (or similar) procedure, and a certain formalism has been developed for this purpose, particularly in relation to decisions on detection. This starts from the assumption that a 'true' answer exists to the question whether an abnormality is present, and compares this with the actual answer given by the imaging procedure. In this way, one can construct the statistical decision matrix shown in table 13.2, covering the four possible situations.

Table 13.2 Statistical decision matrix, showing the four possible situations.

Number of test assessments of presence of abnormality	Number of 'true' assessments of presence of abnormality	
	Yes	No
Yes	True positive (TP)	False positive (FP)
No	False negative (FN)	True negative (TN)

On the basis of this formalism, it is becoming common to use the following terms to indicate the quality of a diagnostic test:

sensitivity (or true positive fraction, TPF)

$$= \frac{\text{number of correct positive assessments}}{\text{number of truly positive cases}}$$

$$= \frac{\text{TP}}{\text{TP} + \text{FN}}$$

specificity (or true negative fraction, TNF)

$$= \frac{\text{number of correct negative assessments}}{\text{number of truly negative cases}}$$

$$= \frac{\text{TN}}{\text{TN} + \text{FP}}$$

$$\text{accuracy} = \frac{\text{number of correct assessments}}{\text{total number of cases}}$$

$$= \frac{\text{TP} + \text{TN}}{\text{TP} + \text{TN} + \text{FP} + \text{FN}}.$$

At this point it is necessary to amplify some points in the above discussion of the objectivity of this kind of analysis. In the first place, particularly if decisions are being made by a human observer (or even by a programmed machine), one must accept the likelihood of bias, deliberate or otherwise. There will always be a certain expectation value for the ratio between normal and abnormal cases, and assumption of an inappropriate ratio will tend to bias the results. More importantly, however, diagnosticians may be strongly influenced in their decision making by knowledge of the consequence of a particular decision. Consider, for example, the hypothetical situations of a diagnostician wishing first of all to screen a population of apparently healthy women for possible signs of breast cancer, and, secondly, to examine a woman with a suspected breast lesion, in support of a decision on whether major surgery should be undertaken. In the first case, the statistical expectation of an abnormality will be very low, but the consequences of a high false-positive rate will be the relatively mild one of an excessive number of patients undergoing further examinations. Thus there will be a valid bias towards achieving high sensitivity at the cost of decreased specificity. In the second case, the expectation of an abnormality will be much higher, but the diagnostician will need to be able to convey to the referring surgeon the degree of confidence that can be attached to the eventual assessment.

Another important qualification to note again is that (as Oscar Wilde remarks) 'truth is rarely pure and never simple': the assumption, implicit in the above matrix, that one can always expect a 'true' assessment of an abnormality is unrealistic. The best that one can usually hope for in the way of a 'definitive diagnosis' is a report on histopathology following surgery, biopsy, or *post mortem* examination. Even this is not always forthcoming, and, when it is, can be subject to considerable uncertainty.

It is thus clear that a simple measure such as sensitivity, specificity, or even accuracy will not be an objective indicator of the quality of decisions available from a particular imaging test procedure. Such an indicator is provided rather better by the so-called 'receiver operating characteristic' (ROC). This is constructed as a plot of true positive fraction (TPF) against false positive fraction (FPF), the individual points on the curve being obtained by repeating the test on a number of occasions with different degrees of bias (or decision threshold) as to the expectation of a positive result. A 'theoretical' set of such ROC curves is illustrated in figure 13.6. The different curves in the figure are indicative of differing quality in the decision process: the diagonal straight line would be the result of a totally uninformative test, whilst lines approaching closest to the FPF $= 0$ and TPF $= 1$ axes are those corresponding to the best performance (Green and Swets 1966, Todd-Pokropek 1981). A practical example of the closely related location ROC (LROC) curves obtained in perceptual tests on radioisotope images is given in figure 13.7.

ROC analysis can be used both to compare the detectability of different kinds of abnormality and to compare performance (in the sense of facilitating detection decisions) either of imaging systems or of their operators, or of both. It can, however, be a very time-consuming procedure, and it is useful to note that equally good results can be produced by a faster rating procedure (Chesters 1982). In this, the observers are required to quantify their degree of certainty about the presence of the abnormality, e.g. within the categories 0–20%, 20–40%, etc.

It is sometimes useful to consider the decision-making procedure in terms of a model in which separation of an abnormal signal from a normal signal is a noise-limited process (figure 13.8). In this, the 'signals' may be any of a large number of potentially quantifiable features of an image (e.g. grey level, smoothness of a lesion boundary),

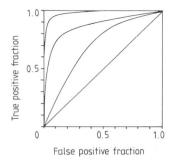

Figure 13.6 'Theoretical' examples of receiver operating characteristics (ROC). Any signal (or test) that generates an ROC curve wholly above and to the left of another is the more detectable.

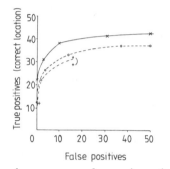

Figure 13.7 Measured ROC curves for an investigation of the effectiveness of different processing and display procedures (in use at three different medical diagnostic 'centres') for identifying both the presence and correct location of a simulated brain 'lesion' that had been inserted mathematically into a normal brain radioisotope scan. Curves of this type have been termed 'location ROC' or LROC curves. (From Houston *et al* (1979).)

and the noise may originate from the observer, the imaging system, and/or variations in the object itself. On this model, one can see rather clearly the effect of placing a decision threshold at some particular value on the scale of signal magnitude. This model also illustrates the fact that decisions (whether observer- or machine-implemented) will often be made on the basis of a number of separate features, for each of which there will be a degree of noise-limited signal separation. In this situation, the decision will be made (whether in the mind of the observer, or in a computer) within the format of a multidimensional feature space (shown as two-dimensional in the figure). A more comprehensive treatment of the subject of image assessment and decision making in diagnostic imaging is given by Goodenough (1977).

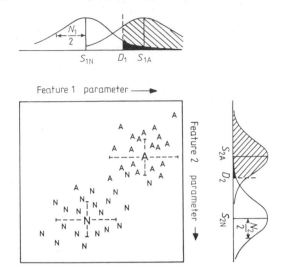

Figure 13.8 Illustration of 'noise'-limited feature separation. The signals for the two different features are characterised by their separations (e.g. $S_{1A} - S_{1N}$) and the widths of their noise spectra (e.g. N_1). Imposing a decision threshold (e.g. D_1) defines true positive fraction (hatched area) and false positive fraction (solid area). Corresponding distributions in two-dimensional feature space are also shown.

REFERENCES

BIBERMAN L M (ed) 1973 *Perception of Displayed Information* (New York: Plenum)

BLACKWELL H R 1946 Contrast thresholds of the human eye *J. Opt. Soc. Am.* **36** 624–43

CHESTERS M S 1982 Perception and evaluation of images *Scientific Basis of Medical Imaging* ed P N T Wells (Edinburgh: Churchill Livingstone) pp237–80

CORNSWEET T N 1970 *Visual Perception* (New York: Academic Press)

COSGROVE D O and McCREADY V R 1982 *Ultrasound Imaging of Liver and Spleen* (New York: Wiley)

GOODENOUGH D J 1977 Assessment of image quality of diagnostic imaging systems *Medical Images: Formation, Perception and Measurement* ed G A Hay (New York: Wiley) pp263–77

GREEN D M and SWETS J A 1966 *Signal Detection Theory and Psychophysics* (New York: Wiley)

HABER R N and HERSHENSON M 1973 *The Psychology of Visual Perception* (New York: Holt, Rinehart and Winston)

HALMSHAW R 1981 Basic theory of the imaging process in the context of industrial radiology *Physical Aspects of Medical Imaging* ed B M Moores *et al* (Chichester: Wiley) pp17–37

HILL C R 1986 Medical imaging *Physical Principles of Medical Ultrasonics* ed C R Hill (Chichester: Ellis Horwood/Wiley) pp262–77

HOUSTON A S, SHARP P F, TOFTS P S and DIFFEY B L 1979 A multi-centre comparison of computer-assisted image processing and display methods in scintigraphy *Phys. Med. Biol.* **24** 547–58

METZ C E and DOI K 1979 Transfer function analysis of radiographic imaging systems *Phys. Med. Biol.* **24** 1079–106

OVERINGTON I 1976 *Vision and Acquisition* (London: Pentech)

PEARSON D E 1975 *Transmission and Display of Pictorial Information* (London: Pentech)

ROSE A 1973 *Vision—Human and Electronic* (New York: Plenum) p12

TODD-POKROPEK A 1981 ROC analysis *Physical Aspects of Medical Imaging* ed B M Moores *et al* (Chichester: Wiley) pp71–94

WEBB S 1987 Significance and complexity in medical images; space variant, texture dependent filtering *Proc. 10th IPMI Conf., Utrecht* ed M Viergever and C N de Graaf (New York: Plenum)

CHAPTER 14

COMPUTER REQUIREMENTS OF IMAGING SYSTEMS

R E BENTLEY AND S WEBB

The type of computer required for image processing and display is, in essence, no different from any other programmable digital computer. The main objective is to present an image on a screen within minutes or less of generating the data. In other words, a fast turn-around time is more important than high throughput. In the particular case of dynamic imaging, where a moving presentation is required, very short calculation times indeed are necessary and, ultimately, processing may be required within one TV frame (40 ms). Computation times are, of course, very heavily dependent on the nature of the problem; image processing, in general, can be extremely demanding of computing resources.

The overall speed of a computer system is dependent not only on the speed of the central processing unit (CPU), but also on the time to transfer image data from one device to another. It is pointless to increase the CPU speed only to find that it takes much longer than the calculation time to transfer an image file from disc to memory. Likewise, if a processor operates on data in one hardware memory and a display device uses a different memory (e.g. a frame store), the time to transfer the image can nullify the gain of the fast processor. For this reason, dual- and multi-ported memories that can be accessed by more than one computing device have been developed.

14.1 SINGLE- VERSUS MULTI-USER SYSTEMS

The relative merits of single-user systems (which may be anything from a simple personal computer (PC) on the desk to an expensive viewing console) and multi-user systems (from a so-called superminicomputer to a mainframe computer) will continue to be debated in many applications of computing for many years. The single-user system is convenient, tailored for the job and maximises speed because it does not have to share resources with other users who may be doing work that has nothing to do with image processing. Examples of such systems that have very general applications are the Apollo, Sun Workstation and VAXstation 2000, although these particular machines may be more appropriate for graphics than imaging. Examples of machines designed more specifically for use in medical imaging are the International General Electric STAR for viewing nuclear medicine images and the Siemens Evaluskop for CT and NMR images.

14.2 GENERATION AND TRANSFER OF IMAGES

Medical images come as a general rule from one of a number of specific devices such as CT scanners, NMR scanners, radioisotope cameras and ultrasound devices as well as radiology with x-rays, described elsewhere in this book. Data will, in general, be in digital form, which means that the operations of image capture and digitisation associated with visual image processing are normally avoided. Less frequently, data may be digitised from analogue images such as x-radiographs but, in the course of time, it is expected that radiological data will also be digital at the point of capture (see §2.7)

Digital data often require to be processed in different computer systems and it is frequently necessary to transfer data from one digital device to another. Data transfers may be purely local, possibly between different computerised devices in the same room, or may involve transmission over long distances for viewing and assessment. The transfer may be done before or after extensive processing. Data transfer may be 'off-line', in which case it involves taking a magnetic medium (tape or floppy disc) from one machine to another, or it may be 'on-line' using direct-wired connection. The data transfer may also be in analogue mode or in digital mode. Analogue transmission uses television technology with video cassettes (off-line) or video cables (on-line). TV technology has the advantage of very high storage density and high transmission rate, but the disadvantage that digital signals may have to be converted to analogue form and then redigitised for processing, with subsequent loss of quality. In practice, the method of transmission will be decided by the available facilities; TV circuits are not generally available to the medical community especially in the form of switchable long-distance networks. However, developments in slow-scan television systems, which work at much lower frame rates than broadcast TV, now make it possible to send a succession of 'still' pictures over ordinary voice-telephone circuits, relatively easily and at low cost (see e.g. Lear *et al* 1987).

14.2.1 Transfer by digital magnetic media

Floppy discs and magnetic tapes are the main media of off-line data transfer. Floppy discs have the advantages of low cost per disc drive, and are especially convenient when the amount of data to be stored is not greater than 1 Mbyte (e.g. about 10 CT slices). Many centres find the concept of one floppy disc per patient easy to handle, even though this means hand searching through boxes of discs. Magnetic tape allows the computer to do the searching, but the search time can be very long if the required data is at the far end of the tape. Magnetic tape has the advantages of greater reliability than a floppy disc and presents less problems of incompatibility, and, although individual floppy discs are inexpensive, the cost per byte of storage is much lower on tape. Two other devices are now becoming available. The first of these is a cartridge tape with data recorded serially on parallel tracks so that there is much reduced search time for any particular data set, especially as these devices are now sufficiently sophisticated to allow searching in both directions. The second device is the optical laser disc, which in its present form allows writing once only but unlimited reads. This clearly gives the fastest access time of all the devices, and the fact that it cannot be erased results in less accidental loss of data and makes it ideal for long-term archiving. Table 14.1 gives a comparison of the order of magnitude characteristics for commonly used devices.

14.2.2 Transfer by hard-wired circuit

Satisfactory as magnetic media may be in many applications, when the two sites are close together, or when the urgency is such that postal delays can be tolerated, there is a strong desire to transmit images across wired physical links. In some cases, a simple link between two specific computers is all that is needed, and the hardware and software to support this is relatively simple. In general, the demand is not to be restricted in this way to a single end-to-end connection but to have access to a fully switched network that will allow unrestricted transmission within the network. A network should provide the capability to connect from any device to any other, if necessary passing through a series of intermediate stages (routing nodes), provide for automatic continuation after interruption and provide an alternative route when the primary route is unavailable. The network must be able to operate between computers of different types and different computer operating systems. It must provide the means for conversion of codes and data formats. It must offer a means of error detection and the facility for retransmission when an error is found. In practice, format conversion procedures are more highly developed for networks (on-line communication) than they are for off-line communication (magnetic media). On-line connection therefore provides a way of overcoming incompatibility problems that arise between systems of different manufacture.

A distinction is made between local area networks (LAN) and wide area networks (WAN), although there is no clear definition of these terms. In general, a LAN, at least in the UK, does not involve the crossing of public property and is generally limited to about 3 km in extent. A WAN, on the other hand, can extend to the whole world.

Table 14.1 Characteristics of exchangeable media for storing medical images[a].

	Storage capacity (number of images)	Average access time (s) (excluding time to load into device)	Transfer time per image (s) (after location on device)	Typical device cost (£)	Typical medium cost (£)	Cost per image (£) (medium only)	Comments
Floppy disc	4	0.25	5	500	2	0.5	Relatively low reliability
Exchangeable hard disc	800	0.03	0.25	10000	500	0.6	Inconvenient because of physical size for long-term storage
Optical (laser) disc	4000	0.15	0.5	10000	500	0.12	Can only be written once; virtually indestructible
Magnetic tape (45 ips at 1600 bpi)	140	300	4	3500	10	0.07	May deteriorate after several years
Magnetic tape (75 ips at 6250 bpi)	550	300	1	7000	10	0.02	May deteriorate after several years
Cartridge tape (serpentine format)	250	30	2.5	7000	25	0.1	Relatively new medium; reliability unknown

[a] Figures assume an image of 512×512 pixels and eight bits per pixel, and magnetic tape is a 2400 ft reel. All data are intended only to give indication of order of magnitude; individual manufacturers' products will vary.

Because of the necessary involvement of telephone authorities, a WAN is often restricted to a transmission speed of no more than 9600 bits per second. At this speed, an array with matrix size 512×512, 8 bits deep, takes at least 3 min to transmit. Kilostream circuits are now available from British Telecom, which work up to 64 000 bits per second when the same image will still take 35 s. Circuits going up to 2 Mbits per second (known as T1 circuits) are now available at great expense in the US and will be available as Megastream in the UK. A 10-fold improvement in the above figures may be obtained by using data-compression techniques. On the other hand, procedures for error correction and retransmission may reduce the figures given above.

In local area networks, higher transmission rates can be used. The two best-known LAN systems are the IBM Token Ring and Ethernet, operating at raw data rates of 4 and 10 Mbits per second, respectively. These systems have high overheads for such things as message addressing that, even in a lightly loaded system with no contention between competing users, the actual throughput may be less than 50% of the specified figure. When there is contention between several applications at the same time, the degradation in performance can be very marked. However, operating at only 10% of the rated figure, a $512 \times 512 \times 8$ image can be transferred over Ethernet in 4 s or so, which becomes feasible for clinical work. It should be realised, however, that transmission in this time assumes that the work load of the two communicating computers is such that their Ethernet interfaces are given sufficient resource to sustain this speed.

14.3 PROCESSING SPEED

Two main components of a computer ultimately limit speed: these are the central processing unit (CPU) and the memory. These elements are constrained by two factors: first, the propagation delay in each gate and, secondly, the signalling velocity between individual elements. Propagation delay depends on the time constant (capacity times resistance) of the element; with silicon devices this sets a lower limit of a few nanoseconds per gate. There is an expectation of a five-fold speed-up using gallium arsenide devices, but their value is as yet unproven. Very high speeds were also promised with Josephson junctions cooled to liquid-helium temperatures but much of the development work in this area has now ceased. Whether the recent announcement of materials that show superconducting properties at higher temperatures will renew interest in this field remains to be seen. Transmission delay is, of course, a function of the velocity of light, and in electrical conductors transmission speed is about $0.7c$. This results in a speed of about 20 cm ns^{-1}, so that higher computing speeds can be attained by making the total device smaller.

Without the promise of faster devices, nothing more can be done to speed up the computation time in a computer having so-called von

Neumann architecture. Such computers are wholly serial and have a bottleneck through which all instructions flow in a predetermined order. They have the advantage of being comparatively easy to program. To obtain faster response, parallel architectures have to be used, and there is now intense activity in research laboratories to produce many different types of parallel or concurrent computer.

Parallelism occurs at several different levels. At the lowest level, it is used simply to manipulate separate bits in the same word when performing arithmetic. For example, in a full adder results are given in parallel by using look-ahead methods not only for individual bits but also for the result of carry and borrow. This level of parallelism has been used in virtually all machines from the 1940s to the present day, although interestingly it has been dropped in some recent massively parallel machines such as the ICL DAP on grounds of cost. (It should be noted that a machine that performs bit-parallel arithmetic still has the classical von Neumann bottleneck for its instruction sequence.) At the opposite extreme, parallelism may be invoked by setting different computers to perform different subtasks and then merging the intermediate results to obtain the required solution. Between these extremes lie a number of other possibilities.

The extent to which parallelism can be used is controlled by the number of dependencies in the problem. In a highly dependent problem, such as a program with many iterations, where each step requires the results of a previous step, a parallel structure will not be any advantage. Fortunately, many image-processing problems have relatively low dependence as they are generally concerned with the identical operation on a very large number of individual picture elements. As a result, parallelism can be very fully exploited.

On the suggestion of Flynn (1972), parallel computer architectures have been divided into two main classes: SIMD (single-instruction multiple-data) and MIMD (multiple-instruction multiple-data). SISD (single-instruction single-data) is the von Neumann class. In a SIMD machine, a number of parallel arithmetic elements are controlled from a single instruction stream and the identical operation is carried out in each element. Examples are the ICL DAP and the CLIP, the latter developed specifically for image-processing applications. They are clearly suitable for many arithmetical operations on multidimensional arrays.

14.3.1 Pipeline processors

A machine with a pipeline is a special case of a SIMD machine in which the same instruction is carried out on different data elements, but in a stepwise manner, not concurrently. This has the great advantage over a truly parallel construction that complex and therefore expensive arithmetic elements do not have to be duplicated. It has the disadvantage that there is an overhead to get the pipe loaded at the beginning and emptied at the end so that for some operations it might be slower than a single-instruction machine.

14.3.2 Multiple-instruction multiple-data (MIMD) machines

Recently, emphasis has tended to centre on MIMD machines as they give a greater degree of programming flexibility than a SIMD machine. The development of the transputer by INMOS has provided a convenient practical tool. The transputer is in itself a conventional programmable computer, with a multitasking capability through a conventional interrupt structure and fast memory on the same chip. Its main power comes, however, from its four very fast input/output ports, which allow transputers to be connected together in a variety of configurations to pass data back and forth. This is in contrast to a more conventional structure where all data between different elements of the same system have to pass through a common highway or bus.

Concurrently with the development of the transputer has been the creation of a programming language OCCAM specifically designed to exploit parallelism. In fact, the transputer has been described as an 'OCCAM engine'. The principle in OCCAM is to consider the program as a means of controlling the flow of data through the system. It is possible to write the program so that it will run on a group of any number of transputers, and, with a configuration procedure, take advantage of as many parallel elements as are available. Thus, in a problem that will break down into many independent parallel elements, the speed will be directly proportional to the number of elements.

Parallel computers are at present at an early stage of development and their methods of use are not well understood. They present many problems to programmers and it seems that the optimum configuration is different for every application. We can, however, expect very extensive application in the image-processing field over the next decade.

14.4 DISPLAY OF MEDICAL IMAGES

Display of medical images is still almost wholly dependent on 50-year-old cathode-ray-tube technology. Only matters of detail have been changed over the years such as wider-angle geometries, improved phosphors and faster scanning rates. Alternative devices based on plasma displays, liquid-crystal electroluminescence and vacuum fluorescence are only just coming into the commercial market, and it is as yet unclear which of these technologies, if any, will eventually supersede the cathode ray tube.

For many medical applications, monochrome (grey-scale) images are adequate and often preferred. Colour scales such as the so-called 'temperature scale' using only a part of the visible spectrum from black to white through reds and yellows is sometimes used as well as scales using the whole spectrum. There is no general agreement on a standard scale that links quantitative level to colour. Many workers believe that intensity is best displayed in monochrome, reserving colour to display other features such as anatomical details or data obtained from another image. Colour can also be used for quantitation as discussed in §6.5.4.

Two methods of moving the CRT spot over the screen are used. For most purposes, a full raster scan is necessary, in which the spot is moved from left to right and from top to bottom as in broadcast television. In some cases, where a line drawing rather than a complete image is adequate, a so-called calligraphic display can be used. This drives the spot only to the points on the screen where it is required to generate bright levels. It may be useful in such fields as molecular modelling, where a line drawing rather than a complete image is required.

In any given application, there are a number of parameters, one of the most important being the required resolution (see also §2.6.3.2). This dictates the television line standard to be adopted. Standard European television uses 625 lines at a frequency of 50 frames per second, but because of two-fold interlacing, there are only 25 complete refreshments of the screen per second. Approximately 40 horizontal lines occur during the period of the frame flyback, so there are in fact only 585 usable lines. This is still adequate to display images of 512×512 pixels. In fact, if it is assumed that the horizontal resolution of the tube is equal to that of the vertical direction, for the normal TV aspect ratio of 4:3, up to 585×768 pixels can be displayed. On the American line standard of 525 lines and 60 frames per second, it is not possible even to display 512 lines and many systems are sold that permit only 480 pixels in the vertical direction.

Another problem is that of flicker, which is disturbing to the human eye, even at 25 Hz. For this reason, the American standard giving an effective 30 Hz refresh rate interlaced is sometimes used in Europe, but great care then has to be taken with power supply smoothing, otherwise a very annoying 5 or 10 Hz beat can be seen on the screen. Another technique for reducing flicker is to employ long-persistence phosphors, but this can then cause an undesirable 'ghosting' effect whenever images are moved about on the screen. Also, there is no satisfactory long-persistent blue phosphor. The only really satisfactory way to avoid flicker is to use a non-interlaced scan, which means going up to a 50 Hz (or 60 Hz) refresh rate. This is in apparent contradiction to the normal view that interlacing reduces flicker; it actually does nothing for a still picture, but in broadcast TV, it may reduce jitter, which is a different phenomenon. The problem in the non-interlaced case is that the video bandwidth has to be doubled, which is a major disadvantage if it is also combined with larger numbers of lines, as discussed in the next paragraph.

Displays offering double the above resolution are reasonably easy to obtain, but systems that use non-standard television often have the disadvantage that it is not then possible to operate additional monitors as 'slaves' off the main system. This greatly restricts the possibility of using auxiliary devices such as hard copy and recording devices. In radiology, which may well become mainly digital in the course of time, it is debated whether displays need to be better than 1024×1024, going up to 2048×2048 or even 4096×4096. The latter is certainly commensurate with the present resolution of 10 line pairs/mm which can be achieved with photographic media. A non-interlaced 1024×1024 system requires a video bandwidth of 24 MHz. Systems working at over 100 MHz have now been developed which should be able to support 2048 × 2048. For even higher resolutions, multiple-gun cathode ray tubes may be required.

The memory associated with a display system, sometimes known as a frame store, is normally dual-ported so that it can be written to by the computer and read from by the TV system. The memory is read out in step with the line and frame timebases, so that panning and scrolling can be performed simply by adding a suitable offset to the start of the timebase. Zooming may be performed geometrically within the analogue electronics of the CRT, or more often by pixel replication at the digital stage. This simply means that the data in a single memory location are read out and applied several times to different spots. By this method, zooming is possible only through integer factors.

14.4.1 Intensity (z axis)

The brightness or intensity of each pixel is determined by the contents of the appropriate memory location. The intensity range depends directly on the number of bits per pixel, so that with 8 bits, 256 different values are possible, with 16 bits ~ 65 000 different levels. For monochrome, these determine grey levels; for colour, it is normal to allocate a certain number of bits to each of the three primary colours.

14.4.2 Look-up tables

Look-up tables (LUT) are one of the least understood, yet most important, features of a display system. Their function is to allow pixel values to be converted into a convenient range of brightness and contrast. Figure 14.1 shows how a narrow range of input intensities may be translated to fill the whole available range of output intensities from fully white to totally black. Instead of using the pixel contents directly to determine the magnitude of the display, they point to a location in the look-up table which contains a value that determines the magnitude of the display. It is analogous to indirect addressing in assembly language programming. Figure 14.2 shows a typical look-up table with eight z planes used for input and with 4 bits of output. The look-up table must have 256 entries, each containing a 4 bit number in the range 0 to 15. To implement the relationship shown in figure 14.1, the contents of all entries in the table up to address 170 will be 0, locations 171 to 180 will cover the range 1 to 14, and locations greater than 180 will have a value of 15.

Confusion exists in the nomenclature of look-up tables, because manufacturers often speak of an n bit look-up table without specifying whether n is the number of bits of input or of output. Note that the number of output bits may be smaller than or greater than the number of input bits. In practice, look-up tables have up to 12 bits of input, then requiring as many as 4096 separate entries. As many as 12 bits are necessary only for CT and NMR images where as many as 11 or 12 bits of significant data may be stored in each pixel. If the full 12 bits are not connected to the table, significant contrast information may be lost.

When using colour, three separate look-up tables are configured for red, blue and green, respectively, and are arranged so that any of the input bits may control any of the output bits for each colour. This results in the implementation shown in figure 14.2 being replicated twice and three tables operated in parallel.

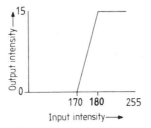

Figure 14.1 Example of output against input in a look-up table set up to concentrate entire output into a narrow range of inputs.

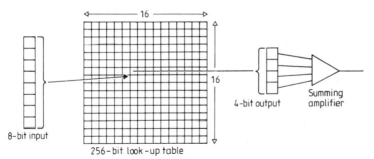

Figure 14.2 Implementation of an eight-bit look-up table using a 256-address memory with four bits of data per address.

14.5 THREE-DIMENSIONAL IMAGE DISPLAY: METHODOLOGY

Tomographic imaging techniques have had a wide impact on diagnostic medicine. Data are generally reconstructed in transaxial planes and viewed as contiguous sets of transaxial data or rearranged into sagittal and coronal distributions. Necessarily it is difficult to display all reconstructed slices in each orientation simultaneously and it has become common, for example, for the observer to view selected slices with reference to position cursors. For most clinical applications, the advantages of viewing tomographic information are self-evident and outweigh the perceptual disadvantages associated with the need to search through sets of data. The computer requirements for such two-dimensional display have already been addressed in preceding sections of this chapter.

In the last few years, interest has grown in the computer production and clinical implementation of 'three-dimensional display' of surfaces generated from tomographic data sets. By this is meant the production of a two-dimensional view or series of views of a three-dimensional pseudo-solid object representation derived from the tomographic data. The term 'three-dimensional display' is in reality a misnomer (with some licence) used to describe this methodology. Three-dimensional display has been achieved at several centres using different computer techniques (see, for example, Udupa 1982, Gibson 1983, Sartoris 1986, Gillespie and Isherwood 1986, Webb *et al* 1986b, 1987). Since the aim here is to illustrate the general principles involved, a system that has been implemented clinically for the display of SPECT data (see Chapter 6) at the Royal Marsden Hospital, Sutton, will be described.

Transaxial data were transferred by magnetic tape from a GE STAR computer on which they were constructed to a VAX 750 computer for production and display of 3D images. The entire set of transaxial data were viewed on a SIGMEX ARGS 7000 display and the sections of the volume(s) of interest were contoured at a threshold set by the user. The threshold was varied until the user was satisfied that the structures of interest were properly contoured. The number of structures outlined in each transaxial frame could be varied. Figure 14.3 illustrates this process for a set of N contours (from N transaxial slices) through the distribution of $^{99}Tc^m$-hexamethylpropyleneamineoxime (HMPAO) in the brain. In the example there is a tumour present that appears on several contours as an indentation.

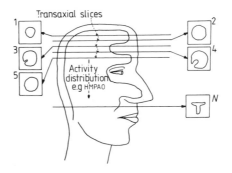

Figure 14.3 Selected contiguous transaxial slices through the activity distribution of interest, and contours found for each slice. (From Webb *et al* (1986b).)

The sets of contours thus produced were submitted to the suite of programs MOVIE.BYU for generating 3D images. MOVIE.BYU is a large, widely distributed, 3D imaging package with many features (Christiansen and Stephenson 1985). The first stage of production of a shaded surface 3D image is the stacking of the transaxial contours as illustrated in figure 14.4. These stacked contours are then connected by a fine mosaic of tiles. The software determines how many nodes on each contour are necessary for a good representation of the surface between pairs of contours. This process is illustrated in figure 14.4, where, for clarity, the mosaic surface between just two adjacent contours is shown. The mosaics can be viewed as 'wire-frame' 3D images, but were more usually shaded corresponding to the surface reflection of light(s) from specified location(s). The production of a shaded surface image from the wire-frame mosaic surface is illustrated in figure 14.5, again for simplicity illustrating the process for just one strip. Where a tile in the mosaic is normal to the line of sight of the observer, a large reflectivity arises and the reflectivity falls as the angle of the tile to the observer increases. In figure 14.5 the amount of light reflected (and hence the brightness ascribed to the reflecting tile) is represented schematically by the number and thickness of the arrowed lines. The shaded surfaces can be made semitransparent for the viewing of otherwise hidden internal structures. Images can be produced in grey tones or in colour, with different colours ascribed to different parts of a complex 3D structure.

Local software was developed to provide a real-time movie of a 3D structure as seen from a series of contiguous locations around the structure. Rotation can be around any chosen axis but, most usefully, following the movements of the gamma camera around the superior–inferior axis.

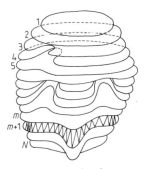

Figure 14.4 Stacking of contours and subsequent mosaic (for clarity, only some of the hidden lines are shown). (From Webb *et al* (1986b).)

Sixteen (256 × 256) 3D images were created corresponding to the 3D view of the structure from equispaced viewing locations in 0–360°. By arranging to flash each of these 3D images sequentially into a (zoomed) 512 × 512 space, an illusion of rotation was achieved. The number of orientations was limited to 16 by the 1024 × 1024 viewing station used.

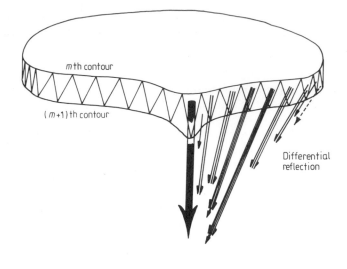

Figure 14.5 Generation of a shaded surface by the mosaics, the degree of shading depending on the angle of each tile to the observer. In the example shown, the light source is at the eye of the observer, but it can be elsewhere. The reflections from just one strip of mosaic are shown. (From Webb *et al* (1986b).)

The rotation is in real time, but is constrained to the orientations for which 3D data have been generated. Ideally for this application, and for those based on x-ray CT data (see next section), joystick control of the viewing direction would be most advantageous, as would real-time translation of the light source to vary the surface shading.

It can easily be appreciated that the computer requirements to achieve three-dimensional display for this or any other technique are demanding. A high-resolution (generally at least 1024×1024 pixels) display device is essential and ideally should be able to be rapidly loaded with data. The software to generate the three-dimensional images generally requires CPU times in excess of an hour and often much longer. It is usually claimed that this is not a real limitation since the tasks for which 3D display have been required are not urgent and otherwise idle computer time, such as night time, may be used. Hardware costs are high, even when general-purpose computers have been used, since, for speedy implementation, these ideally require the use of an array processor. Special-purpose hardware and software for three-dimensional work is available, but costs between $30 000 and $200 000 have been cited (Sartoris 1986). It is thus apparent that, unless three-dimensional display can utilise existing equipment (and by nature such implementations have been somewhat makeshift), the technique, being expensive, is only viable when the clinical information generated can be achieved in no other way and is of extreme importance. We shall highlight some examples in the next section.

14.6 THREE-DIMENSIONAL IMAGE DISPLAY: CLINICAL APPLICATIONS

Two important areas of clinical implementation have received attention in the last few years and the results have been striking. Chronologically first to be explored was the display of x-ray transmission CT data (see Chapter 4) as an aid to surgical procedures. Secondly, the three-dimensional display of the distribution of the uptake of radiopharmaceuticals *in vivo* has been attempted, based on SPECT data (see Chapter 6).

14.6.1 X-ray CT data in 3D

Diagnostic radiology using planar radiographs and x-ray CT traditionally utilises two-dimensional images, and for the vast majority of clinical applications these are entirely adequate. Three-dimensional display will have no impact for routine radiology. Three-dimensional display finds a unique role, however, in the pre-operative planning of a variety of orthopaedic and craniofacial surgical procedures including both pre- and post-operative assessment. Visualisation of bony structure in 3D is relatively straightforward in view of the high linear attenuation coefficient of bone and the ease of defining bony contours by thresholding techniques. The method is useful for craniofacial malformations, investigations of the temperomandibular joint, temporal bone, vertebral column deformities, acetabular fractures and imaging of complex joints such as wrist, ankle and knee. The method is particularly useful in imaging adult hip disease, and by interfacing three-dimensional imaging devices to computer-controlled milling machines, lifesize models for surgical prostheses have been constructed in plastic. The time quoted for this, however, is of the order of days (Sartoris 1986), but it is believed that customised prosthetic surgery may soon be available at some centres. Similarly, models of damaged bony structure, made in this way, serve for rehearsing surgical procedures for bone or joint reconstruction before theatre (Pate *et al* 1987).

Three-dimensional images are generated from either thin contiguous (1.5–2 mm) slices at a 1.5–2 mm spacing for small-scale structure or from overlapping 5 mm slices at 3 mm table increments for large structures such as the pelvis. Typical radiation exposures have been confined to less than 0.01 Gy by reducing the x-ray tube exposure without significant loss of resolution of bony structure (see also §4.5.2).

Three-dimensional display is ideal for evaluation of the spine and analysis of the injured vertebral column. For example, in figure 14.6 retropulsed posterior fracture fragments are seen encroaching on the spinal canal as a result of comminuted fracture of the vertebra. This *can* be visualised from 2D CT sections, but perception is greatly enhanced with 3D display, and the precise position of osseous fragments within the canal can be determined.

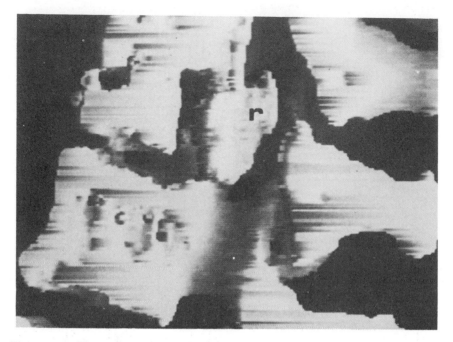

Figure 14.6 Three-dimensional display of the spine depicting encroachment upon the spinal canal by retropulsed posterior fracture fragments (r). (From Sartoris (1986).)

Facial bony structure is also easily visualised. Figure 14.7 shows a cleft palate deformity and multiple congenital anomalies at C1 and C2 due to non-fusion of the three ossification centres of the odontoid process. A small anterior spina bifida defect is also visible in C1.

In figure 14.8 is seen a fracture of the right ilium and superior pubic ramus, which have misaligned on healing of a pathological fracture of the right iliac wing. The patient had Paget's disease of the pelvis and sustained the fracture after a fall.

In view of the digital nature of the data, several additional advantages arise for surgical planning and assessment. It is possible to be very precise about distance measurements, and specific areas of interest for display may be 'cut out' by selectively rejecting overlying structure. Image processing is also possible. Pioneering work in 3D display of x-ray CT data has been performed by Hohne *et al* (1987).

Three-dimensional display of soft-tissue structure from x-ray CT is much more difficult and has only been achieved by intravenous contrast enhancement when vascular structures, blood vessels and intracranial tumours have been investigated or by the injection of intra-articular air (Gholkar and Isherwood 1988). Finally, it is worth noting that 3D display of NMR data has been attempted for quantitating bone-marrow densitometry.

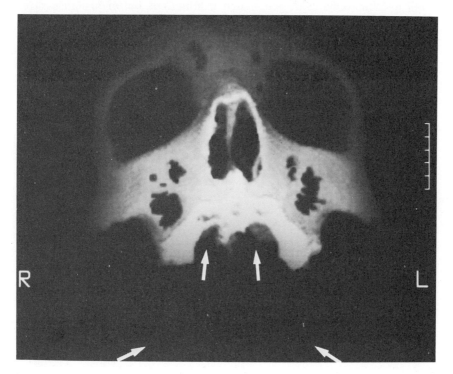

Figure 14.7 Congenital anomalies involving the face and upper cervical spine: antero-inferior view showing cleft palate and anomalous C1 and C2 (arrowed). (From Gillespie and Isherwood (1986).)

14.6.2 SPECT data in 3D

There are several SPECT examinations where viewing the reconstructed data as a three-dimensional image can be beneficial, especially when the facility to rotate the 3D image or view inside it is available. For example, the detection of a space-occupying tumour usually requires the tumour to be seen in at least 2–3 contiguous transaxial slices and in sagittal and coronal views. A 3D display that connects associated structure from contiguous slices can show the exact location of the tumour in relation to normal structure. Secondly, where a tumour is suspected on the surface of an organ (such as sometimes occurs in liver) 3D display of the surface might usefully show features indicative of abnormal pathology. Thirdly, certain diseases of the liver and kidney are associated with changes in the surface texture of the organ, which may be visible on 3D displays (Homma and Takenaka 1985, Merrick *et al* 1980). Fourthly, for diseases (e.g. of the thyroid) treated by administration of a therapy dose of a radionuclide, a knowledge of the functioning volume is desirable (Webb *et al* 1986a, Ott *et al* 1987), and 3D display can act as a guide to selection of the voxels for inclusion in the volume. Finally, conformation radiotherapy might profit from a three-dimensional view of a tumour within normal tissue, and when such a tumour is better visualised from functional imaging than anatomical imaging, 3D display of SPECT data could be beneficial. The use of 3D display, providing a pseudo-tactile image, essentially exploits man's natural in-built ability to perceive a three-dimensional world and interpret two-dimensional pictures of three-dimensional objects.

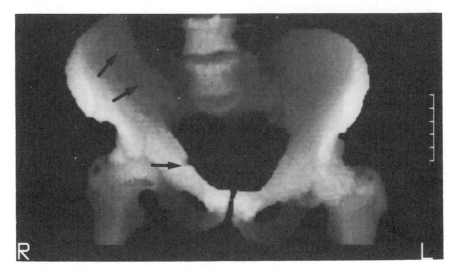

Figure 14.8 Misaligned healing of a pelvic fracture resulting in pelvic asymmetry (right ilium and superior pubic ramus are arrowed). (From Gillespie and Isherwood (1986).)

Figure 14.9 shows a three-dimensional representation of the distribution of ^{99}Tc^m-labelled hexamethylpropylene amine oxime (HMPAO) in the brain of a female patient who had brain metastases from a small-cell lung carcinoma. The images were produced by thresholding the transaxial data at 50% of the maximum reconstructed pixel value. The distribution is shown from three viewpoints (chosen from a set of 16 used to construct a real-time movie). Also shown is the superposition of a wire frame at the 20% contour on the shaded surface distribution at the 50% contour. The uptake of HMPAO in the brain and the surrounding tissues can be clearly seen. The inferior surface of the brain with the inferior aspects of the frontal lobes, the temporal lobes and the occipital region can be clearly identified. In this patient the uptake of HMPAO in the tumour was reduced or absent, and this is seen as a cavity in the three-dimensional reconstruction. It is less easy to visualise the extent of the defect from the corresponding transaxial data (see Webb *et al* 1987).

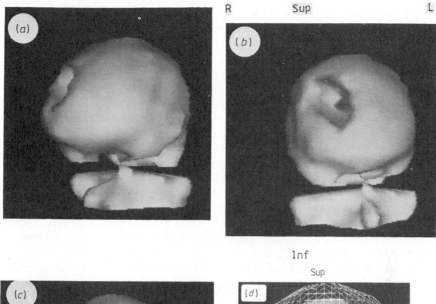

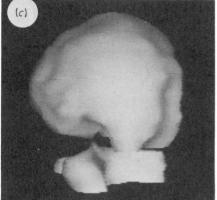

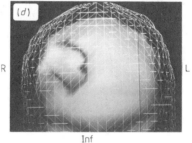

Figure 14.9 HMPAO distribution in brain ((a)–(c)) and wire frame (d) at the 20% contour superposed on shaded surface at the 50% contour. (From Webb *et al* (1986b).)

From the illustrative case it is clear that, despite the difficulties, 3D display of SPECT data is possible and can add clarity of interpretation to inspection of sets of tomographic slices. By rotating or stripping the distributions, additional appreciation of 3D structure arises. The processes of generating 3D images are complex and are certainly not to be undertaken for every case but should be reserved for those occasions where interpretation is difficult from the corresponding 2D images. The user contributes a measure of interactive decision making and thus in a sense 3D images display what the user chooses to see. For diagnosis they should be viewed alongside the 2D data. This is especially important since 3D images contain less information than 2D images, the justification for their production being a potential improved clarity of interpretation.

14.6.3 Other 3D medical applications

Most medical 3D work has concentrated on the display of transmission and emission tomographic data, and the previous two sections have highlighted some examples. Three-dimensional display has also been used for making computer models of human anatomy from contour data obtained from a microtome (a device for preparing thin sections through biological material). The computer, in this respect, replaces the nineteenth-century practice of making wax or wood models from such data. Human anatomical computer models are of use in assisting medical students to appreciate structure without access to numerous cadavers (Cookson *et al* 1987). The same benefit arises for students of dentistry who need to appreciate three-dimensional juxtapositions during their training. It is also worth noting that the need for three-dimensional computer modelling extends well beyond medical applications. For example, a recently announced optical scanner (Bell 1987) can generate anthropomorphic contour coordinates and turn these into 3D pictures on a screen. The system is useful for designing prostheses and is also aiding production of clothing patterns from lines sketched on the 3D display.

Some attempts have also been made to display NMR data in 3D. One particular system uses the technology of the varifocal mirror to display 3D images of the brain and the bifurcation of the left carotid artery (Kennedy and Nelson 1987). The lack of hard copy is a disadvantage and the image segmentation required prior to loading the display is time-consuming even on a mainframe computer. Further interest in 3D display of NMR data is anticipated.

REFERENCES

BELL J 1987 Automation beckons the garment industry *New Scientist* 13 August, 42–3

CHRISTIANSEN H M and STEPHENSON M 1985 *MOVIE.BYU* (Provo, UT: Community Press)

COOKSON J, DYKES E and HOLMAN J 1987 The illusion of solidity *New Scientist* 6 August, 50–3

FLYNN M J 1972 Some computer organisations and their effectiveness *IEEE Trans. Comput.* **C-21** 948–60

GIBSON C J 1983 A new method for the 3D display of tomographic images *Phys. Med. Biol.* **28** 1153–7

GILLESPIE J E and ISHERWOOD I 1986 Three-dimensional anatomical images from computed tomographic scans *Br. J. Radiol.* **59** 289–92

GHOLKAR A and ISHERWOOD I 1988 Three-dimensional computed tomographic reformations of intracranial vascular lesions *Br. J. Radiol.* **61** 258–61

HOHNE K H, RIEMER M and TIEDE U 1987 Volume rendering of 3D tomographic imagery *Proc. 10th Int. Conf. on Image Processing in Medical Imaging*

(*IPMI*), *Utrecht, June 1987* ed M Viergever and C N de Graaf (New York: Plenum)

HOMMA K E and TAKENAKA E 1985 An image processing method for feature extraction of space occupying lesions *J. Nucl. Med.* **26** 1472–7

KENNEDY D N and NELSON A C 1987 Three-dimensional display from cross-sectional tomographic images: an application to magnetic resonance imaging *IEEE Trans. Med. Imaging* **MI-6** (2) 134–40

LEAR J L, PLOTNICK J S, LEE R and ROFF U B 1987 High performance imaging system for digital processing and telephone transmission of nuclear medicine images *Nuklearmedizin* **26** 84–5

MERRICK M V, UTTLEY W S and WILD S R 1980 The detection of pyelonephritic scarring in children by radioisotope imaging *Br. J. Radiol.* **53** 544–56

OTT R J, BATTY V, WEBB S, FLOWER M A, LEACH M O, CLACK R, MARSDEN P K, McCREADY V R, BATEMAN J E, SHARMA H and SMITH A 1987 Measurement of radiation dose to the thyroid using positron emission tomography *Br. J. Radiol.* **60** 245–51

PATE D, RESNICK D, SARTORIS D J and ANDRE M 1987 3D CT of the spine. Practical applications *Diagn. Imaging Int.* **16** (5) 86–94

SARTORIS D J 1986 3D display of CT data; new aid to preop surgical planning *Diagn. Imaging Int.* **2** (3) 26–32

UDUPA J K 1982 Display 82—a system of programs for the display of three-dimensional information in CT data *Medical Image Processing Group Technical Report* MIPG 67, University of Pennsylvania

WEBB S, FLOWER M A, OTT R J, BRODERICK M D, LONG A P, SUTTON B and McCREADY V R 1986a Single photon emission computed tomographic imaging and volume estimation of the thyroid using fan-beam geometry *Br. J. Radiol.* **59** 951–5

WEBB S, McCREADY V R, FLOWER M A, OTT R J and LONG A P 1986b Three dimensional display of functional images generated by single photon emission computed tomography *NUC-Compact; Compact News Nucl. Med.* **17** 323–31

WEBB S, OTT R J, FLOWER M A, McCREADY V R and MELLER S 1987 Three dimensional display of SPECT data *Br. J. Radiol.* **60** 557–62

CHAPTER 15

EPILOGUE

S WEBB AND C R HILL

15.1 INTRODUCTION

This book has described the physics of a variety of approaches to medical imaging. Each chapter has inevitably concentrated attention on a particular approach but, in the real world, things do not work in that way—the different approaches complement each other in the solution of different problems. This chapter tries to put the situation in perspective and also deals with the subject, which is common to them all, of the hazard that may arise from diagnostic use of the radiations on which the different techniques rely.

Even when the physics of medical imaging has been neatly parcelled into discrete sections, as we have done in this volume, it must be said that these segregations are somewhat artificial. They are convenient for our purposes; they represent how a teaching course on medical imaging might be structured, but the divisions are woolly at the edges since so many common features and problems are shared between the imaging modalities. Chapters 2 to 11 presented the major imaging modalities in turn and Chapters 12 to 14 considered some thematic aspects. How, then, might we reasonably end our account?

To start, we might note that the very act of assembling a number of physical imaging techniques under that one label invites *comparison*. Indeed, many conferences have been held with just that intention, whilst at others comparing imaging modalities quickly becomes the topic of discussion. There are a number of reasons why this is important. First, it is inherently important for the person making a diagnosis to request the appropriate images and preferably in some optimum order. For the patient, this is vital also, since the aim of the clinician is to reach a clear diagnosis by the quickest, most accurate, least inconvenient, least harmful and least painful way. The clinical questions being asked thus clearly dictate the choices of imaging method. The choices and order of their execution are generally quite different depending on the nature of these questions and many books (e.g. Sodee and Verdon 1979, Preston *et al* 1979, Simeone 1984) address the problem. A second reason for comparing imaging modalities lies in the need to remember that they rely on different physical principles. What may be a quite impossible task for one method may be solvable by another. The corollary for the physicist is that it may be worth abandoning a line of physical investigation that is becoming progressively more difficult and less rewarding in favour of a quite different approach. For example, we have seen how no amount of struggling with planar x-rays can image deep-seated small-scale disease, whereas this is a comparatively easy problem for x-ray CT or imaging with ultrasound. Thirdly, with only limited financial resources for health care, the questions of what imaging facilities to provide, how many of each and where they should be located are of paramount importance (WHO 1983, 1985). Those charged with making these managerial decisions require detailed knowledge of a comparative nature. Naively, one might imagine that decisions could be based on the known incidences of requesting imaging examinations of various kinds. This would, however, ignore the feedback process whereby it has often been the case that the real clinical demand has only become apparent after the provision of a facility. Demand provision would also ignore the possibility of catering for rarer, but potentially more important, investigations. It is very tempting for the physicist to try to duck out of this question and leave the burden of choice elsewhere. This is wrong, since the physicist can uniquely assess the imaging potential, complementing the clinician, who can translate this into terms of assessing clinical benefit.

The manufacturer also plays an important role, since it has been commonly observed that, however suitable a laboratory imaging tool may be for some purpose, its full potential is rarely realised until a critical mass of people are all working with it. This has been true of x-ray CT, ultrasound and NMR imaging, whilst other techniques still wait their turn.

Even though we have still not strayed far from aspects of imaging to which an understanding of physical principles is the bedrock, we are in dangerous waters. There are aspects of these comparative considerations that are very subjective and which by their nature fuel discussion. It is almost certain that even the limited number of physicists contributing to this volume would find it difficult to be in complete agreement on the role of a particular imaging method. There are too many feedback processes involved, including the biases of their own particular interests and experiences, and the relative novelty of the technique in question; indeed, perhaps the perception of the relative interest in the method by others. Against this background, it is tempting to confess to no consensus and leave the readers to decide for themselves on the roles and relative importances of the subjects covered. In respect of the reader, the bottom line is that we shall do just this. After all, the reader's assessment of medical imaging practice is as valid as ours. Not quite content to rest there, we note that some aspects of the situation are, however, not unlike the problem of consumer choice and, in this spirit, we provide a table of fairly objective comparative information.

The word 'fairly' is meant to imply that even the data in this table are not entirely devoid of argument! To do the job properly, we would have to provide data for a vast number of different investigations even within a particular modality. Table 15.1 shows some of the questions one may need to answer in considering using or providing some investigative tool. Many of these questions have been addressed in detail in earlier chapters. The question of potential hazard, on which we have only briefly touched hereto, is now addressed in a little more detail.

15.2 THE IMPACT OF RADIATION HAZARD ON MEDICAL IMAGING PRACTICE

All the widely used methods of medical imaging entail exposure of the patient, in an abnormal manner, to forms of radiation that are known, in other (but generally different) circumstances, to be harmful. There will thus always in principle be a need to achieve a proper balance between levels of expected or hypothetical risk from a procedure and its anticipated benefit through improved management of the patient. The practical achievement of such a balance is extremely difficult, and will often in a true sense be impossible, because of lack of knowledge about major terms in the equation. On the one hand, virtually all the evidence as to expected hazard is indirect, in the sense that it is derived from levels and conditions of exposure quite different from those appropriate to clinical diagnosis, and it is unclear what should be the basis for making the required interpolation between experimental data and zero dose†. On the other hand, it is generally very difficult, and often impossible, to demonstrate an objective and quantitative relationship between use of a particular imaging investigation and a specific improvement in patient outcome. In this situation, it is necessary to adopt an empirical approach, and to steer on the side of caution.

The literature on *ionising radiation* hazard is enormous, and it would be impossible here even to summarise it. A quantitative feel for the situation can be gained, however, from consensus statements such as that the average risk of inducing a fatal malignancy in a human tissue subjected to a dose of 10 mSv is in the region of 10^{-4} (UNSCEAR 1977), and that 'the overall risk associated with a tissue dose, in the developing embryo or early foetus, of 10 mSv may lie in the range 0–1 per 1000 for all sequelae combined, i.e. serious malformations and cancer put together' (Mole 1979). The message for the conduct of imaging procedures is to keep dose 'as low as reasonably achievable'

---

†The only direct evidence of a quantitative relationship between damage and exposure to diagnostic levels of radiation (ionising or otherwise) is probably that of Bithell and Stewart (1975), who claimed to show a significant excess of early childhood cancer in children who had been exposed to x-rays *in utero* during diagnostic examination of the mother. The validity of even this evidence is, however, questionable (Mole 1979).

(the ALARA principle) and current practice leads to tissue doses per examination that are broadly in the range of 0.1–100 mSv (Shrimpton and Wall 1985). The lower limit here is set by a requirement for a number of recorded events sufficient to avoid intolerable image noise, and there is thus always a major premium on system and detector efficiency. As was seen in Chapter 6, nuclear medicine imaging is an area where this problem presents in a particularly serious form.

NMR *imaging* entails exposure to magnetic and RF electromagnetic fields and, whilst there is no evidence that current practice may be hazardous, the most likely direction from which hazard might arise is an increase in RF frequency and/or field strength to the point of appreciable body heating (see also Chapter 8). NMR is an inherently noise-limited technique and imaging resolution, in particular, is specifically limited by the ratio between the signal derived from a given object voxel and thermal noise arising from the entire sensitive volume of the body being scanned. Thus S/N ratio can in principle be improved by increasing magnetic field strength (and corresponding RF frequency), but the above considerations of possible thermal damage have led to current guidelines of a maximum of 2.5 T and specific absorption rate for RF energy of 0.4 W kg^{-1} for patients over the whole body (NRPB 1983).

The biological basis for these physical limits on exposure conditions arises as follows. The static field limit has been derived by noting that no chromosome aberrations in human lymphocytes nor cellular mutations have been observed after prolonged exposure to static inductions of 1 T. It should, however, be noted that no experiments to test for tumour induction have so far been performed. The whole-body RF energy deposition limit is set at one-tenth the value (4 W kg^{-1}) that would result in a body temperature rise of 1 °C, although even at this value natural thermoregulatory mechanisms (see Chapter 9) should cope. Because the blood supply to different organs and tissues varies, there is an additional local limit to 4 W kg^{-1} in any 1 g of tissue. A third limit has been set on the rate of change of magnetic induction. If the pulse width is greater than about 10 ms, $|dB/dt|$ must be less than 20 T s^{-1}. For pulsewidths $\delta t < 10$ ms, the margin of safety is set by $[dB/dt \ (\text{T s}^{-1})]^2 \ \delta t \ (\text{s}) < 4$.

Finally, the exposure for staff operating equipment (0.02 T whole body; 0.2 T limbs) appears to have been set with due regard to naturally occurring 'recovery' periods between exposures, bearing in mind that even the potential hazards noted above are believed to be reversible. Other potential hazards include the need to screen out patients with metal implants and cardiac pacemakers and the need for care in the use of metal tools near the scanner.

The very widespread use of *ultrasound*, particularly in obstetrics, has directed much attention to questions as to its safe use. Again, there is no direct evidence, or even systematic indirect evidence, that any hazard might arise from its normal diagnostic use (and there is better evidence than for ionising radiation that all known biophysical mechanisms exhibit a threshold). However, under other conditions, and particularly

Table 15.1 Comparison of imaging modalities.

	Planar x-ray	X-ray tomogram	Xeroradiograph	X-ray CT study	Gamma-camera image	Gamma-camera SPECT study	PET study	Rectilinear NM scan	US B-scan and Doppler imaging	Thermogram	NMR image	NMR spectrum	APT	Diaphanograph	Digital radiograph
Anatomy (A) or function (F) measured by basic technique?	A	A	A	A	F	F	F	F	A+F	F	A+F	F	A+F	A	A
Data acquisition time?	<1 s	few s	<1 s	2 min	5 min	20 min	20 min	20 min	1 s	≪1 s	20 min	20 min	≪1 s	<1 s	<1 s
Data reconstruction or processing time?	2 min	2 min	2 min	2 min	<1 s	10 min	10 min	few s	<1 s	≪1 s	2 min	<1 s	few min	<1 s	<1 s
Patient handling time?	5 min	5 min	5 min	30 min	10 min[a]	30 min[a]	30 min[a]	30 min[a]	10 min	10 min	1 h	>1 h	few min	10 min	5 min
Single (S) or multiple (M) images generated	S	S	S	M	S	M	M	S	M	M	M	—	S	M	S
Does imaging require a computer[b]	N	N	N	Y	Y	Y	Y	Y	Y/N	N	Y	Y	Y	N	Y
Can images be further processed easily?[b]	N	N	N	Y	Y	Y	Y	Y	Y	N	Y	Y	Y	N	Y
Minimum number of staff needed to use equipment conveniently	1	1	1	2	1	1	2	1	1	1	2	3	1	1	1

Space requirement[c]	S	S	S	>1	S	L	>1	L	S	S	SB	SB	S	S	L
Is a medical doctor needed for imaging?[b]	N	N	N	N	N	N	N	N	Y/N	N	N	N	N	Y/N	N
How common is equipment? (1(rare)→5(very common))	5	3	2	2	3	2	1	2	4	2	1	1	1	1	2
How many body sites can be studied? (1(few)→5(many))	5	3	2	5	5	4	4	5	3	3	5	3	1	1	2
Spatial resolution? (1(poor)→5(good))	5	5	5	4	3	2	4	2	4	5	4	–	1	2	4
Sensitivity? (1(poor)→5(good))	2	2	2	4	2	3	4	2	3	4	4	2	5	3	2
Inconvenience or unpleasantness to patient (1(little or none)→5(great))	1	2	3	3	2	3	3	2	2	2	4	5	2	1	1
Could the technique be a first-choice investigation?[b]	Y	N	Y	N	Y	N	N	Y	Y	Y	N	N	N	Y	N
Variability of measured parameters (1(little)→5(great))	1	1	1	2	1	1	3	1	3	1	4	4	3	2	2
Typical capital cost (£M)[a]	0.07	0.09	0.05	0.5	0.08	0.2	0.5–2.0	*	0.01–0.1	0.05	1	1.5	*	0.1	0.3

[a] Excluding injection time and wait time.
[b] N = no; Y = yes.
[c] S = small room; >1 = more than one room; L = large room; SB = special building.
* not available

where it is used surgically or therapeutically, ultrasound is clearly capable of inducing change in living systems. Thus the World Health Organisation, for example (Hill and ter Haar 1988), has reviewed the situation and has set out guidelines for its 'appropriate use'. A major criterion of 'appropriateness' in this context is, again, that exposure (measured as peak acoustic pressure amplitude for pulse–echo devices, whose acoustic stress is considered to be a potentially critical mechanism, and as acoustic power for Doppler blood-flow devices that may induce tissue heating) should be 'as low as reasonably achievable' (ALARA). With this proviso, examinations requested on specific medical indications are regarded as entirely appropriate but additional considerations may need to be taken into account when other investigations, such as on volunteers or for routine screening, are to be made. Diagnostic ultrasound is almost invariably used in echo mode for imaging purposes and the effect of stochastic noise here is to limit the range achievable for a given acoustic input amplitude and acoustic frequency (and hence spatial resolving power). This situation has led to the existence of a very wide spread of exposure parameters arising from different machine designs (Duck *et al* 1985), and criteria are only now emerging on how to minimise exposure of the patient in a manner consistent with achieving good diagnostic performance (Hill and ter Haar 1988).

The effects on biological tissue of *time-varying electric fields* have already been reviewed in Chapter 10. At practically utilised frequencies of about 100 kHz, the dominant biological mechanism is heating, since electrolysis is reversible and neural stimulation responds too slowly for damage to occur to brain, heart and muscle. Natural thermoregulatory mechanisms can cope at experimentally used power levels. It is perhaps unfortunate, in view of its lack of adverse biological effect, that applied potential tomography has to date been confined to laboratory measurements.

Diaphanography with red or near-infrared photons is also associated with a low risk (Muirhead and Seright 1984). For this reason, there is renewed excitement at the possibility of this technique at least complementing mammography, which uses ionising radiation (see Chapter 11). At this time, this has not yet occurred.

15.3 ATTRIBUTES AND RELATIVE ROLES OF IMAGING MODALITIES

Returning to table 15.1 we highlight some important attributes that help to determine the roles of the imaging modalities. The modalities divide into two classes: those which essentially give anatomical information and those showing the function of biological tissues. This division breaks down if the anatomical method can generate images fast enough to show changes in anatomy with time (which is then functional information) (e.g. B-scanning the heart). It is also becoming common for B-scan imaging to be combined with Doppler functional measurements and/or parametric functional images in one machine. Those imaging modalities which only show function can sometimes be difficult to interpret in the absence of anatomical landmarks. Although techniques for geometrically registering images from different modalities have been proposed (Gerlot and Bizais 1987), they are not in common use. Some of the methods that yield anatomical information can also yield functional information, generally by the administration of contrast media. For example, blood flow can be measured by computed tomography (Flower *et al* 1985), and radio-opaque barium compounds can be used to image the gastrointestinal tract dynamically with x-rays.

Turning to the time taken for an imaging procedure, it is possible to identify at least three separate components. There is the time it takes to acquire the basic imaging data (which may yet not actually be an image), the time it takes to reduce these data to recognisable images and finally the time the procedure takes as perceived by the patient. The latter will include the time required to prepare for the examination, the intervals during the examination when data are not being acquired (e.g. time taken resetting geometry) and the further 'dead' time after the data are acquired, but before the patient actually leaves the examination room. It is also important to note wide differences in the time needed to 'read' images.

There are several orders-of-magnitude variations in data acquisition time. X-radiography, B-scan ultrasonography, diaphanography and APT are essentially instantaneous data capture procedures. Conversely, most tomographic imaging with radiopharmaceuticals (SPECT and PET) and NMR studies require of the order of 20 min or longer to take data. We have seen that these procedures are noise-limited. Some investigations require little or no processing of the data before the images are available (e.g. gamma-camera planar scintigraphy, B-scan ultrasonography, digital radiography). Others, notably those involving image reconstruction, can take many minutes. How many minutes obviously depends on whether the associated computers utilise fast arithmetic units or array processors and on the skill of optimising the reconstruction programming. Some tomographic scanners always generate multislice data (e.g. gamma-camera-based SPECT) whilst others, although in principle able to make just a single tomogram (e.g. x-ray CT, B-scan ultrasonography, NMR), rarely do so, full studies generally being carried out. The estimates for times in table 15.1 are for full studies rather than single-plane imaging. Finally, there is at least one order-of-magnitude variation in the time perceived by the patient. X-radiography generally requires the shortest attendance (say 5 min) whilst complex procedures such as NMR imaging can take over an hour.

A further way of classifying the modalities into two groups is by multiplicity (M) or singularity (S) of images generated. Again, we have chosen to refer to common practice for a study. The division is also a little arbitrary, since all the modalities labelled 'M' could, of course, be forced to yield just one image; yet they rarely do so, since full advantage is taken of their ability to generate stacking two-dimensional images simulating three-dimensional investigation. Equally, those labelled 'S' are generally requested to form two or more separate images from (for example) orthogonal viewing directions.

The majority of the imaging methods require a computer. This was, of course, a contributing factor to the bunching of developments into the last two decades discussed in Chapter 1. We should not, however, forget that planar diagnostic radiology with x-rays does not have this requirement and, as we saw in Chapter 2, its use probably outnumbers that of other investigative tools. Indeed, a glance down the appropriate column in table 15.1 shows that this, historically the first, modality still retains many attractive features. B-mode ultrasound is also not essentially based on digital computers. Other investigations generally complement x-radiology rather than substitute for it. Since images generated by computer are digital, it follows that they are in a form immediately amenable to further processing. The lack of this ability for radiographs on film or electrostatic plate is a major drawback. Recently, there has been much interest in the possibility of film-less x-radiography (Craig and Glass 1985), which would open up the possibility of further image processing as well as hosts of other advantages. There are corresponding disadvantages of these proposals, and the late 1980s is still firmly the film era. In this volume we have only barely touched on the problems of data compatibility, data compression and storage, and data transmission. These are, however, the rapid growth areas, and some would argue that the growth is not rapid enough in view of the vast quantities of data contained in (particularly tomographic) images.

We see from table 15.1 that most imaging modalities can be operated by just one person. In two instances, B-scan ultrasonography and diaphanography, it is often desirable that this should be a medical doctor, since the information is gathered in real time and the course of action during the investigation may depend on new clinical questions raised at the time. For the others, it is sufficient to arrange a reporting session some time after the images have been taken. That said, an efficient department will often seek a quick medical opinion on single images before the patient departs, in case a further study is requested. When this is not possible (for example, for those modalities generating multiple images), the patient may need to be recalled after the reporting session. It is also the multi-image modalities that generally require more than one person present at the time of data acquisition.

Some of the least common modalities require a great deal of space, although this is not always the case. The commitment of physical space and the capital and revenue financial costs of equipment are clearly factors in determining the number of devices that can be installed. As we intimated earlier, they are not, however, the only deciding factors.

There is a range of applicability by body site. Almost every part of the body can be radiographed, scanned by x-ray CT or NMR or imaged with a gamma camera, although how successful this will be does vary by site. Some probes, e.g. ultrasound, NMR spectroscopy and certain ECT techniques, are not universally applicable, but nevertheless enjoy a wide range of applicability. Other imaging modalities, e.g. xeroradiography, APT and diaphanography, are very specialised tools.

Earlier chapters have shown a wide range of spatial resolution and sensitivity. X-radiology has excellent spatial resolution (less than 1 mm) but is not very sensitive to changes in the parameters being imaged. X-ray CT retains good resolution (1–2 mm) whilst showing an enormous increase in sensitivity. Many PET systems and NMR imagers have comparable resolution to x-ray CT. Digital radiology, being pixel-based, is degraded from film imaging. Gamma cameras have much poorer extrinsic resolution (7–15 mm) whilst SPECT and rectilinear scanning can feature resolution as poor as 2 cm or more. APT currently brings up the tail at around 3 cm spatial resolution. Yet the latter runs away with the prize for best sensitivity (to small changes in water content).

In §15.2 we have already dealt with hazard or perceived risk. To the patient, this may not present as much a problem as the inconvenience of an investigation. The entries in the table are clearly subjective, since what one patient can easily bear may be almost intolerable to another. We have obviously excluded contrast-media investigations in diagnostic radiology, which would clearly be more unpleasant than (say) a chest radiograph and also carries more hazard (De Leonardis and Pearce 1987). Since, if properly explained, most imaging modalities are not particularly unpleasant (with the exception of some more invasive examinations), one might expect low rankings in table 15.1. The higher values generally reflect the increased time required for the patient to attend.

The imaging modalities divide into two roughly equal classes depending on whether the technique would be a first-choice investigative tool, although clearly this is problem-dependent. There is a good correlation that those techniques which are not first-choice tools are also rarer, as one might expect. Surprisingly, perhaps the reverse does not seem to be true. There are several first-choice investigative techniques that are fairly uncommon, for example, xeroradiography and diaphanography. Other reasoning has prevailed. Since it is quite impossible to give more than a succinct generalisation here, the reader is referred elsewhere (e.g. Sodee and Verdon 1979) for an approach to this problem.

Most imaging tools essentially display one property of biological tissue. Some modalities promise more, but it is only NMR and PET imaging that seems to have the flexibility to investigate many properties. X-ray CT and digital radiology can give the composition of tissues, but struggle to do so; ultrasound B-scanning can investigate a variety of elastic properties and APT using multiple frequencies also promises to image more than one electrical property of tissue.

Finally, having tried to look at imaging modalities in this loose common framework, we return to recognising that many of these are 'apples and pears' comparisons. The very emergence of most imaging methods was in response to appreciating that some newly recognised physical property of biological tissue or some metabolic pathway or physiological function could be mapped into an image of diagnostic utility. Such imaging methods build into a formidable arsenal of diagnostic tools, generally complementing each other rather than replacing existing techniques. The search for less hazardous, less invasive investigations goes on. With new technology not only comes better diagnosis and patient management, but imaging that contributes to man's fundamental understanding of human biology. It is not yet a century since the only way to see inside the human body was literally by eye at surgery. Anatomical and functional information is now available on the spatial scale of millimetres. Imaging at the cellular level is still almost impossible—with the exception of ultrasonic microscopy—and we rely on information that is macroscopic by cellular dimensions. The story of medical imaging has not yet reached its final chapter.

REFERENCES

BITHELL J F and STEWART A M 1975 Pre-natal irradiation and childhood malignancy: a review of British data from the Oxford survey *Br. J. Cancer* **31** 271–87

CRAIG J O M C and GLASS H I 1985 The creation of a filmless/digital hospital: the St Mary's experience *Br. J. Radiol.* **58** 803

DE LEONARDIS E A and PEARCE J G 1987 Death during routine outpatient barium enema examination *Appl. Radiol.* (June) 44a–b

DUCK F A, STARRITT H C, AINDOW J D, PERKINS M A and HAWKINS A J 1985 The output of pulse-echo ultrasound equipment: a survey of powers, pressures and intensities *Br. J. Radiol.* **58** 989–1001

FLOWER M A, HUSBAND J E and PARKER R P 1985 A preliminary investigation of dynamic transmission computed tomography for measurements of arterial flow and tumour perfusion *Br. J. Radiol.* **58** 983–8

GERLOT P and BIZAIS Y 1987 Image registration: a review and a strategy for medical applications *Proc. 10th IPMI Meet., Utrecht* ed M Viergever and C N de Graaf (New York: Plenum)

HILL C R and TER HAAR G R 1988 Ultrasound *Non-Ionizing Radiation Protection* 2nd edn (Copenhagen: World Health Organisation)

MOLE R H 1979 Radiation effects on pre-natal development and their radiological significance *Br. J. Radiol.* **52** 89–101

MUIRHEAD A and SERIGHT W 1984 Clinical experience with the diaphanograph machine *Ann. R. College Surg. Engl.* **66** 123–4

NRPB (NATIONAL RADIOLOGICAL PROTECTION BOARD) 1983 Revised guidance on acceptable limits of exposure during nuclear magnetic resonance clinical imaging *Br. J. Radiol.* **56** 974–7

PRESTON K, TAYLOR K J W, JOHNSON S A and AYERS W R (eds) 1979 *Medical Imaging Techniques: A Comparison* (New York: Plenum)

SHRIMPTON P C and WALL B F 1985 Preliminary observations on the relationship between patient exposure and imaging techniques in a sample of British hospitals *Br. J. Radiol.* Suppl. 18, 127–9

SIMEONE J F (ed) 1984 *Co-ordinated Diagnostic Imaging* (New York: Churchill Livingstone)

SODEE D B and VERDON T A (ed) 1979 Correlations in diagnostic imaging *Nuclear Medicine, Ultrasound and Computed Tomography in Medical Practice* (New York: Appleton-Century-Crofts)

UNSCEAR (UNITED NATIONS SCIENTIFIC COMMITTEE ON EFFECTS OF ATOMIC RADIATION) 1977 Sources and effects of ionizing radiation *Report of UN-SCEAR to General Assembly* UN Publ. no. E.77.IX.1

WHO (WORLD HEALTH ORGANISATION) 1983 A rational approach to radiodiagnostic investigations *Technical Report Series* 689 (Geneva: World Health Organisation)

—— 1985 Future use of new imaging technologies in developing countries *Technical Report Series* 723 (Geneva: World Health Organisation)

Index

A-mode display, 347
A-scan, 327, 346
Abdominal ultrasonic imaging, 368
Aberdeen Section Scanner, 12, 153
Absorption of ultrasound, 323
Absorption signal (NMR), 401
Accelerators, 185
Accidental coincidences in PET, 242
Accuracy in imaging, 580
Acoustic field parameters, 320
Acoustic impedance, 322
Acoustic microscopy, 320, 384
Acoustic output measurement, 364
Acoustic radiation fields, 328
Acoustic signal injection for QA, 363
Activation energy of molecule, 406
Activity–time curves, 199, 214, 282
Adiabatic fast passage, 397
Adrenal glands, 298
AGEMA thermovision 870 camera,
    493
Air contrast, 95
Airgaps, 94
Aliasing, 357, 449, 472, 551
Alpha decay, 189
Alzheimers disease, 270
Analogue data transmission, 585
Analogue hard copy, 170, 205
Analogue to digital converter (ADC),
    194
Anger camera, 148, 161
    the first, 11, 148
Anger logic, 168

Angiography by NMR, 435
Angular momentum vector, 391
Angular sampling requirements (CT),
    121, 229
Annihilation, 190, 221
Annihilation coincidence detection
    (ACD), 239
Annular arrays of transducers, 342
Anoxia, 450
Antibody imaging, 304
Antibody labels, 192, 213, 261
Antiparallel gammas, 190, 239
Anti-scatter device, 21
Applied potential tomography (APT),
    15, 513
Apollo workstation, 585
Arthritis, 502
Arithmetic mean isosensitive
    scanning, 211
Array processor, 196, 472
Arrays of transducers, 341
Art and science in imaging, 4
Artefacts, 124, 226, 428, 575
    in NMR images, 428
Arterial disease, 503
Attenuation of light by tissue, 527
Attenuation of ultrasound, 322
Attenuation problem in SPECT, 226,
    231, 233, 236
Attenuation problem in PET, 228,
    240
Autocorrelation function, 564
Autoradiography, 147, 174, 177

Autotuning, 171, 232
Auger electron, 191
Axial (ultrasound) resolution, 334

B-mode display, 348
Background limited infrared
    photoconductor (BLIP), 492
Backing medium for transducer, 339
Backprojection, 114, 226, 513, 535
Backscattering impulse response, 335
Band-limit, 107
Band-limited convolution kernel,
    113, 125
Bandwidth for NMR, 449, 472
Barium contrast, 95
Barium fluoride, 158, 176
Beam hardening artefacts, 124
Beam plot, 364
Beer's Law, 99, 106, 124, 527
Beta continuum, 190
Beta-minus emission, 190
Beta-plus emission, 190
Bialkali photomultipliers, 170
Bifunctional chelate, 192, 261
Biopsy, 373, 525
Bio-electrodes, 511
Biological effects and safety of
    ultrasound, 375
Biological effects of NMR, 474
Biological purity, 265
Birdcage coil, 470
Bismuth Germanate detectors, 138,
    157, 241
Blackwell's experiment, 574
Bladder imaging by ultrasound, 368
Bladder in CT, 129
Bladder tumours, 129
Black body radiation, 490
Bloch equations, 400, 407
Blood–brain barrier, 267
Blood flow, 142, 269, 272, 435, 499,
    520
    artefact in NMR imaging, 428
    imaging by NMR, 435
Blood pump, 214
Blood volume, 142, 269, 272
Blurring, 38, 47
Body temperature, 499

Boltzmann distribution, 399
Bound water and T_1, 407
Brain
    imaging by RI, 267
    imaging by ultrasound, 373
    physiological function, 269
Breast
    imaging by APT, 521
    imaging by CT, 137
    imaging by ionography, 65
    imaging by transillumination, 489,
        524, 528
    imaging by ultrasound, 359, 362,
        371, 380
    imaging by xeroradiography, 60
    imaging by x-rays and film, 37
    phantom, 88
    propagation of ultrasound in, 334
Brightness, 570
Broad focus, 80
Brookhaven National Laboratory,
    10, 12, 123
Bucky factor, 93
Burns, 505
Butterworth filter, 197, 229

C-mode display, 348
Cadmium mercury telluride, 491
Caesium iodide screens, 55
Calcium tungstate scintillator, 11
Calcium tungstate screens, 48
Cancer mortality, 128
Carbon-13 NMR spectroscopy, 451
Carbon-labelled glucose, 273
Carbon-labelled methionine, 274
Cardiac imaging
    by NMR, 425
    in nuclear medicine, 214, 275
    by APT, 519
Cardiac pacemakers and NMR, 474,
    477
Cardiovascular imaging by
    ultrasound, 369
Carrier free radionuclides, 183
Carr–Purcell spin-echo train, 409
Cartridge tape, 586
Catapult grid, 94
Cathode-ray-tube display, 590

Cavitation, 365, 376
Cellular energy balance, 450
Central nervous system imaging, 267
Central processor unit (CPU), 472, 584, 588
Central section theorem, 109
Centre of rotation correction in SPECT, 233, 254
Cerebral blood flow, 269
Cerebral haemorrhage, 269
Cerebral infarction, 269
Cerebral vascular disease, 269
CERN, 172
Cervix, imaging by ultrasound, 367
Characteristic curve, 44, 95, 207
Characteristic impedance, 322
of transducer, 338
Characteristic x-rays, 26, 191
Charge image, 60
Charged-particle accelerators, 185
Charged-particle bombardment, 182, 185
Chelation, 259, 261
Chemical-shift artefact, 428, 431
Chemical-shift imaging, 438
Chemical-shift spectroscopy, 445
Chemotherapy, 128
CHESS, 438
Chest x-ray, 98
Chiral nematic liquid crystals, 505
Cholesteryl nonanoate liquid crystals, 505
Chromatographic techniques, 266
Cine photography, 57
Circularly-polarised RF fields, 395
Classical description of NMR theory, 391
Classical tomography, 86, 99
Cleon scanner, 155, 237
Clinical applications
of APT, 517
of RI imaging, 256
of ultrasound imaging, 365
Clinical thermography, 499
CLIP computer, 589
Clock pulse for echography, 345
Clutter, 336, 352, 575
Coded aperture, 178, 223

Coding messengers in imaging, 537
Coherent radiation, 319, 335, 570
Coherent speckle, 575
Coil design for NMR, 470
Coincidence detection, 157, 173
Collimation of MWPC, 174
Collimator choice, 207, 232
Collimators
for gamma camera, 161, 236
for PET, 157
for rectilinear scanner, 149, 152, 222
for SPECT, 232, 236
Colour flow mapping, 362
Colour scales, 201, 593
Comparisons of imaging modalities, 427, 610
Compound scanning, 336, 348
Compression amplifier, 346
Compton scatter, 92, 144, 178
Compton-scatter camera, 177–9
Compton-scatter tomography, 139
Compton telescope, 177
Computer disk space, 202
Computer memory, 202
Computer peripherals for NMR imaging, 472
Computer requirements of imaging systems, 472, 584
Computer speed, 584, 588
Computers in radionuclide imaging, 193
Computed tomography, 9, 98
Conductivity imaging, 512
Cones, 572
Constant ambient temperature thermography, 500
Constrained deconvolution, 547
Contact scanning (echography), 350
Continuous wave flow detector, 354
Contrast, 27, 51
Contrast/dose compromise, 28, 30
Contrast enhancement, 94
Contrast improvement factor, 93
Contrast in NMR imaging, 439
Contrast in radiography, 99
Contrast media, 95, 99
Contrast resolution, 573

Converging multihole collimator, 164, 222, 232
Conversion electron, 191
Convolution, 112, 327, 335
Convolution and back projection, 111–13
Convolution integral, 538, 540, 564
Convolution theorem, 564
Cooley–Turkey algorithm, 543
Correlation time of molecule, 406
Correlation ultrasonic processing, 328
Couch control by computer, 203
Count-rate performance QA, 253
Coupling, 400
Coupling of spins in NMR spectroscopy, 446
Credit cards and NMR, 474
Crossover, 47
Crosswires, 84
Cryogenics, 147, 176, 491
Cryogens, 479
Crystals for detectors, 148, 157
CT-aided planning and survival, 133
CT number, 111
CT simulator, 137
Cumulative error in measurement of T_2, 410
Cut-away camera head, 236
Cyclotron, 10, 185
Cysts, 524

Data acquisition computing, 194
Data processing computing, 196
Data reconstruction, 196
Data storage, 201
Data transfer, 584–5
Dead time, 169
Deblurring of limited angle ECT, 223
Decay scheme, 188
Decision making, 578
Deconvolution, 199, 220, 538, 541
    in APT, 515
    in SPECT, 228
    in transillumination, 532
    in x-ray CT, 118
Deep vein thrombosis (DVT), 502
De-excitation, 190

Degenerate states, 399
Demand modulation function (DMF), 575
Demodulation, 346
Depth-independent response, 212, 226
Depth of field, 333
Depth-resolved surface coil NMR spectroscopy (DRESS), 459
Detective quantum efficiency, 31, 54
Detectivity, 491
Detector, MWPC, 171
Deuterium imaging by NMR, 438
Diagnostic imaging as a team activity, 1
Diaphanography, 14, 524
Diastole, 215
Differential uniformity, 246
Diffraction field, 330
Diffusion, effect on measurement of T_2, 409
Diffusion imaging by NMR, 436
Digital computer, 584
Digital gamma camera, 204
Digital image processing, 70
Digital radiology, 66, 95, 585
Digital subtraction angiography, 9, 70
Digitisation of ultrasound A-scans, 347
Dipole radiation, 400
Discrete convolution, 115
Discrete Fourier Transform (DFT), 543–4
Disease specific imaging, 303
Dispersion of chemical shifts, 445
Dispersion signal, 401
Display for gamma camera, 170, 200
Display of ultrasound images, 347
Diverging multihole collimator, 165, 223
Divided kidney function, 209
Dixels, 282
DNA sequencing, 175
Doppler equation, limitations, 351
Doppler spectrum, 352
Doppler ultrasound, 327, 351
Dose in CT, 122

Dose reduction, 70
Double-headed scanner, 151, 212
Double-tuned coil, 456
Drift chamber, 146, 177
Drift detector, 177, 180
Drugs, 257, 259
DTPA, 260
Dual-energy radiology, 71
Dual-isotope imaging, 207
Dual-ported memory, 584
Duplex ultrasonic scanners, 361
Dynamic cardiac imaging, 214
Dynamic range of visual system, 571
Dynamic renal imaging, 218
Dynamic scintigraphy, 143, 213

ECG signal, 215
Echo-planar imaging, 425
Echoes for ultrasound, 325
Echography, 343
Eddy currents in superconducting
    magnets, 467, 469
Edge enhancement, 60, 197
Effective dose equivalent, 32
Effective linear attenuation
    coefficient, 80
Effective renal plasma flow, 281
Effective x-ray energy, 80
Elastic modulus, 320, 336
Electrical conductivity of tissue, 510,
    512
Electrical impedance imaging, 15,
    509
Electrical permittivity of tissue, 509
Electric currents; effects on tissue,
    510
Electro-mechanical gantry, 204, 232
Electron capture, 190
Electrophoresis plates, 175
Elliptical gamma-camera orbits, 203,
    236
Emission computed tomography
    (ECT), 143, 153, 221
Emissivity, 490
Emulsion, 41
Energy resolution QA, 249
Endocrine system, 292
Endoscopic ultrasound imaging, 359

Engineering principles of ultrasound
    imaging, 337
Enhanced ultrasonic imaging, 361
Equipotentials, 513
Erythrocytes; importance for
    diaphanography, 526
Ethernet, 588
Europium activated barium
    fluorohalides, 65
Evaluscop (Siemens), 585
Exercise, effect on metabolism, 451
Exercise stress, 217
Excited states, 190
Expert systems, 382
Exponential envelope, 402
Exposure of film, 42
Eye-and-brain, 569
Eye and orbit, imaging by
    ultrasound, 373

F-number, 333
Fab fragments, 304
Fan-beam collimator, 232, 236
Far-field, 330
Fast Fourier transform (FFT), 197,
    543
Fast line-scane imaging, 422
Fast NMR imaging methods, 422,
    469
Fast rotating gradient spectroscopy
    (FROGS), 459
Feature extraction, 382
Ferromagnetic influences in NMR,
    473, 477
Field alignment, 84
Filament, 80
Film cassette, 46
Film gamma, 44
Film gradient, 44
Film granularity, 51
Film latitude, 44
Film resolution, 45
Film sensitivity, 45
Filtered images, 197
Filtered backprojection, 112, 222,
    226, 427
Filtered projection, 114
Filtering images, 197, 233

Filtration of x-ray set, 79
Fine focus, 80
Finite number of imaging classes?, 3
First Pass cardiac study, 194, 217, 276
First Pass extraction brain imaging, 270
First reports of new imaging methods, 7
Fish; electrolocation in, 521
FISP (NMR method), 423
Fission, 183
Fission products, 184
FLASH (NMR method), 423
Flicker, 591
Flicker sensitivity, 577
Floor loading for NMR units, 468, 472
Floppy disks, 202, 585
Flow potential due to NMR, 475
Fluorescent photons, 46
Fluorine imaging *in-vivo* by NMR, 438
Fluorine NMR spectroscopy, 451
Fluoroscopy, 54
Fluorodeoxyglucose (FDG), 272, 438
Focal spot size, 80
Focussed collimator, 155, 222
Focussed grid, 93
Focussed selective excitation, 421
Focussed ultrasound, 333, 338
Foil converters in MWPC, 172
Fog, 43
FONAR, 421
Fourier ghosts, 552
Fourier reconstruction techniques, 109
Fourier representation of convolution integral, 540
Fourier transform, 401, 472, 540, 562
  imaging in NMR, 412
  of sampled and unsampled images, 550
Fourier zeugmatography, 414
Fovea, 572
Frame-mode data, 194, 216
Frame store, 584, 592
Frauhofer zone in acoustic field, 330
Fredholm integral, 542

FREE analysis, 220
Free induction decay, 401
Frequencies in images, 540, 576
Frequency dependence of electrical properties of tissue, 509
Fresnel theory, 328, 330
Fresnel zone in acoustic field, 328, 330
Fresnel zoneplate, 223

Gadolinium doped water, 408
Gadolinium DTPA contrast agent in NMR, 439
Gallium arsenide devices, 588
Gallium citrate, 208, 303
Gallium EDTA, 273
Gallium isotopes, 213, 259
Gamma camera, 148, 161, 535
  comparison with APT, 519
Gamma-camera gantry control, 203
Gamma camera improvements, 170
Gamma camera SPECT, 224
Gamma (film), 44
Gamma-ray imaging; first reports, 10
Gamma-ray scanner; first reports, 11
Gas amplification, 144
Gas avalanche, 174
Gas detectors, 144, 171
Gas-filled chambers, 64
Gastric transit measurement by APT, 518
Gastrointestinal tract, imaging by ultrasound, 360, 368
Gaussian MTF, 542
Geiger–Müller tube, 10
Generations of CT scanners, 101
Generator equations, 187
Generator for radionuclides, 12, 186, 259
Genetic injury, 31
Genetic significant dose (GSD), 31
Geometric-mean isosensitive scanning, 211
Geometric unsharpness, 28, 38
Geometrical efficiency of gamma camera, 161
Geometrical focus of ultrasound, 333

Geometrical region (ultrasound scattering), 324
Germanium detector, 147, 177
Germanium gamma camera, 177
Ghosting (in display), 591
Gibb's phenomenon, 428
GLEEP, 10
Glomerular filtration rate, 219, 281
Glucose metabolism, 273, 279
Gradient field distortions, 433
Gradient in NMR imaging, 412, 468
Grain noise, 576
Grating lobe, 343
Grey scale display, 590
Grids, 92
Grid Bucky factor, 93
Grid contrast-improvement factor, 93
Grid; moving, 94
Grid ratio, 93
Grid selectivity, 93
Grid; stationary, 93
Gynaecological ultrasound imaging, 366
Gyromagnetic ratio, 391

Hahn echoes, 409
Half life, 182
  biological, 192
  effective, 191
Half value layer, 79
Hanning filter, 197
Hardcopy images, 150
Hardcopy images for gamma camera, 170
Harwell, AERE, 10
Hazard, 4, 319, 375, 474, 510, 525, 604, 606
Head coils, 470
Hearing aids and NMR, 474
Heat emission from man, 499
Heat losses in superconducting magnets, 466
Heterodyne Doppler systems, 354
Heterogeneity corrections, 130
High-energy photons, 150, 192, 213
History of imaging developments, 7
Hormones, 259

Hospital equipment affected by presence of NMR units, 473
Hounsfield's work in CT, 9
Huygens' principle, 3, 328
Human *in-vivo* spectroscopy, 449
Human serum albumin, 306
Human tumour NMR spectroscopy, 451
Human visual perception, 567
Hurter and Driffield curve, 44
Hybrid detectors, 175
Hydrophone, 365
Hyperthermia; temperature measurement, 520

IBM token ring, 588
ICL DAP parallel computer, 589
Identification, 576
Ill-conditioning, 545
Ill-posed problems, 120
Image decoherence, 336
Image display, 69, 590
Image-intensifier CT, 136
Image intensifiers, 54, 87
Image perception, 567
Image processing, 534, 538, 539
Image receptors, 40
Image restoration; general theory, 545
Image viewing geometry, 573
Image and object relationship, 534, 535
Imaging at cellular spatial scale, 4, 614
Impedance computed tomography, 513
Impedance imaging, 511
Incomplete data, 110
Independent measurements in APT, 514
Indium antimonide, 491
Indium bleomycin, 301
Indium isotopes, 213, 259
Inflammatory processes, 286, 300, 303, 502
Information transport from object to image, 536–7
Infrared imaging, 488

Infrared imaging systems, 493
Infrared optical components, 493
Infrared photon-detector, 491
Infrared photography, 488
Infrared transillumination, 489, 528
Inorganic scintillator, 145, 175
Instrumentation for NMR, 465
Integrability of images, 537
Integral uniformity, 246
Interfaces for ultrasound reflection, 325
Interlacing, 59, 591
Internal conversion, 191
Interpolation of images, 552, 565
Interpretation of images, 2
Intracavity ultrasound scanners, 359, 367, 368
Intra-operative imaging, 54, 375
Intraventricular haemorrhage– imaging by APT, 520
Intrinsic spatial resolution, 169, 247
Invasive procedures guided by ultrasound imaging, 373
Inverse operator, 545
Inverse square law, 94, 163
Inversion recovery, 404, 408
Iodine contrast, 95
Iodine Iodoamphatamine (IMP), 270
Iodine isotopes, 142, 213, 292
Iodine radiopharmaceuticals, 262
Ionization chamber, 78, 96, 144
Ionography, 63
Iontomat, 96
ISIS, 462
Isomeric states, 191
Isosensitive scanning, 151
Iterative convolution, 231
Iterative image processing, 559
Iterative least-squares technique, 231
Iterative reconstruction in APT, 514
Iterative reconstruction methods, 112, 119, 222, 230

Joints, 285
Josephson junctions, 588
Juke box, 69

Kidney,
  diseases, 219
  excretion, 219
  filtration, 219
  imaging, 209, 214, 218, 279, 368
  perfusion, 219
  transplant, 219
Krebs cycle, 450
Kubelka–Munk model, 528
Kupffer cells, 288
kV meter, 75

LAL test, 266
Large-bore magnets for spectroscopy, 468
Larmor frequency, 392, 395, 398, 400–1, 445
Last axial maximum, 330
Lateral (ultrasound) resolution, 334
Lattice and spins, 406
Lead converters in MWPC, 172
Lead shield in gamma camera, 169
Lead telluride, 491
Lead zirconate titanate (PZT) transducer, 338
Least-squares image processing, 547
Left ventricular ejection fraction (LVEF), 215, 276
Leukocytes (Indium labelled), 303
Light-beam diaphragm, 82, 84
Light field, 84
Light guide in gamma camera, 166
Limitations of CT, 133
Limited angle of ECT, 222, 224
Linear arrays of transducers, 342
Linear RF amplifier for echography, 345
Line integral, 105
Line-pairs test object, 83
Line scanning NMR imaging, 421
Line spread function, 241, 249, 570
Linear accelerator, 185
Linear attenuation coefficient of x rays, 98, 164, 222
Linear imaging equations, 537
Linear probe APT, 520
Linear sampling requirements, 230

Linear scanner, 151
Liquid crystal thermography, 505
Liquid helium cooling, 466
Liquid nitrogen cooling, 466
List-mode data, 194, 216
Liver imaging by planar scintigraphy, 208, 288
Liver imaging by ultrasound, 368
Liver SPECT, 292
Local area networks, 586, 588
Localisation in NMR spectroscopy, 455
Longitudinal relaxation time, 400, 406
Longitudinal-section ECT, 221
Look-up tables, 592
Lorentzian line shape, 401, 408
Low-energy photons, 173
Low-sensitivity NMR, 437
Luminance, 570
Lung diseases and imaging, 286
Lung imaging by APT, 517

M-mode display, 348
Magnet for NMR, 466
Magnetic dipole moment, 391
Magnetic field gradients, 468
Magnetic flux density, 392
Magnetic tapes, 472, 585
Magnetic torque, 392
Magnetic quantum number, 397
Magnetic shielding, 468
Magnification radiography, 40, 82
Main-frame computer, 584
Male sub-fertility, 503
Mammographic unit tests, 88
Mammography, 28, 60
Matching layer for transducer, 338
Mathematical model of contrast, 28
Mathematical model of radiography, 22
Mathematics of image formation and image processing, 534
Matrices for data storage, 194, 233
Matrix representation of discrete imaging, 543
Matrix size in digital radiography, 69

Maximum entropy deconvolution, 549
Megavoltage CT, 138
Meiboom–Gill pulse sequence, 410
Metabolic myopathies, 450
Metabolic processes, 142, 450
Metabolism, 450
Metabolism of radiopharmaceuticals, 258
Metal detector archway, 478
Metastable isotope, 191
mIBG, 299
Microcalcification, 60
Microscopy by NMR, 439
Microscopy, tomographic, 123
Microwave CT, 521
Microwave thermography, 506
MIMD parallel computer, 590
Minicomputers, 193
Minimum resolvable temperature difference (MRTD), 493
Mitochondrial myopathies, 451
Mobile NMR units, 467
Models for the imaging equation, 542
Modulation transfer function (MTF), 28, 56–7, 249, 541, 569–70
Molecular engineering, 304
Molecular frequencies and T_1, 406
Monoclonal antibodies, 304
Motion artefact, 431
Movement effects in ultrasound imaging, 326
Movement unsharpness, 29
MOVIE.BYU, 594
Moving grid, 94
Moving images, 576
Multicrystal PET, 157
Multicrystal scanners (MSPTS), 153, 237
Multigated acquisition (MUGA), 214, 276
Multiple element transducers, 340
Multiple sensitive point NMR imaging, 422
Multiplicity of imaging modalities, 2
Multistep avalanche (MSA) MWPC, 174

Multi-user computer systems, 584
MUP-PET, 242
Muscle energy metabolism, 279, 450
Muscular dystrophy, 451
MWPC detector, 171
MWPC gamma camera, 173
MWPC positron camera, 536, 557
Myocardial imaging, 193, 214, 275

n–γ reaction, 183
Narrow-band noise bursts in NMR, 428
National Physical Laboratory, 10, 75
Natural active emissions for imaging, 5
Natural focus of ultrasound, 333
Near field, 328
Needle-puncture techniques (biopsy), 373
Neighbourhood principle, 536
Neonatal brain imaging by transillumination, 524, 532
Neonatal monitoring; role of APT, 520
Neonatal ultrasound imaging, 359
Networks, 586
Neuro ECAT III, 158
Neutrino, 190
Neutron activation, 182
Neutron capture, 182
Neutron-deficient radioisotopes, 186
Neutron-rich radioisotopes, 184, 190
Ninepoint smoothing, 197
Ninety-degree RF pulse, 402
Niobium-titanium superconductor, 466
NMR
    classical theory of, 391
    data in 3D, 602
    further reading to this volume, 390
    imaging, 389, 412
    historical development, 13, 390
    microscopy, 439
    spectroscopy, 389, 445
    quantum theory of, 397
Nobel prize for principles of NMR, 13
Nobel prize for x-ray CT, 9
Noise, 29, 569

Noise equivalent power (NEP), 491
Noise equivalent temperature difference (NETD), 493
Noise in CT, 122
Noise in imaging equation, 545
Noise in screen–film, 51
Noise in SPECT, 228
Noise in TV, 58
Noise required modulation (NRM), 575
Non-linear imaging equations, 335, 537
Non-linear ultrasound propagation, 337, 364
Non-negativity principle, 537, 570
Non-standard CT scanners, 135
Non uniformity QA, 246
Nuclear fission, 183
Nuclear fuel, 184
Nuclear magnetic resonance – first images, 13
Nuclear medicine, 142
Nuclear reactor, 10, 183
Nuclear-spin quantum number, 397
Nutritional studies with APT, 520
Nutting's law, 43
Nyquist sampling criteria, 115, 551, 576

Oak Ridge, 10
Object and image relationship, 534, 536
Obstetric ultrasound, 366
OCCAM, 590
Oldendorf's laboratory CT, 9
One hundred and eighty degree RF pulse, 404
On-line data correction, 195
Optical density, 42, 52
Optical disc, 69, 472, 586
Optical properties of tissue, 526
Optical transfer function, 570
Organic elements, 259
Oscillating grid, 94
Output of x-ray sets, 78
Ovaries, imaging by ultrasound, 367
Overfielding magnets, 468
Oxygen extraction efficiency, 272

Oxygenation, measurement via NMR, 450

Pacemakers, 474, 477
Pain and trauma, 504
Panning, 592
Pancreas, imaging by ultrasound, 368
Parallel-hole collimator, 161
Parallel-computer architecture, 589
Parallel projections, 101
Paramagnetic species, 406
Paramagnetic spin enhancement
    agents in NMR, 439
Parametric image, 200, 214, 220, 282
Parathyroid, 298, 371
Parent–daughter, 186
Partial-volume artefacts, 124, 235
Patient positioning, 207
Peak potential (x-ray), 75
Penetrameter (x-ray), 75
Penumbra, 29, 38
Performance assessment in RI
    imaging, 245
Performance evaluation, 74
Perfusion (lung), 287
Perfusion study, 287
Peripheral potential gradients, 513
Personal computer (PC), 584
PET, 157, 171, 221, 239, 427
    brain imaging, 272
    image enhancement in, 557
pH measurement by NMR, 450
Phagocytic cells, 288
Phantoms for QA in NM, 247
Phantoms for QA of ultrasonic
    imagers, 362
Phantoms for transillumination, 531
Phase coherence, 402
Phase focussing, 340
Phase steering, 340
Phosphors for image intensifiers, 55
Phosphors for screens, 48
Phosphorus compounds in NMR
    spectroscopy, 450
Photoelectric interaction, 24, 144
Photographic emulsions, 41
Photographic process, 42

Photomultiplier, 11, 145, 157, 167, 195
Photopeak, 145, 150, 168, 207, 232
Photon interactions, 23
Photon transmission through tissue, 24
Photostimulable phosphor, 65
Picture archiving and
    communications systems (PACS), 69
Piezoelectric transducer, 338
Pinhole camera, 80
Pinhole collimator, 164, 223
Pipe-line processors, 589
Pixel depth, 69
Planar scintigraphy, 143, 204
Plane sensitivity QA, 252
Planck's formula, 490
Plastic (PVDF) transducer, 338
Plastic-scintillator detectors, 136
Plumbicon TV, 58
Point methods of NMR imaging, 421
Point process of image formation, 335, 536, 569
Point source response function
    (PSRF), 538
    in SPECT, 226
    in ultrasound imaging, 335
Polar Fourier transform, 112
Polarised RF fields, 394
Polaroid film, 205
Polyclonal antibodies, 303
Population of spin states, 399
Positron annihilation, 190
Positron emission, 190, 193, 239
Positron-emitting
    radiopharmaceuticals, 157
Positron imaging, invention of, 13
Positron range, 190, 239
Post-operative clips and NMR, 479
Post viral fatigue syndrome, 480
Potential (x-ray tube), 75
Powder image, 60
Power spectrum, 29, 564
Power supplies for gradient fields, 469
Precession, 392
Pressure impulse response, 330

Projection radiography, 21
Projection reconstruction for NMR, 412
Projections, 20, 101, 107, 226
Proportional counter, 144
Prostate, imaging by ultrasound, 369
Proteins, 258
Proton CT, 138
Proton density, derivation from NMR signal, 412
Proton density measurement, 411
Proton NMR spectroscopy, 451
Proton spin density, 389
Proton-to-neutron ratio, 183, 190
Pseudoprojections, 231
Pulmonary oedema, 517
Pulse-echo scanning, 335, 343
Pulse pile up, 169
Pulse repetition rate in echography, 345
Pulse sequences for NMR, 402, 472
Pulsed ultrasound, 327
Pulsed-wave flow detector (pulsed Doppler US), 355
Pyloric stenosis, 520
Pyroelectric imaging system, 496
Pyroelectric vidicon TV tube, 489, 496
Pyrogenicity, 265

Q-factor for transducer, 338
Quadrature-phase-demodulator, 354
Quadrature, 401
Quadrature detection, 401, 471
Quality assurance (QA), 74
QA
  in ECT, 254
  of radiopharmaceuticals, 265
  of ultrasonic imaging systems, 362
Quality control in RI imaging, 245
Quality factor (Q) for NMR tuned coils, 470
Quantification in planar scintigraphy, 208
Quantum mechanical description of NMR theory, 397
Quantum mottle, 29, 51, 66
Quenching, 478, 479

Radar, 340
Radiation detectors, 143
Radiation dose in SPECT, PET, TCT, 222
Radiation force balance, 365
Radioactive 'cows', 12, 188
Radioactive decay constant, 182
Radioactive decay equations, 182
Radioactive decay modes, 188
Radioactive half-life, 182
Radiographic contrast, 27, 91
Radiographic mottle, 29, 51, 66
Radiographic quality, 90
Radiographic scatter, 26, 91
Radioimmunoscintigraphy, 307
Radioisotope imaging, 142
Radio-halogens, 259, 262
Radio-metals, 259
Radionuclide generator, 182, 186
Radionuclide production, 182
Radionuclides for imaging, 181, 191
Radiopharmaceutical chemistry, 257
Radiopharmaceutical purity, 265
Radiopharmaceuticals, 181
Radiotherapy, 128
Radon's equation, 9, 112
Rampable magnets for NMR, 467
Random events in PET, 171, 242
Rare-earth screens, 48
Rayleigh region for ultrasound scatter, 326
Reactance of transducer circuit, 339
Real-time scanning (echography), 348
Receiver coils, 470
Receiver operating characteristic (ROC), 580
Receptor binding tracers, 275
Receptor unsharpness, 39
Receptors, 40, 257
Reciprocal space, 540
Reciprocating grid, 94
Recognition, 576
Reconstruction from projections, 108, 225, 379
Recovery of continuous from sampled data, 552
Rectilinear scanner, 148, 222

Reflection of ultrasound, 325
Regions of interest (ROI), 199
Relaxation times, 389, 400, 406
Renal *see* Kidney
Renogram, 220
Repetition times (pulses in NMR),
    404
Rephasing spins, 405
Representation of objects, 535, 539
Resistive magnet, 466
Resistivity of biological tissues, 511
Resolution
    in APT, 514–15
    in x-ray CT, 121
    in diaphanography, 525
    theory of degraded, 539
Resonant and non-resonant
    interactions, 3
Resonant frequency of transducer,
    338
Respiratory artefacts in NMR, 431
Respiratory diseases, 287
Respiratory imaging, 286
Responsivity, 491
Reticuloendothelial system, 288
Retina, 572
Reverberation artefact, 575
Reynolds number in imaging, 546
RF
    power deposition, 396, 451, 476
    power; potential hazard, 476
    pulse, 395
    system for NMR, 469
Ring artefacts, 102
    in NMR, 428
Risk
    from ionising radiation, 31, 606
    in APT, 510, 610
    in diaphanography, 525, 610
    in NMR imaging, 396, 474, 607
    in ultrasound imaging, 607
Rods, 572
Röntgen's discovery of x-rays, 8
Röntgen, Wilhelm, 8, 20
Rose criterion, 30, 574
Rotate-translate CT scanning, 102
Rotating collimator, 223
Rotating frames (NMR theory), 394

Royal Marsden Hospital, 2, 14, 130,
    137, 242, 390, 594
Royal Marsden Hospital CT
    Simulator, 137
Russian CT in the 1950s, 10, 112
Rutherford–Appleton Laboratory,
    172

Saddle shaped coil, 466
Safety and NMR, 474
Safety standards for ultrasound, 364,
    375
Sampling
    in NMR imaging, 442
    in NMR spectroscopy, 449
    of data, 122, 229, 442, 449, 472
    of images, 549
Saturation-recovery pulse sequence,
    402, 408
Scan converter, 347
Scanned projection radiography, 67
Scanner configurations (CT), 101
Scanning acoustic microscope, 384
Scanning methods for ultrasound
    imaging, 348
Scanning slit, 94
Scanning transducer, 340
Scatter detection at receptors, 92–3
Scatter in SPECT, 228
Scatter rejection, 68
Scatter removal, 92
Scatter-to-primary ratio, 23, 91
Scattered coincidence in PET, 244
Scattering centres, 92
Scattering cross section, 25
Scattering of light by tissue, 527
Scattering of ultrasound, 324
Scattering of x-rays, 23, 91
Scintillation detector, 145, 148, 157,
    166
Screen–film combinations, 46
Screening of patient, 55
Screening of population, 55
Screening staff near NMR units, 478
Scrolling, 592
Scrotal imaging, 359, 373, 524
Searle Pho/Con imager, 222
Secular equilibrium, 187

Selective excitation, 413, 441
Selectivity of grid, 93
Selenium, 60
Self-shielding magnets, 468
Selwyn's law, 52
Semi-conductor detector, 146, 176
Semi-opacity requirement, 3
Sensitive point NMR imaging, 421
Sensitivity
  of diagnosis, 579
  of film, 45
  of NMR, why poor?, 399
  of PET, 176, 242
  of screen–film, 47
  of xeroradiography, 60
Separability, 335, 539
Septum, 162
Shadow-shield collimators, 244
Shah function, 550
Shannon sampling theorem, 114
Shield leakage QA, 254
Shielding fields, 446
Shielding magnetic fields, 467, 472
Shimming, 448, 474
Shimming for spectroscopy, 436, 474
Shrapnel injuries, 478
Signal to noise ratio, 30, 569
  in CT, 122, 443
  in deconvolution, 546
  in NMR, 443
  in SPECT, 443
Significance of pixel values, 199
Silicon detector, 146, 180
Silicon wafers, 180
SIMD parallel computer, 589
Simulator CT scanners, 136
Single channel analyser, 150, 168,
  208
Single-section tomography, 153
Single-user computer systems, 584
Site-planning guide for NMR, 472
Siting considerations for NMR, 472
Skeletal imaging, 152, 164, 283
Skin temperature in man, 490, 499
Slant-hole collimator, 223
Slice width in NMR imaging, 441
SLIT DRESS (NMR spectroscopy),
  459

Small-bore NMR systems, 439
Small-parts ultrasound imagers, 359
Smoothing images, 197, 228
Snaking display, 201
Sodium imaging *in-vivo* by NMR, 437
Sodium iodide detector, 146, 148,
  151, 161
Sonar, 320
Sonogram, 354
Software QA, 256
Solenoidal coil, 466
Somatic risk, 31
Space invariance in PET, 241
Space variance and invariance, 335,
  538
Space variant (ultrasound)
  resolution, 335
Spatial distortion, 252
Spatial resolution, 572
  in limited angle ECT, 223
  in PET, 222, 239
  in SPECT, 222, 226, 228
  in ultrasound imaging, 333
  of gamma camera, 161, 169, 212
  QA, 247
Spatially localised NMR spectroscopy,
  445
Speckle, 327, 336, 384
Special collimators for ECT, 223
Special-purpose SPECT systems, 237
Specificity, 579
SPECT, 221, 427, 535
  acquisition parameters, 231
  image enhancement in, 553
  in 3D, 599
  invention of, 11
  processing parameters, 233
  with rotating gamma camera, 224
Spectral broadening, 353
Spectroscopy
  in small-bore NMR, 445
  in-vivo, 449
Spin angular momentum, 397
Spin-echo sequence, 404
Spin lattice relaxation, 406
Spin rephasing, 443
Spin states, 399
Spin temperature, 399

Spin-warp imaging, 415
Spinning top timer, 78
Spleen, imaging by ultrasound, 368
SPRITE detector, 496
Staff dose, 74
Star computer (IGE), 585
Star (spoke) artefacts, 226
Static-field gradients, 468
Static fields; potential hazard, 474
Static planar-scintigraphy, 204
Stationary grid, 93
Statistical distribution of spin states, 399
'Steam train' scanner, 151
Steering transducer, 340
Stefan–Boltzmann law, 490
Sterility, 265
Steroids, 259
Stimulated echoes, 444
Stimulated emission, 400
Stimulated luminescence, 67
Stochastic region for ultrasound scatter, 326
Strength of ultrasound focussing, 333
Strip detector, 147, 177, 180
Structure mottle, 51
Subtraction imaging, 70, 208, 306
Sugars, 259
Sun workstation, 585
Superconductivity at higher temperature, 467
Superconducting magnet, 466
Superficial structures, imaging by ultrasound, 370
Supermini computer, 584
Superposition principle, 537
Surface coils, 456
Susceptibility imaging by NMR, 436
Switched gradients; potential hazard, 475
Synthetic-aperture imaging, 378
Synovitis, 286
Synthetic tracers, 259
System spatial resolution, 247
Systole, 214, 217

T_1 relaxation time, 400, 406
T_1 experimental measurements, 407

T_2 experimental measurements, 409
T_2 relaxation time, 400, 408
T_2^* relaxation time, 409
Target element (reactor), 183
Targetted therapy, 190
Technetium-99m, 188, 192, 259
Technetium-99m; first generator, 12
Technetium complexes, 259
Technetium-essential radiopharmaceuticals, 260
Technetium-labelled DMSA, 302
Technetium-labelled DTPA, 260, 267, 282
Technetium-labelled glucoheptonate, 267
Technetium-labelled HMPAO, 260, 270
Technetium-labelled IDA, 292
Technetium-labelled MDP, 284
Technetium-labelled pyrophosphate, 278
Technetium-labelled red blood cells, 214, 272, 276
Technetium-labelled sulphur colloid, 208, 290
Technetium-tagged radiopharmaceuticals, 260
Telediaphanography, 531
Television, 55, 58, 87, 531, 591
Television bandwidth, 58, 592
Television data transfer, 585
Television resolution, 59, 591
Temperature measurement, 497 by APT, 520
Temperature of lattice, 399, 406
Temperature of spins, 399
Temperature scale for image display, 590
Temporal contrast-sensitivity, 577
Testes, imaging by ultrasound, 359, 373
Test objects for x-radiology, 75
Testicular temperature, 503
Tetrakis (dimethylamino) ethylene (TMAE), 176
Texture, 384
Thallium DDT, 270
Theoretical framework of CT, 100

Thermal detectors, 489
Thermal gradients in man, 499
Thermal neutrons, 183
Thermal radiation, 489
Thermal stress thermography, 500
Thermographic investigations, 499, 502
Thermographic mammography, 499, 505
Thermography, first images, 14
Thermography in oncology, 505
Three-dimensional display, 593
Thyroid, 151, 164, 173, 209, 292, 359, 371
Time after injection, 207
Time constant for vision, 577
Time-gain-control, 345
Time-of-flight PET, 159
Time-of-flight ultrasound tomography, 378
Time-varying field gradients, 469
Timeliness of imaging developments, 4
Timing (x-ray sets), 77
Tissue characterisation using ultrasound, 381
TM-mode display, 348
Tomographic mathematics, 108, 378
Tomographic microscopy, 123
Tomographic planes, 221
Tomographic tests, 86
Tomography
   using x-rays, 100
   using ultrasound, 378
Tomomatic 64, 155
Topical magnet resonance, 421
Toxicity, 265
Tracers, 142, 181
Transducer frequency; choice of, 344
Transducers, 337
Transfer of image data, 585
Transferrin, 260
Transient equilibrium, 187
Transient ischaemic attacks, 269
Transillumination, 489, 525
Transillumination, historical images, 14

Transmission computed tomography (TCT), 222
Transmitter coils, 470
Transverse relaxation time, 400
Treatment planning, 128
Tumour imaging by RI, 300
Transputer, 590
True focus of ultrasound, 333
Tumour localisation by CT, 129
Tumour staging, 129
Tuned coils, 470
Tuning of transducer, 339

Ultrasound
   absorption, 323
   attenuation, 322
   diffraction, 343
   energy, 320
   first images with, 13
   imaging, 319
   imaging parameters, 328
   in general practice, 361
   intensity attenuation coefficient, 322
   interactions in biological tissue, 320
   momentum, 320
   safety of, 375

Ultrasound (*cont.*)
   scatter, 322, 324
   speed, 320
   wave intensity, 321
   wave propagation, 320
Undercouch x-ray tube, 87
Unfocussed grid, 93
Uniformity correction, 233
Uniformity of gamma camera, 246
Unpaired electrons, 406
Unsharpness, 28, 38, 45–6, 56, 62
Urine, 218
Uterus, imaging by ultrasound, 367

Varicosities, 503
Varifocal mirror, 602
Vascular disorders, 502
Vascular disturbance syndromes, 503

Vax computer, 594
Vaxstation 2000, 585
Ventilation study, 287
Ventilation/perfusion study, 288
Ventriculography, 276
Very high field magnets, 468
Vidicon TV, 58, 488
Vision, 567
Vision for survival, 571, 578
Vitamins, 259
Volume imaging by NMR, 426
Volume selective excitation (VSE), 422, 462
Von Neumann architecture, 589

Washout, 270
Waterbath scanning (echography), 350
Water content measurement by APT, 517
Water content of tissues; and T_1, 407
Watches and NMR, 474
Weber ratio, 573
Whittaker–Shannon theorem, 114
Wide-area networks, 586

Wiener filter, 549
Wiener spectrum, 29, 54
Windowed display, 111, 201
Windowed reconstruction, 117
Wire planes in MWPC, 173

Xenon detector, 144, 174
Xenon for imaging, 270
Xeroradiography, 59
Xeroradiography; history, 14
X-ray CT
    in radiotherapy planning, 128
    in 3D, 597
X-ray film, 41
X-ray filters, 35, 37
X-ray imaging, 20
X-ray linear attenuation coefficient, 24, 222
X-ray spectra, 35
X-ray transmission CT, 98
X-ray tube potential, 32, 37
X-ray tubes, 32

Z pulse in gamma camera, 168, 195
Zeugmatography, 414
Zooming, 435, 592